Trophoblast Research
VOLUME 2

CELLULAR BIOLOGY AND PHARMACOLOGY OF THE PLACENTA
Techniques and Applications

Trophoblast Research

Series Editors

Richard K. Miller and Henry A. Thiede

University of Rochester Medical Center
Rochester, New York

Trophoblast Research
VOLUME 2

CELLULAR BIOLOGY AND PHARMACOLOGY OF THE PLACENTA
Techniques and Applications

Edited by
Richard K. Miller and
Henry A. Thiede

University of Rochester Medical Center
Rochester, New York

SPRINGER SCIENCE+BUSINESS MEDIA, LLC

Library of Congress Cataloging in Publication Data

Cellular biology and pharmacology of the placenta.

(Trophoblast research; v. 2)
"Derived from the Tenth Rochester Trophoblast Conference, held October 8–10, 1985, in Rochester, New York " T.p. verso.
Includes bibliographies and index.
1. Placenta—Congresses. 2. Placenta—Effect of drugs on—Congresses. 3. Cytology—Congresses. I. Miller, Richard K. (Richard Kermit), date. II. Thiede, Henry A. III. Rochester Trophoblast Conference (10th: 1985) IV. Series. [DNLM: 1. Planceta—drug effects—congresses. 2. Placenta—metabolism—congresses. W1 TR877 v.2/WQ 212 C393 1985]

QP281.C45   1987   612'.63   87-6934
ISBN 978-1-4757-1938-3   ISBN 978-1-4757-1936-9 (eBook)
DOI 10 1007/978-1-4757-1936-9

Derived from the Tenth Rochester Trophoblast Conference,
held October 8–10, 1985, in Rochester, New York

© 1987 Springer Science+Business Media New York
Originally published by University of Rochester in 1987
Softcover reprint of the hardcover 1st edition 1987

TROPHOBLAST RESEARCH

Trophoblast Research publishes contributions concerning the placenta and the extraembryonic membranes as they relate to embryonic and fetal development and to trophoblastic neoplasia. Original articles, reviews and reports are published in single bound volumes. All articles are peer-reviewed.

The Editorial Office for *Trophoblast Research*:
Department of Obstetrics and Gynecology,
University of Rochester, School of Medicine and Dentistry,
601 Elmwood Avenue, Rochester, New York, USA 14642
716-275-3638

PAST ROCHESTER TROPHOBLAST CONFERENCES

MODERATORS

First Conference 1961   Arthur T. Hertig
                                               Ernest Witebsky

Second Conference 1963   Rupert Billingham
                                               Jack Davies

Third Conference 1965   Donald H. Barron
                                               Claude A. Villee

Fourth Conference 1967   Roy Hertz
                                               Ralph M. Wynn

Fifth Conference 1969   E. J. Quilligan
                                               Kenneth Ryan

Sixth Conference 1971   Donald L. Hutchinson
                                               Frederick P. Zuspan

Seventh Conference 1977   Griff Ross
                                               Ralph M. Wynn

STATE OF THE ART SPEAKERS

Eighth Conference 1980   W. Page Faulk
                                               Maurice Panigel

Ninth Conference 1982   Arnold Klopper
                                               Claude A. Villee
                                               Myron Winick

Tenth Conference 1985   Ross Berkowitz
                                               Peter Johnson
                                               Mont R. Juchau
                                               Frank Young

PREFACE

This volume is the second in the series, *Trophoblast Research* and the tenth report from the Rochester Trophoblast Conferences. Twenty-five years ago Curtis Lund and Henry Thiede held the First Rochester Conference. During the succeeding years, the University of Rochester has had the pleasure of welcoming both basic and clinical scientists who have brought their most exciting, challenging and novel observations concerning all aspects of the placenta to Rochester. Such innovative research has been the foundation for the presentations at the Tenth Conference. As the years have passed, the science of trophoblast research has broadened in range and increased in depth. This is also true of the participants at the Tenth Conference. Two dozen scientists attended in 1961; 179 participants from 15 countries and 22 states were present at the Rochester Hilton from October 8-10, 1985. The diversity of presentations in plenary sessions, posters and workshops represents the multiple backgrounds and training of those scientists who devoted their attention to the critical issues of implantation; biochemistry and physiology of trophoblast function; and technics for studying such tissue function and the richness of materials for comparative species investigation. Note that in Volume 1, Ralph Wynn (1984) reviewed the history of the Rochester Conferences.

The theme for the Tenth Conference was the Cell Biology and Pharmacology of the Placenta. Presentations encompassed all aspects of current scientific study from the molecular biology of trophoblastic neoplasia to methods for altering recurrent spontaneous pregnancy loss. Because of the increased size of the program, the reader will note that the recorded discussion is not included. However, many of the issues raised during the discussions have been incorporated into the text from each contribution. Unfortunately, not all eighty presentations could be included in this volume. The reader is referred to the abstract booklet from the Tenth Rochester Trophoblast Conference for a description of the other equally provocative and exciting presentations.

Included during the conference were workshops emphasizing technics by which one can investigate the trophoblast. Six workshops were held of which three are summarized in the volume. The other three workshops on Receptors (Leaders: Carolyn Coulam, Peter Johnson and Howard Sussman); on Pharmacokinetics (Leaders: David Manchester and John Young); and on Chorionic Villus Biopsy (Leader: Laird Jackson) were intense and educational for all in attendance. However, these sessions were perceived by the session leaders not to produce new data or technics, and therefore these reports were not included. The reader is referred to the workshop reports on Tissue Culture of Trophoblast (Leaders: Robert Hussa and Charles Loke), Technics in Placental Perfusions in Experimental Animals (Leaders: Bruce Kelman and Bernard van Kreel) and Human Placental Perfusions (Leaders: Joseph Dancis and Henning Schneider) where new information has been reported.

The state of the art in any area of science is but a fleeting moment in time. However, Mont Juchau, Ross Berkowitz and Peter Johnson brought enthusiasm and encouragement as well as the latest knowledge in the disciplines of xenobiotic and steroid enzyme regulation, immunobiology of the trophoblast and cell surface receptors. New to the conference this year was the First Rochester Conference

receptors. New to the conference this year was the First Rochester Conference Lecture. This lecture was devised to bring to the conference new and diverse approaches to research, health care and general scientific knowledge not perceived to be among those individuals investigating the placenta. It was a pleasure and an honor to have Frank E. Young, M.D., Ph.D., Commissioner of the Food and Drug Administration of the United States present an address on the subject: The Partnership between a Physician Scientist and the Federal Government in Research and Health Care.

The objectives of the Rochester Trophoblast Conference and the series, *Trophoblast Research* are (1) to provide a forum for the presentation of exciting and innovative research; and (2) to foster an atmosphere within which scientists can exchange ideas, meet new and old colleagues and develop new directions to their creative efforts. If you did not come to Rochester for the Tenth Conference, perhaps in the following pages you will find some helpful ideas. The Rochester Trophoblast Conference was pleased to join with the European Placenta Group for a joint meeting in Rolduc, The Netherlands, September 24-27, 1986. The emphasis of this joint meeting was the vasculature of the placenta. The proceedings of that meeting will be peer-reviewed and published as Volume 3 of *Trophoblast Research*. We also welcome you to attend the Eleventh Rochester Trophoblast Conference to be held from October 9-12, 1988 in Rochester, New York. The theme for the Eleventh Conference will be the Molecular Biology of the Trophoblast. In addition to the major theme, there will be workshops on such topics as Circulation, Recurrent Pregnancy Loss, Cell Culture, Trophoblastic Neoplasia, Technics in Molecular Biology, Advances in Perfusion Technics and others yet to be organized.

Finally, the editors wish to express their gratitude to the Organizing Committee of the Tenth Rochester Trophoblast Conference, who provided the guidance and reviews necessary for the program to be successful. They are Francis Bellino, Carolyn Coulam, Donald Goldstein, Nicholas Illsley, Debabrata Maulik, Eberhard Muechler, Alan Poisner, B.V. Rama Sastry and Bernard van Kreel. We also acknowledge the efforts of Willem Faber, Risa Saltzman, Lisa Smith, Tacey White and John Zongrone, who provided the support necessary for all aspects of the Conference program to be both timely and efficiently conducted. This volume would have not been possible without the devotion and care provided by our conference secretary and editorial assistant, Jacqulyn White with the assistance of Shelley Gordon. Janice Stern and John Matzka from Plenum Publishing Corporation have provided innumerable suggestions during the development of this volume. As with all programs, it is important to acknowledge those who have assisted in the financial aspects of this endeavor. They are the American Cancer Society, Battelle Laboratories, Hoffman-La Roche, Inc., Hybritech Laboratories, the March of Dimes - Birth Defects Foundation, Pennwalt Pharmaceuticals, Ross Laboratories, Upjohn Laboratories and especially the Department of Obstetrics and Gynecology at the University of Rochester.

The editors hope you benefit from the fascinating investigations that follow. If you have any suggestions, please do not hesitate to share them with the editors.

R. K. Miller and H. A. Thiede
Rochester, New York

CONTENTS

CELL BIOLOGY

IMMUNOBIOLOGY

PLACENTAL PHARMACOLOGY

TROPHOBLAST CULTURE

ANIMAL PLACENTAL PERFUSIONS

HUMAN PLACENTAL PERFUSIONS

Contents

CELL BIOLOGY

Trophoblast Research 2:3-15, 1987

BIOCHEMICAL AND IMMUNOLOGICAL ASPECTS OF THE HUMAN TROPHOBLAST CELL SURFACE

- A Review -

Paul D. Webb, Nicholas Hole, P. Jeremy McLaughlin,
Peter L. Stern, and Peter M. Johnson[1]

Pregnancy Immunology Group
Department of Immunology
University of Liverpool
PO Box 147
Liverpool, L69 3BX
United Kingdom

INTRODUCTION

Following blastocyst implantation, the outer layer of human trophoblast rapidly proliferates and several anatomically distinct populations of trophoblast develop in the placenta and extraembryonic membranes as gestation progresses (Hertig et al., 1954; Boyd and Hamilton, 1970; Bulmer and Johnson, 1985). The major trophoblast population covers chorionic villi (or conversely lines intervillous spaces) and is therefore in intimate contact with maternal blood. This villous trophoblast layer is organized into an inner cytotrophoblast and an outer syncytiotrophoblast layer with numerous microvilli on the maternal-facing surface. Also of relevance to this review is extravillous cytotrophoblast which proliferates into maternal uterine tissue from the implantation site. Since high purity preparations of syncytio-trophoblast surface microvillous plasma membrane (StMPM) can be readily obtained from term placentae (Smith et al., 1974; Ogbimi et al., 1979), the majority of biochemical and immunological studies have centered on syncytiotrophoblast.

Trophoblast displays unique nutritional, endocrine, and immunological functions which support the development of the fetal allograft throughout gestation. These diverse properties are reflected at the molecular level by the large number of separate protein components associated with StMPM which can be resolved by two-dimensional gel electrophoresis (Webb et al., 1985). StMPM is a ready source of a number of important cell surface molecules, enabling their characterization and purification from this tissue - for example, the receptors for insulin and transferrin (Newman et al., 1983; Wild, 1983; Ebina et al., 1985). The array of StMPM-associated components has been reviewed in detail elsewhere (Whyte, 1983; Johnson, 1984; Truman and Ford, 1984). This review concentrates on components thought to be relevant in the control of cellular proliferation and evasion of immunological rejection

[1]Correspondence: Professor Peter M. Johnson
Received: 8 October 1985; Accepted: 10 January 1986

by the maternal immune response, and which may also be important in understanding analogous functional characteristics of tumor cells.

StMPM-Associated Glycosylation

The glycosphingolipid (GSL) component of StMPM has received scant attention compared with several other human cell types, but initial reports have indicated this membrane to express an unusually high concentration of gangliosides (sialylated glycolipids). Several are highly sialylated and may be more complex than those normally found in other organs (Cheng and Johnson, 1985). These sialoglycolipids probably account for the strong negative charge demonstrated by StMPM, but of more relevance may be reports that some highly sialylated gangliosides have immunosuppressive properties (Hakamori, 1985). Several complex highly glycosylated neutral GSL are also associated with StMPM (Cheng and Johnson, 1985), and could be analogous to some of the cell surface glycosylated structures shown to be oncodevelopmental antigens and thought to function in cellular growth and differentiation processes (Feizi, 1985; Hakamori, 1985). Thus, GSL undoubtedly contributes to the surface carbohydrate covering of the entire StMPM surface and, along with glycoproteins, will form one layer of interaction with maternal cells. Lectin binding to StMPM has shown that wheat germ agglutinin strongly binds to glycosylated StMPM components throughout gestation, whereas concanavalin A and phytohemagglutinin bind only to first-trimester trophoblast (Johnson et al., 1980); this suggests an alteration in trophoblast glycosylation as pregnancy progresses (Nelson et al., 1978).

Placental-type Alkaline Phosphatase (PLAP)

Many enzymes are expressed by StMPM, but that which consistently attracts most attention is PLAP. This heat-stable sialoglycoprotein isoenzyme can be readily distinguished from other human alkaline phosphatase isoenzymes by biochemical and immunological means, and is an important 'oncotrophoblast' antigen expressed by certain tumors - notably ovarian tumors and seminomas (McDicken et al., 1985; Epenetos et al., 1985a). The availability of monoclonal antibodies (mAbs) specific for PLAP has established this enzyme to be a clinically useful marker for these tumors. Thus, these mAbs have been used to assay trace amounts of PLAP in serum as well as for immunohistological and in vivo radioimmunoimaging tumor localization studies (McDicken et al., 1985; Epenetos et al., 1985b; Critchley et al., 1986). PLAP displays genetic polymorphism, based on electrophoretic mobility in starch gels (Fishman, 1974), but it is not yet proven whether this is the result of differing amino-acid sequences or post-translational structural modifications of the enzyme. There is some evidence of heterogeneity of carbohydrate composition for different PLAP preparations (Ezra et al., 1983). Ultimately, this point will be resolved only by knowledge of the complete primary amino acid sequence of PLAP allotypes. Xenogeneic immunization has produced mAbs with some selective reactivity for different PLAP alleles (McLaughlin et al., 1984). However, these allotypic forms of PLAP do not apparently promote alloimmune responses, since maternal immunity to the paternally-inherited fetal allotype of human PLAP has never been described. This could indicate some form of maternal systemic tolerance to such trophoblast membrane components which is operative in pregnancy, perhaps resulting from the substantial continuous deportation of trophoblastic membrane elements from the placental site into maternal blood.

There are two major forms of PLAP which can be recognized by mAbs. The mAb H317 recognizes only the placental form of the enzyme, whereas the mAbs H315 (Webb et al., 1986) and H17-E2 (McLaughlin et al., 1984) also recognize a placental-like form of alkaline phosphatase. H317-reactive PLAP is expressed by syncytiotrophoblast only after the first trimester of pregnancy (Bulmer and Johnson, 1985), although biochemical studies have shown some expression of a similar enzyme in first trimester placental tissue (Sakiyama et al., 1979). Immunohistology with the mAb H315 has demonstrated some reactivity with early human trophoblast (Bulmer et al., 1984) and, furthermore, this mAb is reactive in enzyme immunoassay with first trimester abortus tissue extracts (McLaughlin, unpublished). H315 has also been employed successfully in identification, by flow cytometry, of exfoliated trophoblast cellular elements in the peripheral blood of women in the first trimester of pregnancy (Covone et al., 1984). It is possible therefore that mAb H317 either recognizes a PLAP epitope which develops as a result of altered post-translational processing of an early PLAP form or detects expression of a different PLAP locus.

Despite the abundance of StMPM-associated PLAP, the biological role of this enzyme has not yet been resolved. Studies with membrane-bound alkaline phosphatase have indicated an interaction between this enzyme and phophatidylinositol (Low and Zilversmit, 1980). Phosphatidylinositol-specific phospholipase C (PI-PLC), purified from certain bacteria, will specifically release alkaline phosphatase in a soluble form from cell membrane preparations (Shukla, 1982). Recent studies with StMPM using phospholipase C isolated from different bacterial sources have also indicated most PLAP to be associated with the inositol phospholipids (Webb, unpublished). Although inositol phospholipids accounts for only a small amount of the lipid in StMPM (Ogbimi and Johnson, 1980), they are of particular significance since phosphatidylinositol and its phosphorylated derivatives play a central role in cellular signal transduction. Thus, binding of growth factors to their cell surface receptors results in the hydrolysis of membrane phosphoinositides with formation of diacylglycerol and inositol triphosphate - these products function as secondary intracellular messengers (for reviews, see Berridge, 1984; Nishizaka, 1984). The cleavage of phosphoinositides is believed to be the result of an endogenous PI-PLC. The increasing concentrations of PLAP released into maternal circulation after the first trimester of pregnancy may therefore be the result of such enzymatic activity on the syncytiotrophoblast membrane, especially since fetal membranes and uterine decidua are known to contain high concentrations of endogenous PI-PLC (Renzo et al., 1981).

PLAP may also play a role in dephosphorylating protein phosphotyrosine at neutral pH (Swarup et al., 1981). Several tyrosine kinases (which are themselves phosphotyrosyl proteins) catalyze protein phosphorylation, including those associated with the platelet-derived growth factor (PDGF), insulin and epidermal growth factor (EGF) cell membrane receptors (see below). A major role for PLAP may thus center on its ability to interact with these phosphotyrosyl-protein kinases in regulation of cellular growth following the first trimester (Risk and Johnson, 1985). However, it must be borne in mind that it is assumed the majority of StMPM-associated PLAP is orientated such that the enzymically active domain is expressed on the extracellular face of the plasma membrane, while the location of kinase activities are at the intracellular face of the plasma membrane. Nevertheless, a further complexity is that 20-30% of StMPM-associated PLAP is resistant to release by PI-PLC (Webb, unpublished). This could result from either inaccessability to the action of PI-PLC or expression of a distinct form of PLAP not associated with phosphatidylinositol. Indeed, analysis of the cytoskeletal components of isolated StMPM vesicle preparations, separated on the basis of insolubility in non-ionic detergents, has shown that 20-30% of PLAP activity remains associated with this cytoskeletal fraction (Webb, unpublished). Thus, there appears to be two populations of PLAP, one associated with the inositol phospholipids of the plasma membrane and the other associated with the submembraneous cytoskeleton.

EGF and Transferrin Receptors

Human StMPM expresses a variety of cell surface receptors, mostly involved in initial events in the selective transfer of ligands (such as IgG) from maternal circulation into fetal tissue (Johnson and Brown, 1981; Wild, 1983). Growth factors and their receptors also constitute a set of regulators of cell proliferation. The receptor for epidermal growth factor (EGF) has been extensively characterized and shown to be a 170kd glycoprotein with a large extracellular domain, a membrane-anchoring sequence, and intracellular domain with intrinsic kinase activity (Hunter, 1984). The kinase activity is tyrosine-specific and induced following EGF binding to the receptor. This receptor is strongly expressed by term StMPM and extracts prepared from whole placentae have shown EGF receptor activity to be expressed as early as 6 wks gestation (Richards et al., 1983; Carson et al., 1983). It is located primarily on syncytiotrophoblast, with the greatest number on microvillous compared with basolateral membrane preparations (Magid et al., 1985; Rao et al., 1985). Another growth factor receptor expressed by StMPM, that for insulin, is similarly equipped with tyrosine kinase activity (Avruch et al., 1982).

Such tyrosine kinase activity is associated with cellular proliferation and activation (Cohen, 1982), although the result of action of growth factors on syncytiotrophoblast receptors is unknown (syncytiotrophoblast being a non-dividing cell reflecting a state of terminal differentiation). It is possible that StMPM binding of such growth factors provides important functional signals affecting cellular differentiation and biosynthetic properties (Carson et al., 1983; Rao et al., 1985). However, higher concentrations of EGF have been reported in cord blood than in maternal blood (Scott et al., 1979). Therefore, binding of EGF to its StMPM receptor may also be the initial step for intracellular transport of this peptide hormone across the syncytiotrophoblast to underlying cytotrophoblast or other fetal tissues (Richards et al., 1983).

The transferrin receptor (like the EGF receptor) is expressed at high concentrations on the surface of human placental syncytiotrophoblast rather than on either villous or extravillous cytotrophoblast. The transferrin receptor is a glycoprotein composed of two identical 95kd subunits linked by a disulfide bridge, and its expression on cell surfaces appears to correlate with cell proliferation (Newman et al., 1983). However, for syncytiotrophoblast (a non-proliferating syncytium), it is assumed that the principle function of the transferrin receptor is the initial molecular recognition of transferrin for the selective transport of iron from mother to fetus

(Johnson et al., 1980). This could, however, be a simplistic view because the total amount of StMPM-associated transferrin receptor may be excess to the requirements of the developing fetus for iron. It has been suggested that this abundance of transferrin receptors would deplete available concentrations of maternal plasma transferrin within placental intervillous spaces and hence impair any local proliferative responses of maternal lymphocytes (Johnson et al., 1980). Furthermore, since transferrin receptors can apparently be selectively endocytosed in the absence of ligand (Watts, 1985), there may be a role for the receptor alone in the regulation of cellular activity.

Oncogenes

Trophoblast has many similarities to transformed cells in its cell surface antigenic phenotype (McLaughlin et al., 1982; Rettig et al., 1985) and ability to proliferate and invade neighboring tissues. Recent studies have shown normal human trophoblastic expression of four separate proto-oncogenes: c-myc, c-sis, c-fos, and c-fms. Transcripts of myc and sis have been localized principally within the cytotrophoblastic shell of early placentae with only weak expression by syncytiotrophoblast (Pfeifer-Ohlson et al., 1984; Goustin et al., 1985). The myc proto-oncogene is expressed by many proliferating cells of diverse origin and encodes a predominantly nuclear protein, while c-sis encodes a protein homologous to the β-chain of the secreted hormone PDGF. The presence of PDGF receptors on trophoblast has also been demonstrated, and an autocrine control of trophoblast growth postulated (Goustin et al., 1985). Both c-fos and c-fms are expressed in human trophoblast tissues, reaching maximal expression at term (Muller et al., 1983; Adamson et al., 1985). There is a wide tissue expression of c-fos, which encodes a 55kd nuclear phosphoprotein. In contrast, c-fms encodes a cell surface protein which demonstrates a greater degree of trophoblast specificity (Muller et al., 1983). Recent studies have indicated that c-fms encodes a 170kd glycoprotein with tyrosyl kinase activity which appears related, and probably identical, to the receptor for colony stimulating factor 1 (CSF1) (Sherr et al., 1985).

HLA Expression and the Immune Response

The various cell surface glycoprotein products (HLA antigens) of class I and II loci of the human major histocompatibility complex (MHC) have important properties relevant to materno-fetal immunology, both as major targets for rejection responses to genetically disparate allografts and as recognition molecules acting to restrict T cell responses to foreign antigens. Thus, class I MHC antigens must be expressed by virally infected cells for the associative recognition of viral antigens and subsequent T cell lysis of infected cells. There have now accumulated many convincing demonstrations that human placental villous trophoblast lacks expression of classical class I or II histocompatibility antigens [HLA-A, B, and DR] (Johnson, 1984; Redman et al., 1984). This is of major importance in provision of villous trophoblast with its unique protection from cytotoxic attack, but also implies that such cells may also allow viral antigen expression safe from immune attack. Indeed, villous trophoblast has been shown to contain virus-like particles, RNA-directed DNA-polymerase activity and to express several endogenous retrovirus-related antigens (Kalter, 1983; Suni et al., 1984a,b; Wahlstrom et al., 1984; Maeda et al., 1985), some of which may be expressed on StMPM (Thiry et al., 1981). Furthermore, 8% of cord sera contain detectable antibodies reactive with retroviral structural protein (Suni et al., 1981).

It has been shown by immunohistology, however, that subpopulations of extravillous cytotrophoblast in the placental bed and chorion laeve throughout gestation react strongly with several mAbs recognizing monomorphic determinants of class I HLA antigens and also express β_2-microglobulin (Redman et al., 1984; Bulmer and Johnson, 1985; Kabawat et al., 1985). Not all such mAbs reactive with class I HLA demonstrated this property (Hsi et al., 1984; Wells et al., 1984) and, significantly in the context of maternal immune recognition, there has been a failure to demonstrate expression by extravillous cytotrophoblast of allotype-specific HLA determinants (Redman et al., 1984). The same distinction between villous and extravillous trophoblast in expression of a class I MHC antigen also occurs in hydatidiform mole and ectopic tubal pregnancy (Sunderland et al., 1985; Earl et al., 1985). These results have been interpreted as representing extravillous trophoblastic expression of some hitherto unidentified class I-like MHC antigen (Redman et al., 1984; Bulmer and Johnson, 1985). It is possible that these are either truncated HLA molecules or products of genes analogous to those of the Qa-Tla region of the mouse which contains many relatively non-polymorphic class I MHC genes (Winoto et al., 1982). There is evidence that the human MHC contains several class I genes in addition to those identified as HLA-A,B,C (Strachan at al., 1984). Northern blotting analysis using a cDNA probe homologous to HLA-A,B,C sequences has shown a low level of mRNA transcripts in a human placental cytotrophoblast preparation ($<0.3\%$ of lymphocyte transcripts) (Kawata et al., 1984). However, the specific probes and the hybridization conditions used are central to interpretations of this approach and the application of different specific oligonucleotide probes to class I MHC gene loci is required.

It is known that gamma-interferon (Rosa and Fellous, 1984) and certain steroidal hormones (Klareskog et al., 1980) can enhance cellular HLA expression, whereas prostaglandins (Synder et al., 1982) may reduce cellular HLA expression. Such factors may be extremely important in influencing local regulation of HLA gene expression in uteroplacental tissues. Recently, it has been shown that gamma-interferon enhances class I MHC antigen expression in a weakly HLA-positive choriocarcionoma cell line whereas it did not induce expression in an HLA-negative choriocarcinoma cell line (Anderson and Berkowitz, 1985). It remains to be clarified which factors are important in trophoblastic transcriptional control of genes encoding, HLA, PLAP, proto-oncogenes, and cell surface receptor structures.

Developmental tumor cell types, in particular teratocarcinomas, have also been of use in providing representatives of early embryonic cell types for biochemical and immunological studies (Stern, 1983, 1984; Stern et al., 1986). There is considerable overlap in surface antigenic phenotype of embryonal carcinoma (EC) cells and several normal embryonic cell types, including trophoblast (Stern, 1984; Stern et al., 1986). Serological and cell-mediated immune studies have provided evidence that, while these EC cells do not express classical class I MHC antigens, they may express novel class I MHC type molecules (Bell and Stern, 1985; Aspinall and Stern, 1985); these are possibly products of the Qa-Tla region. Several EC cell lines have now been isolated from human teratocarcinoma tumors and shown to retain the ability to differentiate into several different cell types. The stem cells do not express HLA or β_2-

microglobulin while the differentiated cell types are HLA-positive (Thompson et al., 1984; Andrews, 1984).

It is also of interest that human EC cells express the trophoblast-leukocyte common cell surface antigen recognized by the mAb H316 (Stern et al., 1986); this antigen appears related to the TLX antigen system described by McIntyre and Faulk (1982). Heteroantisera raised against trophoblast membrane antigens inhibit the in vitro correlate of cellular immune response to an allograft, the mixed lymphocyte culture reaction (McIntyre and Faulk, 1979). These antisera have been shown to bind to lymphocytes in a specific, but HLA-independent, manner and are thought to recognize TLX antigens (McIntyre and Faulk, 1982). The mAb H316 was raised against isolated StMPM and reacts with a cell surface antigen common to all populations of human trophoblast as well as peripheral blood leukocytes, possibly recognizing a monomorphic determinant of TLX (McLaughlin et al., 1982; Bulmer and Johnson, 1985). Biochemical studies have now shown H316-reactive molecules to be glycoproteins of approximate MW's of 55kd and 65kd (Stern et al., 1986). Comparative studies of different placentae have, however, shown some size heterogeneity of their H316-reactive material. Thus, all populations of fetal trophoblast express an apparently polymorphic non-HLA cell surface antigen system that might influence events in the materno-fetal immunogenetic relationship by signaling or modulating protective maternal antibody and local infiltrating cellular responses, the failure of which could contribute to recurrent pregnancy failure.

ACKNOWLEDGEMENTS

This work was supported by grants from the Medical Research Council (UK), Cancer Research Campaign, North West Cancer Research Fund and Imperial Cancer Research Fund.

REFERENCES

Adamson, E.D., Meek, J., and Edwards, S.A. (1985) Product of the cellular oncogene, c-fos, observed in mouse and human tissues using an antibody to a synthetic peptide. *EMBO. J.* 4, 941-947.

Anderson, D.J. and Berkowitz, R.S. (1985) Gamma-interferon enhances expression of class I MHC antigens in the weakly HLA-positive human choriocarcinoma cell line BeWo but does not induce MHC expression in the HLA-negative choriocarcinoma cell line JAr. *J. Immunol.* 135, 2498-2501.

Andrews, P.W., Damjanov, I., Simon, D., Banting, G.S., Carliu, C., Dracopoli, N.C., and Fogh, J. (1984) Pluripotent embryonal carcinoma clones derived from the human teratocarcinoma cell line Tera2. *Lab. Invest.* 50, 147-162.

Aspinall, R. and Stern, P.L. (1985) Analysis of the xenogeneic T-cell response to murine H-2 negative embryonal carcinoma cells. *Immunol.* 54, 549-557.

Avruch, J., Nemenoff, R.A., Blackshear, P.J., Pierce, M.W., and Osathanondh, R. (1982) Insulin-stimulated tyrosine phosphorylation of the insulin receptor in detergent extracts of human placental membranes. *J. Biol. Chem.* 257, 15162-15166.

Bell, S.M. and Stern, P.L. (1985) Rat natural killer cells and cytotoxic T cell lysis of H-2 negative murine embryonal carcinomal cells. *Eur. J. Immunol.* 15, 59-65.

Berridge, M.J. (1984) Inositol trisphosphate and diacylglycerol as second messengers. *Biochem. J.* 220, 345-360.

Boyd, J.D. and Hamilton, W.J. (1970) *The Human Placenta.* W. Heffer and Sons Ltd., Cambridge.

Bulmer, J.N., Billington, W.D., and Johnson, P.M. (1984) Immunohistologic identification of trophoblast populations in early human pregnancy with the use of monoclonal antibodies. *Am. J. Obstet. Gynecol.* 148, 19-26.

Bulmer, J.N. and Johnson, P.M. (1985) Antigen expression by trophoblast populations in the human placenta and their possible immunobiological relevance. *Placenta* 6, 127-140.

Carson, S.A., Chase, R., Wep, E., Scommegna, A., and Benveniste, R. (1983) Ontogenesis and characteristics of epidermal growth factor receptors in human placenta. *Am. J. Obstet. Gynecol.* 147, 932-939.

Cheng, H.M. and Johnson, P.M. (1985) A description of human placental syncytiotrophoblast membrane glycosphinoglipids. *Placenta* 6, 229-238.

Cohen, P. (1982) The role of protein phosphorylation in neural and hormonal control of cellular activity. *Nature* 296, 613-620.

Covone, A.E., Mutton, D., Johnson, P.M., and Adinolfi, M. (1984) Trophoblast cells in peripheral blood from pregnant women. *Lancet* ii, 841-843.

Critchley, M., Brownless, S., Patten, M., McLaughlin, P.J., Tromans, P.M., and Johnson, P.M. (1986) Radionuclide imaging of epithelial ovarian tumours with ^{123}I-labelled monoclonal antibody (H317). *Clin. Radiol.* 37, 107-112.

Earl, U., Wells, M., and Bulmer, J.N. (1985) The expression of major histocom-patibility complex antigens by trophoblast in ectopic tubal pregnancy. *J. Reprod. Immunol.* 8, 13-24.

Ebina, Y., Ellis, L., Jarnagin, K., Edery, M., Graf, L., Clauser, E., Ou, J.H., Masiarz, F., Kan, Y.W., Goldfine, I.D., Roth, R.A., and Rutter, W.J. (1985) The human insulin receptor cDNA: the structural basis for hormone-activated transmembrane signalling. *Cell* 40, 747-758.

Epenetos, A.A., Munro, A.J., Tucker, D.F., Gregory, W., Duncan, W., MacDougall, R.H., Faux, M., Travers, P., and Bodmer, W.F. (1985a) Monoclonal antibody assay of serum placental alkaline phosphatase in the monitoring of testicular tumours. *Br. J. Can.* 51, 641-644.

Epenetos, A.A., Snook, D., Hooker, G., Begent, R., Durbin, H., Oliver, R.T.D., Bodmer, W.F., and Lavender, J.P. (1985b) Indium-111 labelled monoclonal antibody to placental aklaline phosphatase in the detection of neoplasms of testis, ovary, and cervix. *Lancet* ii, 350-353.

Ezra, E., Blacher, R., and Udenfriend, S. (1983) Purification and partial sequencing of human placental alkaline phosphatase. *Biochem. Biophys. Res. Commun.* 116, 1076-1083.

Fishman, W.H. (1974) Perspectives on alkaline phosphatase isoenzymes. *Am. J. Med.* 56, 617-650.

Goustin, A.T., Betsholtz, C., Pfeifer-Ohlsson, S., Persson, H., Rydnert, J., Bywater, M., Homgren, G., Heldin, C.-H., Westermark, B., and Ohlsson, R. (1985) Co-expression of the sis and myc proto-oncogenes in developing human placenta suggests autocrine control of trophoblast growth. *Cell* 41, 301-312.

Hakomori, S.-I. (1985) Abberant glycosylation in cancer cell membranes as focused on glycolipids. Overview and perspectives. *Cancer Res.* 45, 2405-2414.

Hertig, A.T., Rock, J., and Adams, E.C. (1956) A description of 34 human ova within the first 17 days of development. *Am. J. Anat.* 98, 435-494.

Hsi, B.-L., Yeh, C.J.-G., and Faulk, W.P. (1984) Class I antigens of the major histocompatibility complex on cytotrophoblast of the human chorion laeve. *Immunol.* 52, 621-629.

Hunter, T. (1984) The epidermal growth factor receptor gene and its product. *Nature* 311, 414-415.

Johnson, P.M. (1984) Immunobiology of the human trophoblast. In: *Immunological Aspects of Reproduction in Mammals*, (ed.), D.B Crighton, Butterworth Press, London, pp. 109-131.

Johnson, P.M. and Brown, P.J. (1981) Fc γ receptors in the human placenta. Placenta 2, 355-370.

Johnson, P.M., Brown, P.J., and Faulk, W.P. (1980) Immunobiological aspects of the human placenta. In: *Oxford Reviews of Reproductive Biology*, vol. 2 (ed.), C.A. Finn, Oxford University Press, Oxford, pp. 1-40.

Kabawat, S.E., Moustafi-Zadeh, M., Driscoll, S.G., and Bhan, A.K. (1985) Implantation site in normal pregnancy: A study with monoclonal antibodies. *Am. J. Pathol.* 118, 76-84.

Kalter, S.S. (1983) Viral expression in the trophoblast. In: *Biology of Trophoblast*, (eds.), Y.W. Loke and A. Whyte, Elsevier, Amsterdam, pp. 627-662.

Kawata, M., Parnes, J.R., and Herzenberg, L.A. (1984) Transcriptional control of HLA-A,B,C antigens in human placental cytotrophoblast isolated using trophoblast and HLA-specific monoclonal antibodies and the fluorescence-activated cell sorter. *J. Exp. Med.* 160, 633-651.

Klareskog, L., Forsum, U., and Peterson, P.A. (1980) Hormonal regulation of the expression of Ia antigens on mammary gland epithelium. *Eur. J. Immunol.* 10, 958-963.

Low, M.G. and Zilversmit, D.B. (1980) Role of phosphatidylinositol in attachment of alkaline phosphatase to membranes. *Biochem.* 19, 3913-3918.

McDicken, I.W., McLaughlin, P.J., Tromans, P.M., Luesley, D.M., and Johnson, P.M. (1985) Detection of placental-type alkaline phosphatase in ovarian cancer. *Br. J. Cancer* 52, 59-64.

McIntyre, J.A. and Faulk, W.P. (1979) Antigens of human trophoblast. *J. Exp. Med.* 149, 824-833.

McIntyre, J.A. and Faulk, W.P. (1982) Allotypic trophoblast-lymphocyte cross reactive (TLX) cell surface antigens. *Human Immunol.* 4, 27-35.

McLaughlin, P.J., Cheng, H.M., Slade, M.B., and Johnson, P.M. (1982) Expression on cultured human tumour cells of placental trophoblast membrane antigens and placental alkaline phosphatase defined by monoclonal antibodies. *Int. J. Cancer* 30, 21-26.

McLaughlin, P.J., Travers, P.J., McDicken, I.W., and Johnson, P.M. (1984) Demonstration of placental and placental-like alkaline phosphatases on non-malignant human extracts using monoclonal antibodies in enzyme immunoassay. *Clin. Chim. Acta* 137, 341-348.

Maeda, S., Yonezawa, K., and Yachi, A. (1985) Serum antibody reacting with placental syncytiotrophoblast in sera of patients with autoimmune diseases - a possible relation to type C RNA retrovirus. *Clin. Exp. Immunol.* 60, 645-653.

Magid, M., Nanney, L.B., Stoscheck, C.M., and King, L.E. (1985) Epidermal growth factor binding and receptor distributions in term human placenta. *Placenta* 6, 519-526.

Muller, R., Tremblay, J.M., Adamson, E.D., and Verma, I.M. (1983) Tissue and cell type-specific expression of two human c-onc genes. *Nature* 304, 454-456.

Nelson, D.M., Enders, A.C., and King, B.F. (1978) Cytological events involved in glycoprotein synthesis in cellular and syncytial trophoblast of human placenta. An electron microscope autoradiographic study of [3-H] galactose incorporation. *J. Cell. Biol.* 76, 418-429.

Newman, R., Schneider, C., Sutherland, R., Vodinelich, L., and Greaves, M. (1982) The transferrin receptor. *Trends Biochem. Sci.* 7, 397-400.

Nishizuka, Y. (1984) The role of protein kinase C in cell surface signal transduction and tumour promotion. *Nature* 308, 693-698.

Ogbimi, A.O. and Johnson, P.M. (1980) Further immunological and chemical characterization of human syncytiotrophoblast microvillous plasma membrane-associated components. *J. Reprod. Immunol.* 2, 99-108.

Ogbimi, A.O., Johnson, P.M., Brown, P.J., and Fox, H. (1979) Characterisation of the soluble fraction of human syncytiotrophoblast microvillous plasma membrane-associated proteins. *J. Reprod. Immunol.* 1, 127-140.

Pfeifer-Ohlsson, S., Goustin, A.S., Rydnert, J., Bjersing, L., Wahlstrom, T., Stehlin, D., and Ohlsson, R. (1984) Spatial and temporal pattern of cellular myc oncogene expression in developing human placenta: Implications for embryonic cell proliferation. *Cell* 38, 585-596.

Rao, C.V., Ramani, N., Chegini, N., Stadig, B.K., Carman, F.R., Woost, P.G., Schultz, G.S., and Cook, C.L. (1985) Topography of human placental receptors for epidermal growth factor. *J. Biol. Chem.* 260, 1705-1710.

Redman, C.W.G., McMichael, A.J., Stirrat, G.M., Sunderland, C.A., and Ting, A. (1984) Class I MHC antigens on extravillous trophoblast. *Immunol.* 52, 457-468.

Renzo, G.C.D., Johnston, J.M., Okazaki, T., Okita, J.R., MacDonald, P.C., and Bleasdale, J.C. (1981) Phosphatidylinositol-specific phospholipase C in fetal membranes and uterine decidua. *J. Clin. Invest.* 67, 847-856.

Rettig, W.J., Cordon-Carlo, C., Koulos, J.P., Lewis, J.L., Oettgen, H.F., and Old, L.J. (1985) Cell surface antigens of human trophoblast and choriocarcinoma defined by monoclonal antibodies. *Int. J. Cancer* 35, 469-475.

Richards, R.C., Beardmore, J.M., Brown, P.J., Molloy, C.M., and Johnson, P.M. (1983) Epidermal growth factor receptors on isolated human placental syncytiotrophoblast plasma membrane. *Placenta* 4, 133-138.

Risk, J.M. and Johnson, P.M. (1985) Antigen expression by human trophoblast and tumor cells: models for gene regulation? In: *Contributions in Gynecology and Obstetrics*, S. Karger, AG, Basel, in press.

Rosa, F. and Fellous, M. (1984) The effect of gamma-interferon on MHC antigens. *Immunol. Today* 5, 261-262.

Sakiyama, T., Robinson, J.C., and Chou, J.Y. (1979) Characterization of alkaline phosphatases from first-trimester human placentae. *J. Biol. Chem.* 254, 935-938.

Scott, I.V., Bardsley, W.G., Gregory, H., and Tindall, V.R. (1979) Human placental diamine oxidase concentration and urogastrone: a possible role in polyamine metabolism and fetal growth. In: *Proceedings of the 8th International*

Symposium on Clinical Enzymology, (eds.), A. Berlina and L. Galzingera, Piccini Medical Books, Padua and London, pp. 563- 578.

Sherr, C.J., Rettenmier, C.W., Sacca, R., Roussel, M.R., Look, A.T., and Stanley, E.R. (1985) The c-fms proto-oncogene product is related to the receptor for the mononuclear phagocyte growth factor, CSF-1. *Cell* 41, 665-676.

Shukla, S.D. (1982) Phosphatidylinosiotol specific phospholipase C. *Life Sci.* 18, 1323-1335.

Smith, N.C., Brush, M.G., and Luckett, S. (1974) Preparation of human placental villous surface membranes. *Nature* 252, 302-393.

Snyder, D.S., Beller, D.I. and Unanue, E.R. (1982) Prostaglandins modulate macrophage Ia expression. *Nature* 299, 163-165.

Stern, P.L. (1983) Serological and cell-mediated immune recognition of teratocarcinomas. In: *Current Problems in Germ Cell Differentiation*, (eds.), A. McLaren and C.C. Wylie, Cambridge University Press, pp. 157-173.

Stern, P.L. (1984) Differentiation antigens of embryonal carcinoma cells and embryos. *Brit. Med. Bull.* 40, 218-223.

Stern, P.L., Beresford, N., Thompson, S., Johnson, P.M., Webb, P.D., and Mole, N. (1986) Characterization of the human trophoblast-leukocyte antigenic molecules defined in a monoclonal antibody. *J. Immunol.*, in press.

Strachan, T., Sodoyer, R., Damotte, M., and Jordan, B.R. (1984) Complete nucleotide sequence of a functional class I HLA gene, HLA-A3: Implications for the evolution of HLA genes. *EMBO. J.* 3, 887-894.

Sunderland, C.A., Redman, C.W.G., and Stirrat, G.M. (1985) Characterization and localization of HLA antigens on hydatidiform mole. *Am. J. Obstet. Gynecol.* 151, 130-135.

Suni, J., Narvaren, A., Wahlstrom, T., Aho, M., Pakkanen, R., Vaheri, A., Copeland, T., Cohen, M., and Oroszlan, S. (1984a) Human placental syncytiotrophoblast Mr 75,000 polypeptide defined by antibodies to a synthetic peptide based on a cloned human endogenous retroviral DNA sequence. *Proc. Natl. Acad. Sci., USA*, 81, 6197-6201.

Suni, J., Narvanen, A., Wahlstrom, T., Lehtovirta, P., and Vaheri, A. (1984b) Monoclonal antibody to human T-cell leukemia virus p19 defines polypeptide antigen in human choriocarcinoma cells and syncytiotrophoblast of first trimester placentas. *Int. J. Cancer* 33, 293-298.

Suni, J., Wahlstrom, T., and Vaheri, A. (1981) Retrovirus p30-related antigen in human syncytiotrophobalst and IgG antibodies in cord sera. *Int. J. Cancer* 28, 559-566.

Swarup, G., Cohen, S., and Garberg, D.L. (1981) Selective dephosphorylation of proteins containing phosphotyrosine by alkaline phosphatase. *J. Biol. Chem.* 256, 8197-8201.

Thiry, L., Loke, Y.W., Hard, R.C., Sprecher-Goldberger, S., and Buekens, P. (1981) Heterologous antiserum to human syncytiotrophoblast membrane is cytotoxic to retrovirus-producing cells and to some cancer cell lines. *Am. J. Reprod. Immunol.* 1, 241-245.

Thompson, S., Stern, P.L., Webb, M., Walsh, F.S., Engstrom, W., Evans, E.P., Shi, W.-K., Hopkins, B., and Graham, C.F. (1984) Differentiation of neuron-like cells and other cell types from cloned human teratoma cells cultured in retinoic acid. *J. Cell Sci.* 72, 37-64.

Truman, P., and Ford, H.C. (1984) The brush border of the human term placenta. Biochim. Biophys. Acta 779, 139-160.

Wahlstrom, T., Nieminen, P., Narvanen, A., Suni, J., Lehtoviria, P., Saskela, A., and Vaheri, A. (1984) Monoclonal antibody defining a human syncytiotrophoblast polypeptide immunologically related to mammalian retrovirus structural protein p30. *Placenta* 5, 465-474.

Watts, C. (1985) In situ ^{125}I-labelling of endosome proteins with lactoperoxidase conjugates. *EMBO. J.* 3, 1965-1970.

Webb, P.D., Evans, P.W., Molloy, C.M., and Johnson, P.M. (1985) Biochemical studies of human placental microvillous plasma membrane proteins. *Am. J. Reprod. Immunol. Microbiol.* 8, 113-119.

Webb, P.D., McLaughlin, P.J., Risk, J.M., and Johnson, P.M. (1986) Isolation of placental-type alkaline phosphatase associated with human syncytiotrophoblast membranes using monoclonal antibodies. *Placenta*, in press.

Wells, M., Hsi, B.-L., and Faulk, W.P. (1984) The expression of class I antigens of the major histocompatibility complex on the cytotrophoblast shell cells of the placental basal plate. *Am. J. Reprod. Immunol.* 6, 167-174.

Whyte, A. (1983) Biochemistry of the human syncytiotrophoblast plasma membrane. In: *Biology of Trophoblast*, (eds.), Y.W. Loke and A. Whyte, Elsevier, Amsterdam, pp. 513-529.

Wild, A.E. (1983) Trophoblast cell surface receptors. In: *Biology of Trophoblast*, (eds.), Y.W. Loke and A. Whyte, Elsevier, Amsterdam, pp. 472-505.

Winoto, A., Steinmetz, M., and Hood, L. (1983) Genetic mapping in the major histocompatbility complex by restriction enzyme polymorphisms. Most mouse class I genes map to the Tla complex. *Proc. Natl. Acad. Sci., USA*, 80, 3425-3429.

Trophoblast Research 2:17-27, 1987

IDENTIFICATION OF DISTINCT RECEPTORS FOR NATIVE AND ACETYLATED-LOW-DENSITY LIPOPROTEIN IN HUMAN PLACENTAL MICROVILLI

Eliane Alsat[1], Francoise Mondon[1], André Malassiné[1],
Regis Rebourcet[2], Sonia Goldstein[2], and Lise Cedard[1,3]

[1]Maternité Baudelocque
INSERM U. 166
123 Bld de Port-Royal
75014 Paris, France

[2]Hopital St-Antoine
CNRS UA 524
27 rue Chaligny
75012 Paris, France

INTRODUCTION

Human placental progesterone biosynthesis is essentially dependent upon the maternal cholesterol (Hellig et al., 1970) delivered from plasma low-density lipoprotein (LDL), the major cholesterol-carrier protein in the human (Winkel et al., 1980a; Winkel et al., 1981; Simpson and MacDonald, 1981; Gwynne and Strauss, 1982). Cultured trophoblastic cells take up and degrade LDL by a receptor mediated endocytosis process (Winkel et al., 1980b), similar to that initially described by Goldstein and Brown (1974) for human fibroblasts in culture. Placental binding sites specific for LDL have been identified simultaneously, in a crude membranous fraction (Cummings et al., 1982) and on microvillous membranes, the effective site of exchange between the maternal blood and the syncytiotrophoblast (Alsat et al., 1982). Further investigations have allowed the characterization of the LDL binding sites, which are detected on microvillous membranes as early as the 6th wk of pregnancy (Alsat et al., 1984). Their presence on term placental microvilli has been confirmed by an ultrastructural study of villi incubated with dense electron conjugates: 80% of ferritin-LDL or colloidal gold-LDL are bound to sites sparsely distributed over the membrane surface of the microvilli (Malassiné et al., 1984).

However, during pregnancy, the maternal plasma LDL are increased and changed in composition (Williams et al., 1976), the potential occurence in the placenta of receptors for modified LDL, so-called "scavenger receptors", has been investigated. For this purpose, acetylated-LDL (acetyl-LDL) which present a modification of about 20% of their lysine residues and an increased net negative charge, were used after iodination, as a ligand for the binding to microvilli isolated for mature and immature trophoblast. The binding characteristics of ^{125}I-acetyl-LDL were compared to those

[3]To whom correspondence should be addressed.
Received: 10 October 1985; Accepted: 15 March 1986

previously reported for the [125]-I-LDL binding. The results obtained implied the presence on mature and immature placental microvilli of distinct binding sites specific for native LDL and for acetyl-LDL. We have subsequently investigated by the ligand-immunoblotting method of Towbin et al. (1979) whether these binding sites for LDL and for acetyl-LDL may be detected as two distinct receptor-proteins.

MATERIALS AND METHODS

- **Placental microvillous membranes** were prepared from either term placentae or immature placentae. Term placentae were collected immediately after elective cesarean section or vaginal delivery from healthy mothers. Immature placentae were obtained from early gestation (8-12 wks of amenorrhoea), or mid-gestation (20-22 wks of amenorrhoea) in mothers undergoing legal abortion by vacumm-curettage or therapeutic prostaglandin abortion. The placental microvilli were isolated by cold saline extraction and centrifugation as described elsewhere (Alsat et al., 1982; 1984). The purity of the preparations were assessed by the analysis of specific microvillous membrane enzymes (17 to 25-fold enrichment in alkaline phosphatase) and by ultrastructural studies (placental microvilli were found as largely predominant constituents in the preparations).

- **Lipoproteins**. Human LDL (d.1.024-1.050 g/ml), VLDL (d < 1.006 g/ml), HDL (d.1.070-1.21 g/ml) and HDL_3 (d.1.125-1.215 g/ml) were prepared from normolipidaemic plasma by sequential ultracentrifugation (Havel et al., 1955). The acetylation of LDL was performed with acetic anhydride (Goldstein et al., 1979). The efficiency of acetylation was established by amino-acid analysis (modification of about 20% of lysine residues) and by agarose gel electrophoresis (increased mobility). Native LDL and acetyl-LDL were labeled with ^{125}I-Na, according to the method of Bilheimer et al., (1972).

- **Binding assays** of ^{125}I-LDL and ^{125}I-acetyl-LDL were determined as previously described (Alsat et al., 1982) according to Basu et al. (1978). Specific binding was calculated as difference between the binding in the absence or presence of an excess of unlabeled ligand (≈ 0.5 mg/ml).

- **Electrophoresis and ligand-immunoblotting of membrane proteins**. Microvillous membrane pellets were solubilized with Triton X 100, then subjected to SDS-PAGE and electrophoretic transfer to nitrocellulose sheets, under conditions described by Beisiegel et al. (1982). After transfer, strips were first incubated with the ligand (LDL or acetyl-LDL) and subsequently with rabbit antiserum to LDL or to acetyl-LDL. The visualization was achieved by incubation with a second antibody (peroxydase- conjugated sheep anti-rabbit IgG) according to Rebourcet et al. (1986).

RESULTS

^{125}I-LDL and ^{125}I-Acetyl-LDL binding to placental microvilli

The binding equilibrium is rapidly reached at 4°C, and remains constant for up to 80 min for both LDL and acetyl-LDL. The specific binding of each ligand, as a function of lipoprotein concentration, is saturable (Figure 1). Term placental microvilli exhibit high affinities for the binding of both ligands with apparent K_D

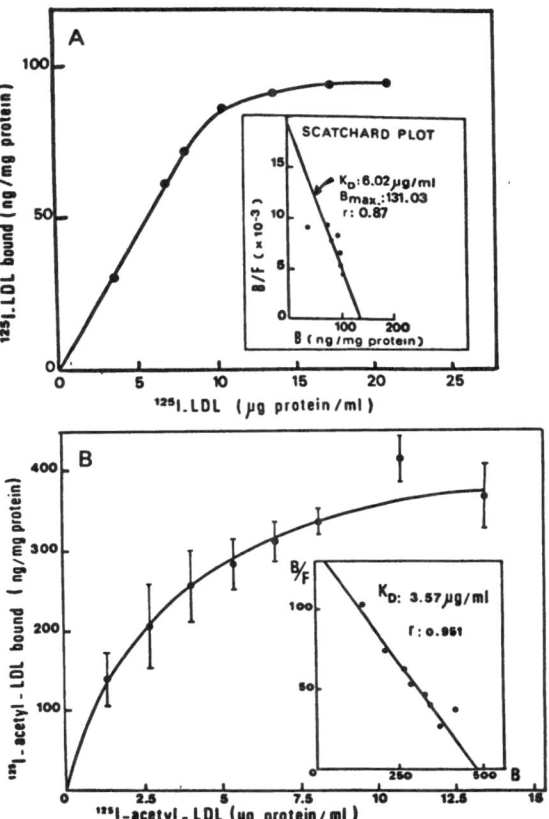

Figure 1. Saturation curves for the specific binding of ^{125}I-LDL (A) and ^{125}I-acetyl-LDL (B) to term placental microvilli. The specific binding was obtained by subtracting nonspecific binding from total binding. Each point represents the mean value of triplicate determinations for a representative experiment (A) or for 3 different microvillous preparations (B). Scatchard plots obtained from these data are shown in insets.

values of 6 µg protein/ml for ^{125}I-LDL and 3.5 µg protein/ml for ^{125}I-acetyl-LDL, as calculated by Scatchard plots (insets). It is of note that the affinity is about 2-fold higher for ^{125}I-acetyl-LDL than for ^{125}I-LDL. In the same way, for a given microvillous preparation, the binding capacity is higher for ^{125}I-acetyl-LDL than for ^{125}I-LDL.

The specificity of ^{125}I-LDL and ^{125}I-acetyl-LDL binding is assessed by competition studies with unlabeled LDL, acetyl-LDL, VLDL and HDL. Increasing concentrations of unlabeled competing lipoproteins inhibits the specific binding of ^{125}I-LDL (A) or ^{125}I-acetyl-LDL (B) (Figure 2). Unlabeled native LDL or acetyl-LDL effectively compete with the homologous iodinated lipoprotein, whereas HDL and VLDL are markedly less effective or totally ineffective. The specificity of the binding

Alsat et al.

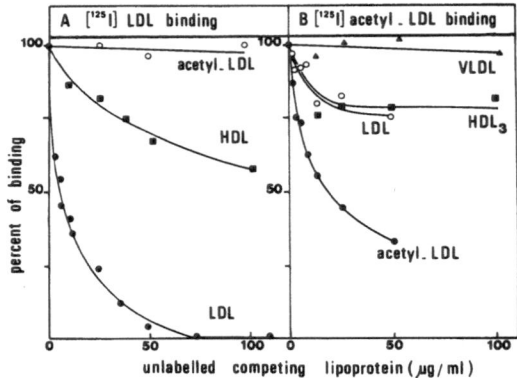

Figure 2. Effect of increasing concentrations of unlabeled lipoproteins on the binding of 125I-LDL (A) and 125I-acetyl-LDL (B) to term placental microvilli. Each point represents the mean value of triplicate determinations and is corrected for the nonspecific binding.

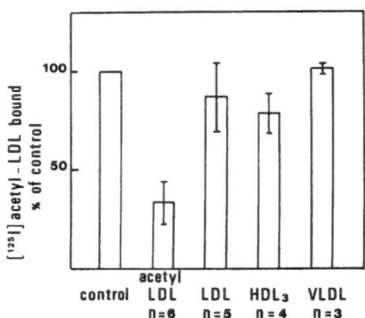

Figure 3. Comparison of different unlabeled lipoproteins to compete with 125I-acetyl-LDL for the binding to term placental microvilli. Incubations were performed in the absence (control) or presence of 50 µg/ml unlabeled acetyl-LDL, or 49.2 µg/ml unlabeled native LDL, or 50.4 µg/ml unlabeled HDL$_3$ or 52.5 µg/ml unlabeled VLDL. Values are corrected for nonspecific binding and represent the means $\pm$ SD of the indicated number of assays.

sites for LDL or for acetyl-LDL has been determined with microvilli isolated from several term placentae. A significant inhibition of 66.3 $\pm$ 11.2% (mean $\pm$ SD) for 125I-acetyl-LDL binding is observed only when unlabeled acetyl-LDL ($\simeq$ 50 µg/ml) are present in the incubation media (Figure 3). These data indicate that microvilli isolated from human term placenta possess distinct binding sites for native LDL and for acetyl-LDL.

Table 1

Comparison of the ^{125}I-LDL and ^{125}I-acetyl-LDL binding to microvilli isolated from placentae obtained at different gestational ages. The binding capacity (Bmax) and the affinity (K_D) for each ligand, are obtained form Scatchard analyses. The values of Bmax are means with the extreme values indicated in brackets, and the values of K_D are means$\pm$SD; n indicates the number of different microvillous preparations tested in each age group.

Gestational age of placental microvilli (weeks of amenorrhoea)	^{125}I-LDL binding		^{125}I-acetyl-LDL binding	
	Bmax	K_D	Bmax	K_D
	(ng/mg protein)	(µg/ml)	(ng/mg protein)	(µg/ml)
8 - 10	225 (17 - 1170) (n = 15)	6.98 $\pm$ 0.83 (n = 5)	702.5 (483 - 954) (n = 6)	2.46 (n = 1)
20 - 22	115 (27 - 410) (n = 7)	6.57 $\pm$ 0.81 (n = 5)	157.8 (n = 1)	N.D.
term	270 (24 - 1165) (n = 19)	6.12 $\pm$ 1.32 (n = 6)	335.1 (159 - 502) (n = 7)	3.63 $\pm$ 1.16 (n = 7)

The specific binding sites for LDL or for acetyl-LDL are sensitive to proteolytic digestion: a 30 min. exposure of placental microvilli to increasing concentrations of pronase is followed by a progressive decrease in the binding capacity (50-60 µg/ml of pronase suppresses the binding of ^{125}I-LDL and ^{125}I- acetyl-LDL).

The binding data for both ligands at different gestational periods are compared in Table 1. Immature placental microvilli are able to bind specifically ^{125}I-LDL and ^{125}I-acetyl-LDL, as early as the 6th wk of pregnancy. The affinity for ^{125}I-LDL remains constant throughout the pregnancy, as indicated by similar K_D values obtained for all of the gestational age groups (6.98 ± 0.83, 6.57 ± 0.81 and 6.12 ± 1.32 µg protein LDL/ml for early, mid-term and term gestation microvilli, respectively). The binding capacity (Bmax) for ^{125}I-LDL exhibits large variations in each age group, without significant differences among them, although the mid-term microvilli possess a smaller number of binding sites than early and term microvilli, when determined under identical experimental conditions. For ^{125}I-acetyl-LDL, the Bmax values are more closely related in each age group, and early placental microvilli possess a higher number of binding sites for acetyl-LDL, than term microvilli ($p < 0.001$, Student's t-test). Moreover, as reported above for term microvilli, we have observed in early placenta a higher number of high affinity binding sites for ^{125}I-acetyl-LDL than for ^{125}I-LDL.

Ligand-immunoblotting of membrane proteins

The results obtained by ligand-immunoblotting are shown on Figure 4. The incubation with LDL, as ligand, and then with the homologous antibody reveals a LDL receptor-protein with an apparent molecular weight of 160,000 (lane 1); a weaker band is observed in the 120 KDa region. Blotting with acetyl-LDL and then with specific antiserum to acetyl-LDL leads to the detection of a 200 KDa protein, corresponding to the acetyl-LDL receptor (lane 4). The antiserum to acetyl-LDL does not detect the binding of native LDL (lane 2) and conversely, the antiserum to LDL does not detect the binding of acetyl-LDL (lane 3). In simultaneous control experiments, ligand immunoblotting was performed with membranes from cells known to possess only the LDL receptor, such as bovine adrenal cells and human skin fibroblasts, or also the acetyl-LDL receptor, such as J774 macrophage-like cells (data not shown).

DISCUSSION

The ability of microvillous membranes isolated from human placenta, to specifically bind LDL and acetyl-LDL, as early as the sixth week of pregnancy, is demonstrated. The binding sites for each ligand are distinct, as shown by the competition studies with unlabeled lipoproteins: when LDL and acetyl-LDL binding are compared, higher maximal binding capacity and high affinity are reported for acetyl- LDL binding sites. Moreover, ligand immunoblotting enables us to detect two different receptor proteins for LDL and for acetyl-LDL, of 160 KDa and 200 KDa, respectively, molecular weights similar to those reported in other cells (Beisiegel et al., 1982; Dresel et al., 1984).

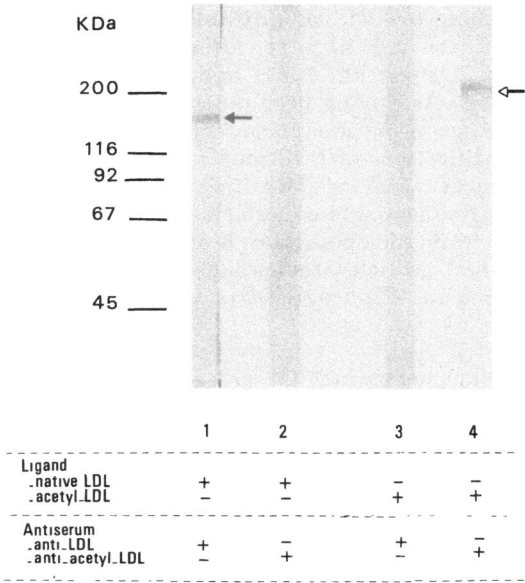

KDa

200 —

116 —
92 —

67 —

45 —

	1	2	3	4
Ligand				
native LDL	+	+	–	–
acetyl-LDL	–	–	+	+
Antiserum				
anti-LDL	+	–	+	–
anti-acetyl-LDL	–	+	–	+

Figure 4. Ligand immunoblotting of partially purified LDL and acetyl-LDL receptors from term placental microvillous membranes. After solubilization and DEAE-cellulose chromatography, the proteins were separated by the SDS-PAGE and transferred electrophoretically to nitrocellulose sheets. The strips were incubated with ligand, then with a rabbit antiserum, as indicated, and the visualization was achieved with a peroxydase-conjugated sheep anti-rabbit IgG.

The presence of the classical LDL receptor in human steroidogenic organs is well known; thus, it has been described and related to the steroid production in human corpus luteum (Carr et al., 1981), in human fetal adrenals (Ohashi et al., 1981) and in human fetal testes (Carr et al., 1983). However, the presence on placental membranes, of the "scavenger receptor" is rather unexpected; until now, essentially human monocyte-macrophages have been reported to possess both LDL and modified-LDL receptors (Brown et al., 1980; Fogelman et al., 1980; Traber et al., 1980). Therefore, the main question which arises from the presence of such binding sites for acetyl-LDL in our preparations is their possible contamination with membranes of other cells known to possess receptors for modified-LDL, i.e., placental macrophages or blood monocytes. Several lines of evidence argue against this possibility: the method used for the preparation of microvilli includes multiple washings and centrifugations, but does not involve tissue homogenization. Moreover, the absence of such contaminants was assessed by repeated ultrastructural studies, as previously reported (Alsat et al., 1982, 1984, 1985). Even though such contaminants may be present and escape our investigations, their amount would be too low to explain the binding capacity for acetyl-LDL obtained in the microvillous membranes.

Modifications leading to LDL recognizable by the "scavenger receptor" of macrophages, has been reported, in vitro, after incubation of native LDL with endothelial cells (Henriksen et al., 1981) and, in vivo, in human aortic cells (Clevidence et al., 1984). As during pregnancy, the maternal plasma LDL are modified (in charge, composition and/or conformation), we speculate that these naturally occurring modifications of LDL may explain the presence, on placental microvilli, of the "scavenger receptors", identified in this study with LDL modified by chemical treatment. Preliminary studies performed with LDL isolated from maternal peripheral blood, and from retroplacental blood, are in agreement with this hypothesis; indeed, the LDL particles from pregnant women are able to compete, like unlabeled acetyl-LDL, with ^{125}I-acetyl-LDL for the binding to the placental "scavenger receptor".

This current study demonstrates the presence in placental microvilli of both types of LDL receptors, described for the first time in steroidogenic cells. These LDL receptors may be related to the high amount of cholesterol required by the human placenta during pregnancy, for its cellular growth and its intensive progesterone synthesis.

SUMMARY

Distinct binding sites for native LDL and for acetyl-LDL were characterized in microvilli isolated from term placenta. For each ligand, the binding sites are saturable, specific and sensitive to proteolysis. Scatchard analysis of the binding data for term placenta reveals a 2-fold higher affinity for ^{125}I-acetyl-LDL than for ^{125}I-LDL (app. K_D values of 3.63 ± 1.16 µg/ml and 6.12 ± 1.32 µg/ml, respectively).

Immature placental microvilli bind specifically ^{125}I-LDL and ^{125}I-acetyl-LDL, as early as the 6th wk of pregnancy. While the number of binding sites for LDL (expressed per mg of microvillous proteins) is similar in microvilli from early or term placenta, the number of binding sites for acetyl-LDL is 2-fold higher ($p < 0.001$) in early than in term microvilli.

These distinct binding sites for either LDL or acetyl-LDL have been visualized by ligand-immunoblotting, as two different receptor proteins with molecular weights of 160 KDa and 200 KDa, respectively.

Results in the present study demonstrate that placental microvilli possess distinct receptors for native and modified-LDL. Whether receptors for modified-LDL are implied in uptake and release of cholesterol in the human placenta, remains to be determined. Nevertheless, the presence of these high affinity receptors for modified-LDL, as early as the 6th wk of pregnancy, allows us to suggest that they may contribute to the intensive supply of cholesterol during pregnancy for placental cellular growth and progesterone synthesis.

ACKNOWLEDGEMENTS

The authors gratefully acknowledge Miss Martine Berthelier for her technical assistance and Mrs. Michèle Verger for her skillful secretarial assistance. This study was partially supported by grants from INSERM (CRE no 84.70.04).

REFERENCES

Alsat, E., Bouali, Y., Goldstein, S., Malassiné, A., Laudat, M.H., and Cedard, L. (1982) Characterization of specific low-density lipoprotein binding sites in human term placental microvillous membranes. *Mol. Cell. Endocrinol.* 28, 439-453.

Alsat, E., Bouali, Y., Goldstein, S., Malassiné, A., Berthelier, M., Mondon, F., and Cedard, L. (1984) Low-density lipoprotein binding sites in the microvillous membranes of human placenta at different stages of gestation. *Mol. Cell. Endocrinol.* 38, 197-203.

Alsat, E., Mondon, F., Rebourcet, R., Berthelier, M., Erlich, D., Cedard, L., and Goldstein, S. (1985) Identification of specific binding sites for acetylated low-density lipoprotein in microvillous membranes from human placenta. *Mol. Cell. Endocrinol.* 41, 229-235.

Basu, S.K., Goldstein, J.L., and Brown, M.S. (1978) Characterization of the low-density lipoprotein receptor in membranes prepared from human fibroblasts. *J. Biol. Chem.* 253, 3852-3856.

Beisiegel, U., Schneider, W.J., Brown, M.S., and Goldstein, J.L. (1982) Immunoblot analysis of low-density lipoprotein receptors in fibroblasts from subjects with familial hypercholesterolemia. *J. Biol. Chem.* 257, 13150-13156.

Bilheimer, D.W., Eisenberg, S., and Levy, R.I. (1972) The metabolism of very low-density lipoproteins. I. Preliminary in vivo and in vitro observations. *Biochim. Biophys. Acta* 260, 212-221.

Brown, M.S., Basu, S.K., Falck, J.R., Ho, Y.K., and Goldstein, J.L. (1980) The scavenger cell pathway for lipoprotein degradation: specificity of the binding site that mediates the uptake of negatively-charged LDL by macrophages. *J. Supramol. Struct.* 13, 67-81.

Carr, B.R., Sader, R.K., Rochelle, D.B., Stalmach, M.A., McDonald, P.C., and Simpson, E.R., (1981) Plasma lipoprotein regulation of progesterone biosynthesis by human corpus luteum tissue in organ culture. *J. Clin. Endocrinol. Metab.* 52, 875-881.

Carr, B.R., Parker, C., Ohashi, M., McDonald, P.C., and Simpson, E.R. (1983) Regulation of human fetal testicular secretion of testosterone: low density lipoprotein-cholesterol and cholesterol synthesized de novo as steroid precursor. *Am. J. Obstet. Gynecol.* 146, 241-247.

Clevidence, B.A., Morton, R.E., West, G., Dusek, D.M., and Hoff, H.F. (1984) Cholesterol esterification in macrophages, stimulation by lipoproteins containing ApoB isolated from human aortas. *Arteriosclerosis* 4, 196-207.

Cummings, S.W., Hatley, W., Simpson, E.R., and Ohashi, M. (1982) The binding of high and low-density lipoproteins to human placental membrane fractions. *J. Clin. Endocrinol. Metab.* 54, 903-908.

Dresel, H.A., Weigel, I.O.H., Schettler, G., and Via, D.P. (1984) A simple and rapid anti-ligand enzyme immunoassay for visualization of low-density lipoprotein membrane receptors. *Biochim. Biophys. Acta* 795, 452-457.

Fogelman, A.M., Shechter, I., Seager, J., Hokom, M., Child, J.S., and Edwards, P.A. (1980) Malondialdehyde alteration of low-density lipoproteins leads to cholesteryl ester accumulation in human monocyte-macrophages. *Proc. Natl. Acad. Sci. USA* 77, 2214-2218.

Goldstein, J.L., Ho, Y.K., Basu, S.K., and Brown, M.S. (1979) Binding sites on macrophages that mediates uptake and degradation of acetylated LDL producing massive cholesterol deposition. *Proc. Natl. Acad. Sci. USA* 76, 333-337.

Goldstein, J.L., and Brown, M.S. (1974) Binding and degradation of low-density lipoprotein by cultured human fibroblasts. *J. Biol. Chem.* 249, 5153-5162.

Gwynne, J.T., and Strauss III, J.F. (1982) The role of lipoproteins in steroidogenesis and cholesterol metabolism in steroidogenic glands. *Endocr. Rev.* 3, 299-329.

Havel, R.J., Eder, H.A., and Bragdon, J.H. (1955) The distribution and chemical composition of ultracentrifugally separated lipoproteins in human serum. *J. Clin. Invest.* 34, 1345-1353.

Hellig, H., Gattereau, D., Lefebvre, Y., and Bolte, E. (1970) Steroid production from plasma cholesterol. I. Conversion of plasma cholesterol to placental progesterone in humans. *J. Clin. Endoc. Metab.* 30, 624-631.

Henriksen, T., Mahoney, E.M., and Steinberg, D. (1981) Enhanced macrophage degradation of LDL previously incubated with cultured endothelial cells: Recognition by receptors for acetylated-LDL. *Proc. Natl. Acad. Sci. USA* 78, 6499-6503.

Malassiné, A., Goldstein, S., Alsat, E., Merger, Ch., and Cedard, L. (1984) Ultrastructural localization of low-density lipoprotein binding sites on the surface of the syncytial microvillous membranes of the human placenta. *IRCS Med. Sci.* 12, 166-167.

Ohashi, M., Carr, B.R., and Simpson, E.R. (1981) Effects of adrenocorticotropic hormone on low-density lipoprotein receptors of human fetal adrenal tissue. *Endocrinol.* 108, 1237-1242.

Rebourcet, R., Alsat, E., Mondon, F., Berthelier, M., Pastier, D., Cedard, L., and Goldstein, S. (1986) Detection of different receptors for native and acetylated low-density lipoproteins in human placental microvilli by ligand immunoblotting. *Biochimie*, in press.

Simpson, E.R. and McDonald, P.C. (1981) Endocrine physiology of the placenta. *Ann. Rev. Physiol.* 43, 163-188.

Traber, M.G., and Kayden, H.J. (1980) Low-density lipoprotein receptor activity in human monocyte derived-macrophages and its relation to atheromatous lesions. *Proc. Natl. Acad. Sci. USA* 77, 5466-5470.

Towbin, H., Staehelin, T., and Gordon, J. (1979) Electrophoretic transfer of proteins from polyacrylamide gels to nitrocellulose sheets: procedure and some applications. *Proc. Natl. Acad. Sci USA* 76, 4350-4354.

Williams, P.F., Simons, L.A., and Turtle, J.R. (1976) Plasma lipoproteins in pregnancy. *Hormone Res.* 7, 83-90.

Winkel, C.A., Snyder, J.M., McDonald, P.C., and Simpson, E.R. (1980a) Regulation of cholesterol and progesterone synthesis in human placental cells in culture by serum lipoproteins. *Endocrinol.* 106, 1054-1060.

Winkel, C.A., Gilmore, J., McDonald, P.C., and Simpson, E.R. (1980b) Uptake and degradation of lipoproteins by human trophoblastic cells in primary culture. *Endocrinol.* 107, 1892-1898.

Winkel, C.A., McDonald, P.C., Hemsell, P.G., and Simpson, E.R. (1981) Regulation of cholesterol metabolism by human trophoblastic cells in primary culture. *Endocrinol.* 109, 1084-1090.

Trophoblast Research 2:29-43, 1987

INFLUENCE OF THE LIPID ENVIRONMENT ON INSULIN BINDING TO PLACENTAL MEMBRANES FROM NORMAL AND DIABETIC MOTHERS

Gernot Desoye and Peter A.M. Weiss

Department of Obstetrics and Gynecology
University of Graz
A-8036 Graz, Austria

INTRODUCTION

It is generally accepted that the function of membrane proteins is affected by the fluidity of the membrane (Sandermann, 1978; Kates and Kuksis, 1980; Shinitzky et al., 1980; Chapman, 1983). Temperature (Lee, 1977), the protein to lipid content (Shinitzky et al., 1980) and the lipid composition (Phillips et al., 1969; Borochov et al., 1979) are the major determinants of membrane fluidity. Effects of fluidity on receptor specificity and affinity have been reported for the thyrotropin receptor (Mehdi et al., 1977; Lee et al., 1978), for the serotonin receptor (Heron et al., 1980) and for the insulin receptor (Amatruda and Finch, 1979; Grunfeld et al., 1981; McCaleb and Donner, 1981; Ginsberg et al., 1981; Gould et al., 1982; Bar et al., 1984). The studies directed to the modulation of membrane proteins by the lipid environment have been accomplished on cultured cells by dietary manipulations leading to a modification of either the fatty acid composition or phospholipid headgroups. Corresponding studies with isolated membranes used physical techniques which altered the bulk fluidity of the membrane or by treatment of the membranes with phospholipases (Gould and Ginsberg, 1984). In the present study we chose a different approach. The affinity of insulin receptors from various tissues has been repeatedly shown to be altered in diabetes mellitus (Andreani et al., 1981). We analyzed the insulin receptors in placental membranes from normal and diabetic mothers and correlated the receptor affinities with parameters which are known to determine membrane fluidity. Thus, we did not study the insulin receptor system in an artificially altered lipid environment but investigated how the receptor affinity is affected by an in vivo modification of the membrane as a result of a pathological state.

MATERIALS AND METHODS

Neuraminidase (EC3.2.1.18) from vibrio cholerae (Behringwerke, Marburg, FRG). Phospholipase-A2 (Plase A2, EC 3.1.1.4) from porcine pancreas, phospholipase-C (Plase C, EC 3.1.4.3) from bacillus cereus, phospholipase-D (Plase D, EC 3.1.4.4) from peanut, all other enzymes, substrates, bovine serum albumin (BSA) and 4-methyl 12, 6-di-tert-butylphenol (BHT) were obtained from SIGMA Chemie (Taufkirchen, FRG). Monoiodinated (A14Tyr-^{125}I)-insulin and (4-^{14}C)-cholesterol

Received: 8 October 1985; Accepted : 29 March 1986.

were from Amersham (England), insulin from Novo (Vienna, Austria). Fatty acid methyl-ester (FAME) standards were obtained from Larodan Fine Chemicals (Malmo, Sweden) and from Pharmacia (Uppsala, Sweden). Boronfluoride-methanol and all other reagents and solvents were from Merck (Darmstadt, FRG) at the highest available purity.

Preparation of Membranes

Human full term placentae were collected within 30 min after delivery, dissected from amnion and chorion and washed free of blood in 50 mM Tris/HCl, pH 7.4. About 60-80 g of tissue were dissected from the center of cotyledons and washed again. Membranes were prepared according to Haour and Bertrand (1974). All steps were carried out at 4°C. Protein (prot) content was measured by the method of Lowry et al. (1951) using BSA as standard.

Membrane Enzymes and Sialic Acids

5'-nucleotidase (EC 3.1.3.5) was measured as described by Arkejsteijn (1976), alkaline phosphatase (EC 3.1.3.1) according to Bowers and McCombe (1966), beta-glucuronidase (EC 3.2.1.31) by the method of Fishman et al. (1967) with the modification of 1 hr incubation at 56°C, and isocitrate-dehydrogenase (EC 1.1.1.42) was determined according to Bowers (1959). Sialic acids were analyzed by the thiobarbituric acid method of Warren (1963).

Insulin Receptor Analysis

Placental membranes (approx. 0.2 mg protein/tube) were incubated in 50 mM Tris/HCl, pH 7.4, at 4°C with monoiodinated insulin (30000 cpm) in the presence of increasing concentrations of unlabeled insulin, ranging from 1.65 pM to 49.5 nM. For each of the 22 different concentrations of insulin at least 3 parallel incubations were performed. Total assay volume was 0.5 ml. After 20 hr, the membrane suspensions were diluted with 2 ml incubation buffer and centrifuged. The supernatant containing unbound insulin was discarded. Non-specific binding was determined in the presence of 4.95 µM unlabeled insulin. Radioactive disintegrations of bound tracer were counted in a 16 channel counter (Nuclear Enterprises NE1600) at approx. 65% counting efficiency. Degradation of insulin was measured by precipitation with tricholoracetic acid.

Treatment with Phospholipases

Membranes were preincubated with Plase A2 (100 U/mg prot.), Plase C (25 U/mg prot.) and Plase D (100 U/mg prot.) for 30 min at 37°C in KRP (130 mM NaCl, 4.2 mM KCl, 1.2 mM $MgSO_4$, 0.75 mM $CaCl_2$, and 10 mM Na_2HPO_4, pH 7.6). Insulin binding was then determined as outlined above with 33 pM and 3.3 nM unlabeled insulin. Non-specific binding was measured in the presence of 330 nM insulin.

Data Reduction

Analysis of insulin binding was performed by fitting of data in terms of the actual measured quantity (cpm) according to a non-cooperative model comprising one class of binding sites described by a non-linear regression function. Pre-estimates of parameters were entered into the function and residuals were minimized by least squares methods utilizing computer facilities (UNIVAC 1100/81) at the University of Graz.

Extraction of Lipids

Placental tissue (approx. 5 g) and membranes (approx. 40 mg prot.) were extracted according to Folch et al. (1957) with modifications described by Veerkamp and Broekhuyse (1976). Recovery of $(4-^{14}C)$-cholesterol added in tracer amounts was usually 93-96%. Extracts were stored frozen in benzene-methanol (4:1 v/v) at -40°C.

Analysis of Phospholipid-Phosphorus and Cholesterol

The cholesterol (chol) content was determined after evaporation of the extract and resolving in 2-propanol by the method of Siedel et al. (1981). Phospholipids (PL) were precipitated with trichloroacetic acid (1.2 M) and the amount of phosphorus was determined by the method of Fiske and Subbarow (1926). Total phospholipid content was calculated by multiplying the lipid phosphorus in mg/dl by 25.

Preparation of FAMEs and Gas Chromatography

FAMEs were prepared by treatment of lipids with boronfluoride in methanol (14% w/v) as described by Morrison and Smith (1964) in the presence of 0.001% BHT as antioxidant (Wren and Szczepanowska, 1964). Methyl esters were separated using a Hewlett-Packard 5840A gas chromatograph equipped with a flame ionization detector on a 50 m WCOT glass capillary column (Chrompack, Middleburg, Netherlands) packed with CP Sil 88. Peak areas were measured with a Hewlett-Packard 5840A electronic integrator. Methyl-tricosanoate (C23:0) (100 ng) was used as internal standard.

Statistical Evaluations

For the purpose of intergroup comparison we used student's t-test for all data with the modification for unknown, unequal variances. Analysis of linear correlation was performed and the significance of the coefficient of correlation r to be different from zero was tested. Significances were determined at a level of 5% or less.

RESULTS

The membranes prepared from 80-100 g of wet tissue contained about 200 mg membrane protein with a relative amount of 2.1 mg protein per g placental tissue. This extraction corresponds to a protein yield of about 3% of the starting material (Table 1). The relative content of sialic acids was about three-fold higher in the membrane preparation compared to the crude placental homogenate. Enzyme activities of 5'-nucleotidase, alkaline phosphatase and beta-glucuronidase were all considerably enriched (Table 1), whereas the activity of isocitrate-dehydrogenase was decreased more than two-fold in the preparation.

Table 1

Characterization of crude placental homogenates and membrane suspensions of 27 placentae (mean ± sd).

	Crude Homogenate	Membrane Suspension	Enrichment (%)
Protein (mg/g Placenta)	72.0 ± 14.2	2.13 ± 0.56	3.08 ± 1.12
Sialic acids (n Mole/mg Prot.)	13.5 ± 3.5	37.1 ± 10.3	292 ± 111
5'-Nucleotidase (mU/mg Prot.)	23.3 ± 10.9	162 ± 95	860 ± 670
Alkaline Phosphatase (mU/mg Prot.)	86.1 ± 44.0	1173 ± 884	1540 ± 1150
β-Glucuronidase (U/mg Prot.)	69 ± 28	210 ± 78	330 ± 160
Isocitrate-Dehydrogenase (mU/mg Prot.)	16.3 ± 10.3	7.9 ± 8.9	46.4 ± 56.9

Enrichment (%) expresses the amount of the different components in the final membrane suspension compared with crude homogenate (100%).

Table 2

Association constants for insulin (K) and insulin receptor concentration (R) at 4°C of placental membranes from normal and diabetic mothers expressed as mean ± sd.

	K (l/Mole) $\times 10^8$	R (Mole/g prot.) $\times 10^{-10}$
Normal (10)	8.4 ± 4.1	3.4 ± 2.0
Diabetics (17)	9.6 ± 5.8	4.7 ± 2.4
Subgroup I (11)	12.6 ± 5.0	3.1 ± 1.9
Subgroup II (6)	4.1 ± 1.4[x]	7.7 ± 3.1[x]

Figures in parentheses represent the number of placentae analyzed.
[x] $p < 0.01$ vs normal.

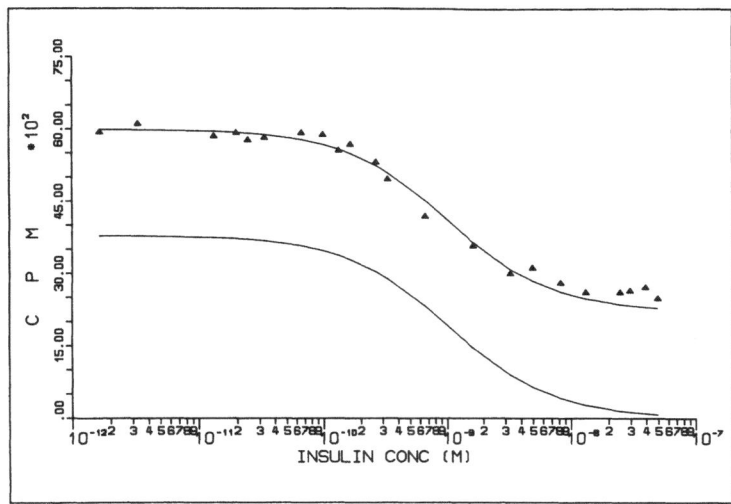

Figure 1. (^{125}I)-insulin binding to placental membranes displaced by varying concentrations of unlabeled insulin. The triangles represent the experimental data, each is the mean of at least 3, usually 9, parallel measurements. The upper line reflects the theoretically calculated (cf. Methods) curve for total binding, the lower curve shows the corresponding specific binding. Results of the experiment shown are: K:1.08 x 10^9 (l/Mole), 2.22 x 10^{-10} Mole receptors/g prot. and total error sum of squares: 1.30 x 10^6.

The analysis of insulin binding to the membranes was performed as competitive assay with unlabeled insulin over a concentration range of five orders of magnitude. The large number of data points was necessary due to the large variations of results at insulin concentrations below 0.1 nM. An example of a typical assay is shown in Figure 1. Degradation of insulin was less than 2% at 4°C. The association constants (K) and receptor concentrations (R) were approximately 9.10^8 l/Mole and 4.10^{-10} Mole/g prot., respectively, and were similar in the normal and diabetic group (Table 2). However, an arbitrary splitting of the diabetics into subgroups according to the individual K-values above (I) or below (II) the mean value of the normal group resulted in significant differences in some parameters (Table 2 and Table 4). In spite of a 50% higher value for K in subgroup I the difference did not reach statistical significance (t = 1.731, threshold value for p < 0.05: t = 1.740) due to large variations of individual K-values. The receptor concentration was comparable to the normal group whereas subgroup II exhibited a 126% gain in receptors/mg protein (p < 0.01).

Separation of the fatty acids in the lipid extracts as methyl esters by GC revealed palmitic acid (C16:0; 22.1 ± 1.6%) and arachidonic acid (C20:4; 19.5 ± 2.2%) as major constituents and a high portion of polyunsaturated fatty acids (PUFA) (42.4 ± 2.2%). Other fatty acids at more than 10% abundance were stearic acid (C18:0; 13.1 ± 1.0%), oleic acid (C18:1; 12.5 ± 0.8%) and linoleic acid (C18:2; 10.8 ± 1.3%). The

Table 3

Fatty acid composition of placental membranes from 10 normal mothers.

Fatty Acid	Area %
16:0	22.1 ± 1.6
16:1	1.5 ± 0.3
18:0	13.1 ± 1.0
18:1	12.5 ± 0.8
18:2	10.8 ± 1.3
22:0	5.3 ± 1.6
20:4	19.5 ± 2.2
22:4	1.2 ± 0.3
24:1	1.4 ± 0.2
22:6	3.2 ± 0.6
Saturated	41.4 ± 1.9
Unsaturated	57.9 ± 2.3
Monounsaturated	15.4 ± 1.0
Polyunsaturated	42.4 ± 2.2
Saturated/unsaturated	0.72 ± 0.07

Figures represent area percent (mean ± sd). Fatty acids with less than 1% abundance are not shown.

fatty acid with the highest number of carbon atoms was nervonic acid (C24:1) at 1.4% abundance. No cerotic acid (C26:0) could be detected.

A comparison of the cholesterol, total phospholipid and total FAME content relative to the amount of protein does not show significant differences between the groups (Table 4), nor does the portion of polyunsaturated fatty acids. However, the ratio of total fatty acids to cholesterol is increased by 20% in subgroup I and reduced by 30% in subgroup II. The amount of PUFA relative to the phospholipid content parallels this finding with smaller differences compared to normal, 15% for group I and 10% for group II (Table 4).

Table 4

Analysis of cholesterol (chol), total phospholipids (pl), fatty acids as methyl esters (FAME), polyunsaturated fatty acids (PUFA) and corresponding ratios in lipid extracts from normal and diabetic mothers (mean ± sd).

	Normal	Diabetics	
		I	II
µg chol/mg prot	74.9 ± 20.8	69.5 ± 51.0	103.0 ± 41.8
µg pl/mg prot	438 ± 62	498 ± 165	476 ± 81
µg FAME/mg prot	156 ± 51	146 ± 106	134 ± 44
µg FAME/µg chol	1.99 ± 0.63	2.38 ± 0.86	1.39 ± 0.32§§+
ng FAME/µg pl	344 ± 75	486 ± 188x	332 ± 71§
ng PUFA/µg pl	163.1 ± 50.0	187.3 ± 88.2	145.6 ± 33.6
pl/chol (m/m)	3.0 ± 0.8	2.5 ± 0.5	2.1 ± 0.5++§
PUFA %	42.4 ± 2.2	44.2 ± 4.5	43.8 ± 0.8+

x $p < 0.05$ normal vs I; + $p < 0.05$, ++ $p < 0.01$, normal vs II; § $p < 0.05$ I vs II

Table 5

Binding characteristics (association constant K, receptor concentration R) of 5 placental membrane preparations from normal mothers after preincubation with different phospholipases and buffer (control) expressed as mean ± sd.

	K (l/Mole) x10^8	R (Mole/g prot.) x10^{-10}
Control	1.7 ± 0.7	4.1 ± 1.1
Plase A2	1.8 ± 0.6	3.3 ± 0.7
Plase C	2.0 ± 0.7	3.0 ± 0.7
Plase D	1.9 ± 0.7	3.6 ± 1.1

Individual results of the lipid and fatty acid analysis have been linearly correlated with the association constant K of insulin binding. K was most strongly dependent upon the PL/Chol-ratio, upon the PL/protein-ratio and upon the ratio PUFA/PL as shown in Figure 2. Additionally, K was correlated with the portion of unsaturated fatty acids ($r = 0.418$, $2\,p < 0.05$) and inversely correlated with the portion of saturated fatty acids ($r = -0.410$, $2\,p < 0.05$), among which K depended on palmitic acid ($R = 0.414$, $2\,p < 0.05$).

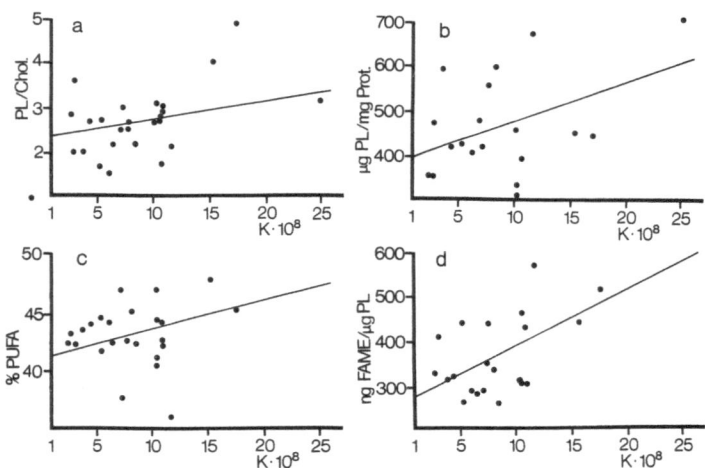

Figure 2. Correlation of insulin binding affinity (K; l/Mole) with a) molar PL/Chol-ratio, b) μg PL/mg prot., c) % polyunsaturated fatty acids and d) ng FAME/μg PL. Experimental values exceeding the range of the axis are not shown. $2\,p < 0.05$ for all correlation coefficients.

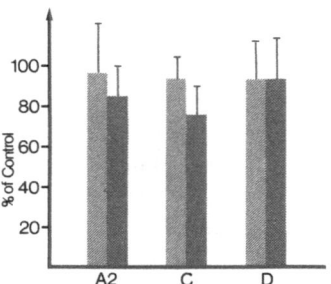

Figure 3. Specific binding relative to control of insulin at 330 pM (hatched bars) and 33 nM (dotted bars) unlabeled insulin after preincubation of placental membranes with phospholipases A2, C, and D. The means of 5 experiments with different membrane preparations and one standard deviation are shown.

Five different membrane preparations from normal mothers were treated with phospholipases. This treatment resulted in a reduction of specific binding compared with control at 330 pM and 33 nM unlabeled insulin, respectively. The binding at 33 nM insulin was 85% ($p < 0.05$), 76% ($p < 0.05$) and 94% of control for Plase A2, C, and D, respectively. Binding data were analyzed by the model fitting procedure (cf. Methods). The results exhibited a loss of binding sites with receptor concentrations of 80% (Plase A2), 74% (Plase C) and 86% (Plase D) of control. The affinities were increased by 7% (Plase A2), 20% (Plase C) and 13% (Plase D). None of these results reached statistical significance (Table 5).

DISCUSSION

Human placenta is a rich source of insulin receptors and is the tissue of origin for the isolation of the receptor (Williams and Turtle, 1979; Harrison and Itin, 1980). Autoradiography with ^{125}I-insulin and electron microscopy using insulin-ferritin conjugates demonstrated that the insulin binding sites are located on the microvillous membrane of the syncytiotrophoblast (Nelson et al., 1978; Whitsett and Lessard, 1978). Numerous studies comprising other tissues and circulating cells have been directed at elucidating the effects of pathological states on the insulin receptor properties with conflicting results concerning the placenta. Posner (1973) has shown that insulin binding was unaffected by maternal diabetes. This report, however, does not specify the diabetes with respect to its severity. Harrison et al. (1979) examined a microsomal-membrane fraction from human placentae of insulin dependent diabetic mothers and demonstrated a significant decrease in the membrane concentration of the receptors to about half the normal. Duran-Garcia et al. (1979) reported a decrease in the number of insulin receptors from gestational diabetic mothers.

In our study, we failed to find pronounced differences in the binding behavior when the total collective of diabetic mothers was compared with the group of the normal (Table 2). However, the diabetic group comprised subjects with gestational diabetes and overt diabetes mellitus of different severity. A detailed analysis pertinent to the question how the severity of maternal diabetes affects the insulin binding behavior is outside the scope of this article.

For the purpose of our present study we divided the diabetic group into two subgroups based on the value found for the receptor affinity compared with the mean of the normal group. Those with a higher affinity constituted subgroup I, whereas the others formed subgroup II (Table 2).

Several factors which are known to determine membrane fluidity were analyzed in all three groups. These factors include the amount of phospholipid and cholesterol relative to protein (chol/prot, pl/prot), the ratio of phospholipid to cholesterol (pl/chol), the amount of fatty acids relative to cholesterol and phospholipid, the portion of polyunsaturated fatty acids (PUFA) and their amount relative to phospholipid (Table 4).

Results obtained show that an increase in receptor affinity is paralleled by a decrease of the cholesterol to protein ratio and by a concomitant increase of the amount of fatty acids relative to cholesterol and phospholipids and of the polyenoic acid/phospholipid ratio. The portion of PUFA is higher in both diabetic subgroups whereas the pl/chol-ratio is lower in both groups. In order to rule out a possible bias of the data due to the variations within the groups which might come up when comparing means, we further attempted to correlate the various factors with the affinity constants found for the 27 individual membrane preparations. As shown in Figure 2, the pl/chol-ratio as well as the portion of PUFAs, significantly correlated with the affinity. Furthermore, the coefficients of linear correlation for the amount of phospholipid relative to protein as well as for the amount of fatty acids relative to phospholipid significantly differ from zero. Thus, we conclude that enrichment with polyunsaturated fatty acids of placental membranes results in an increased affinity for insulin binding, unless other determinants of membrane fluidity predominate and revert the effect of PUFAs.

Studies reported in the literature which modified the fatty acid composition of cultured cells by dietary means do not show uniform behavior of the receptors. Ginsberg et al. (1981) found that Friend erythroleukemia cells, enriched in unsaturated fatty acids, had an increased number of insulin receptors and a concomitant decreased receptor affinity. Similar results were found by the same author (Ginsberg et al., 1982) in Ehrlich ascites cells. When the partially purified insulin receptor was inserted into the unilamellar liposomes (Gould et al., 1982), the similar qualitative results were observed. In a recent report, Bar et al. (1984) failed to demonstrate endothelial cells changes in the binding of insulin or the insulin-like growth factor despite extensive fatty acyl changes. It is difficult to explain the discrepancies in our findings compared to the studies with cultured cells. Unfortunately, the authors listed data only for the fatty acid composition, nothing is mentioned about the other factors relevant for membrane fluidity.

The current concepts of biomembrane fluidity (Shinitzky and Henkart, 1980; Chapman, 1983; Stubbs and Smith, 1984) assume that the penetration of molecules such as cholesterol results in a more rigid lipid lattice of the membrane, whereas an increase in the degree of fatty acid unsaturation and an increasing amount of phospholipid relative to protein promote a greater freedom of molecular motion within the membranes. Our data, which were derived from the analysis of lipids in three groups of membranes with different association constants as well as the corresponding analysis of linear correlation, can consistantly be interpreted in terms of membrane fluidity.

Furthermore, we studied the influence of phospholipid modification by treatment with the various phospholipases and found a marked decrease in specific binding for phospholipases A2 and C whereas experiments with phospholipase D exhibited only minor changes (Figure 3). Analysis of these data by our computer modeling procedure reveals a loss of binding sites and an increase in the receptor affinity. These affinity changes were considerable for Plases C and D, respectively, and probably reflect alterations in electrostatic interactions between ligand and the receptor subsequent to phospholipid headgroup cleavage. The reason why the affinity was only slightly increased (7%) after Plase A2 preincubation might be that the cleaved fatty acid chains had not been removed from the membrane lipid phase by incubation with bovine serum albumin. Although one has to keep in mind that the affinity data obtained are based on the fitting of only two concentrations of unlabeled insulin, one might speculate that the membrane destablization, i.e., increased membrane fluidity, by Plase A2 leads to an increased receptor binding.

Thus, it is tempting to conclude that as the receptor molecules gain more freedom of molecular motion in the membrane, the association with the ligand becomes stronger. In terms of a molecular picture, it is assumed that by an increased membrane fluidity the translational movements and/or oscillations of the receptor molecules or parts of it vertically to the plane of the membrane are facilitated. The more frequent the insulin binding sites are exposed to the surrounding water phase, the higher the probability of molecular contacts between the binding sites of receptor and ligand which thermodynamically results in a higher association constant.

This interpretation varies from the hypothesis of Shinitzky and Inbar (1976) and Borochov et al. (1979). These authors are in agreement with our view that the membrane proteins are able to move more easily in the membrane as the membrane becomes more fluid. However, they further provide evidence for the conclusion that the membrane proteins would be displaced deeper within the membrane thus impeding the ligand binding. The experiments suggesting this scheme have been performed at 37°C whereas we studied insulin binding at 4°C. It seems conceivable that the behavior of membrane lipids and their interaction with membrane proteins differs below and above the transition temperature range of the lipid phase. This could imply than below the transition temperature either the membrane fluidity is differently modulated by the several parameters we measured leading to a decrease in affinity with increasing membrane fluidity or that the lateral movements are more restricted than those perpendicular to the membrane. The authors favor the latter explanation. Below the transition temperature, the membrane becomes more densely packed. In terms of fluidity, the density of the membrane is the primary general feature which determines the absorption of kinetic energy by the microenvironment via molecular collisions (Lands, 1980).

It is proposed that the increase in membrane density is not isotropic and that the gain in lateral molal density predominates. This would favor translational kinetic energy to be mainly absorbed by lateral collisions. Unfortunately, data which result from direct measuring translational membrane fluidity are not available. These measurements are undoubtedly necessary to further substantiate or reject this interpretation.

SUMMARY

How the association constants for insulin binding to its receptor are affected by various parameters known to modify membrane fluidity was investigated. Placental membranes were analyzed from normal and diabetic mothers in order to cover a large range of affinity values. The strongest linear correlations with the binding affinity were observed with the phospholipid/cholesterol-ratio ($r = 0.467$, 2 $p < 0.05$), the amount of phospholipid relative to protein ($r = 0.449$, 2 $p < 0.05$) and the ratio of fatty acids to phospholipids ($r = 0.452$, 2 $p < 0.05$). Preincubation of the membranes with phospholipases A2, C, and D resulted in a reduction of receptor concentration and a gain in affinity.

These results are interpreted in terms of membrane fluidity and are consistent with an increase in the mobility of the insulin receptor leading to an increased affinity.

ACKNOWLEDGEMENTS

The authors are indebted to J. Haas for his support in statistical problems, to H. Hofmann for providing us with the placentae and to P. Puerstner for his technical help with the gas chromatograph. The technical assistance of M. Wertanzl and A. Klimpfinger and the secretarial work of A. Novotny is gratefully acknowledged.

REFERENCES

Amatruda, J.M. and Finch, E.D. (1979) Modulation of hexose uptake and insulin action by cell membrane fluidity. *J. Biol. Chem.* 254, 2619-2625.

Andreani, D., DePirro, R., Lauro, R., Olefsky, J.M., and Roth, J. (eds.) (1980) *Current Views on Insulin Receptors*, New York, Academic Press.

Arkejsteijn, C.L.M. (9176) A kinetic method of serum 5'-nucleotidase using stabilised gluatmate dehydrogenase. *J. Clin. Chem. Clin. Biochem.* 14, 155-158.

Bar, R.S., Dolash, S., Spector, A.A., Kaduce, T.L., and Figard, Ph.D. (1984) Effects of membrane lipid unsaturation on the interactions of insulin and multiplicating stimulating activity with endothelial cells. *Biochim. Biophys. Acta* 804, 466-473.

Borochov, H., Abbott, R.E., Schachter, D., and Shinitzky, K. (1979) Modulation of erythrocyte membrane proteins by membrane cholesterol and lipid fluidity. *Biochem.* 18, 251-255.

Bowers, G.N., Jr. (1959) Measurement of isocitric dehydrogenase activity in body fluids. *Clin. Chem.* 5, 509-518.

Bowers, G.N., Jr., and McComb, R.B. (1966) A continuous spectrophotometric method for measuring the activity of serum alkaline phosphatase. *Clin. Chem.* 12, 70-89.

Chapman, D. (1983) Biomembrane fluidity: The concept and its development. In: *Membrane Fluidity in Biology*, Vol. 2, New York, Academic Press, pp. 5-42.

Duran-Garcia, S., Nieto, J.G., and Cabello, A.M. (1979) Effect of gestational diabetes on insulin receptors in human placenta. *Diabetologia* 16, 87-91.

Fishman, W.H., Kato, K., Anstiss, C.L., and Green, S. (1967) Human serum beta-glucuronidase; its measurement and some of its properties. *Clin. Chim. Acta* 15, 435-447.

Fiske, C.H. and Subbarow, Y. (1926) The colorimetric determination of phosphorus. *J. Biol. Chem.* 66, 375-400.

Folch, J., Lees, M., and Stanley, G.H.S. (1957) A simple method for the isolation and purification of total lipids from animal tissue. *J. Biol. Chem.* 226, 497-509.

Ginsberg, B., Brown, T.J., Simon, I., and Spector, A.A. (1981) Effect of the membrane lipid environment on the properties of insulin receptors. *Diabetes* 30, 773-780.

Ginsberg, B.H., Jabour, J., and Spector, A.A. (1982) Effect of alterations in membrane lipid unsaturation on the properties of the insulin receptor of Ehrlich ascites cells. *Biochim. Biophys. Acta* 690, 157-164.

Gould, R.J., Ginsberg, B.H., and Spector, A.A. (1982) Lipid effects on the binding properties of a reconstituted insulin receptor. *J. Biol. Chem.* 257, 477-484.

Gould, R.J. and Ginsberg, B.H. (1984) Biochemistry and analysis of membrane phospholipids: Application to membrane receptors. In: *Membranes, Detergents, and Receptor Solubilization*, (eds.), J.C. Venter and L.C. Harrison, New York, Alan R. Liss, pp. 65-83.

Grunfeld, C., Baird, K.L., and Kahn, C.R. (1981) Maintenance of 3T3-L1 cells in culture media containing saturated fatty acids decreases insulin binding and insulin action. *Biochem. Biophys. Res. Commun.* 103, 219-226.

Haour, F., and Bertrand, J. (1974) Insulin receptors in the plasma membranes of human placenta. *J. Clin. Endocrinol. Metab.* 38, 334-337.

Harrison, L.C., Billington, T., Clark, S., Nichols, R., East, I., and Martin, F.I.R. (1977) Decreased binding of insulin by receptors on placental membranes from diabetic mothers. *J. Clin. Endocrinol. Metab.* 44, 206-209.

Harrison, L.C. and Itin, A. (1980) Purification of the insulin receptor from human placenta by chromatography on immobilized wheat germ lectin and receptor antibodies. *J. Biol. Chem.* 255, 12066-12072.

Heron, D.S., Shinitzky, M., Hershkowitz, M., and Samuel, D. (1980) Lipid fluidity markedly modulates the binding of serotonin to mouse brain membranes. *Proc. Natl. Acad. Sci. USA* 77, 7463-7467.

Kates, M. and Kuksis, A., (eds.) (1980) *Membrane Fluidity: Biophysical Techniques and Cellular Regulation*, Clifton, New Jersey, The Humana Press.

Lands, W.E.M. (1980) Fluidity of membrane lipids. In: *Membrane Fluidity. Biophysical Techniques and Cellular Regulation*, (eds.), M. Kates and A. Kuksis, Clifton, New Jersey, The Humana Press, pp. 69-73.

Lee, A.G. (1977) Lipid phase transitions and phase diagrams. I. Lipid phase transitions. *Biochim. Biophys. Acta* 472, 237-281.

Lee, G., Consiglio, E., Habig, W., Dyer, S., Hardegree, C., and Kohn, L.D. (1978) Structure-function studies of receptors for thyrotropin and tetanus toxin. Lipid modulation of effect or binding to glycoprotein receptor component. *Biochem. Biophys. Res. Commun.* 83, 313-320.

Lowry, O.H., Rosenbrough, N.J., Farr, A.L., and Randall, R.J. (1951) Protein measurement with the folin phenol reagent. *J. Biol. Chem.* 193, 265-275.

McCaleb, M.L. and Donner, D.B. (1981) Affinity of the hepatic insulin receptor is influenced by membrane phospholipids. *J. Biol. Chem.* 256, 11051-11057.

Mehdi, S.Q., Nussey, S.S., Shindelman, J.E., and Kriss, J.P. (1977) Influence of lipid substitution on thyrotropin-receptor interactions in artificial vesicles. *Endocrinol.* 101, 1406-1412.

Morrison, W.R. and Smith, L.M. (1964) Preparation of fatty acid methyl esters and dimethylacetals from lipids with boronfluoride-methanol. *J. Lipid Res.* 5, 600-608.

Nelson, D.M., Smith, R.M., and Jarett, L. (1978) Nonuniform distribution and grouping of insulin receptors on the surface of human placental syncytial trophoblast. *Diabetes* 27, 530-538.

Phillips, M.C., Williams, R.M., and Chapman, D. (1969) On nature of hydrocarbon chain motions in lipid liquid crystals. *Chem. Phys. Lipids* 3, 234-244.

Posner, B.I. (1973) Insulin receptors in human and animal placental tissue. *Diabetes* 23, 209-217.

Sandermann, H., Jr. (1978) Regulation of membrane enzymes by lipids. *Biochim. Biophys. Acta* 515, 209-237.

Shinitzky, M. and Inbar, M. (1976) Microviscosity parameters and protein mobility in biological membranes. *Biochim. Biophys. Acta* 43, 133-149.

Shinitzky, M. and Henkart, P. (1980) Fluidity of cell membranes - current concepts and trends. *Int. Rev. Cytol.* 60, 121-147.

Shinitzky, M., Borochov, H., and Wilbrandt, W. (1980) Lipid fluidity as a physiological regulator of membrane transport and enzyme activities. In: *Membrane Transport in Erythrocytes*, (eds.), H.H. Ussing and J.O. Wieth, Copenhagen, Munksgaard, pp. 91-107.

Siedel, J., Schlumberger, H., Klose, S., Ziegenhorn, J., and Wahlefeld, A.W. (1981) Improved reagent for the enzymatic determination of serum cholesterol. *J. Clin. Chem. Clin. Biochem.* 19, 838-839.

Stubbs, C.D. and Smith, A.D. (1984) The modification of mammalian membrane polyunsaturated fatty acid composition in relation to membrane fluidity and function. *Biochim. Biophys. Acta* 779, 89-137.

Veerkamp, J.H. and Broekhyse, R.M. (1976) Technique for the analysis of membrane lipids. In: *Biochemical Analysis of Membranes*, (ed.), A.H. Maddy, London, Chapman and Hall, pp. 252-282.

Warren, L. (1963) Thiobarbituric acid assay of sialic acids. *Meth. Enzymol.* 6, 463-466.

Whitsett, J.A. and Lessard, J.L. (1978) Characteristics of the microvillus brush border of human placenta: Insulin receptor localization in brush border membranes. *Endocrinol.* 103, 1458-1468.

Williams, P.F., and Turtle, J.R. (1979) Purification of the insulin receptor from human placental membranes. *Biochim. Biophys. Acta* 579, 367-374.

Wren, J.J. and Szczepanowska, A.D. (1964) Chromatography of lipids in presence of an antioxidant, 4-methyl-2,6-di-tert-butylphenol. *J. Chromatog.* 14, 405-410.

Trophoblast Research 2:45-60, 1987

RENIN RELEASE FROM HUMAN CHORIONIC TROPHOBLASTS IN VITRO: THE ROLE OF CYCLIC AMP AND PROTEIN KINASE C

Alan M. Poisner, Promila Agrawal, and Roselle Poisner

Department of Pharmacology,
Toxicology and Therapeutics
University of Kansas
College of Health Sciences
Kansas City, Kansas 66103, USA

INTRODUCTION

Trophoblasts from human chorion laeve contain renin (Poisner et al., 1981b), HCG (Poisner et al., 1983a) and progesterone (Tonkowicz and Poisner, 1985). Most of the renin in the chorion is present in the form of prorenin (Poisner et al., 1981b). Preliminary work indicated that renin release is dependent on external calcium and can be inhibited by agents which block calcium influx (Poisner et al., 1983b; 1984a,b). This suggested that renin release may be mediated through the common calcium-dependent exocytosis pathway. Progesterone secretion from these cells is enhanced by dibutyryl cyclic AMP and agents which increase cyclic AMP (Tonkowicz and Poisner, 1985). Cyclic AMP has also been suggested as a mediator of renin secretion from the kidney (Fray, 1980), but the role of calcium is controversial (Chen and Poisner, 1976; Park et al., 1981; Hinko et al., 1984). Another mediator of secretion in a number of cells is the diacyl glycerol-activated protein kinase (C-kinase), which is also activated by the tumor promoter phorbol esters (Castagna et al., 1982). The present studies were carried out to determine if cyclic AMP and protein kinase C may also function as mediators of renin secretion from chorionic trophoblasts. Preliminary studies have been reported (Poisner, 1984; Poisner et al., 1984b).

MATERIALS AND METHODS

The chorion from term pregnancies was separated from the amnion and digested with collagenase and DNAse as previously described (Poisner et al., 1982, 1986). Cells (2-10 x 10^5/well) were plated in 24-well multiwell plates for 5-13 days in 1.0 ml CMRL-1066 medium containing 6% NuSerm and penicillin/streptomycin (100 units/100 μg/ml), with medium changes at 2-3 day intervals. The monolayer was confluent at 3-5 days with no further increase in cell number.

Received: 8 October 1985; Accepted: 15 May 1986

Culture Conditions

At the time of the experiment, the wells were drained, rinsed with saline, and 1.0 ml of control or test media was added to each well. All conditions were done in quadruplicate wells. Media used include CMRL-1066 containing NuSerm (6%), bovine serum albumin (0.1%) or a balanced salt solution of the following composition: NaCl, 134 mM, KCl, 5.4 mM; $CaCl_2$; 1.0 mM; $MgCl_2$, 0.5 mM; HEPES buffer (7.4), 20 mM; Dextrose, 1 mg/ml; BSA, 1 mg/ml. In some experiments, the medium also contained soybean trypsin inhibitor (SBTI) and bacitracin. After incubation, the medium was removed and assayed immediately or frozen for later assay. For assay of cell content, the wells were rinsed with saline and then the cells were sonicated in 1.6 ml saline with a Branson sonicator.

Analytical Methods

Renin was measured (after trypsin activation) by radioimmunoassay of angiotensin I generated from sheep substrate, as described before (Poisner et al., 1981a). The standard curve ranged from 1.5 to 100 µUnits/ml. None of the agents used in the experiments interfered with the assay at the concentrations present. Cyclic AMP was assayed by radioimmunoassay, utilizing the Immunonuclear RIA Kit. Protein was measured by the Bradford dye method, using ovalbumin as standard. All samples were analyzed in duplicates.

Chemicals and Media

NuSerm was obtained from Collaborative Research, phospholipase C was obtained from CalBiochem. All other reagents were obtained from Sigma Chemicals. Standard renin was obtained from the National Institutes for Biological Standards, Holly Hill, London, UK.

Expression of Results

Results are expressed in most cases as release of substances per well with the mean and standard error of the mean indicated. All conditions were tested in quadruplicate wells and at least two plates were used for each type of experiment. Representative experiments are indicated. Statistical testing employed analysis of variance with $p < 0.05$.

RESULTS

Release of Renin in Response to Agents Influencing cyclic AMP

Cholera toxin (CT) (20 ng/ml) and forskolin (50 µM), which are known to activate adenyl cyclase, caused an increase in renin release, as did methyl isobutyl xanthine (MIX) (30 µM), an inhibitor of phosphodiesterase (Figure 1). CT and MIX showed a dose dependent stimulation of renin release, with high concentrations producing less release (Figure 2). The inhibition of renin release by high concentrations of MIX and CT is probably due to toxic effects of excessive stimulation. This has been found in other biological systems and should alert investigators to try a variety of concentrations before concluding that an agent is merely inhibitory.

8-bromo-cyclic AMP, as described below, was also effective in stimulating renin secretion.

Release of cyclic AMP From Trophoblasts

The combination of MIX plus forskolin caused a time-dependent release of cyclic AMP without altering renin release in 120 min (Figure 3). Isoproternol and epinephrine also increased the release of cAMP, but were not as effective as forskolin (Figure 4).

Release of Renin in Response to Agents Acting on Protein Kinase C

PMA, which is known to simulate diacyl glycerol in activating protein kinase C, also caused the release of renin from trophoblasts and was synergistic with CT and MIX (Figures 5 and 6). PMA was also synergistic with 8-bromo-cyclic AMP (BrcAMP) in producing renin release (Figure 7). Renin release was also brought about by phospholipase C and this response was potentiated by BrcAMP (Figure 8).

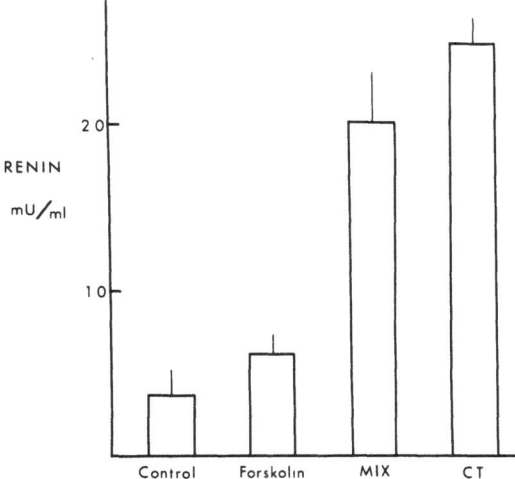

Figure 1. Stimulation of renin release from chorionic trophoblasts by forskolin (F), methyl isobutyl xanthine (MIX) and cholera toxin (CT). 2×10^5 cells were cultured for 5 days in CMRL/NuSerm and then exposed for 48 hr to fresh media containing no additions (control), forskolin (50 μM), MIX (30 μM) or CT (20 ng/ml). Total renin released in the final 48 hr is shown. Responses to drug treatments are significantly greater than the control.

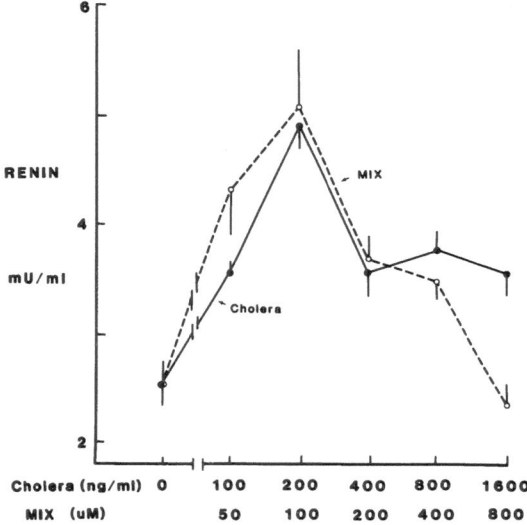

Figure 2. Dose response relationship of stimulation of renin release by cholera and MIX. 5×10^5 cells were cultured for 6 days and then exposed for 72 hr to different concentrations of cholera or MIX.

Effect of Calcium on Renin Release

The release of renin was reduced as the extracellular calcium concentration was decreased. Low calcium, which is known to increase cell permeability, also caused an increase in the leakage of LDH (Figure 9). Renin release was also reduced by inhibitors of calcium influx (gadolinium and verapamil) and by an inhibitor of calcium-calmodulin activity (trifluoperazine) (Figures 10 and 11).

Effect of a Lipoxygenase Inhibitor on Renin Release

Nordihydroguaretic acid (NDGA), an inhibitor of the lipoxygenase pathway of arachidonic acid metabolism, has been shown to inhibit HCG secretion and stimulate progesterone secretion from choriocarcinoma cells (Ilekis and Benveniste, 1983). NDGA was found to inhibit renin release in response to MIX and CT (Figure 12).

DISCUSSION

The cellular control of renin secretion from chorionic trophoblasts is poorly understood. Mediators of secretion in other cells include calcium, cyclic AMP, arachidonic acid metabolites, and metabolites of phospholipid breakdown, including diacyl glycerol. The present study was designed to examine the effects of agents known to influence these mediators on the secretion of renin from chorionic trophoblasts.

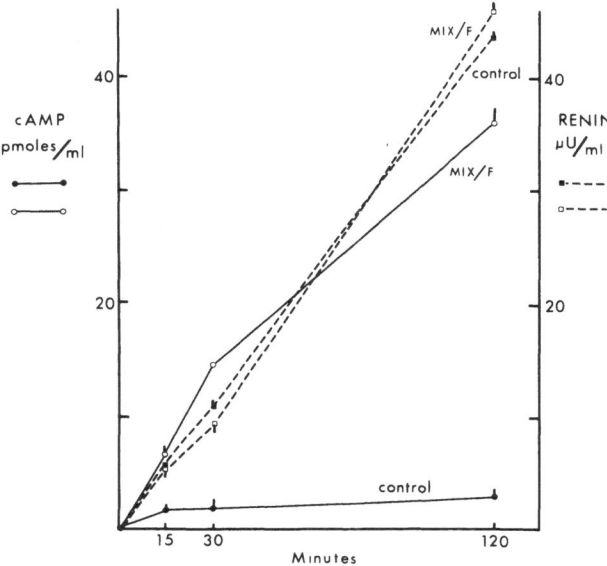

Figure 3. Simulation of cyclic AMP release (cAMP) from chorionic trophoblasts by forskolin plus MIX. 4x10[5] cells were cultured for 5 days, rinsed with saline, and then exposed to a balanced salt solution.(see Materials and Methods) with or without forskolin (100 μM) plus MIX (1.0 mM). At various times from 15 to 120 min, the media was removed and assayed for cAMP and renin. The circles indicate that cAMP release was increased by drug treatment; while the squares show that there was no effect on renin release within 120 min.

    The results provide evidence pointing to adenyl cyclase-cyclic AMP in the control of renin secretion. Agents known to increase adenyl cyclase, cholera toxin (CT), and forskolin caused an increase in renin secretion, with a time delay suggesting that renin secretion reflects the state of synthesis. This conclusion is similar to that relating the release of HCG from malignant trophoblasts (Benveniste et al., 1978). Similarly, MIX, an inhibitor of phosphodiesterase, which should increase cyclic AMP within the cells, also increased renin release. Finally, 8- bromo-cyclic AMP, also enhanced renin secretion. Cyclic AMP was released from trophoblasts upon exposure to forskolin plus MIX and also the adrenergic agents, epinephrine and isoproterenol. Taken together, these results strongly support the importance of cyclic AMP as a mediator of chorionic renin release. It is expected from results in other tissues that the cyclic AMP effect is mediated through a cyclic nucleotide-dependent protein kinase. Indeed, cyclic AMP-stimulated protein kinase (A-kinase) has been demonstrated in human placenta (Moore et al., 1984).

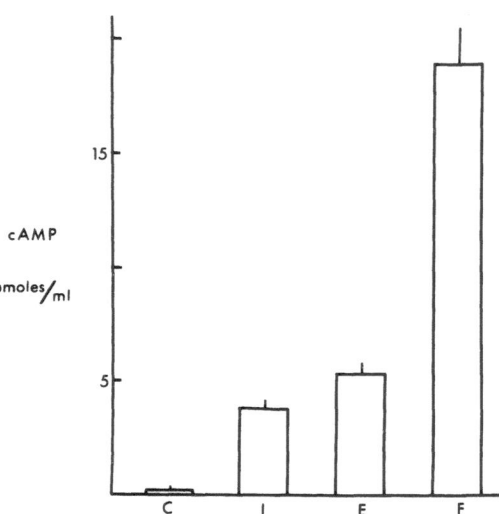

Figure 4. Stimulation of cAMP release from chorionic trophoblasts by isoproterenol (Iso), epinephrine (Epi), and forskolin (F). 10×10^5 cells were cultured for 10 days, rinsed with saline, and then exposed to a balanced salt solution for 30 min. The concentrations used were: Iso, 10 μM, Epi, 10 μM; F, 100 μM. All responses to drug treatments are significantly greater than the control.

        Evidence for a diacyl glycerol-stimulated protein kinase (C-kinase) in the regulation of chorionic renin secretion has also been found. Thus, renin secretion is enhanced by phorbol ester (PMA) which is known to simulate diacyl glycerol in activating protein kinase C (Castagna et al., 1982). Preliminary experiments indicate that phorbol dibutyrate (PDB) is much weaker than PMA in promoting renin secretion. PDB has also been shown to be less active than PMA in a vareity of other biological systems. Phospholipase C, which releases diacyl glycerol from membrane phospholipids, also stimulates renin secretion. There is a profound synergism between the effects of PMA and agents acting on the cyclic AMP system (CT, MIX, and BrcAMP). Similarly, the effects of phospholipase C and BrcAMP on renin secretion were super additive. Although there is evidence in some tissues that PMA can enhance the production of cyclic AMP (Perchellet and Boutwell, 1980; Brostrom et al., 1983), in other tissues PMA has been found to reduce cyclic AMP accumulation (Brostrom et al., 1982; Heyworth et al., 1984). The synergism between PMA and BrcAMP suggests that the interaction between PMA and CT or MIX on renin secretion is beyond the step of cyclic AMP accumulation. Previous evidence has implicated cyclic AMP in chorionic progesterone secretion (Tonkowicz and Poisner, 1985). However, preliminary results show that PMA inhibits CT-induced progesterone secretion. This demonstrates that the cellular mechanisms for secretion of renin and progesterone can be dissociated. PMA has been reported to inhibit synthesis of progesterone and testosterone in ovarian and testicular cells (Welsh et al., 1984). The inhibition of PMA in these cells was seen in response to cholera toxin

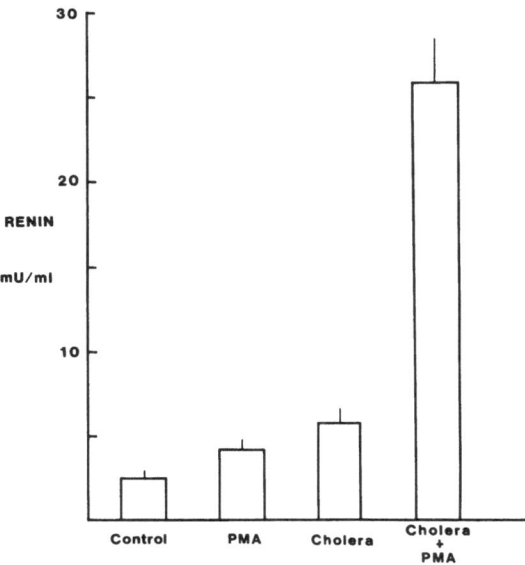

Figure 5. Synergistic effect of PMA and cholera toxin (CT) on renin release from chorionic trophoblasts. 5×10^5 cells were cultured for 10 days and then exposed for 72 hr to CMRL/BSA media containing no additions (control), PMA (10 ng/ml), CT (100 ng/ml) or CT plus PMA. The effects of PMA and CT are significantly greater than the control and less than the combination of PMA plus CT.

and to dibutyryl cyclic AMP and therefore is presumably exerted at a stage distal to cyclic AMP generation. PMA and phospholipase C also cause secretion of prolactin from cultured anterior pituitary cells (Koike et al., 1985). In these cells, as well as others (see Nishizuka et al., 1984), protein kinase C activation and calcium mobilization are synergistic in inducing secretion.

Calcium is commonly found to be required for secretion in a variety of systems. As indicated before, there is evidence for and against it as a cellular stimulant for renal renin secretion. Different conclusions regarding the role of calcium most likely depend on the experimental systems employed and the time course of the protocols. It is clear that there are differences between short term effects and those occurring over longer periods of time (Lester and Rubin, 1977; Hinko et al., 1984). Chorionic trophoblasts exposed to low calcium showed a decrease in renin secretion over a 24 hr period which was opposite to the effect on lactic dehydrogenase (LDH). Since LDH leakage is a marker of increased cell permeability, this evidence suggests that renin is not handled by the cells as a free cytoplasmic enzyme, but rather may have some special packaging. Other results from our laboratory using digitonin also point to different handling of renin and LDH (Poisner, unpublished). Agents which interfere with calcium influx include gadolinium, a lanthanide element known to prevent

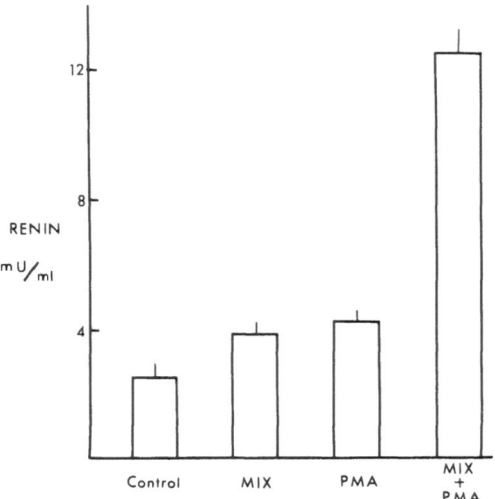

Figure 6. Synergistic effect of PMA and MIX on renin release from chorionic trophoblasts. 5×10^5 cells were cultured for 10 days and then exposed for 72 hr to CMRL/BSA media containing no additions (control), MIX (100 µM), PMA (10 ng/ml), or MIX plus PMA. The effects of MIX are significantly greater than the control and less than the combination of PMA plus MIX.

calcium influx, and verapamil, a organic calcium channel blocker. Both of these agents, which are known to inhibit secretion in other systems, reduced chorionic renin secretion. Finally, trifluoperazine, a calcium-calmodulin inhibitor, also blocked renin secretion. These results indicate that interference with calcium influx or utilization inhibit chorionic renin secretion. Since the diacyl glycerol-activated protein kniase is a calcium dependent enzyme (Castagna et al., 1982), this points to one possible site of action of calcium. Also, since there is a calcium-calmodulin activated protein kinase in the placenta which is inhibited by trifluoperazine (Moore et al., 1984), this is another possible site of action of calcium in promoting renin secretion. Other possible sites of action of calcium include the adenyl cyclase system and various phospholipases.

Arachidonic acid metabolites are known to stimulate secretion in a number of systems. Although indomethacin, an inhibitor of cyclooxygenase, has no effect on HCG secretion from malignant trophoblasts, NDGA, an inhibitor of lipoxygenase inhibits basal and stimulated secretion at 10 and 25 µg/ml (Ilekis and Benveniste, 1984). In the present study, NDGA was found to inhibit basal and stimulated release of renin from chorionic trophoblasts at 10 µg/ml. Other results indicate that indomethacin does not inhibit and that PGE_2, a stimulant of renal renin secretion, does not stimulate chorionic renin secretion (Cooke et al., 1986; Poisner, unpublished).

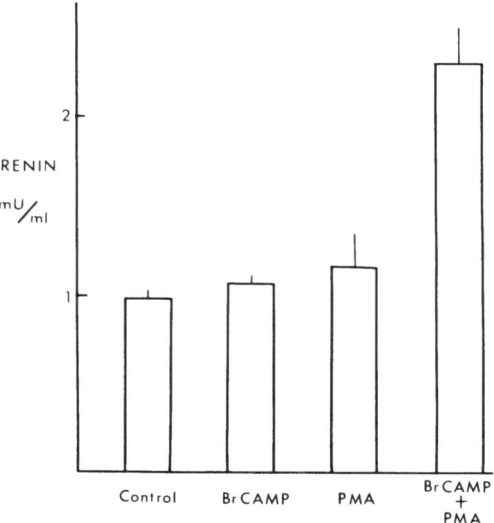

Figure 7. Synergistic effect of PMA and 8-bromo-cyclic AMP (BrcAMP) on renin release from chorionic trophoblasts. 4×10^5 cells were cultured for 4 days and then exposed for 24 hr to fresh media containing PMA (10 ng/ml), BrcAMP (1.0 mM), or PMA plus BrcAMP. The effects of PMA and BrcAMP are significantly greater than the control and less than the combination of PMA plus BrcAMP.

Current views on interactions between cellular mediators suggest that phospholipid breakdown (associated with membrane excitation) yields diacyl glycerol and inositol phosphates which lead to further changes in protein kinase C and cellular calcium. A further metabolite of membrane phospholipids is arachidonic acid. This can be further metabolized by the cyclooxygenase or lipoxygenase pathways. Primary messengers which activate adenyl cyclase initiate the chain of events which ultimately leads to protein kinase A activation. As described elsewhere (Nishizuka et al., 1984), there are many possible interactions of these cellular mediators. Renin release from chorionic trophoblasts appears to behave as a monodirectional system with respect to protein kinase C and protein kinase A, in the terminology of these authors, where the two messengers produce similar cellular responses.

The only clues concerning primary messengers for regulation of chorionic renin secretion comes from the evidence on adrenergic agents activating cyclic AMP release. Trophoblasts could be exposed to catecholamines emanating from amniotic fluid or maternal or fetal circulation. Further studies will be needed to determine if this is coupled to renin secretion and if other hormones can act on an adenyl cyclase-coupled secretory pathway. The ability of physiological agents to activate phospholipid breakdown (and subsequent activation of protein kinase C) also remains to be determined.

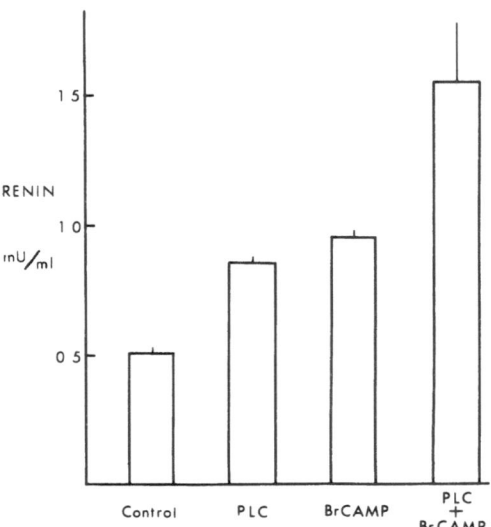

Figure 8. Synergistic effect of phospholipase C (PLC) and 8-bromo-cyclic AMP (BrcAMP) on renin release from chorionic trophoblasts. 5x10⁵ cells were cultured for 3 days and then exposed for 48 hr to CMRL/BSA medium containing no additions (control), PLC (0.2 Units/ml), BrcAMP (0.1 mM), or PLC plus BrcAMP. The effects of PLC and BrcAMP are significantly greater than the control and less than the combination of PLC plus BrcAMP.

In addition to discerning the cellular mechanisms for regulation of chorionic renin secretion, it will be important to elucidate the physiological and pathological significance of this substance. There is evidence that chorionic renin may be secreted into the maternal circulation (Hsueh et al., 1982) and it may be surmised that most of the renin in amniotic fluid also comes from this source. Renin also has been found in a trophoblastic tumor (hydatidiform mole) (Poisner et al., 1984c). Recent studies indicate that the trophoblasts possess angiotensin receptors and release LHRH-like activity on exposure to angiotensin II (Poisner and Poisner, 1985). LHRH is synthesized initially in the placenta as a larger molecular weight precursor (Gautron et al., 1981; Nikolics et al., 1985) and current sudies in our laboratory indicate that a higher molecular weight form of LHRH is the major species of the hormone released from chorionic membranes. Since most of the renin in both normal and abnormal trophoblasts is also in the form of the inactive precursor, it is possible that a variety of peptides in the placenta are not processed as completely as they are in other tissues.

SUMMARY

Trophoblasts from human chorion laeve possess an adenyl cyclase coupled to renin release. This is supported by the release of renin in response to cholera toxin (CT), methyl isobutyl xanthine (MIX), and 8-bromo-cyclic AMP (BrcAMP). Cyclic

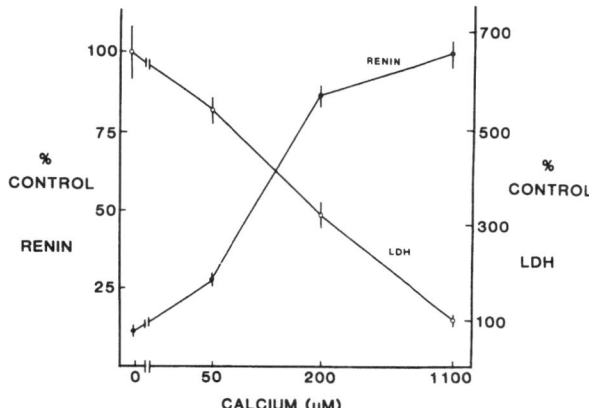

Figure 9. Inhibition of renin release and stimulation of LDH release from chorionic trophoblasts by low calcium. 5x10⁵ cells were cultured for 3 days, rinsed with saline and then exposed to balanced salt solution with varying concentrations of calcium. The results are expressed as a percentage of the amount of renin or LDH present in the control wells (calcium = 1100 μM). As the calcium concentration was reduced from 1100 to 0 mM, LDH release went from 100% to 680%, while renin release went from 100% to 10%. The calcium concentration is positively correlated with renin release and negatively correlated with LDH release.

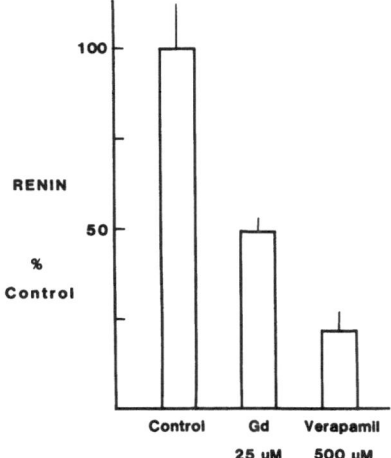

Figure 10. Inhibition of renin release from chorionic trophoblasts by gadolinium and verapamil. 5x10⁵ cells were cultured for 5 days and the exposed for 24 hr to fresh medium containing gadolinium (Gd) (25 μM) or verapamil (500 μM), known inhibitors of calcium influx. Renin release is expressed as a percentage of that found in the control wells. Gd and verapamil produce a significant inhibition of renin release.

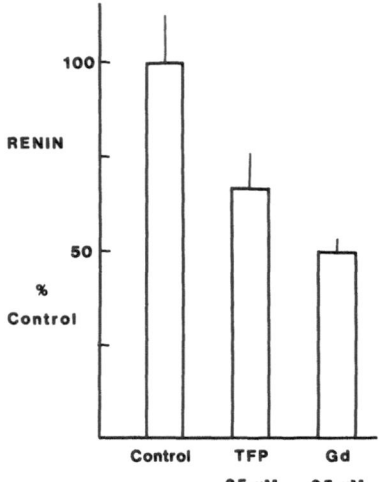

Figure 11. Inhibition of renin release from chorionic trophoblasts by trifluoperazine (TFP). 5×10^5 cells were cultured for 5 days and the exposed for 24 hr to fresh medium containing TFP (25 μM) or gadolinium (25 μM). Renin release is expressed as Figure 10. TFP and Gd produce a significant inhibition of renin release.

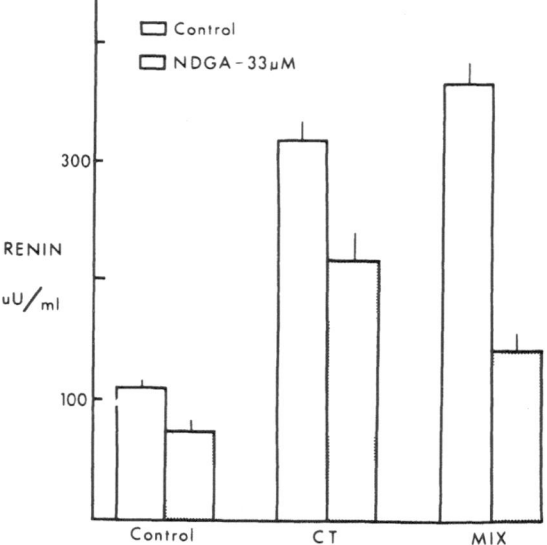

Figure 12. Inhibition of renin release from chorionic trophoblasts by an inhibitor of lipoxygenase, nordihydroguairetic acid (NDGA). 2×10^5 cells were cultured for 13 days and then exposed for 24 hr to CMRL/BSA media. Drug concentrations used were: cholera toxin (CT), 1.0 μg/ml; MIX, 1.0 mM; and NDGA, 33 μM (10 μg/ml). CT and MIX cause a significant stimulation of renin release and NDGA produces a significant inhibition under all conditions.

AMP is released from trophoblasts in response to forskolin, cholera toxin, MIX, and isoproterenol. Renin release is also enhanced in the presence of phorbol myristate acetate (PMA) and phospolipase C, which suggests the presence of a membrane receptor linked to protein kinase C. The simultaneous addition of PMA with CT or MIX causes a superadditive effect, suggesting synergism between the cAMP-activated protein kinase (A kinase) and the diacyl glycerol-activated kinase (C kinase). Simultaneous treatment with BrcAMP and phospholipase C also produces a synergistic response. The release of renin is dependent on extracellular calcium and is inversely correlated with the release of LDH. An inhibitor of lipoxygenase, NDGA, decreases renin release is response to CT and MIX. The results suggest that renin release from trophoblasts is similar to secretion from a number of cell types, utilizing A kinase and C kinase, dependent on calcium, and modulated by arachidonic acid metabolites. The primary signals for renin release and its physiological significance remain to be elucidated.

ACKNOWLEDGEMENTS

This research was supported by NIH grant HD-16552. We wish to acknowledge the help of Dr. Charles King in obtaining the clinical material.

REFERENCES

Benveniste, R., Speeg, K.V., Long, A., and Rabinowitz, D. (1978) Concanavalin-A stimulates human chorionic gonadotropin (hCG) and hCG-alpha secretion by human choriocarcinoma cells. *Biochem. Biophys. Res. Commun.* 84, 1082-1087.

Brostrom, M.A., Brostrom, C.O., Brotman, L.A., Lee, C.S., Wolff, D.J., and Geller, H.M. (1982) Alterations of glial tumor cell Ca^{2+} metabolism and Ca^{2+}-dependent cAMP accumulation by phorbol myristate acetate. *J. Biol. Chem.* 257, 6758-6765.

Brostom, M.A., Brostrom, C.O., Brotman, L.A., and Green, S.S. (1983) Regulation of Ca^{2+}-dependent cyclic AMP accumulation and Ca^{2+} metabolism in intact pituitary tumor cells by modulators of prolactin production. *Molec. Pharmacol.* 23, 399-408.

Castagna, M., Takai, Y., Kaibuchi, K., Sano, K., Kikkawa, U., and Nishizuka, Y. (1982) Direct activation of calcium-activated, phospholipid-dependent protein kinase by tumor-promoting phorbol esters. *J. Biol. Chem.* 257, 7847-7851.

Chen, D.S. and Poisner, A.M. (1976) Direct stimulation of renin release by calcium. *Proc. Soc. Exper. Biol. Med.* 152, 565-567.

Cooke, S.F., Craven, D.J., and Symonds, E.M. (1986) Prostaglandin E_2 and the release of renin from human chorion. *Clin. Exp. Hypertension*, in press.

Fray, J.C. (1980) Stimulus-secretion coupling of renin. *Circulat. Res.* 47, 485-492.

Gaurton, J.P., Pattou, E., and Kordon, C. (1981) Occurrence of higher molecular weight forms of LHRH in fractionated extracts from rat hypothalamus, cortex and placenta. *Molec. Cell. Endocrinol.* 24, 1-15.

Heyworth, C.M., Whetton, A.D.. Kinsella, A.R., and Houslay, M.D. (1984) The phorbol ester, TPA inhibits glucagon-stimulated adenylate cyclase activity. *FEBS Letters* 170, 38-42.

Hinko, A., Franco-Saenz, R., and Mulrow, P.J. (1984) Biphasic alteration of renin release by calcium. *Proc. Soc. Exp. Biol. Med.* 175, 454-457.

Hsueh, W.A., Luetscher, J.A., Carlson, E.J., Fraze, W., and McHargue, A. (1982) Changes in active and inactive renin throughout pregnancy. *J. Clin. Endocrinol. Metab.* 54, 1010-1016.

Ilekis, J. and Benveniste, R. (1983) The arachidonic acid pathway and the secretion of chorionic gonadotropin (hCG) and progesterone by the cultured human choriocarcinoma JEG-3 cell. *Trophoblast Res.* 1, 289-300.

Koike, K., Judd, A.M., Yasumoto, T., and MacLeod, R.M. (1985) Calcium mobilization potentiates prolactin release induced by protein kinase C activators. *Molec. Cell. Endocrinol.* 40, 137-143.

Lester, G.E. and Rubin, R.P. (1977) The role of calcium in renin secretion from the isolated cat kidney. *J. Physiol.* 269, 93-108.

Moore, J.J., Baker, J.V., and Whitsett, J.A. (1983) cAMP-dependent protein kinase and cAMP, cGMP and calcium stimulated phosphorylation in human placenta: comparison of cytosol and membrane fractions. *Trophoblast Res.* 1, 185-196.

Nikolics, K., Mason, A.J., Szonyi, E., Ramachandran, J., and Seeburg, P.H. (1985) A prolactin-inhibiting factor within the precursor for human gonadotropin-releasing hormone. *Nature* 316, 511-517.

Nishizuka, Y., Takai, Y., Kishimoto, A., Kikkawa, U., and Kaibuchi, K. (1984) Phospholipid turnover in hormone action. *Rec. Prog. Horm. Res.* 40, 301-341.

Park, C.S., Han, D.S., and Fray, J.C.S. (1981) Calcium in the control . of renin secretion: Ca^{2+} influx as an inhibitory signal. *Am. J. Physiol.* 240, F70-F74.

Perchellet and Boutwel (1980) Enhancement of isobutyl-1-methylxanthine and cholera toxin of 12-O-tetradecanoylphorbol-13-acetate stimulated cyclic nucleotide levels and ornithine decarboxylase activity in isolated epidermal cells. *Cancer Res.* 40, 2653-2660.

Poisner, A.M. (1984) Effect of cholera toxin, methyl-isobutyl-xanthine, phorbol myristate acetate, sodium butyrate and vinblastine on renin secretion from cultured human chorionic trophoblasts. *J. Cell. Biol.* 99 (4, Part 2), 225a.

Poisner, A.M. and Poisner, R.B. (1985) Angiotensin receptors and release of LHRH activity from human chorion laeve. *Proc. 67th Endocrinol. Soc. Mtg.* 33.

Poisner, A.M., Johnson, R.L., Hanna, G., and Poisner, R. (1981a) Activation of renin in human amniotic fluid and placental membranes. In: *Heterogeneity of Renin and Renin Substrate*, (ed.), M.P. Sambhi, Amsterdam: Elsevier North Holland, pp. 335-347.

Poisner, À.M., Wood, G.W., Poisner, R., and Inagami, T. (1981b) Localization of renin in trophoblasts in human chorion laeve at term pregnancy. *Endocrinol.* 109, 1150-1155.

Poisner, A.M., Wood, G.W., and Poisner, R. (1982) Release of inactive renin from human fetal membranes and isolated trophoblasts. *Clin. Exp. Hypertension* A4, 2007- 2017.

Poisner, A.M., Cheng, H.C., Wood, G.W., and Poisner, R. (1983a) Storage and release of renin and hCG in trophoblast from human chorion laeve. *Trophoblast Res.* 1, 279-288.

Poisner, A.M., Wood, G.W., and Poisner, R. (1983b) Stimulation and inhibition of renin release from trophoblasts cultured from human chorion laeve. *Fed. Proc.* 42, 455.

Poisner, A.M., Poisner, R., and Agrawal, P. (1984a) Effects of calcium and heavy metals on release of renin and LDH from human chorionic trophoblasts. *Fed. Proc.* 43, 582.

Poisner, A.M., Poisner, R.B., and Tonkowicz, P.A. (1984b) Renin and progesterone secretion from human trophoblasts: role of cyclic AMP and calcium. *Proc. 7th Int. Congr. Pharmacol.*, p.1920.

Poisner, A.M., Cheng, H.C., Tomita, T., Poisner, R., and King, C.R. (1984c) Report of a hydatidiform mole containing renin, inactive renin, progesterone and HCG. *Proc. 7th Int. Congr. Endocrinol.*, p. 1251.

Poisner, A.M., Poisner, R., and Agrawal, P. (1985) Synergistic effects between cholera toxin and phorbol myristate acetate in producing HCG secretion from a malignant trophoblastic cell line. *Fed. Proc.* 44, 1737.

Poisner, A.M. and Poisner, R (1986) Methods for studying renin secretion from cultured trophoblasts and superfused chorion. In: *In Vitro Methods for Studying Secretion*, (ed.), A.M. Poisner and J.M. Trifaro, Amsterdam: Elsevier North Holland, in press.

Tonkowicz, P.A. and Poisner, A.M. (1985) Evidence for cyclic-AMP mediated progesterone secretion from human chorionic trophoblasts. *Endocrinol.* 116, 646-650.

Welsh, Jr., T.H., Jones, P.B.C., and Hsueh, A.J.W. (1984) Phorbol ester inhibition of ovarian and testicular steroidogenesis in vitro. *Cancer Res.* 44, 885-892.

Trophoblast Research 2:61-69, 1987

AMNIONIC ASSOCIATION WITH CHORION AND PROSTAGLANDIN E₂ PRODUCTION BEFORE AND AFTER LABOR

John A. McCoshen

Department of Obstetrics, Gynecology
and Reproductive Sciences
University of Manitoba
59 Emily Street
Winnipeg, Manitoba
Canada R3E 0W3

INTRODUCTION

Myometrial contractions and cervical ripening associated with the spontaneous initiation of labor appear to be related to alterations in production and metabolism of a variety of prostaglandins by intrauterine tissues (Embrey, 1969; Karim et al., 1969; Karim, 1975; Shepherd, 1976; and Wingerup et al., 1978). Furthermore, while temporal changes in steroidogenesis by both mother and fetus may be involved in the natural termination of pregnancy via prostaglandins (Blocke et al., 1984; Khan-Dawood and Dawood, 1984; and Mitchell et al., 1982), much attention is now being focused on other possible factors as controlling agents of labor-related prostaglandin synthesis. Specifically, the fetal membranes and decidualized endometrium of pregnancy are suspected important elements in the complex process of labor. Both inhibitory and stimulatory factors of prostaglandin production by amnion have been identified in amniotic fluid, amnionic, chorionic, and decidual tissues (Karim, 1975; Mitchell et al., 1984; Rhenstrom et al., 1983; and Saeed and Mithcell, 1982). Suggestions that the fetal kidney contributes to the amniotic fluid pool of prostaglandins via fetal urine have been proposed (Casey et al., 1983) as well as evidence that fetal urine contains a factor(s) stimulatory to fetal membrane prostaglandin production and the initiation of labor (Strickland et al., 1983). Thus, the hypothesis that the fetus is instrumental in initiating the onset of labor and delivery may have credence in view of the propensity of fetal membranes to produce high concentrations of a variety of prostaglandins. Furthermore, amnionic cells obtained after spontaneous labor onset and vaginal delivery produce more prostaglandins as compared to cells obtained prior to labor (Olson et al., 1983). During the course of labor, amniotic fluid prostaglandins E₂ and F₂α substantially increase in concentration (Manabe et al., 1983; Nieder and Augustin, 1983).

The use of prostaglandins, specifically prostaglandin F₂α (PGF), and prostaglandin E₂ (PGE), administered by intravenous infusion, injection into amniotic fluid or as vaginal suppositories, have been used to induce first trimester abortion (Bygdeman, 1984). Currently, PGE is used in a number of centres to stimulate

Submitted: 10 October 1985; Accepted: 10 May 1986

cervical ripening in preparation for labor induction (Lorenz et al., 1984; Nimrod et al., 1984). Furthermore, spontaneous rupture of fetal membranes, amniotomy, and mechanical distension of the amniotic sac facilitate the initiation of labor and have been associated with alterations in prostaglandin production and sensitivity of uterine contractility (Guzick and Winn, 1985; Manabe et al., 1982; Manabe et al., 1983; and Mitchell et al., 1977).

While a strong link exists between the prostaglandins and labor, it is as yet unclear as to whether or not changes in prostaglandin levels actually initiate labor or simply result from the onset of labor. Furthermore, prostaglandin production by individual components of intrauterine tissue may require interaction between amnion and the underlying villous chorion and reflected choriodecidua. Thus, the present study sought to identify alterations in prostaglandin production by term fetal membranes relative to associations with underlying chorion laeve or villous chorion, and to mode of delivery relative to non-labor and post-labor conditions.

MATERIALS AND METHODS

Placentae with attached umbilical cord and reflected membranes were used in these studies. All tissues were washed in cold (4°C) phosphate buffered saline at pH 7.4 to remove residual blood. Sections of umbilical cord were obtained no less than 10 cm away from the placental junction. Five 2 cm lengths were cut from each of 5 umbilical cords obtained from 5 elective repeat term cesarean deliveries (38-39 weeks) and 5 term vaginal deliveries (39-41 weeks) following spontaneous onset of labor. Each section of umbilical cord was weighed, placed in 18 ml of synthetic amniotic fluid (SAF) at pH 7.4 (Schwartz et al., 1977) and pre-incubated for 2 hr at 20°C to allow elution of endogenous PGE. The medium was changed at 1 and 2 hr, respectively. Subsequently, the tissues were incubated at 37°C for 4 hr following a 15 minute temperature equilibration period. Aliquots of 0.5 ml were removed at 0, 2, and 4 hr with addition of an equal volume of SAF to maintain a constant volume of 18 ml.

From an additional 15 cesarean and 15 vaginal deliveries, amnion was removed from the reflected chorion laeve and from the placenta. Paired discs (8 cm^2) of reflected and placental amnion were mounted on 22 mesh stainless steel screens and secured within perfusion apparati as previously described by McCoshen et al. (1981). Each chamber was filled to 18 ml with SAF and these tissues were treated in a manner identical to that described above for umbilical cord.

From an additional 5 cesarean and 4 vaginal deliveries, paired membranes consisting of reflected amnion (A), choriodecidua (CD), and amnion retained in cellular contact with choriodecidua (ACD) were prepared as described above. During the course of both pre-incubation at 20°C and incubation at 37°C, aliquots of incubation media were removed from each of the fetal and maternal sides of each membrane. In all instances, samples of incubation medium were quick-frozen in acetone/dry ice (-70°C) and stored frozen until time of radioimmunoassay (RIA) for PGE.

All membranes were incubated in medium circulated with humidified filtered air. The membranes were tested for leakage by first adding 9 ml of SAF to the fetal chamber and observing the maternal chamber for 10 min. Each membrane was weighed following incubation, placed flat and secured to wax cards, fixed in 10% neutral buffered formalin and subsequently stained as whole mounts in 1% toluidine blue. RIA of PGE was performed according to the method of Dubin et al. (1979). All results were analyzed by multiple analysis of variance (MANOVA) to test treatment effects of type of membrane, membrane side and mode of delivery and incubation time. Multiple two-tailed t-tests and analysis of correlation were also used for testing specific variables.

RESULTS

Total PGE produced by reflected (R) and placental (P) amnion after 4 hr of incubation at 37°C is presented in Figure 1. The comparison of amnion from non-labor cesarean and post-labor vaginal deliveries is also illustrated. Placental amnion from cesarean deliveries produced significantly more PGE (mean $\pm$ SEM) than reflected amnion (1123 $\pm$ 153 ng/ml vs 207 $\pm$ 29 ng/ml) whereas amnion from vaginal deliveries produced 2257 $\pm$ 274 ng/ml vs 730 $\pm$ 96 ng/ml) respectively. Significant differences were identified between location and mode of delivery as summarized in Table 1. Comparisons of PGE production by the fetal and maternal sides of amnion (A), choriodecidua (CD), and composite amnion plus choriodecidua (ACD) from the reflected location is presented in Figures 2-4, respectively. Statistical analyses of these data (MANOVA) identified significant differences in total PGE produced between cesarean and vaginally delivered amnion (F = 4.29; d.f.1:48; P < 0.05) and choriodecidua (F = 37.91; d.f. 1:48; P < 0.001). A significant difference between fetal and maternal PGE production by ACD was found for composite membrane from vaginal deliveries (F = 4.22; d.f. 1:48; P < 0.05) but not from cesarean deliveries. Umbilical cord PGE production is presented in Figure 5. No significant differences in PGE production by cord from cesarean versus vaginal deliveries were found, however, total productivity by umbilical cord from each mode of delivery was substantial being almost 4000 pg/ml in each instance.

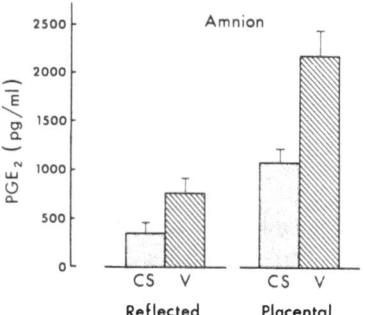

Figure 1. Net PGE (pg/ml) $\pm$ SEM produced over 4 hr incubation at 37°C per 8 cm[2] of amnion. Total PGE production by placental amnion is significantly (P < 0.001) greater than that of reflected amnion. In each instance, vaginally (V) delivered membrane is significantly (P < 0.001) more productive than cesarean delivered (CS) membrane. (n = 30 for each mean.)

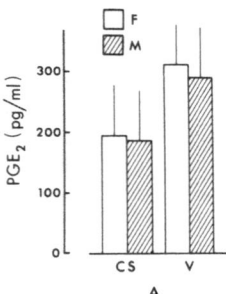

Figure 2. Net PGE ± SEM production by reflected amnion (A) after 4 hr of incubation at 37°C. PGE production by vaginally (V) delivered amnion is significantly (P<0.05) greater than cesarean (CS) delivered membranes. No difference between PGE concentrations on the fetal (F) and maternal (M) sides concentrations was found. (n=10, CS; n=8, V, for each mean.)

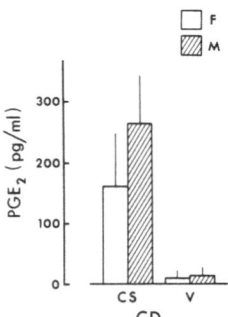

Figure 3. Net PGE ± SEM production by reflected choriodecidua (CD) from vaginal (V) deliveres is significantly (P<0.001) less than cesarean (CS) delivered membranes. In each of CS and V tissues, fetal (F) and maternal (M) PGE concentrations are equivalent. (n is the same as Figure 2.)

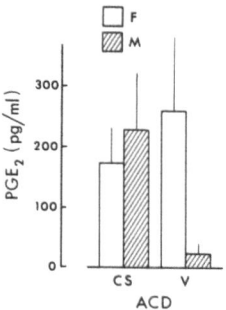

Figure 4. Net PGE $\pm$ SEM production by composite amnion plus choriodecidua (ACD). PGE concentrations on the fetal (F) or amnionic side of each of cesarean (CS) and vaginally (V) delivered ACD is equal to that of respective A membrane alone. On the maternal (M) side, PGE is equivalent to the M side of CD from CS and V deliveries. A significant (P < 0.05) difference exists for PGE concentrations between F and M sides of ACD membranes from V deliveries but not CS deliveries. (n is the same as Figure 2.)

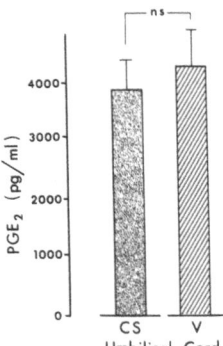

Figure 5. Net PGE $\pm$ SEM production by 2 cm lengths of umbilical cord from CS and V deliveries. No significant difference exists between cord PGE production from each mode of delivery. (n = 10 for each mean.)

DISCUSSION

The results of these studies identify a significant difference in production of PGE by amnion relative to location in the gestational sac. The locational associations include reflected amnion derived from contact with chorion laeve and placental amnion derived from contact with villous chorion. Furthermore, significant increases in amnionic PGE production from each source was found for those tissues obtained at parturition following spontaneous labor onset. However, umbilical cord from each

Table 1

Summary of multiple range test of amnionic PGE production according to location in the gestational sac (R, reflected; P, placental) and mode of delivery (CS, caesarean; V, vaginal)

*Comparison	*Comparison			
	**LSD	Probability	***SNK	Probability
CS: R vs P	10.45	<0.001	14.78	<0.01
V: R vs P	17.40	<0.001	24.61	<0.01
R: CS vs V	5.96	<0.001	8.43	<0.01
P: CS vs V	12.91	<0.001	18.26	<0.01

*   Error mean square = 115528, n = 30 for each mean.
**   Least significant difference.
***   Student-Newman-Keul's Test.

mode of delivery produced equivalent PGE concentrations. Except for the point of junction of the cord and placenta, umbilical cord has no direct association with either chorion laeve or villous chorion. Thus, we suggest that the association of amnion with the underlying chorion may in part be responsible for the difference in PGE production by reflected and placental amnion.

In those studies that considered the fetal and maternal sides of suspended amnion, choriodecidua and composite amnion plus choriodecidua, it was also found that reflected amnionic PGE production increased following parturition subsequent to spontaneous labor onset. However, choriodecidua displayed almost no net PGE production over time following vaginal delivery as compared to those membranes obtained at cesarean section prior to labor. Confirmaton of this finding is seen in those membranes consisting of composite amnion plus choriodecidua in that fetal PGE increased between the two modes of delivery whereas on the maternal side, representing the choriodecidual component, PGE production was significantly reduced following vaginal delivery. These findings suggest that functional changes may take place in choriodecidua from a time prior to labor until parturition, that result in a substantial reduction in the ability of choriodecidua to produce PGE. On the other hand, in all instances amnion, regardless of location or chorionic associ-ation, displayed a net increase in PGE productivity at parturition.

Prostaglandin E_2 represents the major prostanoid produced by amnionic membrane (Mitchell et al., 1978). Differences in amnionic PGE production according to mode of delivery have been described by others where dispersed amnionic cells were utilized (Olson et al., 1983; Skinner and Challis, 1985). Furthermore, chorion and decidua appear to have the ability to produce PGE and PGF as the principle prostanoids. The assumption that high levels of both PGE and PGF in amniotic fluid gain access to the myometrium via the reflected fetal membrane, and subsequently participate in myometrial contractions, may not be correct in view of the significant decline in PGE on the maternal side of choriodecidua and amnion plus choriodecidua. However, these results do not deny a role for prostaglandin from amnion or amniotic fluid in myometrial contractions, since it is possible that PGE is rapidly metabolized by the choriodecidua (Casey and MacDonald, 1983), possibly to PGF. Skinner and Challis (1985) have recently reported that chorionic and decidual prostaglandin metabolite 13, 14-dihydro-15-keto prostaglandin $F_2\alpha$ substantially increases following spontaneous onset of labor.

These current studies suggest that amnionic PGE production may be influenced by the underlying chorion. Furthermore, while amnionic PGE production is increased at parturition following spontaneous onset of labor, choriodecidual PGE production is substantially decreased. Thus, these results, while confirming increased amnionic PGE production during the course of labor, fail to identify the choriodecidua as a route of transport or a source of PGE to the underlying myometrium. These findings may suggest that PGE is important around the time of labor onset but not as labor progresses. Alternatively, the prostaglandin requirements of myometrium may change subsequent to labor onset in favor of another prostanoid(s), or possibly a requirement for PGE may diminish during active labor.

SUMMARY

In summary, differential production of PGE has been found for amnion and choriodecidua that appears related to both chorionic associations and the circumstances of non-labor and post-labor delivery. The results further suggest that choriodecidual PGE productivity is diminished at parturition while that of amnion is increased. Thus, reciprocal changes in PGE productivity by the reflected membranes occur that may in part be an expression of physiologic interactions between the fetal membranes occurring during labor.

REFERENCES

Block, B.S.B., Liggines, G.C., and Creasy, R.K. (1984) Preterm delivery is not predicted by serial plasma estradiol or progesterone concentration measurements. *Am. J. Obstet. Gynecol.* 150, 716.

Bygdeman, M. (1984) The use of prostaglandins and their analogues for abortion. *Clin. Obstet. Gynaecol.* 11, 573.

Casey, M.L., Carter, S.I., and Mitchell, M.D. (1983) Origin of prostanoids in human amniotic fluid: The fetal kidney as a source of amniotic fluid prostanoids. *Am.J. Obstet. Gynecol.* 147, 547

Casey, M.L. and MacDonald, P.C. (1983) Pisseaktis in human amniotic fluid. Presented to the 65th Annual Meeting of the Endocrine Society, San Antonio, Texas. *Abstract* 600.

Dubin, N.H., Ghodgaonkar, R.B., and King, T.M. (1979) Role of prostaglandin production in spontaneous and oxytocin-induced uterine contractile activity in in vitro pregnant rat uteri. *Endocrinol.* 105, 47.

Embrey, M.P. (1969) The effect of prostaglandins on pregnant human uterus. *J. Obstet. Gynaecol. Br. Common.* 76, 683.

Guzick, D.S. and Winn, K. (1985) The association of chorioamnionitis with preterm delivery. *Obstet. Gynecol.* 65, 11.

Karim, S.M.M., Trussell, R.R., Hillier, K., and Patel, R.C. (1969) Induction of labor with prostaglandin F_{2a}. *J. Obstet. Gynaecol. Br. Common.* 76, 169.

Karim, S.M.M. (1975) *Prostaglandins and Reproduction*, pp. 150-178, University Park Press, Baltimore, Maryland.

Khan-Dawood, F.S. and Dawood, M.Y. (1984) Estrogen and progesterone receptor and hormone levels in human myometrium and placenta in term pregnancy. *Am. J. Obstet. Gynecol.* 150, 501.

Lorenz, R.P., Botti, J.J., Chez, R.A., and Bennett, N. (1984) Variations of biologic activity of low-dose prostaglandin E_2 on cervical ripening. *Obstet. Gynecol.* 64, 123.

Manabe, Y., Manabe, A., and Takahashi, A. (1982) F prostaglandin levels in amniotic fluid during balloon-induced cervical softening and labor at term. *Prostaglandins* 23, 247.

Manabe, Y., Okazaki, T., and Takahashi, A. (1983) Prostaglandin E and F in amniotic fluid during stretch-induced cervical softening and labor at term. *Gynecol. Obstet. Invest.* 15, 343.

McCoshen, J.A., Chudasama, S., and Tyson, J.E. (1981) Differential responsiveness of cells of human amniotic epithelium to ferritin and ^{125}I-prolactin in vitro. *Placenta* (Suppl. 3), 33.

Mitchell, M.D., Flint, A.P.F., Bibby, J., Brunt, J., Arnold, J.M., Anderson, A., and Turnbull, A. (1977) Rapid increases in plasma prostaglandin concentrations after vaginal examination and amniotomy. *Br. Med. J.* 2, 1183.

Mitchell, M.D., Bibby, J., Hicks, B.R., and Turnbull, A.C. (1978) Specific production of prostaglandin E_2 by tissues from human uterus and fetoplacental unit. *Prostaglandins* 15, 377.

Mitchell, B., Cruickshank, B, McLean, D., and Challis, J. (1982) Local modulation of progesterone production in human fetal membranes. *J. Clin. Endocrinol. Metabol.* 55(6), 1237.

Mitchell, M., MacDonald, P.C., and Casey, M.L. (1984) Stimulation of prostaglandin E_2 synthesis in human amnion cells maintained in monolayer culture by a substance(s) in amniotic fluid. *Prosta. Leuko. Med.* 15, 399.

Nieder, J. and Augustin, W. (1983) Increase of prostaglandin E and F equivalents in amniotic fluid during late pregnancy and rapid PGF elevation after cervical dilatation. *Prosta. Leuko. Med.* 12, 289.

Nimrod, C., Currie, J., Yee, J., Dodd, G., and Persaud, D. (1984) Cervical ripening and labor induction with intracervical triacetin base prostaglandin E gel: A placebo-controlled study. *Obstet. Gynecol.* 64, 476.

Olson, D.M., Skinner, K., and Challis, J.R.G. (1983) Prostaglandin output in relation to parturition by cells dispersed from human intrauterine tissues. *J. Clin. Endocrinol. Metabol.* 57(4), 694.

Rehnstrom, J., M. Ishikawa, F. Fuchs, and A-R. Fuchs (1983) Stimulation by myometrial and decidual prostaglandin production by amniotic fluid from term, but not mid-trimester pregnancies. *Prostaglandins* 26(6), 973.

Saeed, S.A. and Mitchell, M.D. (1982) Stimulants of prostaglandin biosynthesis in human fetal membranes, uterine decidua vera and placenta. *Prostaglandins* 24(4), 475.

Schwartz, A.L., Forster, C.S., Smith, P., and Liggins, G.C. (1977) Human amniotic metabolism. I. In vitro maintenance. *Am. J. Obstet. Gynecol.* 127, 470.

Shepherd, J., Sims, C., and Craft, I. (1976) Extra amniotic E_2 and the unfavorable cervix. *Lancet* ii, 709.

Skinner, K. and Challis, J.R.G. (1985) Changes in the synthesis and metabolism of prostaglandins by human fetal membranes and decidua at labor. *Am. J. Obstet. Gynecol.* 151, 519.

Strickland, D.M., Saeed, S.A., Casey, M.L., and Mitchell, M.D. (1983) Stimulation of prostaglandin biosynthesis by urine of the human fetus may serve as a trigger for parturition. *Science* 220, 521.

Wingerup, L., Andersson, K-E., and Ulmsten, U. (1978) Ripening of the uterine cervix and induction of labor with prostaglandin E_2 in viscous gel. *Acta Obstet. Gynecol. Scand.* 57, 403.

Trophoblast Research 2:71-83, 1987

ARACHIDONIC ACID METABOLISM IN THE HUMAN PLACENTA

Matthew P. Rose, Murdoch G. Elder, and Leslie Myatt[1]

Institute of Obstetrics and Gynaecology
Hammersmith Hospital
Du Cane Road
London W12 0HS, United Kingdom

INTRODUCTION

The synthesis of arachidonic acid metabolites via the cyclooxygenase enzyme pathway in the pregnant human uterus is well documented (Keirse, 1978). There is ample evidence for the participation of these metabolites in implantation (Pakrasi and Dey, 1982), maintenance of placental blood flow (Wallenberg, 1981) and in the initiation and maintenance of parturition (Goldberg and Ramwell, 1975). Arachidonic acid is also metabolized via lipoxygenase enzymes to hydroxy-eicosatetraenoic acids (HETEs) and leukotrienes (LTs) and, via cytochrome P_{450} enzymes, to stable epoxides (Chacos et al., 1982), HETEs and vicinal diols (Oliw and Oates, 1981). There is evidence to suggest that fetal membranes may be able to synthesize hydroxyfatty acids in addition to prostaglandins in humans (Kinoshita and Green, 1980; Saeed and Mitchell, 1982) and in rabbits (Elliott et al., 1984). However the synthesis of lipoxygenase products by the human placenta is not well documented.

We have studied the metabolism of arachidonic acid in vitro by human placental cells obtained during the first trimester of pregnancy, when the utero-placental circulation is being established, or obtained at term. Alterations in arachidonic acid metabolism have been suggested to be etiological factors in pregnancies complicated by fetal growth retardation or pre-eclampsia (Stuart et al., 1981; Goodman et al., 1982; Ogburn et al., 1984). Pathological features of these complications such as formation of thrombi within the utero-placental circulation (Sheppard and Bonnar, 1976) may be increased by an imbalance of arachidonic acid metabolites affecting platelet function and vascular tone (Jogee et al., 1983).

The metabolism of arachidonic acid supplied exogenously in cell culture medium and of that released from intracellular sources was studied as there is evidence that different prostaglandins may arise from the two sources of substrate (Coene et al., 1982). The spectrum of arachidonic acid metabolites formed was analyzed by reverse- phase high performance liquid chromatography (RP-HPLC).

[1]To whom correspondence should be addressed.
Received: 8 October 1985; Accepted: 15 June 1986

MATERIALS AND METHODS

Human placental tissue was obtained at first trimester termination of pregnancy (6 experiments) or following delivery by spontaneous labor or elective cesarean section (not in labor) at term from normal pregnancies (4 experiments each). Cells were dispersed enzymatically and grown in monolayer culture using the methods described by Jogee et al. (1983). After dispersion with trypsin (0.5%, Difco Bacto 1/250, 37°C, 1 hour) cells were collected by filtration through sterile gauze and pelleted by centrifugation (150 x g, 10 min.). The cells were washed by resuspending in tissue culture medium (TC-M199 + 10% horse serum, Gibco), repelleted and finally resuspended in culture medium (0.5×10^6 ml^{-1}). Suspensions (10 ml) were plated out in tissue culture dishes and incubated in a humidified atmosphere (5% CO_2 in air, 37°C).

After 24 hr the supernatant was discarded, and red blood cells and other non-adherent cells were removed by washing with sterile phosphate buffered saline (3 x 10 ml Dulbeccos A, Oxoid Ltd). Tissue culture medium (10 ml) containing (^{3}H)-arachidonic acid (0.2 µCi ml^{-1}, 5, 6, 8, 9, 11, 12, 14, 15 - ^{3}H(N)-arachidonic acid, Amersham, UK) was added to the monolayer cultures which were then incubated for another 24 hr. After this time the medium was removed and stored at -20°C prior for analysis. The cultures were washed with sterile PBS and fresh culture medium (10 ml), without (^{3}H)-arachidonic acid, was added. After an additional 24 hr, medium was collected and stored at -20°C. Continuing viability of the cells was determined using the trypan blue exclusion test.

In order to control for auto-oxidation culture medium (10 ml) containing (^{3}H)-arachidonic acid (0.2 µCi ml^{-1}) was incubated in the absence of cells for 24 hr and treated in the same way as culture supernatants. In separate experiments, first trimester placental cells were incubated with (^{3}H)-arachidonic acid as described previously (controls) with additional cultures being performed with addition of indomethacin (10^{-6}M) or nordihydroguaiarietic acid (NDGA) (5×10^{-4}M), α-tocopherol (10 µg/ml), α-naphthoflavone (1-100 µg/ml) or SKF 525A (1-100 µg/ml) for the final 24 hr of incubation.

The extraction and subsequent analysis of arachidonic acid metabolites was performed using reverse phase adsorption chromatography on octadecyl (C_{18}) silica columns. This involves adsorption of metabolites onto the hydrophobic C_{18} solid support and their elution in order of decreasing polarity (i.e. most polar first) by decreasing the polarity of the eluting solvent or solvent mixtures. Initial extraction was performed on minicolumns (C_{18} Sep-Paks) followed by fractionation on an analytical column.

C_{18} Sep-Paks were prepared by wetting with acetonitrile (10 ml, HPLC grade, Rathburn Chemicals Ltd) and washing successively with distilled water (10 ml), petroleum spirit (10 ml, May and Baker), methylformate (10 ml BDH) and 80% ethanol (10 ml). Duplicate samples of culture supernatants (10 ml) were pooled, acidified to pH 3.0 with formic acid and loaded with a glass syringe onto prepared C_{18} Sep-Paks which were then washed with distilled water (10 ml) and petroleum spirit (10 ml).

Oxygenated metabolites of arachidonic acid were eluted with methyl formate (5 ml) into siliconized glass test tubes. Methyl formate was reduced under oxygen free nitrogen and the residue was redissolved in 1 ml of HPLC solvent 1. Prior to separation, the sample for analysis was filtered through Millipore solvent resistant filters (0.45 μM).

Prostaglandins, prostaglandin metabolites and di- and mono-HETEs were separated on a reverse phase Waters Fatty Acid Analysis Column (Millipore, UK, Ltd) using three isocratic solvent mixtures at a flow rate of 1 ml min^{-1}. Solvent 1 consisted of acetonitrile: 40 mM triethylamine formate pH 3.15 (TEAF) 30/70 v/v for 25 min. Solvent 2 consisted of acetonitrile: TEAF 50:50 v/v for 25 min. and solvent 3 acetonitrile 100% for 8 min. Solvent composition and flow rate were controlled by a Spectra-Physics SP 8700 solvent delivery system and solvents were degassed with helium. Eluting solvent was collected into liquid scintillation vials (0.5 ml) using a Pharmacia Frac 100 fraction collector and radioactivity was determined by liquid scintillation counting.

The fatty acid analysis column was calibrated with standard radiolabeled or unlabeled metabolites fractionated under the conditions described above (Table 1). (^{3}H)-6-keto-PFG$_1$α and (^{3}H)-TxB$_2$ were obtained from New England Nuclear, (^{3}H)-PGF$_2$α, (^{3}H)-PGE$_2$, (^{3}H)-13,14-dihydro-15-keto-PGE$_2$ and (^{3}H)-LTB$_4$ were obtained from Amersham International. (^{14}C)-5-, 8-, 9-, 11-, 12-, and 15-HETEs were gifts from Dr. P. Woolard, Institute of Dermatology, London. Unlabeled metabolites were obtained from Sigma Chemical Co., Ltd. and their elution positions determined by continuous U.V. monitoring.

RESULTS

Culture of human placental cells in vitro yields four cell types we have previously described (Jogee et al., 1983). The majority of cells appears to be trophoblast in origin from both first trimester and term tissue. There was no appreciable change in cell viability (greater than 95%) throughout the experiment. The profiles of radiolabeled arachidonic acid metabolites obtained were compared with the retention data for authentic standards (Table 1) and radioactivity per peak calculated by summation.

All of the control experiments showed a major peak of radioactivity eluting in the column void volume perhaps representing short chain fatty acids. Lesser amounts of radioactivity were observed in peaks co-chromatographing with PGI$_2$ metabolites, TxB$_2$, LTB$_4$, and HETEs. The extent of auto-oxidation was 3.6 $\pm$ 0.94% (n=4) of the total radioactivity incubated. In all the control experiments, the radioactivity in these peaks was integrated and compared with that observed in supernatants of cell cultures, in order to assess the significance of metabolism by cells.

Representative product profiles obtained following incubation of first trimester, elective cesarean section or term spontaneous delivery placental cell monolayer cultures with medium containing (^{3}H)-arachidonic acid are shown in Figures 1a, b, and c, respectively. The conversion of added (^{3}H) arachidonic acid to polar products by placental cells in culture was 12.9 $\pm$ 3.7, 11.0 $\pm$ 5.2, and 2.9 $\pm$ 2.7%

Table 1

Retention data for arachidonic acid metabolites on reverse-phase high performance
liquid chromatography

Compound	t_R (mins.)	Peak width (mins.)
6-keto-PGF$_1$	8.0	7.0-9.0
2, 3-dinor-6-keto-PGF$_1$*	8.15	Not determined
13, 14-dihydro-6, 15-diketo-PGF$_1$*	Not detectable	
6-keto-PGE$_1$*	7.7	Not determined
TxB$_2$	11.0	10.5-12.0
PGF$_2$	13.0	12.5-14.5
15-keto-PGF$_2$*	18.0	Not determined
13, 14-dihydro-15-keto-PGF$_2$*	22.0	Not determined
PGE$_2$	16.5	15.5-18.0
15-keto-PGE$_2$*	20.5	Not determined
13, 14-dihydro-15-keto-PGE$_2$	27.0	26.0-28.5
PGD$_2$*	19.0	Not determined
PGA$_2$*	29.0	Not determined
PGB$_2$*	31.8	Not determined
LTB$_4$	34.0	34.0-35.0
5-HETE	35.5, 43.0	34.5-36.5, 42.0-44.0
8-HETE	42.5	41.5-43.0
9-HETE	44.0	43.0-44.5
11-HETE	42.5	43.0-44.0
12-HETE	35.5, 42.5	35.0-36.0, 41.5-43.0
15-HETE	42.0	41.5-43.0
Arachidonic acid	58.0	57.5-59.0

* = Non-radioactively labelled compounds.

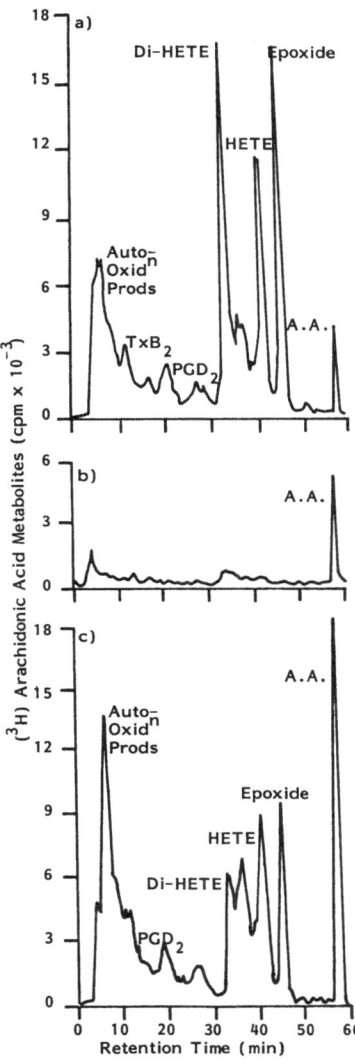

Figure 1.   Metabolites of exogenously added (^{3}H) arachidonic acid from a) first trimester; b) term elective cesarean section, or c) term spontaneous labor placental cells. Tissues were enzymatically dispersed and cells grown in monolayer culture at 0.5 x 10^6 ml^{-1} and incubated with (^{3}H) arachidonic acid (0.2 µCi ml^{-1}) for 24 hr at 37°C. Oxygenated metabolites were extracted from 10 ml culture supernatant and analyzed by reverse phase HPLC on a fatty acid analysis column 0.5 ml fractions of effluent were collected and radioactivity counted.   Retention times of products were compared with authentic standards.

by first trimester, term spontaneous labor and term cesarean section, respectively. There appeared to be no significant difference compared with controls in the amount of auto-oxidation products formed and which chromatographed in the column void volume. While the metabolites co-chromatographing with Thromboxane B_2, PGD_2, and PGE were found the major metabolites co-chromatographed with the dihydroxy fatty acid standard Leukotriene B_4 (shown as diHETE), with monoHETEs and as a less polar compound (retention time 45-46 mins.) which did not chromatograph with any of the standards available. The total radioactivity per metabolite was determined by integration and was compared with that produced by auto-oxidation of (^{3}H)-arachidonic acid. The first trimester placental cells (Figure 2a) appeared to be more metabolically active than term cells obtained following spontaneous labor (Figure 2b). The least active were cells obtained by elective cesarean section prior to labor and the amount of metabolites produced were not significantly different from control (data not shown).

In addition to conversion of exogenously added (^{3}H)-arachidonic acid placental cells could also incorporate it intracellularly. The incorporation was 76.4 ± 3.1, 24.5 ± 9.0, and $8.6 \pm 1.4\%$ for first trimester, term spontaneous labor and term

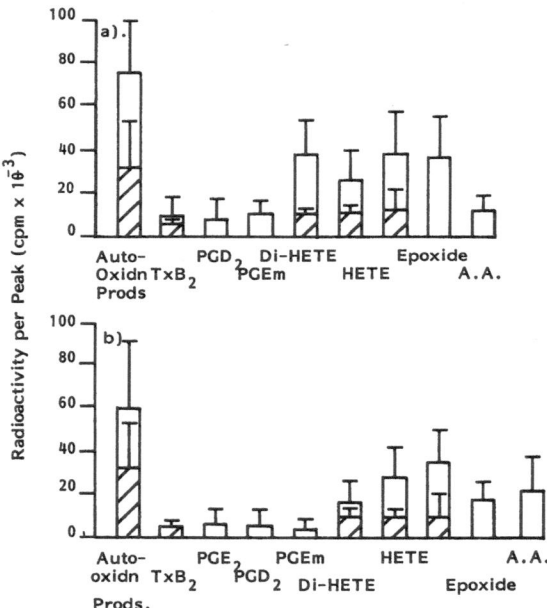

Figure 2. Quantitation of (^{3}H) arachidonic acid metabolites produced by a) first trimester and b) term spontaneous labor placental cells. The radioactivity co-chromatographing with known standards (open bars) was summated and compared with that found in control incubations in the absence of cells (closed bars). The mean $\pm$ SD is shown for 6(a) and 4(b) separate experiments.

cesarean section, respectively. Following removal of medium containing (^{3}H)-arachidonic acid and replacement with fresh medium not containing radioactivity for the final 24 hr of incubation the products of this metabolism of endogenously incorporated arachidonic acid could be studied (Figure 3). The proportion of total incorporated radioactivity released was 3.9 $\pm$ 1.0, 7.8 $\pm$ 1.8, and 6.9 $\pm$ 2.5%, respectively and its distribution among the metabolites is shown in Figure 4. Only compounds co-chromatographing with the diHETE and monoHETE standards and the non-polar compound at 45-46 min were seen. Again cells obtained at elective cesarean section (not in labor) did not produce appreciable quantities of metabolites (data not shown) when compared to first trimester cells (Figures 3a and 4a) and cells obtained following term spontaneous labor (Figures 3b and 4b).

Addition of either indomethacin (10^{-6}M), the antioxidant α-tocopherol (10 μg/ml) or the cytochrome P_{450} inhibitor metyrapone did not affect the profile of products released from this endogenous substrate by first trimester placental cells (data not shown). Preliminary experiments, however, showed that the lipoxygenase inhibitor NDGA (500 μM) inhibited production of all the products chromatographing with retention times greater than 30 min (Figure 5) and that the cytochrome P_{450} inhibitor SKF 525A gave inhibition (1-100 μg/ml) of these products with the most pronounced effect being on the metabolites chromatographing at 45-46 min (Figure 6).

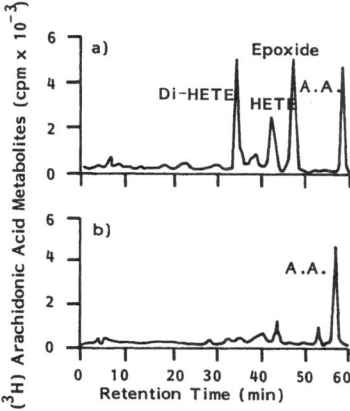

Figure 3. Metabolites of (^{3}H) arachidonic acid released from endogenous pools in a) first trimester or b) term spontaneous labor placental cells. Tissues were enzymatically dispersed, plated at 0.5 x 10^6 cells ml^{-1} and incubated with (^{3}H) arachidonic acid for 24 hr at 37°C to allow incorporation into endogenous pools. The medium was then exchanged for fresh medium not containing (^{3}H) arachidonic acid and cells were incubated for a further 24 hr. Medium was then removed and oxygenated metabolites extracted and analyzed by reverse phase HPLC.

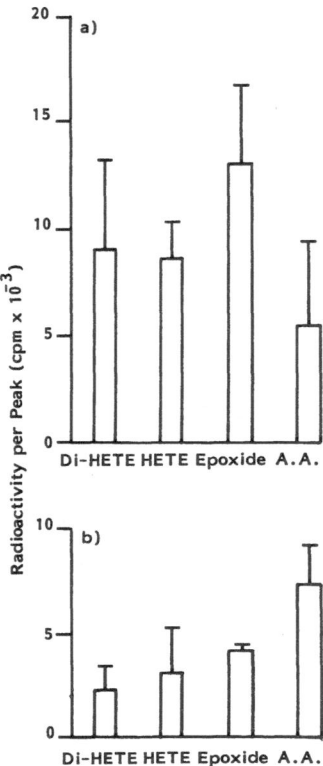

Figure 4. Quantitation of (^{3}H) arachidonic acid metabolites produced by a) first trimester or b) term spontaneous labor placental cells from endogenous pools. The radioactivity co-chromatographing with known standards was summated. The mean ± SD of 6(a) or 4(b) separate experiments are shown.

DISCUSSION

Monolayer cell culture offers an accurate and reproducible technique for studying the metabolic fate of arachidonic acid. Radiolabeled arachidonic acid added to the medium may be either directly metabolized via cyclooxygenase, lipoxygenase, and cytochrome P_{450} enzymes or incorporated into cellular phospholipids for subsequent release by phospholipases and consequent metabolism. This latter pathway may be the intrinsic pathway for the metabolism of arachidonic acid in most tissues since phospholipase action is purportedly the rate limiting step in prostaglandin synthesis (Flower and Blackwell, 1976). The profile of products observed after 24 hr incubation with medium containing (^{3}H)-arachidonic acid probably represents contributions from both direct metabolism and that following incorporation into phospholipids. However, during the 24 hr following removal of added arachidonic acid, metabolites must be formed from the endogenous pool.

The major products of arachidonic acid metabolism by placental cells during both time periods appeared to be compounds co-chromatographing with mono- and di-

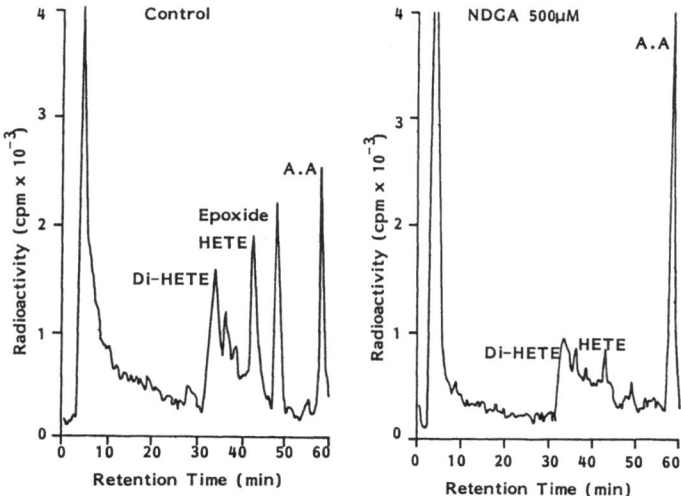

Figure 5. The effect of the lipoxygenase inhibitor NDGA on (^{3}H) arachidonic acid metabolism by first trimester placental cells. First trimester placental cells were incubated with (^{3}H) arachidonic acid for 24 hr at 37°C and then this medium replaced by fresh medium not containing (^{3}H) arachidonic acid either with (NDGA) or without (control) 500 μM NDGA for a further 24 hr incubation at 37°C. Oxygenated metabolites were separated by reverse phase HPLC. A representative profile is shown.

HETEs. These can be characterized by comparison of their retention times on RP-HPLC and those of authentic standards. Synthesis of monohydroxylated compounds has been described previously in intrauterine tissues (Kinoshita and Green, 1980) and 12-HETE production has been reported in both human (Saeed and Mitchell, 1982) and rabbit amnion (Elliott et al., 1984). In common with many other studies, the conversion of added arachidonic acid was low, being at most 9.3% in first trimester placental cells after allowance for auto-oxidation of arachidonic acid.

First trimester placental cells appear to be more metabolically active than those obtained at term. This may represent a physiological difference whereby first trimester trophoblast is actively invading decidua and spiral arteries in addition to having exchange functions. The apparent lack of activity in term placental cells obtained prior to spontaneous labor is intriguing but may reflect the general stimulus of the parturition process. Under the conditions of culture reported placental cells released only up to 7.8 ± 1.8% of the total incorporated arachidonic acid as oxygenated metabolites. This contrasts with other cells such as macrophages, which when stimulated, may release up to 50% of the intracellular arachidonic acid (Scott et al., 1980). The reason for the low release reported in this study could be that the placental cell cultures were not specifically stimulated and therefore the metabolism of arachidonic acid was at a basal rate.

The compounds co-chromatographing with the LTB$_4$ standard in the solvent system used could only be tentatively identified as di-HETEs since the various isomers of LTB$_4$ may also chromatograph in this position and different solvent

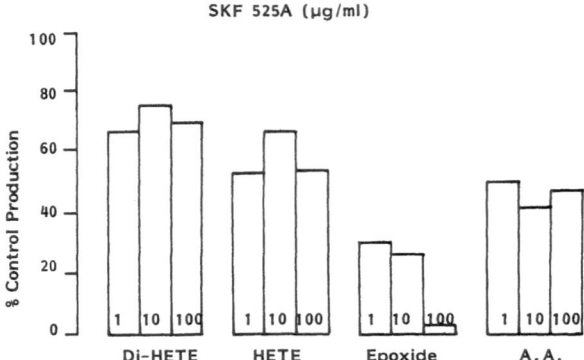

Figure 6. The effect of the cytochrome P_{450} inhibitor SKF 525A on (^{3}H) arachidonic acid metabolism by first trimester placental cells. First trimester placental cells were incubated with (^{3}H) arachidonic acid for 24 hr at 37°C before replacement of medium with fresh medium not containing (^{3}H) arachidonic acid either with or without SKF 525A at 1, 10, or 100 ug/ml. Metabolites co-chromatographing with known standards were summated and quantities expressed as a percentage of amounts formed in control incubations not containing SKF 525A.

systems would be required for their resolution. The relatively non-polar compound (elution time 45-46 mins) did not co-chromatograph with any of the authentic standards. On the basis of its acid stability and relative polarity this compound may be a fatty acid epoxide produced by cytochrome P_{450} enzyme as described by Chacos et al. (1982). The placenta contains cytochrome P_{450} enzymes which may hydroxylate steroids (Meigs and Ryan, 1968, 1971) and it also contains an epoxide hydrolase (Juchau and Namburg, 1974) which could convert epoxides to more polar vicinal diols. These compounds would have a polarity similar to LTB_4 and may therefore chromatograph with LTB_4.

Neither indomethacin, a cyclooxygenase inhibitor, nor the general anti-oxidant α-tocopherol altered the product profile seen from endogenous substrate. As these products appear mainly to be mono and diHETE products, possibly from the lipoxygenase pathway, this finding for indomethacin may not be surprising. The lack of effect of α-tocopherol may simply be a reflection of the concentration used (10 µg/ml).

Of the cytochrome P_{450} enzyme inhibitors used metyrapone had no effect but may not be a specific inhibitor of the human placental enzyme. SKF 525A gave inhibition of metabolism with dose-dependent effects on the putative epoxide at 45-46 min. The apparent inhibition of the products co-chromatographing with the diHETE and monoHETE standards implies that they too may initially arise from an epoxidation step and perhaps be vicinal diols. NDGA, a lipoxygenase inhibitor, appeared to prevent formation of all the less polar products at 500 µM.

Prostaglandins were only formed in the presence of exogenously added (^{3}H-arachidonic acid. This may be methodologically important as most studies of arachidonic acid metabolism involve the addition of exogenous substrate. Coene et al. (1982) have previously shown that rabbit mesothelial cells metabolise exogenously added arachidonic acid and endogenously derived arachidonic acid to different prostaglandin products. A similar situation may be operating in the first trimester placenta where lipoxygenase and putative cytochrome P_{450} products appeared to be the only metabolites synthesized from endogenously derived arachidonic acid. Prostaglandins were only formed in small amounts from added exogenous arachidonic acid when cells were in culture. Both intracellular mechanisms and extra-cellular signals may exist in vivo to direct phospholipid-derived substrate to either lipoygenase or cyclo-oxygenase pathways. The data reported here may only represent the situation in the absence of extracellular signals.

The physiological role of HETEs and LTs in reproductive tissues have not been fully investigated. They may be vasoactive (Letts et al., 1984), have immunological effects (Samuelsson, 1983) and may be involved in muscle contraction (Lewis et al., 1983). Fatty acid epoxides have been implicated in the action of LHRH in the pituitary as an intermediate in the release of LH (Snyder et al., 1983). Similar mechanisms regulating peptide release may be present in the placenta as lipoxygenase products have been associted with the regulation of HCG release (Ilekis and Benveniste, 1983).

The finding that lipoxygenase products are the major arachidonic acid metabolites produced by human placental cells in the first trimester necessitates further invesitgation to determine their roles at this time. Further structural identification of the compounds produced by placental tissue at different stages of gestation may help in the elucidation of their roles.

SUMMARY

Monolayer cultures of human placental cells were incubated with (^{3}H)-arachidonic acid which was converted to oxygenated metabolites or incorporated into cellular phospholipid pools for subsequent release and metabolism. The arachidonic acid metabolites were separated by reverse phase HPLC. The major products of metabolism were diHETEs co-chromatographing with LTB_4 and monoHETEs plus an unidentified product less polar than HETEs which could be a cytochrome P_{450} product. Only these products appeared to be formed from phospholipid derived substrate whereas exogenously added (^{3}H)-arachidonic acid also gave rise to small amounts of cyclooxygenase product. This preponderance of lipoxygenase and cytochrome P_{450} metabolites in first trimester placenta indicates that they may play roles in trophoblast invasion, vascularization or the immunology of pregnancy.

REFERENCES

Capdevila, J., Marnett, L.J., Chacos, N., Prough, R.A., and Estabrook, R.W. (1982) Cytochrome P_{450} - dependent oxygenation of arachidonic acid to hydroxy-eicosatetraenoic acids. *Proc. Natl. Acad. Sci. USA* 79, 767-770.

Chacos, N., Falck, J.R., Wischrom, C., and Capdevila, J. (1982) Novel epoxides formed during the liver cytochrome P_{450} oxidation of arachidonic acid. *Biochem. Biophys. Res. Commun.* 104, 916-922.

Coene, M.-C., Van Hove, C., Claeys, M., and Herman, A.G. (1982) Arachidonic acid metabolism by cultured mesothelial cells. Different transformations of exogenously added and endogenously released substrate. *Biochem. Biophys. Acta* 710, 437-445.

Elliott, W.J., McLaughlin, L.L., Block, M.H., and Needleman, P.W. (1984) Arachidonic acid metabolism by rabbit fetal membranes of various gestational ages. *Prostaglandins* 27, 27-36.

Flower, R.J. and Blackwell, G.J. (1976) The importance of phospholipase A2 in prostaglandin biosynthesis. *Biochem. Pharmacol.* 25, 285-291.

Goldberg, V.J. and Ramwell, P.W. (1975) Role of prostaglandins in reproduction. *Physiol. Rev.* 55, 325-351.

Goodman, R.P., Killam, A.P., Brash, A.R., and Branch, R.A. (1982) Prostacyclin production during pregnancy. Comparison of production during normal pregnancy and pregnancy complication by hypertension. *Am. J. Obstet. Gynecol.* 142, 817-822.

Ilekis, J. and Benveniste, R. (1983) The arachidonic acid pathway and the secretions of chorionic gonadotrophins (hCG) and progesterone by cultured human choriocarcinoma JEG-3 cells. *Trophoblast Res.* 1, 289-300.

Jogee, M., Myatt, L., and Elder, M.G. (1983) Decreased prostacyclin production by placental cells in culture from pregnancies complicated by fetal growth retardation. *Br. J. Obstet. Gynaecol.* 90, 247-250.

Jogee, M., Myatt, L., Moore, P., and Elder, M.G. (1983) Prostacyclin production by human placental cells in short term culture. *Placenta* 4, 219-230.

Juchau, M.R. and Namburg, M.J. (1974) Studies on the biotransformation of naphthalene-1,2-oxide in fetal and placental tissue of humans and monkeys. *Drug Metab. Disp.* 2, 380-385.

Keirse, M.J.N.C. (1978) Biosynthesis and metabolism of prostaglandins in the pregnant human uterus. *Adv. Prostag. Throm. Res.* Vol. 4, (eds.), F.Coceani and P.M. Olley, Raven Press, New York, pp. 87-102.

Kinoshita, K. and Green, K. (1980) Bioconversion of arachidonic acid to prostaglandins and related compounds in human amnion. *Biochem. Med.* 23, 183-197.

Letts, L.G., Cirino, M., Lord, A., and Yuski, P. (1984) The actions of leukotrienes on blood vessels in vitro. *Prostaglandins* 28, 602-604.

Lewis, R.A., Lee, C.W., Levine, L., Morgan, R.A., Weiss, J.A., Drazen, J.M., Oh, H., Hoover, D., Corey, E.J , and Austen, K.F. (1983) Biology of the C-6-sulphidopeptide leukotrienes. *Adv Prostag. Throm. Res.* 11, 15-26.

Meigs, R.A. and Ryan, K.J. (1968) Cytochrome P_{450} and steroid biosynthesis in the human placenta. *Biochem. Biophys. Acta* 165, 476-482.

Meigs, R.A. and Ryan, K.J. (1971) Enzymatic aromatization of steroid I effects of oxygen and carbon monoxide on the intermediate steps of estrogen biosyntehsis. *J. Biol. Chem.* 246, 83-87.

Ogburn, P.L., Williams, P.P., Johnson, S.B., and Holman, R.T. (1984) Serum arachidonic acid levels in normal and pre-eclamptic pregnancies. *Am. J. Obstet. Gynecol.* 148, 5-9.

Oliw, E.H. and Oates, J.A. (1982) Oxygenation of arachidonic acid by hepatic microsomes of the rabbit mechanism of biosynthesis of two vicinal dihydroxyeicosatrienoic acids. *Biochim. Biophys. Acta* 666, 327-340.

Pakrasi, P.L. and Dey, S.K. (1982) Prostaglandins in the uterus: Modulation by steroid hormones. *Prostaglandins* 26, 991-1009.

Saeed, S.A. and Mitchell, M.D. (1982) Formation of arachidonate lipoxygeanse metabolites by human fetal membranes, uterine decidua vera and placenta. *Prostaglandins, Leukotrienes and Med.* 8, 635-640.

Samuelsson, B. (1983) Leukotrienes: A new class of mediators of immediate hypersensitivity reactions and inflammation. *Adv. Prostag. Throm. Res.* 11, 1-13.

Scott, W.A., Zriker, J.M., Hammill, A.L., Kempe, J., and Cohn, Z.A. (1980) Regulation of arachidonic acid metabolites in macrophages. *J. Exp. Med.* 152, 324-335.

Sheppard, B.L. and Bonnar, J. (1976) The ultrastructure of the arterial supply of the human placenta in pregnancy complicated by fetal growth retardation. *Br. J. Obstet. Gynaec.* 83, 948-959.

Snyder, G.P., Capdevila, J. Chacos, N., Manna, S., and Falck, J.R. (1983) Action of luteinizing hormone-releasing hormone: involvement of novel arachidonic acid metabolites. *Proc. Natl. Acad. Sci USA* 80, 3504-3507.

Stuart, M.J., Sunderji, S.G., Yambo, T., Clark, D.A., Allen, J.B., Elrad, J.B., and Slott, J.H. (1981) Decreased prostacyclin production. A characteristic of chronic placental insufficiency syndromes. *Lancet* I, 1126-1128.

Wallenberg, H.C.S. (1981) Prostaglandins and the maternal placental circulation: Review and perspectives. *Biol. Res. Preg.* 2, 15-22.

Trophoblast Research 2:85-93, 1987

DOES HUMAN TROPHOBLAST AFFECT DECIDUAL CELL FUNCTION DURING GESTATION?

Ljiljana Vićovac, Mirjana Vučković, and Olga Genbačev

INEP - Institute of Endocrinology, Immunology
and Nutrition, Banatska 31-b
11080 Zemun, Yugoslavia

INTRODUCTION

This study was undertaken to investigate the possible role of human trophoblast in the regulation of protein and prolactin (PRL) production and secretion by human decidua of early pregnancy.

The process of decidualization involves the transformation of endometrial stromal fibroblast elements to epitheloid appearing cells. Decidual endometrial cells have been demonstrated to synthesize PRL regardless of whether their origin is normal decidua from early or term pregnancy (Riddick et al., 1978; Golander et al., 1979; Healy et al., 1979), decidua obtained from endometrial currettage following the removal of tubal ectopic pregnancies (Maslar et al., 1980), or endometrial samples taken during the late luteal phase of the normal menstrual cycle (Maslar and Riddick, 1979; Maslar et al., 1980).

The factors that control and regulate PRL production by human decidual tissue are largely unknown at present. It has recently been reported that some placental polypeptide(s) and human chorionic gonadotropin (hCG) (Handwerger et al., 1983) stimulate PRL production by term decidua.

To determine whether protein and PRL synthesis and secretion by first trimester decidua are affected by one or more factors of trophoblast origin, the effects of placental conditioned media (PCM) and hCG were studied.

MATERIALS AND METHODS

Tissues

Decidual tissue was obtained after legally performed abortions (6-8 wks of gestation). A pool of tissue was made for each experiment from material collected from 8-10 patients. The tissue was extensively washed in phosphate buffered saline (PBS, 0.01 M, pH 7.3), minced to 2 by 3 mm in essential medium (MEM) containing 125 mIU/ml of penicillin, 125 μg/ml of streptomycin and 0.25 μg/ml of nystatin, and washed in the same medium two more times. Placental tissue used for preparation of

Received: 8 October 1985; Accepted: 1 June 1986

PCM was obtained after legal first trimester abortions and a pool of 8-10 placentae was used for each experiment. For preparation of PCM_2 a pool of 3 term placentae obtained after spontaneous vaginal delivery was used.

Batch Incubations

Preparation of PCM

Media were prepared by incubating tissue explants (0.5 g/5 ml medium) for 3 hr in MEM. Conditioned media from first trimester (PMC_1) and term placentae (PCM_2) were decanted from the tissue, centrifuged at 2500 x g for 30 min and the supernatant was filter-sterilized. Placental conditioned media were assayed for essential amino acids, glucose, and protein concentration (Lowry et al., 1951). Leucine concentration in control Eagle's MEM (Torlak, Yugoslavia) was 72 mg/l, in PCM_1 79 mg/l, in PCM_2 81 mg/l. Media were corrected for initial glucose concentration (0.55 mM) and osmolarity, and the pH checked routinely.

The amount of hCG, estradiol (E_2), progesterone (PRG), and alpha fetoprotein (AFP) was determined by specific RIAs using INEP-VINCA kits (Yugoslavia).

In selected experiments, hCG was absorbed from PCM_1 by affinity chromatography (Hudson and Hay, 1976) using specific anti-hCG antibodies coupled to Sepharose 4B.

Incubation Procedure

First trimester decidual tissue was preincubated in MEM overnight (18 hr), then transferred to PCM_1 or PCM_2 and incubated for 3 and 6 hr.

In another group of experiments, decidual tissue was incubated in Eagle's MEM supplemented with hCG (Ayerst, USA) in final concentrations of 100, 250, 500, and 1000 IU/ml.

In order to study total protein synthesis, 2.5 µCI of uniformly labeled ^{14}C-leucine (Radiochemical Center, Amersham, Searle, England, specific activity 344 µCI/nmol) was added to MEM.

Processing of Tissue and Incubation Media

The incubation was terminated by immersing the Erlenmeyer flasks into an ice bath. Following incubation, the media and tissue fragments were filtered through two layers of cheese cloth; the filtrate was centrifuged at 2000 x g for 10 min, the pellet was discarded. The supernatant fraction was recentrifuged at 10000 x g and kept frozen at -30°C until use.

The tissue was removed from the cheese cloth and homogenized with a Teflon-coated pestle for 3 min in a 10 ml grinding vessel which was immersed in ice. PBS, pH 7.2, was used. The crude homogenate was centrifuged at 2000 x g for 15 min; the pellet was discarded, and the supernatant fraction was recentrifuged at 10000 x g for 15 min.

For the determination of ^{14}C-protein in tissue extracts and incubation medium, aliquots of 0.1 ml of incubation medium and 10000 x g tissue supernatant fraction were precipitated with 10% cold trichloroacetic acid (TCA), filtered, and washed twice with 2 ml of 5% TCA through Sartorius membrane filter (SM 11406), (Genbacev et al., 1981). The radioactivity was counted in Bray's solution in a liquid scintillation counter (Model SI 4000, Intertechnique).

The method described by Lowry et al. (1951) was used for total protein determination. Highly purified bovine serum albumin was used as a standard.

PRL Radioimmunoassay (RIA)

Prolactin concentration in the incubation medium and tissue extract 10000 x g supernatant was measured by a double antibody RIA as described by Sinha et al. (1973). Prolactin for iodination (NIAMDD-hRPL-I-6, AFP-2284 C2), PRL antiserum (AFP- C 11580), and the PRL reference preparation (AFP-2312C) used as a standard were kindly supplied by Dr. Salvatore Raiti, Director of the National Hormone and Pituitary Program, USA. Goat anti-rabbit gamma globulin antiserum was prepared in our laboratory. Prolactin concentrations are expressed as ng/0.5 g wwt.

Statistical Analysis

The results are expressed as the mean $\pm$ SEM. The differences were evaluated statistically using the unpaired student's T-test.

RESULTS

Two groups of experiments were performed to study the possible effects of trophoblast on decidual cells of early gestation.

Table 1

Protein and hormone concentrations in PCM_1 and PCM_2. To obtain PCM_1 and PCM_2, placental tissue (0.5 g wwt) was incubated for 3 hr in 5 ml Eagle's MEM.

	total protein mg/ml	beta-hCG IU/ml	PRG ng/ml	E_2 pg/ml	AFP µg/ml
PCM_1	2.1 $\pm$ 0.3	60–100	20–40	60–100	1.5–2.5
PCM_2	2.2 $\pm$ 0.3	0.5–3	40–60	100–150	0.6–1.5

AFP = alpha Fetoprotein; E_2 = Estradiol; PRG = Prolactin; beta hCG = beta Human Chorionic Gonadotropin

Placental-conditioned Media (PCM)

The first group of experiments was conducted to elucidate whether any factor(s) secreted by human first or third trimester trophoblast affect(s) protein synthesis and PRL secretion by decidual explants.

Placental-conditioned media was prepared by incubating tissue explants from first trimester (PCM_1) and term placentae (PCM_2) in MEM for 3 hr. Hormone content and protein concentration of PCM are shown in Table 1. Decidual explants were exposed to PCM after 18 hr preincubation in MEM.

The incorporation of ^{14}C-leucine into total proteins, after incubation of decidual explants for 3 and 6 hr in PCM_1 and PCM_2, is shown in Figure 1 (A and B). A significant inhibition ($p < 0.001$) of total protein synthesis was noted with PCM_1 after

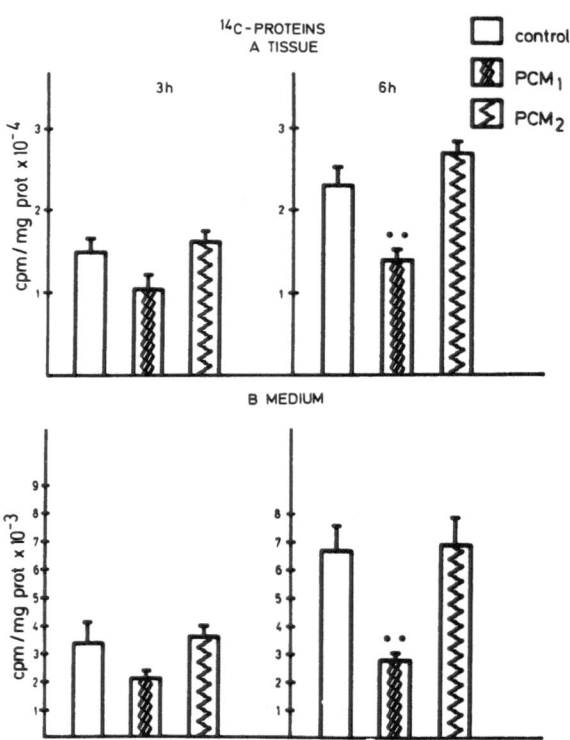

Figure 1. Effect of PCM_1 and PCM_2 on protein synthesis by decidual tissue explants (A-tissue; B-medium). After an 18 hr preincubation, decidual explants were incubated for 3 and 6 hr in control medium or placental conditioned media with ^{14}C-leucine. The results are expressed in cpm/mg protein. Each bar represents the mean results from 5 incubation flasks $\pm$ S.E.M. Statistically significant differences for $p < 0.05$ and $p < 0.001$ are marked by one and two asterisks, separately.

6 hr, both in the tissue (A) and the medium (B). Incubation in PCM_2 was without effect.

The major difference in hormone content between PCM_1 and PCM_2 is the hCG concentration. To test the hCG effect on protein synthesis, PCM_1 was subjected to affinity chromatography on a column with anti-hCG antibody coupled to Sepharose 4B. The hCG concentration in PCM_1 after absorption ranged from 0.01 - 0.05 IU/ml. The initial hCG concentration in PCM_1 was 87 IU/ml.

Incubation in hCG-free PCM_1 did not affect total protein synthesis by decidual explants (Figures 2A and B).

Incubation of decidual explants in the absorbed PCM_1 supplemented with exogenous hCG (100 IU/ml) resulted in protein synthesis inhibition comparable to that induced by nonabsorbed PCM_1 (Figures 2A and B).

The total amount of PRL in the tissue and medium was not affected by incubation in PCM (Figure 3).

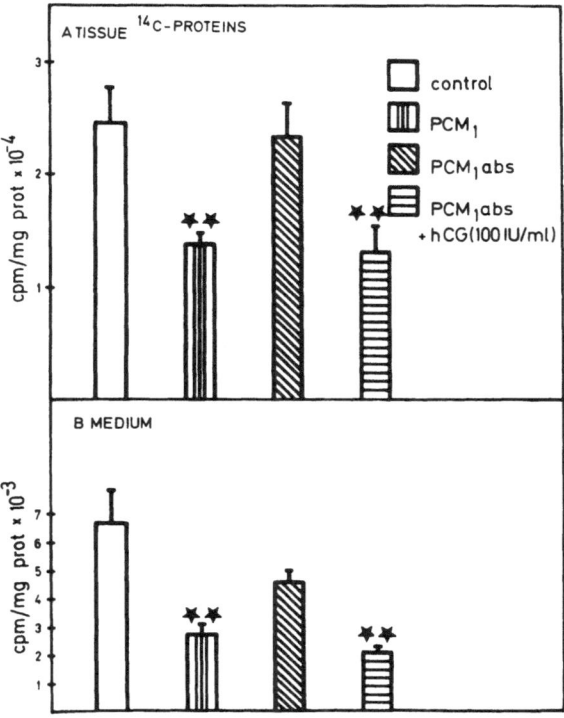

Figure 2. The effect of PCM_1, hCG-free PCM_1 and hCG-free PCM_1 supplemented with exogenous hCG (100 IU/ml) on protein synthesis by decidual tissue explants (A-tissue; B-medium). Results are expressed in cpm/mg protein. Bars and statistical significance as for Figure 1.

Effect of hCG

The effect of hCG added to MEM incubation media on ^{14}C-leucine incorporation and PRL tissue and medium concentration was studied in the second group of experiments (Figure 4).

Human chorionic gonadotropin was added in final concentrations of 250, 500, and 1000 IU/ml to 18 flasks (6 flasks for each concentration) and compared with 6 hCG- free controls. In the first group (Figure 4A) ^{14}C-leucine and hCG were added to leucine-free medium at the beginning of a 3 hr incubation. ^{14}C-leucine incorporation into total proteins decreased gradually with the increase in the dose of hCG. However, a statistically significant decrease ($p < 0.05$) was observed only with the highest hCG concentration (Figure 4B).

If the tissue was preincubated for 3 hr and then transferred to the hCG containing medium and incubated for another 3 hr, the observed effect of hCG was more pronounced.

DISCUSSION

As hCG production by human trophoblast is maximal during the first trimester of pregnancy and since hCG is readily available to adjacent decidua, it was reasonable to assume that hCG might participate in the regulation of decidual cell metabolism and PRL production in early gestation. It has been demonstrated recently by

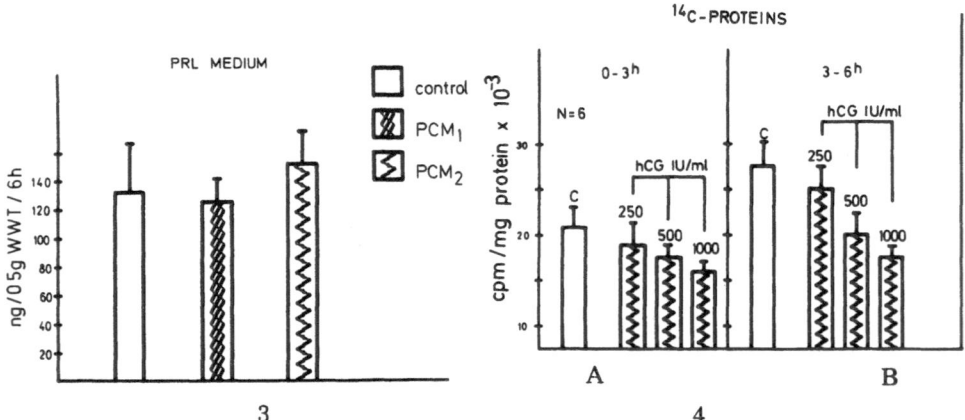

Figure 3. The effect of PCM on release of decidual PRL. Decidual tissue explants preincubated for 18 hr were exposed to PCM$_1$ and PCM$_2$ for an additional 6 hr. Media were assayed for PRL using RIA. Results are expressed in ng/0.5 g wwt. Each bar is the mean from 5 incubation flasks.

Figure 4. The in vitro effect of hCG on total protein synthesis by human endometrium of early pregnancy. The effect of hCG treatment during the first (A) and second (B) incubation interval is compared. Results are expressed in cpm/mg of protein. Statistically significant differences as for Figure 1.

Rosenberg et al. (1984) that hCG in a dose range from 10-100 IU/ml stimulates PRL production by term decidua in vitro.

Handwerger et al. (1983) have shown that term PCM contains a heat stable protein of molecular weight greater than 100,000 daltons that selectively stimulates PRL synthesis and secretion by term decidual explants. The same authors failed to demonstrate any hCG effect on PRL production with much lower doses (1-5 IU/ml) of hCG than those used by Rosenberg et al. (1984).

To our knowledge, the effect of hCG on decidual cells of early gestation has not been reported. Two types of experiments were done to consider this problem.

In the first group of experiments, decidual tissue explants were incubated in PCM which was collected after 3 hr of incubation of placental explants of the first (PCM_1) and the third (PCM_2) trimester of pregnancy. The main difference in hormone content of PCM_1 and PCM_2 was the hCG concentration which was 60-100 IU/ml and 0.5-3 IU/ml, respectively.

The obtained results indicate that some factor(s) released by first trimester trophoblast inhibit(s) protein synthesis by decidual explants of the same gestational age. Incubation in term PCM had no effect on protein synthesis.

In order to test whether hCG is one of the factors responsible for the observed inhibition of protein synthesis, PCM_1 was absorbed with specific anti-hCG antibodies coupled to Sepharose 4B. The concentration of hCG in absorbed PCM_1 was 0.01-0.05 IU/ml, i.e., 50 times lower than in term PCM. Incubation in absorbed PCM_1 did not affect total protein synthesis by decidual explants. The addition of hCG in the concentration of 100 IU/ml to absorbed PCM_1 (hCG concentration in native PCM_1 was 87 IU/ml) restored the observed inhibitory effect on protein synthesis.

These data indicate that hCG, alone or in combination with some factor(s) secreted by first trimester trophoblast, inhibit(s) protein synthesis in decidual explants that have been maintained out of homeostatic control, in tissue culture for 18 hr, prior to exposure to placental-conditioned media.

Prolactin secretion was unaffected by incubation in PCM_1. However, a slight stimulation was observed during incubation in PCM_2.

Handwerger et al. (1983) reported that term placenta produces factors that stimulate PRL release in term decidua. Our findings seem to indicate that a similar, but less pronounced, effect can be obtained with early gestation decidua.

In the second group of experiments, decidual tissue explants were incubated in MEM supplemented with hCG in concentrations of 250, 500, and 1000 IU/ml. Lower doses of hCG, comparable to those recovered in PCM_1 were ineffective. Human chorionic gonadotropin (500 and 1000 IU/ml) inhibited total protein synthesis and did not modify PRL release.

These data indicate that other factors secreted by first trimester trophoblast regulate, synergistically with hCG, decidual protein synthesis. The dose of exogenous hCG in MEM needs to be 5 times greater to induce the same effect on protein synthesis as PCM_1.

On the basis of the results obtained, we suggest that the interrelationship between decidualized endometrium and growing trophoblast in early pregnancy may restrict trophoblast invasion on the one hand, and control proliferation of endometrium on the other.

SUMMARY

This study was undertaken to investigate the possible effects of trophoblast on decidua of early gestation.

Decidual tissue explants were incubated in conditioned media (PCM) from cultured placental explants of early (PCM_1) and term (PCM_2) gestation. A significant inhibition ($p < 0.001$) of total protein synthesis was noted with PCM_1, in both tissue and medium. Incubation in PCM_2 was without effect on protein synthesis, but slightly stimulated PRL release. Placental conditioned medium of first trimester placentae absorbed with specific anti-hCG antibodies did not affect total protein synthesis by decidual explants. The addition of hCG in the concentration of 100 IU/ml to absorbed PCM_1 (hCG concentration in unabsorbed PCM_1 was 87 IU/ml) restored the observed inhibitory effect on protein synthesis.

These data indicate that hCG, alone or in combination with some factor(s) secreted by first trimester trophoblast, inhibit(s) protein synthesis in decidual explants of the same gestational age and does not affect PRL secretion.

REFERENCES

Genbacev, O., Cemerikic, B., Movsesijan, M., and Sulovic, V. (1981) Protein synthesis by human fetal lung. *Am. J. Obstet. Gynecol.* 139, 41-46.

Golander, A., Hurley, T., Barrett, J., and Handwerger, S. (1979) Synthesis of prolactin by human decidua in vitro. *J. Endocrinol.* 82, 263-267.

Handwerger, S., Barry, S., Markoff, E., Barrett, J., and Conn, P.M. (1983) Stimulation of the synthesis and release of decidual prolactin by a placental polypeptide. *Endocrinol.* 112, 1370-1374.

Healy, D.L., Kimpton, W.G., and Muller, H.K. (1979) The synthesis of immunoreactive prolactin by decidua-chorion. *Br. J. Obstet. Gynaecol.* 86, 307-313.

Hudson, L. and Hay, F.C. (1976) *Practical Immunology*, (eds.) L. Hudson and F.C. Hay, Oxford, Blackwell Scientific Publications, pp. 192-194.

Lowry, O.H., Rosenbrough, N.J., Farr, A.L., and Randall, R.J. (1951) Protein measurement with the folin phenol reagent. *J. Biol. Chem.* 193, 265-276.

Maslar, I.A. and Riddick, D.H. (1979) Prolactin production by human endometrium during the normal menstrual cycle. *Am. J. Obstet. Gynecol.* 135, 751-759.

Maslar, I.A., Kaplan, B.M., Luciano, A.A., and Riddick, D.H. (1980) Prolactin production by the endometrium of early human pregnancy. *J. Clin. Endocrinol. Metabol.* 51, 78-83.

Riddick, D.H., Luciano, A.A., Kusmik, W.F., and Maslar, I.A. (1978) De novo synthesis of PRL by human decidua. *Life Sci.* 23, 1913-1922.

Rosenberg, S.M. and Bhatnager, A.S. (1984) Sex steroid and human chorionic gonadotropin modulation of in vitro prolactin production by human term decidua. *Am. J. Obstet. Gynecol.* 148, 461-465.

Sinha, Y.N., Selby, P.W., Lewis, V.J., and Vanderland, W.P. (1973) A homologous radioimmunoassay for human prolactin. *J. Clin. Endocrinol. Metabol.* 36, 509-513.

Trophoblast Research 2:95-120, 1987

S-ADENOSYL-L-METHIONINE MEDIATED ENZYMATIC METHYLATIONS IN THE PLASMA MEMBRANES OF THE HUMAN TROPHOBLAST

Stanley L. Barnwell and B.V. Rama Sastry[1]

Department of Pharmacology
Vanderbilt University
School of Medicine
Nashville, Tennessee 37232 USA

INTRODUCTION

Enzymatic methylations, in the presence of S-adenosyl-L-methionine (SAM) as a methyl donor, play a significant role in several cellular functions. During the past decade, three of these SAM-mediated methylations have received special attention: (a) stepwise conversion of membrane phosphatidylethanolamine (PE) to phosphatidyl-N-methylethanolamine (PME), phosphatidyl-N,N-dimethylethanolamine (PMME) and phosphatidylcholine (PC) by two phospholipid N-methyltransferases (PMT I and II) in several tissues (Hirata et al., 1978; Hirata and Axelrod, 1978a,b, 1980; Crews et al., 1980; Sastry et al., 1981a, b, 1982; Sastry and Janson, 1983; Jaiswal et al., 1983), (b) formation of protein carboxymethlyesters (PCME) by protein carboxymethylase (PCM) in several cell systems (Gagnon et al., 1978, 1979; Alder, 1979; Sastry et al., 1983), and (c) methylation of endogenous fatty acids by fatty acid carboxymethylases (FACM) in several tissues (Zatz et al., 1981; Engelsen and Zatz, 1982; Stephan and Sastry, 1984, 1985).

Membrane phospholipid N-methylation has been implicated in the coupling of beta adrenergic receptor and adenylate cyclase (Hirata et al., 1979); membrane transduction of receptor mediated biosignals (Axelrod and Hirata, 1981); ribosomal insertion of cytochrome P-450 into rat liver endoplasmic reticulum during induction by phenobarbital and 3-methylcholanthrene (Sastry et al., 1981b), rat liver regeneration (Jaiswal et al., 1982), modification of cholinergic responses of the rat hemidiaphragm (Sastry et al., 1982), and human sperm motility (Sastry and Janson, 1983). Protein carboxymethylation has been implicated in the control of the direction of rotation of bacterial flagella (Alder, 1979; Stock and Koshland, Jr., 1979), sperm motility (Gagnon et al., 1979; Sastry and Janson, 1983), and neurosecretory processes (Diliberto et al., 1979; Eiden et al., 1979). The physiological significance of fatty acid carboxymethylation is not established. Fatty acid methylesters may alter membrane fluidity and thereby alter membrane function and activities of membrane-bound enzymes.

[1]To whom correspondence should be addressed.
Submitted: 6 October 1985; Accepted: 14 June 1986

Enzymatic phospholipid N-methylation is known to occur in placental tissues (Barnwell and Sastry, 1981a, b; Welsch et al., 1981; Sastry et al., 1983). Further, it has been demonstrated that increasing the intracellular levels of S-adenyosyl-L-homocysteine (SAH), an inhibitor of phospholipid N-methylation, inhibits the uptake of α-aminoisobutyric acid (AIB) by human placental villus (Barnwell et al., 1981b; Sastry et al., 1983). SAH is not a selective inhibitor of enzymatic methylation; it inhibits all SAM-mediated enzymatic methylations. A question arises as to which one of the SAM-mediated enzymatic methylations plays a significnat role in the uptake of AIB by placental villus. Amino acid uptake systems occur in the plasma membrane. Therefore, plasma membrane from human placental villus was prepared and analyzed for the above three groups of enzymes involved in SAM-mediated methylations. According to these investigators, (a) the relative distribution of PMT enzymes and amino acid carrier systems, PCM and FACM in the plasma membrane and placental villus homogenates and (b) depression of AIB uptake by inhibition of PMT enzymes indicate that plasma membrane phospholipid methylation plays a significant role in the uptake of amino acids by the human trophoblast.

METHODS

Standard Phospholipids

Several phospholipids were collected for identification of the methylated phospholipids. Phosphatidylethanolamine, phosphatidyl-N-methylethanolamine, and phosphatidyl-N,N-diemthyllethanolamine were obtained from the Grand Island Biological Co. (Grand Island, NY). These phospholipids were derived from egg phosphatidylcholine by the exchange of bases in the presence of phopholipase D. Synthetic α,β-dipalmitoyl-γ-phosphatidylethanolamine and its N-methylated derivatives were obtained from the Calbiochem-Behring Corp. (La Jolla, CA).

Radiolabeled Compounds

S-Adenosyl-L-[methyl-^{3}H]methionine (SAM) (64.6 Ci/mmole) was purchased from the New England Nuclear Corp. (Boston, MA). This preparation of SAM was diluted with 100 mM Tris-glycylglycine buffer to obtain concentrations lower than 1.0 μM. Unlabeled SAM was added to labeled SAM to obtain concentrations higher than 2 μM. Unlabeled SAM was purchsed from Boehringer-Mannheim (Indianapolis, IN). 2-amino(1-^{14}C)isobutyric acid (61 Ci/mmol) and L-[Methyl-^{14}C]methionine (50 Ci/mmole) were purchased from Amersham Corp. (Arlington Heights, IL).

Chemicals Used For the Formation of S-adenosyl-L-homocysteine, An Inhibitor of Enzymatic Methylations

Erthro-9-(2-hydroxy-3-nonyl)adenine (EHNA) was supplied by Burroughs Wellcome Co., Research Triangle Park, NC. L-Homocysteine thiolactone (L-HCT) and adenosine were obtained from Sigma Chemial Co. (St. Louis, MO). SAH was formed in the placental cells along the pathways shown in Figure 2 when the tissue was incubated with adenosine, L-HCT and EHNA.

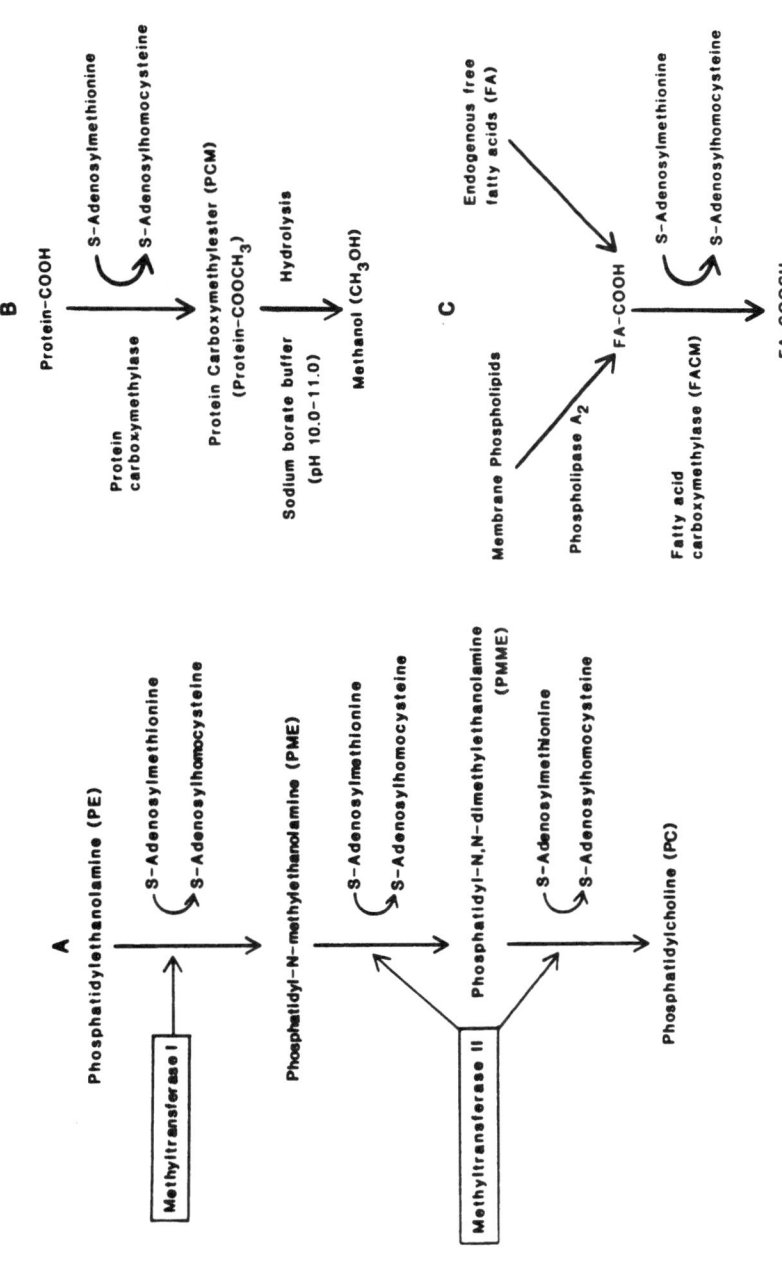

Figure 1. Metabolic pathways for enzymatic methylations: A. Stepwise methylation of phosphatidylethanolamine (PE) to phosphatidylcholine (PC). B. Methylation of protein carboxyl groups and liberation of methanol by the hydrolysis of protein carboxymethyl esters. C. Methylation of fatty acids forming methyl esters.

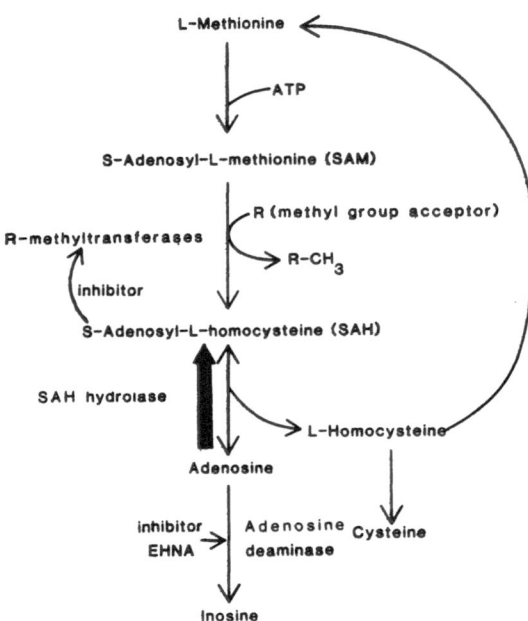

Figure 2. Metabolic pathways for the formation of S-adenosyl-L-methionine (SAM) and S-adenosyl-L-homocysteine (SAH). Intracellular concentrations of SAH are increased by inhibiting metabolism of adenosine by adenosine deaminase with EHNA and increasing the rate of the reverse reaction by SAH hydrolase by supplying adenosine and L-homocysteine thiolactone.

Preparation of Human Placental Plasma Membrane

Full term human placentae (gestation period, 38-40 wk) were obtained within 30 min of either vaginal or Cesarean section delivery at Vanderbilt University Hospital, placed on ice and brought to the laboratory for preparation of plasma membranes. Placental plasma membranes were prepared as previously described by Smith et al. (1974, 1977) with some modifications. The basal and chorionic plates were trimmed away from the placenta, and the remaining villus tissue (100 g) was cut into 10 mm diameter pieces and placed in Krebs-Ringer bicarbonate buffer (KRB) gassed with 95% O_2/5% CO_2. The KRB solution was prepared with the following composition: NaCl, 118.50 mM; KCl, 4.87 mM; $CaCl_2$, 2.54 mM; $MgSO_4$, 1.19 mM; KH_2PO_4, 1.19 mM; $NaHCO_3$, 24.88 mM; dextrose, 121.11 mM. It was maintained at pH 7.4

All of the procedures subsequent to the dissection of the villus tissue were performed in the cold room. The tissue was rinsed four times with 100 mM $CaCl_2$

followed by three rinses with phosphate buffered saline, pH 7.4. The tissue was transferred to a beaker containing 150 ml of the phosphate buffered saline and gently stirred with a magnetic stirring bar for 30 min. During this step, plasma membrane exfoliated from the syncytiotrophoblast cells and was suspended in the medium. The villus tissue was removed by filtering through nylon mesh (0.2 mm mesh). The filtered fluid containing plasma membrane was centrifuged for 10 min at 800 x g to remove blood cells and large tissue fragments. The supernatant fluid was centrifuged for 30 min at 10,400 x g to remove remaining tissue fragments. The supernatant fluid was aspirated off and centrifuged for 30 min at 110,000 x g. The plasma membrane pellet was washed with phosphate buffered saline to remove soluble protein and resuspended in 50 μM Tris glycylglycine buffer, pH 8.1. The tissue was stored at -20°C.

For characterization of the PMT enzyme, sufficient batches of plasma membrane preparations were combined so that all analyses of the enzymes were performed on a uniform batch of plasma membranes. The final preparation contained about 280 μg of protein per 50 μl. Proteins were determined by the method of Lowry et al. (1951).

Characterization of Human Placental Plasma Membrane

The activities of alkaline phosphatase and 5'-nucleotidase were each determined using a homogenate of whole human placental villus tissue and the plasma membrane from the same placenta. The alkaline phosphatase assay determined the formation of p-nitrophenol by its absorbance at 410 nM (Bowers and McComb, 1966). p-Nitrophenol was formed as a cleavage product of p-nitrophenylphosphate in the presence of alkaline phosphatase. Alkaline phosphatase activity was expressed as nmol of p-nitrophenol formed per min per mg protein. 5'-Nucleotidase activity was determined by measuring the hydrolysis of [2-^{3}H]-adenosine 5'-phosphate (Avruch and Wallach, 1971). It was expressed as nmoles of product, [2-^{3}H]-adenosine, formed per min per mg protein.

Assay for Phospholipid N-Methyltransferases in Human Placental Plasma Membrane

The assay procedures for PMT enzymes were similar to those described by Hirata et al. (1978), Sastry et al. (1981a), and Sastry and Janson (1983) with minor modifications. The total volume of the reaction mixture was 70 μl and was composed of three solutions. The first solution contained 50 μl (280 μg of protein) of the membrane preparation. The second solution of 10 μl was used for the addition of inhibitors. The third solution of 10 μl contained the substrate, S-adenosyl-L- [methyl-^{3}H]-methionine (SAM). The reaction was stopped by adding 0.5 ml of ice cold 15% (w/v) trichloroacetic acid (TCA). The tubes containing the reaction mixture were centrifuged at 9300 x g for 10 min in a Sorvall RC-58 refrigerated centrifuge. The supernatant fluid was aspirated off the pellet. Three ml of chloroform/methanol/ hydrochloric acid (2/1/0.02, by volume) were added to the pellet, and the reaction mixture was shaken for 10 min. The chloroform-methanol-HCl extract was washed twice by shaking for 10 min with 2 ml of 0.1 M KCl in 50% aqueous methanol. The upper aqueous phase was aspirated off each time. One half ml of the chloroform phase was transferred to a counting vial and evaporated to dryness. Evaporation removed the tritium-labeled methanol, a metabolite of the radioactive SAM. Ten ml of Aquasol (New England Nuclear, Boston, MA) were added, and radioactivity was measured. The remaining chloroform phase was dried with anhydrous sodium sulfate and used for characterization and estimation of the radioactive metabolites. The amount of tritium-labeled methyl groups incorporated into PME, PMME, and PC were determined by thin layer chromatography. Unless otherwise specified, data were expressed in terms of the product formed.

In experiments using low concentrations of SAM, undiluted radioactive SAM was used. The experiments using high concentrations of SAM required dilution of the radioactive SAM with nonradioactive SAM. Therefore, the specific activity of the SAM was lower in experiments using high concentrations of SAM than in experiments using low concentrations of SAM. In experiments where inhibitors were not used, the second solution was eliminated and the reaction mixture volume was 60 μl. The reaction was started by adding the labeled substrate to the mixture of the first and second solutions. The reaction mixture was incubated for a specified time at 37°C in a shaking water bath.

Thin Layer Chromatography of Phospholipids

The chloroform extracts of phospholipids were evaporated under a stream of nitrogen and dissolved in 50 μl of chloroform/methanol/HCl. The samples were applied on one inch strips of Silica gel G phase (Uniplate, Analtech, Inc., Newark, DE). Chromatograms were developed in a solvent system of propionic acid/n-propyl alcohol/chloroform/water (2/2/1/1, by volume). The solvent front moved about 6.25 inches. Quarter inch cuts were collected into counting vials, 10 ml of Aquasol were added and radioactivity was counted Authentic samples of phospholipids were applied to the plates with the chloroform extracts. The phospholipids on the plate were visualized with iodine vapor.

Measurement of S-Adenosyl-L-Homocysteine in Intact Isolated Human Placental Villus

Villus tissues from full term human placentae were used for measuring S-adenosyl-L-homocysteine (SAH) levels. Placentas were perfused with KRB to remove fetal blood. The chorionic and basal plates were trimmed away, and the remaining villus tissue was cut into 0.1 to 0.3 g pieces. The villus tissue was washed with KRB and placed in a beaker with KRB. The beaker was warmed to 37°C and gassed with 95% O_2/5% CO_2. Inhibitors were added to this incubation medium whenever required. Approximately 0.1 to 0.3 g of tissue was sampled out for determination of SAH. After the incubation in the presence or absence of drugs, the tissue was homogenized in cold 10% (w/v) TCA and centrifuged at 10,000 x g for 10 min at 4°C. The supernatant fluid was removed and extracted three times with diethyl ether to remove the TCA. The resulting aqueous fluid, free from TCA, was used for the assay of SAH.

The procedure for high performance liquid chromatography of SAH was adapted from Glazer and Hartman (1980) and Glazer and Peale (1978). SAH was detected by UV absoprtion at 254 nm and identified by the retention time at 20 min, using purified standards.

Measurement of Phospholipid N-Methyltransferase Activities in Intact Isolated Human Placental Villus

Villus tissue from full term human placentae were prepared as described before (Barnwell and Sastry, 1983; Sastry et al., 1983). The villus tissue was preincubated for 90 min in the KRB. During the preincubation, the tissue was continuously oxygenated with 95% O_2/5% CO_2 and kept at 37°C. At the end of the preincubation, approximately 100 mg of tissue was placed in a test tube containing 400 µl of KRB. The methylation reactions were begun by adding 100 µl of ^{3}H-SAM (final concentration, 1.2 µM) or 100 µl of ^{3}H-methionine (final concentration, 2.56 µM). The reactions were allowed to proceed for 1 hr. At the end of the incubation period, 1 ml of cold, 15% TCA was added to the tubes, and the tissue was homogenized. Extraction and identification of phospholipids in the homogenized tissue were performed as described in the procedure using human placental plasma membrane.

Assay of Protein Carboxymethylase Activity in Homogenates of Intact Isolated Human Placental Villus and Human Placental Plasma Membrane

The assay for PCM was adapted from the method described by Diliberto and Axelrod (1974) and Sastry and Janson (1983). The villus tissue homogenate and the plasma membrane resuspended in KRB were used as the source of the enzyme.

Protein carboxymethylase transfers methyl groups from radioactive SAM to carboxyl groups on proteins. It was measured by the radioactive methanol formed after cleavage of the methyl groups from the carboxyl moieties of proteins by sodium borate buffer. The radioactive methanol was extracted into the organic phase containing isoamyl alcohol and toluene.

To assay for PCM, 20 µl of tissue (or a suspension of the plasma membrane) homogenate was placed in a test tube with 10 µl of 0.5 M sodium phosphate buffer, pH 6.0, containing 100 mg/ml of human serum albumin. The albumin was included as a methyl acceptor protein (MAP). The exogenous MAP kept the endogenous substrate from being rate limiting. The reaction was begun by adding 20 µl of a stock solution of ^{3}H-SAM. This stock solution consisted of 200 µl of commercially available SAM (64.4 Ci/mmol, 1 mCI in 2 ml) with 40 µl of 0.5 M sodium phosphate buffer, pH 6 and 10 µl of 0.01 M sodium hydroxide, pH 10. After the SAM solution was added to the reaction tube, the reaction proceeded for 15 min at 37°C in a shaking water bath. At the end of the incubation, 250 µl of 0.5 M sodium borate buffer with 1% methanol, pH 10-11, and 3 ml of cold isoamyl alcohol: toluene mixture (2:3, v/v) were added.

The test tubes were capped and placed on ice for 15 min, spun on a vortex mixer and the organic and aqueous phases were allowed to separate for 15 min. The test tubes were uncapped and centrifuged at 3000 x g for 10 min in a refrigerated centrifuge. One ml of the top layer was added to the scintillation vial containing 10 ml of Aquasol. This vial was immediately counted in a liquid scintillation counter. Another 1 ml aliquot of the top layer was added to a scintillation vial and evaporated before 10 ml of Aquasol was added. The vial was then counted. The difference in the radioactivities of these two vials gives the radioactive methanol formed. In blanks, boiled homogenates or plasma membranes were used.

Uptake of the Nonmetabolizable Amino Acids, Alpha-aminoisobutyric Acid by the Isolated Placental Villus

Details of AIB uptake by isolated placental villus have been described in several published papers (Rowell and Sastry, 1978, 1981; Barnwell and Sastry, 1983). The present studies use the same procedures.

Statistics

Results are expressed as means and standards errors wherever possible. Straight lines and initial linear velocities were established with linear regression analysis by the method of least squares. The significance of the difference between the mean values was calculated by Student's t test. A result was considered significant if its p value was less than 0.05.

RESULTS

Characterization of Human Placental Plasma Membranes Preparation

Alkaline phosphatase and 5'-nucleotidase are marker enzymes for placental surface membranes. The activity of these enzymes was enriched about 25-fold in the human placental plasma membrane as compared to the whole placental villus homogenate (Table 1). This enrichment of alkaline phosphatase and 5'-nucleotidase was similar to the enrichment found by other workers for human placental plasma membranes (Smith et al., 1977; Fant, 1980).

Identification of Methylated Phospholipids from Human Placental Plasma Membrane Incubated with ^{3}H-SAM

Methylated phospholipids formed after plasma membrane had been incubated with ^{3}H-SAM were identified by thin layer chromatography (Figure 3). Placental plasma membrane incubated with 0.5 µm ^{3}H-SAM for 30 min formed PME, PMME, and PC (Figure 3A). The proporation of the total radioactivity in the chloroform extract containing the phospholipids was 10% as PME, 27% as PMME, and 50% as PC. About 10% of the radioactivty was present at the solvent front, which possibly was FAME. About 3% of the radioactivity was not identified. Steady state levels for the endogenous occurrence of PME and PMME are not significant. Therefore, the number of radioactive methyl groups incorporated into PME, PMME, and PC should be 1, 2, and 3, respectively. The specific activity of newly formed PMME was twice and the specific activity of newly formed PC was three times as high as PME. Therefore, all radioactive PC and PMME were formed from PME, the total amount of product PME formed was equal to the sum of moles of PME, PMME, and PC. Similarly, the total amount of PMME formed was equal to the sum of moles of PMME and PC. From these sums, the relative distribution of products can be calculated. This distribution converted to a product molar ratio of 1.00:76:0.41 for PME, PMME, and PC, respectively. Placental plasma membranes incubated with 200 µM SAM formed PC as the major product (Figure 3B). Approximately 85% of the radioactivity in the chloroform extract was in PC.

Table 1

Enzyme activities in human placental plasma membrane and villus tissue

Enzyme activity or product formed	Units for enzyme activity[a]	Villus fraction[b]		Ratio PM/H
		Plasma membrane (PM)	Homogenate (H)	
Alkaline phosphatase	nmoles/mg/min	857 ± 71	34.8 ± 2.2	24.6[c]
5'-Nucleotidase	nmoles/mg/min	2550 ± 100	101 ± 13	25.2[c]
Phospholipid-N-methyltransferase I (PME formation)	fmoles/mg/30 min	260 ± 6	166 ± 7	1.57[c]
Phospholipid-N-methyltransferase II (PC formation)	fmoles/mg/30 min	971 ± 10	1590 ± 32	0.61
Protein carboxy-methylase (PCME formation)	pmoles/mg/30 min	7.8 ± 0.2	27.8 ± 0.2	0.28
Fatty acid carboxymethylase[d] (FACM formation)	fmoles/mg/30 min	31.4 ± 0.7	147 ± 6	0.21

[a]   All activities are expressed per mg protein.

[b]   All values are means ± standard error for 10 values.

[c]   This ratio indicates significant enrichment of the enzyme activity in the plasma membrane ($p < 0.001$).

[d]   The products are assumed to be nonpolar lipid methyl esters (Stephan and Sastry, 1984, 1985). They have to be identified by strict chemical criteria. It is assumed that one mole of the methyl group from SAM is transferred to each molecule of the nonpolar lipid or an acceptor molecule.

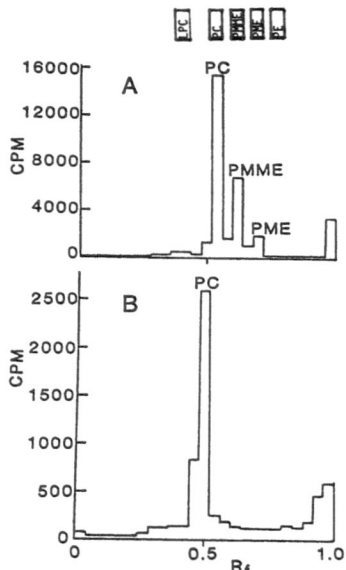

Figure 3. Separation and identification of methylated phospholipids by thin layer chromatography. TLC was used to separate phosphatidyl N-methylethanolamine (PME), phosphatidyl N,N-dimethylethanolamine (PMME) and phosphatidylcholine (PC). Placental plasma membranes were incubated for 30 min with (A) 0.5 µM S-adenosyl-L-(methyl-³H)-methionine (SAM) or (B) 200 µM SAM. The peaks for formation of PME, PMME, and PC were identified by comparison with chromatographic patterns of authentic samples of PME, PMME, PC, and lysolecithin (LPC) visualized with iodine stains. The relative positions of these standards are shown in the rectangles above Figure 3A. Chromatograms were developed with a solvent system containing propanol/propionic acid/chloroform/water (2/2/1/1, by volume). The solvent front advanced 6.25 inches from the origin. The ordinate represents the radioactivity (cpm) present in each section of TLC. The peaks of methylated phospholipids in Figures 3A and 3B were not corrected for specific activities. The endogenous concentrations of PME and PMME were negligible. The radioactive PME, PMME, and PC were formed by methylation of membrane PE. The ratio of radioactive methyl groups incorporated into PME, PMME, and PC were 1, 2, and 3, respectively.

Enzymatic Formation of PME and PC by the Plasma Membrane as a Function of Time

The formation of PME and PC by the plasma membrane with time are shown in Figure 4. When the concentration of SAM was 1 µM, the formation of PME was linear for 30 min of the incubation period (Figure 4A). There appeared to be a deviation in the time course for formation of PME beyond 30 min. During 0-30 min, PME was the major product in the reaction medium. By about 30 min, threshold concentrations of PME accumulated in the medium and was further methylated to form PMME and PC.

Therefore, SAM was possibly utilized during 30-60 min at a rate different from that during 0-30 min. A new set of reaction conditions were established at 30 min, which was indicated by a shoulder or a deviation from linearity in time-velocity plot (Figure 4A). These biphasic time velocity curves were observed at only low SAM concentrations. Therefore, a reaction time of 30 min, when the formation of PME was linear, was used in all further experiments.

At a high concentration of SAM (Figure 4B), the rate of formation of PC was linear for 60 min.

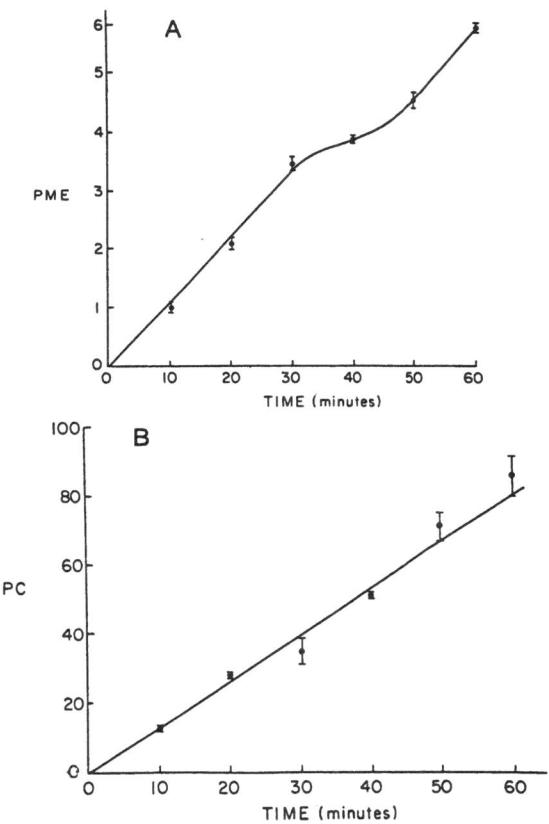

Figure 4. Enzymatic methylation of human placental plasma membrane phosphatidylethanolamine as a function of time. Enzymatic synthesis of (A) phosphatidyl N-methylethanolamine (PME) and (B) phosphatidyl-choline (PC) were measured. The ordinate represents pmol of product formed per mg protein. PME synthesis was measured when the concentration of S-adenoysl-L-(methyl-^{3}H)-methionine (SAM) concentration was 1 μM. PC synthesis was measured when the concentration of SAM was 200 μM. Vertical lines represent standard errors of the mean. Each point represents the mean of ten values.

Effect of pH on the Enzymatic Formation of PME and PC by the Plasma Membrane

To provide evidence for or against the presence of more than one PMT in the plasma membrane, the formation of PME and PC was measured as a function of pH (Figure 5). The formation of PME was measured while the concentration of SAM was 1 μM. The formation of PC was measured which the concentration of SAM was 200 μM. The rate of formation of PME was the highest when the pH was about 8.6. The curve for its formation as a function of pH was symmetrical. The rate of formation of PC was highest at about pH 10.4. The formation of PC as a functiun of pH demonstrated a shoulder at about pH 8.6, the optimum pH for PME formation.

Effect of Magnesium Ion on the Ezymatic Formation of PME and PC

The effect of magnesium ion on the formation of methylated phospholipids is shown in Figure 6. The presence of magnesium ion did not significantly alter the formation of PME when the concentration of SAM was 1 μM. Magnesium ion concentration as high as 10 mM did not have an effect on the formation of PME. When the concentration of magnesium ion was increased to 2.5 mM, there was a 47% increase in the formation of PC. As the concentration of magnesium ion was increased to 5.0 mM and 7.5 mM, there were no significant changes in the rate of PC formation compared to its rate of formation at 2.5 mM magnesium ion.

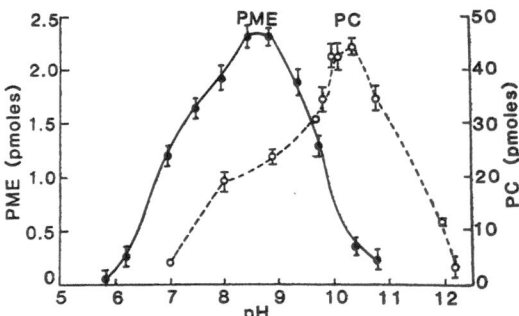

Figure 5. Enzymatic formation of phosphatidyl N-methylethanolamine (PME) of phosphatidylcholine (PC) in human placental plasma membranes as a function of pH. The ordinate represents pmol of product per mg protein per 30 min incubation. Formation of PME was measured when the concentration of S-adenosyl-L-(methyl-^{3}H)-methionine (SAM) concentration was 0.5 μM. Formation of PC was measured when the concentration of SAM concentration was 200 μM. Two buffer systems were used to regulate pH. Sodium phosphate buffer (50 mM) was used for pH less than 8.0 and glycine-NaOH buffer (50 mM) was used for pH higher than 8.0. Vertical lines represent standard errors of the mean. Each point represents the mean of ten values.

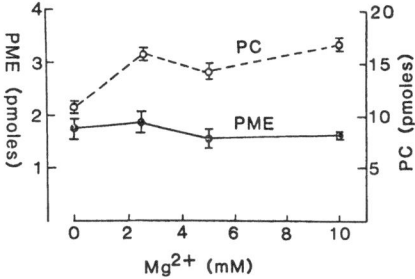

Figure 6. Effect of magnesium ion on the enzymatic formation of methylated phospholipids by human placental plasma membrane. The formation of phospholipids was measured in the presence of various concentrations of magnesium. Prior to the experiment, the plasma membrane was dialyzed for 16 hr at 4°C in 1 liter of 50 mM Tris-glycylglycine buffer (pH 8.1) containing EDTA (0.1 mM), with two changes of buffer medium. The formation of phosphatidyl-N-methyl-ethanolamine (PME) was measured at 1 µM S-adenosyl-L-(methyl-^{3}H) methionine (SAM). The formation of phosphatidylcholine (PC) was measured at 200 µM SAM. Exogenous magnesium chloride stimulated the formation of PC ($p < 0.001$) whereas there was no significant effect on the formation of PME. Formation of phospholipids is expressed in picomoles of product/mg protein/30 min. Each point is mean $\pm$ standard error from ten values.

Enzymatic Formation of PME at Low SAM Concentrations by the Placental Plasma Membrane

Placental plasma membranes were incubated with various concentrations of SAM and the formation of PME was measured. As the concentration of SAM was increased from 0.25 µM to 3.4 µM, increasing amounts of PME were formed. These low concentrations of SAM were selected because other PMTs which synthesized PME from PE have a K_m of about 1.0 µM (Hirata et al., 1978; Crews et al., 1980; Sastry et al., 1981a). The rate of incorporation of radioactive methyl groups into phospholipids was linear during the first 30 min of incubation of placental plasma membrane. The curve showing the rate of formation of PME as a function of SAM concentration was approximately a rectangular hyperbola. Double reciprocal plots using the initial linear velocity for the formation of PME and SAM concentration gave a K_m of 3.5 µM for the enzymatic formation of PME (Figure 7A).

Enzymatic Formation of PMME and PC by the Plasma Membrane at High SAM Concentrations

In order to evaluate whether more than one enzyme was involved in the stepwise methylation of PE, human placental plasma membranes were incubated with high concentrations of SAM. Concentrations of SAM between 20 µM and 200 µM were selected because the K_m of other PMTs that converted PME to PMME and PC was high (Sastry et al., 1981a; Crews et al., 1980; Hirata et al., 1978). At these concentrations of SAM, the rate of incorporation of radioactive methyl groups into phospholipids was linear for 30 min. The curves demonstrating the rates of formation of PMME and PC as a function of substrate concentrations approximated a rectangular hyperbola. Double reciprocal plots using the initial linear velocities for the formation of PMME or PC and SAM concentrations were used to determine the K_ms. The K_m for the formation of both PMME and PC was 20 µM (Figure 7B).

Activity of PMT Enzymes in the Plasma Membrane

Homogenates of whole human placental villus and plasma membrane from the placental villus were examined to compare their abilities to form PME in the presence of a low concentration of SAM and PC in the presence of a high concentration of SAM (Table 1). At a SAM concentration of about 1 µM, the whole placental villus homogenate formed significantly less PME per mg protein during the 30 min incubation period than the plasma membrane. At a higher concentration of SAM (50

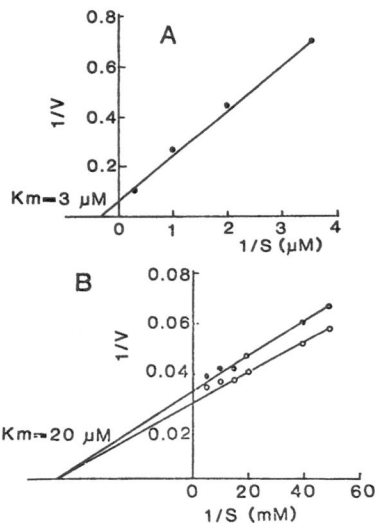

Figure 7. Double-reciprocal plots of the formation of phospholipids by human placental plasma membranes were incubated with various concentrations of S-adenosyl- L-(methyl-3H)-methionine (SAM). The initial velocities of product formation were measured during an incubation period of 30 min. The formation of phosphatidyl-N-methylethanolamine (PME) was measured at low concentrations of SAM (A). An apparent K_m of 3 µM for the substrate SAM was observed. The formation of phosphatidylcholine (PC) and phosphatidyl-N,N-dimethylethanolamine (PMME) was measured at high concentrations of SAM (B). An apparent K_m of 20 µM for the substrate SAM was observed for formation of either PC or PMME. Each point is a mean from ten values.

µM), the whole placental villus homogenate formed significantly more PC per mg protein than the plasma membrane.

Inhibition of the Enzymatic Formation of Methylated Phospholipids

S-Adenosyl-L-homocysteine (SAH) is a competitive inhibitor of SAM mediated methylations. The effect of SAH on formation of PME and PC by human placental plasma membrane during a 30 min incubation was studied (Table 2). About 33% of the PMT I was inhibited at 20 nM SAH which increased to 70% at 200 nM SAH. The I50 for the inhibition of PME formation at 1 µM SAM was 86.5 nM SAH. About 11% of PMT II was inhibited at 20 nM SAH which increased to 64% at 200 nM SAH. When the SAM concentration was 50 µM, the I50 for the inhibition of PC formation was 152.5 nM.

Identification of Methylated Phospholipids from Intact, Isolated Human Placental Villus Incubated with SAM or L-Methionine

Intact isolated human placental villus (0.1-0.3 g) was incubated with SAM or L-[methyl-^{3}H]-methionine. The methylated phospholipids were identified by thin layer chromatography (Figure 8). The tissue incubated with 1.2 µM SAM for 60 min formed PME, PMME, and PC. The proportion of the total radioactivity in the chloroform extract was 15% PME, 14% PMME, and 19% PC. This distribution converted to a mole ratio of 1.00:0.46:0.21 for PME, PMME, and PC, respectively. In addition to the three major peaks representing PME, PMME, and PC, about 25% of the radioactivity in the chloroform extract migrated with the solvent front (R_f 0.85-1.00) during the thin layer chromatography procedure. Such a huge solvent front did not occur during enzymatic methylation of PE in plasma membrane (Figure 3).

The products which appeared at R_f 0.85-1.00 were not identified. In studies on other tissues, one methyl group of SAM was transferred to one molecule of an acceptor. This transfer of methyl groups was found to be enzymatic and was inhibited by inhibitors of enzymatic methylations (Stephan and Sastry, 1984, 1985). One category of methyl group acceptors is fatty acids, mainly palmietic and oleic acids. These unidentified enzymatic methylations were expressed as methyl groups transferred from SAM (Table 1).

Protein Carboxymethylase Activity in Homogenates of Intact Isolated Human Placental Villus and Human Placental Plasma Membrane

The presence of PCM was studied in homogenates of intact isolated human placental villus and human placental plasma membranes. This enzyme also uses SAM as a substrate to methylate carboxyl groups on proteins. PCM was present in both whole tissue homogenates and the plasma membrane fraction (Table 1). The PCM activity of the plasma membrane fraction is about 23% of that of the whole homogenate.

Table 2

Inhibition of enzymatic methylations by S-adenosyl-L-homocysteine
in vivo and in situ[a]

Enzyme (Product)	Substrate (μM)	Enzyme Source	Inhibitors (μM)	Inhibition (%) $M \pm E$[b]
PMT I (PME)	SAM (1.0)	Plasma membrane	SAH (0.02)	33 ± 5
PMT I (PME)	SAM (1.0)	Plasma membrane	SAH (0.2)	79 ± 3
PMT II (PC)	SAM (50)	Plasma membrane	SAH (0.02)	11 ± 6
PMT II (PC)	SAM (50)	Plasma membrane	SAH (0.2)	64 ± 3
PMT I (PME)	SAM (1.0)	Plasma membrane	HCT (100) + Adenosine (100)	60 ± 3
PMT II (PC)	SAM (51)	Plasma membrane	HCT (100) + Adenosine (100)	38 ± 3
PMT I (PME)	SAM (1.2)	Villus	HCT (100) + Adenosine (100)	87 ± 2
PMT I	Methionine (2.6)	Villus	HCT (100) + Adenosine (100)	75 ± 7
PCM (PCME)	SAM (2.4)	Plasma membrane	SAH (0.2)	78 ± 6
PCM (PCME)	SAM (2.4)	Villus	HCT (100) + Adenosine (100)	75 ± 1
FACM (FAME)	SAM (1.2)	Villus	HCT (100) + Adenosine (100)	85 ± 3

[a]   To form S-adenosyl-L-homocysteine (SAH) in situ, human placental villus was
incubated with adenosine and L-homocysteine thiolactone (HCT). HCT is formed
by the reverse reaction of L-HCT hydrolase. The reaction mixture also contains
erythro-9-(3-(2-hydroxynonly))-adenine (10 μM) to inhibit adenosine deaminase
and to conserve adenosine.

[b]   Mean ± standard error from 10 values.

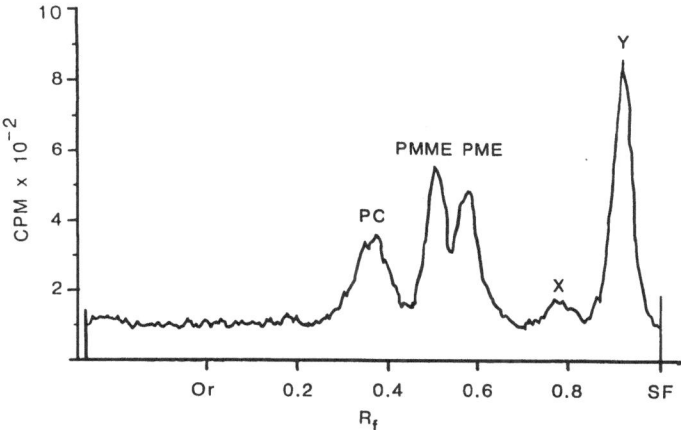

Figure 8. Formation of methylated phospholipids by intact isolated human placental villus. The tissue was incubated for 1 hr in the presence of 1 µM S-adenosyl-L-(methyl-^{3}H)-methionine (SAM). The phospholipids were extracted and separated by thin layer chromatography. The products were identified by co-migration with authentic phosphytidylcholine (PC), phosphatidyl-N,N-dimethylethanolamine (PMME), and phosphatidyl-N-methylethanolamine (PME) visualized with iodine stains. The peaks 'X' and 'Y' were not identified. All incubations with intact isolated human placental villus were performed in Krebs Ringer bicarbonate buffer, pH 7.4. The above tracing of thin layer radiochromatogram was obtained using Packard Radiochromatogram Scanner and a gaseous medium of helium (98.7%) and butane (1.3%) as described by Sastry et al. (1982).

Relationship Between Inhibition of PMT Enzymes in the Plasma Membrane and Depression of AIB Uptake by the Placental Villus

Adenosine (100 µM), L-HCT (100 µM), and EHNA (10 µM) were found to be weak inhibitors of PMT I (26-28%) and PMT II (7-27%) of the plasma membrane (Table 3). This inhibition increased to 60% inhibition of PMT I and 38% inhibition of PMT II when all three agents were present in the reaction medium (Table 3). This increased inhibition is possible due to formation of SAH from adenosine and L-HCT in the presence of SAH hydrolase (SAHase). SAHase is considered to be a cytoplasmic enzyme in placenta (Hershfield, 1979). However, its binding to plasma membrane and its entrapment by plasma membrane vesicles cannot be evaluated from the available data. Under the present conditions, plasma membrane did seem to form SAH to reach a concentration of about 150 nM to account for the inhibition.

Since PMT I is the rate limiting enzyme in stepwise methylation, its activity in the villus with and without inhibitors, was studied. About 75 to 87% of PMT I was inhibited with SAM or L-methionine as the substrate when the inhibitors were present in the medium. While adenosine metabolism was inhibited by EHNA (10 µM,

Table 3

Effect of adenosine, erythro-9-(3-(2-hydroxynonyl))-adenine (EHNA) and L-homocysteine thiolactone (L-HCT) on phospholipid methylation in human placental plasma membranes and uptake of alpha-amino isobutyric acid by placental villus

Drug	Percent of Control[a] Formation of		Inhibition of AIB Uptake % of Control[b]		SAH formed in situ[c]
	PME[a]	PC	30 min	2 hr	
Control	100 ± 4	100 ± 6	100 ± 5	100 ± 5	N.S.
Adenosine (100 µM)	74 ± 3	93 ± 7	113 ± 6	125 ± 7	N.S.
EHNA (10 µM)	72 ± 4	85 ± 7	106 ± 5	111 ± 6	N.S.
L-HCT (100 µM)	73 ± 4	73 ± 7	97 ± 5	87 ± 5	N.S.
Adenosine + EHNA + L-HCT (SAH formation)[d]	40 ± 2	62 ± 2	69 ± 3	59 ± 3	+ + +

a   Plasma membranes were incubated for 30 min with S-adensoly-L-methionine in the presence or absence of drugs. At the end of the incubation period, phospholipids were extracted and formation of PME and PC was measured. Each value is a mean ± standard error from 10 values. Product formation is significantly lower ($p < 0.05$) from control except when PC formation is measured while adenosine is present. The concentration of substrate S-adenosyl-L-methionine (SAM) was 1 µM when PME was measured. It was 51 µM when PC was measured.

b   The uptake by the isolated placental villus was measured. AIB concentration 0.1 mM. The depression of AIB uptake was significant ($p < 0.05$) with adenosine + EHNA + L-HCT.

c   S-Adenosyl-homocysteine (SAH) formed within the cell was measured by HPLC.

d   SAH does not enter cells. It was formed in situ from adenosine and L-HCT. EHNA is an inhibitor of adenosine deaminase which conserves adenosine. Under the present experimental conditions, tissue concentrations of SAH were found to be 220 nM.

an inhibitor of adenosine deaminase), SAH was formed from adenosine (100 µM) and HCT (100 µM). Under these conditions, about 40% of AIB uptake was inhibited.

DISCUSSION

There is considerable pharmacological interest in the transport of amino acids across placenta because several of the drugs of abuse including nicotine and other components of tobacco smoke, morphine, and cocaine have been implicated to interfere with the mechanisms of amino acid transport and thereby contribute to intrauterine fetal growth retardation (Sastry et al., 1983; Sastry, 1984). Several processes which are related to the plasma membrane of the trophoblast including acetylcholine (ACh) release (Sastry et al., 1976, 1977; Sastry and Sadavongvivad, 1979), phospholipid N-methylation and influx of Ca^{++} ions (Sastry et al., 1983) seem to be interrelated for efficient uptake of amino acids by placental villus (Sastry et al., 1983). Phospholipid N-methylation in the plasma membrane affects Ca^{++} fluxes which, in turn, influence ACh release as well as amino acid uptake. This is the first report characterizing PMT enzymes in human placental plasma membrane and their relationship to the amino acid transport in placenta.

The cell surface of syncytiotrophoblast, which is in contact with maternal blood, has numerous microvilli. The plasma membrane prepared in this study has many of the characteristics of syncytial trophoblastic microvillus membrane (Smith, 1974; Carlson et al., 1976; Smith et al., 1977; Ruzycki et al., 1978). It is presumed that the membrane structures arise by abscission and rapid resealing of the enveloping membrane of microvilli. The present preparation was enriched in enzymes characteristic of surface membranes of syncytiotrophoblast and other cells and tissues. For example, the present preparation was enriched by about 25 times in alkaline phosphatase and 5'-nucleotidase.

The data in this report suggest that there are at least two PMT enzymes in the placental plasma membrane. The first enzyme, PMT I, converted membrane bound PE to PME. It had a low K_m of 3.5 µM and a pH optimum of 8.6. Mg^{++} had no effect on PME formation. The second enzyme, PMT II, converted PME into PMME and PC. This enzyme had a K_m of 20 µM for the formation of both PMME and PC and pH optimum of 10.4. The activity of this enzyme was increased by Mg^{++} (2.5 mM).

There is a significant enrichment (1.6 times) of PMT I activity in the plasma membrane when compared to its activity in the villus homogenate. The activity of PMT II in the plasma membrane was lower than that in the villus homogenate. The ratio of PMT II/PMT I in the villus homogenate was 9.6 which decreased to 3.7 in the plasma membrane. These observations indicate that part of PMT II is solubilized and lost during preparation of the plasma membrane. Similarly, PMT IIs of the adrenal microsomes and liver microsomes were solubilized and the solubilized enzymes utilized exogenous PME and PMME to form PC (Hirata et al., 1978; Sastry et al., 1981a).

The activity of protein carboxymethylase (PCM) was much lower in the plasma membrane fraction than the whole villus tissue homogenate. It could not be discerned if the decreased activity of the enzyme in the plasma membrane was due to a loss of the enzyme during isolation or whether the enzyme had a lower activity in the plasma membrane as compared to other subcellular fractions of the cell. The distribution of PCM within the cell varies among different tissues. In the adrenal medulla, the enzyme was almost entirely cytoplasmic whereas in the ox brain and mouse pancreas approximately 30% of the enzyme was membrane bound (Gilliland et al., 1981). The cyotplasmic PCM activity was probably not present in the plasma membrane preparation since the membrane was washed before being assayed. Some of the enzyme activity might have come from contaminating membranes in the plasma membrane preparation. The preparation of human placental plasma membranes did not totally exclude other intracellular membranes (Smith et al., 1977). These other intracellular membranes could have been a source of PCM. It is also possible that some cytoplasmic PCM was trapped during abscission and resealing of the enveloping membrane of microvilli. Further, exogenous methyl acceptor protein (MAP), exogenous human albumin, was used for the assay of PCM. All the conditions tend to increase the value of PCM in the plasma membrane. Even then, the activity of PCM in the plasma membrane was about 27% of that in the homogenate. These observations are suggestive that PCM may have a more significant role in the whole villus tissue than in the plasma membrane.

There was an unidentified peak at R_f 0.85-9.0 on TLC when the phospholipids from the whole villus homogenates were methylated. This peak possibly represents nonpolar lipids, mainly methylesters of fatty acids fromed by FACM (Zatz et al., 1981; Stephan and Sastry, 1984, 1985). These products have yet to be identified. However, their formation was negligible compared to the methylated polar lipids (phospholipids) by the plasma membrane in which amino acid transport systems are located. Therefore, FACM is not of significant importance for the transport of AIB by the plasma membrane.

Several investigations indicated that stepwise methylation of PE to PC affected various cell properties. These effects on cell function appeared to be mediated through changes in the microviscosity of the membrane and translocation of phospholipids from one side of the membrane bilayer to the other side (Hirata and Axelrod, 1978a, b; 1980). Whether these same processes occurred in the human placental plasma membrane is not known. The data from the present studies of placental tissue did suggest that phospholipid N-methylation is involved in trophoblast function. In the placenta, phospholipid N-methylation appears to exert some effect on amino acid transport. Increasing the intracellular level of SAH blocked phospholipid N-methylation. Amino acid uptake by the placental villus was also inhibited by these same conditions. The blockade of amino acid transport was not complete. When phospholipid N-methylation was blocked more than 60%, amino acid uptake decreased only about 40%. There might have been a basal level of amino acid uptake which was not regulated by phospholipid N-methylation. The increased SAH levels inhibited PCM and FACM in addition to phospholipid N-methylation. The distribution of PMT I, PMT II, PCM, and FACM, as well as the amino acid carrier systems (e.g., gamagutamyltranspeptidase) indicated that phospholipid N-methylation is possibly more intimately involved in the amino acid transport across the plasma membrane of the trophoblast than the other two enzymatic methylations.

The activity and kinetics of membrane bound enzymes, transport systems and their regulators were affected by the membrane lipid composition (Kaduce et al., 1977; Shinitzky, 1984). According to Yuli et al. (1982), there was no optimal membrane fluidity for the peak uptake of AIB by mouse fibroblasts. During phospholipid N-methylation, there was transient accumulation of PME in the lipid bilayer of the membrane. PME accumulation in the membrane alters membrane fluidity (Sastry et al., 1981b) which influences the lateral microdisplacement of transport systems and interactions of amino acids with their transport systems. Therefore, phospholipid N-methylation may influence amino acid transport through the plasma membrane by regulating its fluidity.

SUMMARY

In recent years, several enzymatic reactions mediated by S-adenosyl-L-methionine (SAM) have been described. Of these, (a) stepwise conversion of membrane phosphatidylethanolamine (PE) to phosphatdiyl-N-methylethanolamine (PME) and phosphatidylcholine (PC) by phosphilipid N-methyltranferases (PMT I and II), (b) formation of protein carboxymethylesters (PCME) by protein carboxymethylase (PCM), and (c) methylation of endogenous fatty acids by fatty acid carboxymethylases (FACM) have received considerable attention because they have been implicated in several cellular functions and in biosignal transfer across membranes. Inhibition of SAM-mediated enzyme reactions in intact human placental villus has depressed the active uptake of alpha-aminoisobutyric acid (AIB) by this tissue. Amino acid uptake systems are known to be localized in the plasma membrane. Therefore, the plasma membrane of the human trophoblast was prepared and SAM-mediated methylations were studied to evaluate as to which one of the above enzymes might play a role in depressing the AIB uptake.

The plasma membrane of the human full term placental villus was prepared by exfoliation of microvilli from syncytiotrophoblast and differential centrifugation. Alkaline phosphatase and 5'-nucleotidase are marker enzymes for placental surface membranes. Their activities were enriched by about 25-fold in the plasma membrane as compared to the villus homogenate. The plasma membrane contained enzyme systems which formed PME (PMT I: K_m 4< μM, pH optimum 8.6) and PC (PMT II: K_m > 20 μM, pH optimum 10.5). PMT I was enriched in the plasma membrane. Part of the PMT II (about 40%) was solubilized and possibly lost during the preparation of the plasma membrane. Even then, PMT II was about 4 times that of PMT I. The activities of PCM and FACM were considerably lower in the plasma membrane than those in the villus homogenate. These investigations suggest that plasma membrane phospholipid N-methylation plays a significant role in the uptake of amino acids by the human trophoblast. Transient accumulation of PME in the membrane lipid bilayer during phospholipid N-methylation may alter membrane fluidity which, in turn, influences the lateral microdisplacement and alignment of amino acid carrier systems in the membrane and their interactions with amino acids.

ACKNOWLEDGEMENTS

This investigation was partially supported by grants from the Council for Tobacco Research USA, Inc., and by United States Public Health Service Research Grants ES-03172, HD-10607, and RR-05424. One of the authors (SLB) was supported by Pharmaceutical Manufacturers Association Foundation, Inc. Medical Student Research Fellowship in Pharmacology-Clinical Pharmacology, predoctoral traineeship from the National Institutes of Health Grant GM-07628 from the

National Institutes of General Medical Sciences, and the Vivian Allen Fund Fellowship for Combined M.D./Ph.D. studies from Vanderbilt University School of Medicine.

REFERENCES

Alder, J. (1979) The role of methylation of proteins in chemotaxis. In: *Transmethylation*, (ed.), E. Usdin, R.T. Borehardt, and C.R. Creveling, New York, Amsterdam and Oxford, Elsevier/North Holland, pp. 505-510.

Avruch, J. and Wallach, D.F.H. (1971) Preparation and propertises of plasma membrane and endoplasmic reticulum fragments from isolated rat fat cells. *Biochim. Biophys. Acta* 233, 334-347.

Axelrod, J. and Hirata, F. (1981) Phospholipid methylation and receptor mediated transmission of biological signals through membranes. In: *Drug Receptors and Their Effectors*, (ed.), N.J.M. Birdsall, London and Basingstoke, MacMillan Publishers, Ltd., pp. 51-58.

Barnwell, S.L. and Sastry, B.V.R. (1981a) The role of enzymatic phospholipid methylation on amino acid uptake in human placental villus. *Fed. Proc.* 40, 667.

Barnwell, S.L. and Sastry, B.V.R. (1981b) Depression of the active uptake of alpha-aminoisobutyric acid by the isolated placental villus during inhibition of phospholipid methylation. *Proc. Eighth Internat. Cong. Pharmacol., Japanese Pharmacological Society and Science Council of Japan, Abstracts*, p. 814.

Barnwell, S.L. and Sastry, B.V.R. (1983) Depression of amino acid uptake in human placental villus by cocaine, morphine and nicotine. *Trophoblast Res.* 1, 101-120.

Bowers, G.N. and McComb, R.B. (1966) A continuous spectrophotometric method for measuring the activity of serum alkaline phosphatase. *Clin. Chem.* 12, 70-89.

Carlson, R.W., Wada, H.G., and Sussman, H.H. (1976) The plasma membrane of human placenta: Isolation of microvillus membrane and characterization of protein and glycoprotein in subunits. *J. Biol. Chem.* 251, 4139-4146.

Crews, F.T., Hirata, F., and Axelrod, J. (1980) Identification and properties of methyl-transferases that synthesize phosphatidylcholine in rat brain synaptosomes. *J. Neurochem.* 34, 1491-1498.

Diliberto, E.J., Jr. and Axelrod, J. (1974) Characterizaion and substrate specificity of a carboxymethylase in the pituitary gland. *Proc. Natl. Acad. Sci. USA* 71, 1701-1704.

Diliberto, E.J., Jr., O'Dea, R.F., and Viveros, O.H. (1979) The role of protein carboxy-methylase in secretory and chemotactic eukarotic cells. In: *Transmethylation*, (eds.) E. Usdin, R.T. Borchardt, and C.R. Creveling, New York, Amsterdam, Oxford, Elsevier/North Holland, pp. 529-538.

Eiden, L.E., Borchardt, R.T., and Rutledge, C.O. (1979) Protein carboxymethylation in neurosecretory processes. In: *Transmethylation*, (eds.) E. Usdin, R.T. Borchardt, and C.R. Creveling, New York, Amsterdam, Oxford, Elsevier/North Holland, pp. 539-546.

Engelsen, S.J. and Zatz, M. (1982) Stimulation of fatty acid methylation in human red cell membranes by phospholipase A_2 activation. *Biochim. Biophys. Acta* 711, 515-520.

Fant, M.E. and Harbison, R.D. (1981) Syncytiotrophoblast membrane vesicles: A model for examining the human placental cholinergic system. *Teratol.* 24, 187-199.

Gagnon, C., Bardin, C.W., Strittmater, W., and Axelrod, J. (1979) Protein carboxy-methylation in the parotid gland and the male reproductive system. In: *Transmethylation*, (eds.) E. Usdin, R.T. Borchardt, and C.R. Creveling, New York, Amsterdam, Oxford, Elsevier/North Holland, pp. 521-528.

Gagnon, C., Viveros, O.H., Diliberto, E.J., Jr., and Axelrod, J. (1978) Enzymatic methylation of carboxyl groups of chromaffin granule membrane proteins. *J. Biol. Chem.* 253, 3778-3781.

Gilliland, E.L., Turner, N., and Steer, M.L. (1981) The effect of ethionine administration and choline deficiency on protein carboxymethylase activity. *Biochem. Biophys. Acta* 672, 280-287.

Glazer, R.I. and Hartman, K.D. (1980) Evidence that inhibitory effect of adenosine, but no cordyceptin, on the methylation of nuclear RNA is mediated by S-adenosylhomocysteine hydrolase. *Mol. Pharmacol.* 18, 483-490.

Glazer, R.I. and Peale, A.L. (1978) Measurement of S-adenosyl-L-methionine levels by SP-sephadex chromatography. *Analyt. Biochem.* 91, 516-520.

Hershfield, M.S., Kredicch, N.M., Small, W.C., and Fredericksen, M.L. (1979) Suicide-like inactiviation of human S-adenosylhomocysteine hydrolase by analogs of adenosine. In: *Transymethylation*, (eds.) E. Usdin, R. Borchardt, and C.R. Creveling, New York, Amsterdam, Oxford, Elsevier/North Holland, pp. 173-180.

Hirata, F. and Axelrod, J. (1980) Phospholipid methylation and biological signal transmission. *Sci.* 209, 1082-1090.

Hirata, F. and Axelrod, J. (1978a) Enzymatic synthesis and rapid translocation of phosphatidylcholine by two methyltransferases in erythrocyte membranes. *Proc. Natl. Acad. Sci. USA* 75, 2348-2352.

Hirata, F. and Axelrod, J. (1978b) Enzymatic methylation of phosphatidylethanolamine increases erythrocyte membrane fluidity. *Nature* 275, 219-220.

Hirata, F., Viveros, O.H., Diliberto, E.M., Jr., and Axelrod, J. (1978) Identification and properties of two methyltranferases in the conversion of phosphatidylethanolamine to phosphatidylcholine. *Proc. Natl. Acad. Sci. USA* 75, 1718-1721.

Hirata, F., Strittmater, W.J., and Axelrod, J. (1979) Beta-adrenergic receptor agonists increase phospholipid methylation, membrane fluidity and beta- adrenergic receptor-adenylate cyclase coupling. *Proc. Natl. Acad. Sci. USA* 76, 368-372.

Jaiswal, R.K., Sastry, B.V.R., and Landon, E.J. (1982) Changes in microsomal membrane microviscosity and phospholipid methyltransferases during rat liver regeneration. *Pharmacol.* 24, 355-365.

Jaiswal, R.K., Landon, E.J., and Sastry, B.V.R. (1983) Methylation of phospholipids in microsomes of the rat aorta. *Biochim. Biophys. Acta* 735, 367-379.

Kaduce, T.L., Awad, A.B., Fontenelle, L.J., and Spector, A.A. (1977) Effect of fatty acid saturation on alpha-aminoisobutryic acid transport in Ehrlich ascites cells. *J. Biol. Chem.* 252, 6624-6630.

Lowry, O.H., Rosenbrough, N.J., Farr, A.L., and Randall, R.J. (1951) Protein measurement with the Folin phenol reagent. *J. Biol. Chem.* 193, 265-275.

Ruzycki, S.M., Kelley, C.K., and Smith, C.H. (1978) Placental amino acid uptake IV. Transport by microvillus membrane vesicles. *Am. J. Physiol.* 234, C27-CX35.

Rowell, P.P. and Sastry, B.V.R. (1978) The influence of cholinergic blockade on the uptake of alpha-aminoisobutyric acid by isolated human placental villi. *Toxicol. Appl. Pharmacol.* 45, 79-93.

Rowell, P.P. and Sastry, B.V.R. (1981) Human placental cholinergic system: Depression of the uptake of alpha-aminoisobutyric acid in isolated human placental villi by choline acetyltransferase inhibitors. *J. Pharm. Exp. Therap.* 216, 232-238.

Sastry, B.V.R. (1984) Amino acid uptake by human placenta: Alterations by nicotine and tobacco smoke components and their implications on fetal growth. In: *Developmental Neuroscience: Physiological, Pharmacological and Cliical Aspects*, (eds.) F. Caciagli, E. Biacobini, and R. Paoletti, Amsterdam, New York and Oxford, Elsevier Science Publishers, pp. 137-140.

Sastry, B.V.R. and Janson, V.E. (1983) Depression of human sperm motility by inhibition of enzymatic methylation. *Biochem. Pharmacol.* 32, 1423-1432.

Sastry, B.V.R., Barnwell, S.L., and Moore, R.D. (1983) Factors affecting the uptake of alpha-amino acids by human placental villus: Acetycholine, phospholipid methylation, Ca^{++} and cytoskeletal organization. *Trophoblast Res.* 1, 81-100.

Sastry, B.V.R. and Sadavongvivad, C. (1979) Cholinergic systems in non-nervous tissues. *Pharmacol. Rev.* 30, 65-132.

Sastry, B.V.R., Olubadewo, J.O., and Boehm, F.H. (1977) Effects of nicotine and cocaine on the release of acetylcholine from isolated human placental villi. *Arch. Inter. Pharm. Ther.* 229, 23-36.

Sastry, B.V.R., Olubadewo, J., Harbison, R.D., and Schmidt, D.E. (1976) Human placental cholinergic system: Occurrence, distribution and variation with gestational age of acetylcholine in human placenta. *Biochem. Pharmacol.* 25, 425-431.

Sastry, B.V.R., Owens, L.K., and Janson, V.E. (1982) Enhancement of the responsiveness of the rat diaphragm by L-methionine and phospholipid methylation and their relationship to aging. *J. Pharmacol. Exp. Therap.* 221, 629-636.

Sastry, B.V.R., Statham, C.N., Axelrod, J., and Hirata, F. (1981a) Evidence for two methyltranferases involved in the conversion of phosphatidylethanolamine to phosphatidylcholine in the rat liver. *Arch. Biochem. Biophys.* 21, 762-774.

Sastry, B.V.R., Statham, C.N., Meeks, R.G., and Axelrod, J. (1981b) Changes in phospholipid methyltransferases and membrane microviscosity during induction of rat liver microsomal cytochrome P-450 by phenobarbital and 3-methylcholanthrene. *Pharmacol. (Basel)* 23, 211-232.

Shinitzky, M. (1984) Membrane fluidity and cellular functions. In: *Physiology of Membrane Fluidity*, (ed.), M. Shinitzky, Boca Raton, Florida, CRC Press, Inc., Volume 1, pp. 1-52.

Smith, C.H., Nelson, D.M., King, B.F., Donohue, T.M., Ruzyski, S.M., and Kelley, L.K. (1977) Characterization of the microvillus membrane preparation from human placental syncytiotrophoblast: A morphologic, biochemical and physiologic study. *Am. J. Obstet. Gynecol.* 128, 190-196.

Smith, N.C., Brush, M.G., and Luckett, S. (1974) Preparation of human placental villus surface membrane. *Nature* 252, 302-303.

Stephan, C.C. and Sastry, B.V.R. (1984) Separation of products of enzymatic methylations in rat kidney membranes and human placental villus. *Pharmacologist* 26, 194.

Stephan, C.C. and Sastry, B.V.R. (1985) Enzymatic formation of fatty acid methylesters by rat renal microsomes in the presence of S-adenosyl-L-methionine. *Fed. Proc.* 44, 1112.

Stock, J.B. and Koshland, D.E., Jr. (1979) Identification of a methyltransferase and a methylesterase as essential genes in bacertial chemotaxis. In: *Transmethylation*, (eds.) E. Usdin, R.T. Borchardt, and C.R. Creveling, New York, Amsterdam and Oxford, Elsevier/North Holland, pp. 511-520.

Welsch, F., Wenger, W.C., and Stedman, D.B. (1981) Choline metabolism in placenta: evidence for the biosynthesis of phosphatidylcholine in microsomes via the methylation pathway. *Placenta* 2, 211-221.

Yuli, I., Incerpi, S., Luly, P., and Shinitzky, M. (1982) Insulin stimulation of glucose and amino acid transport in mouse fibroblasts of elevated membrane microviscosity. *Experientia* 38, 1114-1115.

Zatz, M., Dudley, P.A., Kloog, Y., and Markey, S.P. (1981) Nonpolar lipid methylation: Biosynthesis of fatty acid methylesters by rat lung membranes using S-adenosylmethionine. *J. Biol. Chem.* 256, 10028-10032.

IMMUNOBIOLOGY

Trophoblast Research 2:123-137, 1987

RECENT ADVANCES IN UNDERSTANDING THE IMMUNOBIOLOGY OF GESTATIONAL TROPHOBLASTIC DISEASE

- A Review -

Ross S. Berkowitz[1], Donald P. Goldstein, and Deborah J. Anderson

Fearing Research Laboratory
New England Trophoblastic Disease Center
Division of Gynecologic Oncology
Brigham and Women's Hospital
Department of Obstetrics and Gynecology and Pathology
Harvard Medical School
Boston, Massachusetts, USA

Gestational trophoblastic disease, including hydatidiform mole, invasive mole, and choriocarcinoma, comprise a group of interrelated diseases which arise from pregnancy and have varying potentials for local invasion and distant spread. It is now well established that hydatidiform mole may precede the development of persistent gestational trophoblastic tumors including gestational choriocarcinoma (Bagshawe, 1969; Goldstein and Berkowitz, 1982). The results of clinical management of molar pregnancy and gestational trophoblastic tumors has improved considerably over the past three decades due to advances in understanding of the pathology, natural history, and endocrinology of these trophoblastic diseases. In 1956, Li et al. inaugurated a new era in the therapy of gestational choriocarcinoma when they reported curing three women with metastases by treatment with methotrexate. Gestational trophoblastic tumors are now known to be curable with chemotherapy even in the presence of widely metastatic disease. The effective treatment of trophoblastic tumors with chemotherapy represents one of the most dramatic successes in the management of human malignancy.

As the results of clinical management of gestational trophoblastic disease has greatly improved, our knowledge of the immunobiology of gestational trophoblastic disease has also moved forward. During the past decade, our understanding of the genesis of complete molar pregnancy has been advanced considerably by cytogenetic studies. Several investigations have also been undertaken to study immunological interactions between the maternal host and gestational trophoblastic disease which is a foreign allograft of paternally-derived tissue. This manuscript will review recent advances in our understanding of the immunobiology of gestational trophoblastic disease, and suggest new directions for further research.

[1]To whom correspondence should be addressed: Brigham and Women's Hospital, 75 Francis Street, Boston, Massachusetts 02115
Received 10 October 1985; Accepted 1 December 1985

Cytogenetics of Complete Molar Pregnancy

Kajii and Ohama (1977) studied the chromosomal patterns of 20 complete molar pregnancies and found a karyotype of 46, XX in all cases. However, the mode of inheritance of the molar chromosomes was examined in seven cases using Q- and R-band polymorphisms as markers and the molar chromosomes were completely of paternal origin in all cases. The paternal origin of molar chromosomes was subsequently confirmed in studies by Wake et al. (1978) and Lawler et al. (1979). To further elucidate the mechanism of androgenesis, Yamashita et al. (1979) studied 13 complete moles and lymphocytes from each parent for HLA specificities (A and B loci). Molar tissues expressed homozygous HLA A and B specificities identical to one paternal haplotype. These data suggest that a complete hydatidiform mole develops from an ovum which has been fertilized by a haploid sperm which then duplicates its own chromosomes. The authors speculated that the ovum nucleus was either inactivated or absent. Jacobs et al. (1980) performed cytogenetic studies of 24 complete moles and the parent using enzyme markers and Q- and C-banding techniques. Twenty-one moles had a 46,XX karyotype and molar chromosomes were clearly of paternal origin in 22 cases. The karyotype in the remaining three moles included one 46,XY, one with an additional chromosome and one with a variable count with a mode of 85,XXXX. Analysis of the enzyme markers was strongly supportive of a haploid sperm origin for most of the 46,XX moles.

While most complete moles have a 46,XX karyotype and are homozygous, about 10 percent of hydatidiform moles have a 46,XY constitution and are heterozygous. Surti et al. (1979) demonstrated that the molar chromosomes were entirely of paternal origin in a case of 46,XY complete mole using polymorphic chromosomal and enzyme markers. Another case of 46,XY complete mole was analyzed by Pattillo et al. (1981) and cytogenetic studies indicated that this 46,XY mole was derived from fertilization of an ovum by two spermatozoa.

Infrequently, complete hydatidiform moles have been found to have a karyotype other than 46,XX or 46,XY such as 45,X (Berkowitz et al., 1982). Therefore, complete molar pregnancy appears to be the morphologic expression of a variety of aberrant chromosomal patterns.

It is important to emphasize that in general all the chromosomes in a complete mole are paternal in origin. A complete hydatidiform mole is therefore a complete allograft and should stimulate a vigorous immunologic response from the maternal host.

Genetic studies of complete moles, invasive moles and choriocarcinoma suggest that the heterozygous complete mole may have a more malignant potential than the homozygous counterpart. Wake et al. (1984) determined that 21 complete moles were homozygous and 5 complete moles were heterozygous using chromosomal and phosphoglucomutase 1 polymorphisms. While only 1 of the 21 patients with homozygous moles developed postmolar trophoblastic disease, 3 of the 5 patients with heterozygous moles developed postmolar trophoblastic disease. Furthermore, Wake et al. (1981) studied three cases of postmolar choriocarcinoma and found that all three were heterozygous. Davis et al. (1984) found that 14 of 19 cases of choriocarcinoma were Y-chromatin positive, thereby indicating the presence of gamete heterozygosity.

The postmolar choriocarcinoma cell line E1Fa was also shown to be heterozygous because it had a Y chromosome (Sasaki et al., 1982). Wake et al. have speculated that the homozygous mole is at a proliferative disadvantage because of homozygosity for recessive mutations at certain loci necessary for autonomous cell proliferation. However, there is still limited data concerning the malignant potential of the heterozygous complete mole. Fisher and Lawler (1984) did not observe an increased incidence of postmolar trophoblastic disease in patients with heterozygous moles (postmolar trophoblastic disease in 2 of 14 patients with homozygous moles and 1 of 4 patients with heterozygous moles).

Immunopathology of the Molar Implantation Site

The molar implantation bed is a site of intimate contact between molar and maternal tissue. Studies have therefore been undertaken to determine whether a host humoral or cellular immune response could be detected at the molar implantation site.

The implantation sites from 11 complete molar pregnancies were examined by direct immunofluorescence for the deposition of immunoglobulin and complement (Berkowitz et al., 1982). None of the patients had a history of diabetes mellitus, hypertension, renal disease, arthritis, systemic lupus erythematosus, or toxemia and none exhibited fever or infection either at the time of molar evacuation or postoperatively. None of the ten patients with their first molar pregnancy had detectable immunoglobulin or complement deposition at the implantation site. Therefore, first molar pregnancies do not appear to induce a vigorous host humoral immune response at the implantation site at the time of clinical presentation. However, the one patient in the study with a second molar pregnancy had marked immunoglobulin and complement deposition at the implantation site. Immunofluorescent localization of immunoglobulin (IgG, IgM) and complement (C3) was confined to the decidual arteries in that patient and was not associated with endovascular trophoblastic deposits. The implantation site in that patient was also notable for the presence of a perivascular lymphocytic infiltrate.

Cellular infiltration at the molar implantation site has been studied recently using monoclonal antibodies that recognize human lymphocyte subsets and macrophages (Kabawat et al., 1985). Tissue samples were obtained from three evacuated moles, sharp curettings from ten molar implantation sites and sharp curettings from nine normal placental implantation sites obtained at the time of first and second trimester therapeutic abortion. The chorionic villi from the three evacuated complete moles had no detectable infiltrating inflammatory cells. In contrast, in the molar implantation site there was a moderate increase in inflammatory cells as compared to normal pregnancy. The molar implantation sites had a 5 fold increased infiltration of T cells, 75% of which were T_4,+, Leu-3a+ cells (helper/inducer T cells). The T4+ Leu 3a+ cells tended to aggregate around the implantation site blood vessels. The pattern of T cell infiltration at the molar implantation site is similar to that described in the immunologic rejection of human skin allografts (Bhan et al., 1982). T_4+, Leu-3a+ cells may induce a delayed hypersensitivity type of reaction at the molar implantation site, causing vascular changes with resulting ischemia and subsequent abortion or rejection of the molar allograft.

Circulating Immune Complexes and Complete Molar Pregnancy

Circulating immune complexes were measured in patients with complete molar pregnancy to investigate the potential host humoral immune response to the molar allograft (Berkowitz et al., 1983c). Circulating immune complex levels were measured by the polyethylene glycol turbidity assay, a nonspecific antigen assay. Circulating immune complex levels were in the normal range in 27 (87%) of 31 patients at the time of molar evacuation. Three of the four patients with elevated immune complex levels had concurrent illness including viral hepatitis, renovascular hypertension and pre-eclamptic toxemia. Eighteen patients were followed with serial measurements until gonadotropin remission was attained and all 18 patients developed increased circulating immune complex levels as they entered remission. Circulating immune complex values remained elevated during gonadotropin remission from 6 to 16 wk (mean = 11.5 wk) and then declined to initial values.

There are at least two potential explanations for why immune complex levels may increase at gonadotropin remission. Firstly, as the patient approaches remission, the circulating antigen level diminishes. The declining antigen load may provide a more favorable antigen: antibody ratio for the detection and/or formation of circulating immune complexes. Secondly, immune complex levels may increase at the time of remission due to increased host antibody production.

Circulating immune complexes from patients with complete molar pregnancy have been fractionated and characterized to identify the antigen components (Lahey et al., 1984). An identifiable antigenic component of circulating immune complexes was found to be paternal HLA antigen. The latest eluting immune complex fraction inhibited paternal lymphocyte lysis using anti-HLA antisera against the husband's HLA tissue type. In each case, this fraction contained no immunoglobulin or beta-2 microglobulin and was antigenically crossreactive with only one of the husband's HLA haplotypes. Lawler et al. (1974) has previously reported the presence of anti-HLA antibodies in the sera of primigravid women following hydatidiform mole. Four cases in 31 primigravid women with hydatidiform mole had anti-HLA antibodies specific for the husband's HLA specificities. The maternal host with a complete molar pregnancy is therefore sensitized to paternal HLA antigens.

HLA Antigen Expression and Complete Molar Pregnancy

Foreign HLA antigens normally induce humoral and cellular immune responses that result in the rejection of foreign tissues. The sensitization of the maternal host to paternal HLA antigens may therefore be important in the immunobiology of complete molar pregnancy.

Extracts of molar tissue have been shown to contain paternal HLA antigens in cytotoxicity-blocking assays (Yamashita et al., 1981). In the five molar tissues that were studied, all HLA specificities that were expressed by the moles were of paternal origin. The localization of HLA antigens in molar chorionic villi has been determined in immunofluorescence tests (Berkowitz et al., 1983a). HLA A,B,C (Class I) antigens were detected on the stromal cells of molar chorionic villi, but not on the villous trophoblast. HLA DR (Class II) antigens were not detectable in either the villous stromal cells or villous trophoblast. Sunderland and associates (1985) confirmed these findings and further observed that the proliferating extravillous trophoblast expressed HLA Class I antigens. HLA Class I antigens were detectable on the extravillous trophoblast using four different antibodies to monomorphic determinants but not with antibodies to the appropriate polymorphic HLA A or B type. Therefore, HLA antigens expressed by molar tissue are similar to those expressed on normal first trimester human placenta (Faulk and Temple, 1976; Sunderland et al., 1981) and may be modified or non-classical MHC antigens.

There are at least three potential mechanisms by which the maternal host with a complete molar pregnancy may be sensitized to paternal HLA antigens. First, the proliferating extravillous trophoblast, which infiltrates the decidua and expresses HLA Class I antigen, may sensitize the maternal host. Second, if the villous trophoblast layer is disrupted, HLA-positive villous stromal cells may be exposed to the maternal circulation. Third, stromal cells may release soluble HLA antigens. If this is the case, the molar villous fluid that bathes the stromal cells should contain soluble paternal HLA antigen. However, in a recent study, cell-free villous fluid from healthy and necrotic molar villi did not contain detectable soluble HLA Class I antigens as determined in immunoprecipitation and Western blot experiments (Berkowitz et al., 1984). Therefore, HLA sensitization of the maternal host is probably not attributable to the shedding of large amounts of HLA antigen by viable or degenerating HLA-positive stromal cells.

HLA Antigen Expression and Gestational Choriocarcinoma

Several lines of evidence indicate that gestational choriocarcinoma cells express HLA Class I antigens. Anti-HLA antibodies specific for paternal HLA specificities have been detected in the sera of patients with choriocarcinoma (Shaw et al., 1979). Circulating immune complexes from a patient with gestational choriocarcinoma were also found to contain paternal HLA antigen (Lahey et al., 1984). Sasagawa et al. (1984) observed by immunoperoxidase immunohistologic tests that a subpopulation of trophoblastic cells in gestational choriocarcinoma express HLA Class I antigen. Furthermore, immunoprecipitation studies by Trowsdale et al. (1980) demonstrated that the BeWo human choriocarcinoma cell line actively synthesizes HLA Class I antigens in vitro. Surface HLA-A,B,C antigen expression by choriocarcinoma cell lines has been shown to be rate limited by the amount of HLA-A,B,C mRNA and not by beta-2 microglobulin mRNA which is present in excess (Kawata et al., 1984).

We have recently investigated the expression of major histocompatibility complex (MHC) antigens by gestational choriocarcinoma cell lines in the presence of phytohemagglutinin-activated lymphocyte culture supernatants (PHA-ALCS) and high doses of affinity purified human gamma-interferon (Anderson and Berkowitz, 1985). Immunofluorescence assays demonstrated morphologic changes and a marked increase in surface expression of class I MHC antigens in BeWo choriocarcinoma cell cultures following a 7 day exposure to PHA-ALCS or high doses (1,000 and 5,000 µg/ml) of gamma-interferon. All BeWo cultures were negative for class II antigens in these assays and another choriocarcinoma cell line, Jar, did not demonstrate class I or II MHC expression under any of the experimental conditions. Surface binding to MHC monoclonal antibodies was quantitated by radioimmunoassay. Two- to four-fold more W6/32 and BBM.1 (monoclonal antibodies specific for HLA-A,B,C and beta-2 microglobulin) bound to BeWo cells that were exposed to ALCS or gamma-interferon than BeWo cells grown in control media. ALCS or gamma-interferon-treated BeWo cells did not reproducibly express class II antigens as detected by our panel of monoclonal antibodies directed against various types of human class II antigens. Jar choriocarcinoma cell cultures did not express quantitatively elevated levels of either class I or II MHC antigens before or after incubation in ALCS or gamma-interferon. Immunoprecipitation of metabolically-labeled HLA molecules was performed to rule out the possibility that increased levels of MHC detected in BeWo cultures was due to binding of lymphocyte MHC to the choriocarcinoma cells. Depending on experimental conditions, two- to six-fold more ^{35}S-labeled class I heavy chain and beta-2 microglobulin was precipitated from ALCS and gamma-interferon-treated BeWo cells than from untreated BeWo cells. The heavy chain bands precipitated with W6/32 from BeWo cultures were close if not identical in molecular weight to the 44 KD band immunoprecipitated by W6/32 from a human lymphoblastoid cell line. A number of laboratories have demonstrated enhanced expression of MHC following exposure of lymphoid and non-lymphoid cells to gamma-interferon (Heron et al., 1978; Fellous et al., 1982). Enhancement of MHC antigen expression in trophoblast cells by lymphokines may have biological and clinical significance. It is possible that rapidly progressing and fatal gestational choriocarcinomas are of the non-HLA inducible type. It would be of interest to investigate whether gamma-interferon induces or enhances expression of paternal HLA in normal or molar trophoblast.

It has been suggested that the development and progression of persistent gestational trophoblastic tumors may be favored by histocompatibility between the patient and her partner. If the patient and her partner are histocompatible, the trophoblastic tumor that bears paternal antigens might not be immunogenic in the maternal host. The intensity of the host's immunologic response may depend upon the immunogenicity of the trophoblastic tumor.

However, histocompatibility between patients and their partners does not appear to be a prerequisite for the development and persistence of gestational trophoblastic tumors. We (Berkowitz et al., 1981) recently studied the frequency distribution of HLA-A,B,C (54 specificities) and DR (7 specificities) in 29 patients with gestational choriocarcinoma and their partners as well as in a matched control population of healthy couples. All 29 patients achieved complete remission with chemotherapy. The HLA-antigen frequency was normal in the patients and their partners and there was no abnormal HLA-antigen sharing between patients and partners. Most patients were not histocompatible with their partners and the frequency of histocompatibility between patients and partners did not differ from that of control couples. Klouda et al. (1972), Mittal et al. (1975), and Yamashita et al. (1981) also observed no increase in HLA-compatible couples among patients with gestational trophoblastic tumors. The potential relationship between histocompatibility and choriocarcinoma was also studied by Lawler et al. (1971) and Lewis and Terasaki (1971) who examined term pregnancies in which the child antecedent to the choriocarcinoma was available for HLA testing. HLA compatible children were detected in only 1 of 18 patients by Lawler et al. (1971) and in 2 of 15 patients by Lewis and Terasaki (1971).

Table 1

Reactivity of monoclonal antibodies to oncofetal antigens with BeWo and Jar
choriocarcinoma cell lines and human fibroblast cells in immunofluorescence assays*

Monoclonal Antibody	Source	Tissue of origin	BeWo	Jar	Human Fibroblast
DU-PAN-2	Daasch et al	Pancreatic Cancer	–	–	–
19-9	Magnani et al	Pancreatic Cancer	–	–	–
MOV 2	Miotti et al	Ovarian Cancer	+/–	+/–	–
OC133	Berkowitz et al	Ovarian Cancer	+/–	+/–	–
115D8	Hilkens et al	Milk fat globule membrane	+	+	–
OC125	Bast et al	Ovarian Cancer	+	+	–
SSEA-1	Solter and Knowles	Murine Teratocarcinoma	+	+	–
126G5	Centocor Co.	Breast Cancer	+	+	–
126E7	Centocor Co.	Breast Cancer	+	+	+
125B4	Centocor Co.	Breast Cancer	+	+	+
123D8	Centocor Co.	Breast Cancer	+/–	–	+/–
125B5	Centocor Co.	Breast Cancer	–	–	–
126E7	Centocor Co.	Breast Cancer	+	+	+
130B9	Centocor Co.	Breast Cancer	+	+	+
3D6	Centocor Co.	Anti-CEA	–	–	–
5E9-1E9	Centocor Co.	Anti-CEA	–	–	–
OVTL3	Centocor Co.	Ovarian Cancer	+/–	+/–	–
DF3	Kufe et al	Breast Cancer	+/–	+/–	–

* The BeWo and Jar human gestational choriocarcinoma cell lines were
obtained from Dr. R.A. Pattillo (Medical College of Wisconsin, Milwaukee, WI)
and fibroblast cell cultures were established from normal adult human lung
(Pattillo and Gey, 1968; Pattillo et al., 1971).

The HLA system may influence the natural history of rapidly progressive and fatal gestational trophoblastic tumors. Morgensen et al. (1969) have observed that histocompatibility between patients and their partners at certain HLA sites was more common than expected with metastatic trophoblastic disease. Tomoda et al. (1976) have noted different patterns of HLA compatibility between patients and their partners in drug-sensitive and resistant choriocarcinoma. Drug-resistant chorio-carcinoma was associated with increased histocompatibility between the patient and her husband. Histocompatibility between the patient and her partner may reduce the immunogenicity of the trophoblastic tumor and thereby lessen the host's immunologic response. The development and progression of poor prognosis trophoblastic tumors may therefore be facilitated by tumor-host (partner-patient) histocompatibility.

Relationships Among Placenta, Complete Mole and Gestational Choriocarcinoma

The relationships among normal placenta, complete mole, and choriocarcinoma are not clearly understood. The normal trophoblast is highly proliferative fulfilling many of the morphologic criteria for malignancy, may invade maternal decidua and may be transported via maternal vascular channels to the pulmonary vasculature (Bernirschke and Driscoll, 1967). Choriocarcinoma is an aggressive trophoblastic malignancy with a propensity for early metastasis and in the absence of therapy has a predictably fatal outcome (Goldstein and Berkowitz, 1982; Bagshawe, 1969). Complete hydatidiform mole may also be invasive and metastatic and in some cases (5%) gives rise to choriocarcinoma (Berkowitz and Goldstein, 1981). However, it is still unclear whether complete mole is a true trophoblastic neoplasm, a premalignant phenotype, or a variant of missed abortion. Kato et al. (1982) studied the behavior of hydatidiform mole after transplantation into nude mice. Unlike choriocarcinoma, molar tissue became avascular and atrophic following transplantation and thereby did not behave like a trophoblastic neoplasm. Takeuchi (1980) has proposed that hydatidiform mole develops because persistent production of blocking IgG antibody protects trophoblastic growth after the demise of the embryo. Patients with molar pregnancy have increased levels of blocking IgG antibody as compared to patients with spontaneous abortion.

Molar trophoblast is more like normal placental trophoblast than choriocarcinoma in the expression of certain products. Placental and molar trophoblast express alpha 2-macroglobulin, placental protein 5, and trophoblast-specific membrane antigen, whereas these three substances are not expressed on the trophoblast of invasive mole or choriocarcinoma in immunoperoxidase studies (Saksela et al., 1981; Seppala et al., 1979; Loke et al., 1980). Both alpha 2-macroglobulin and placental protein 5 are known to inhibit proteases such as plasmin and trypsin; therefore expression of alpha 2-macroglobulin and placental protein 5 on normal placental trophoblast may be involved in the regulation of trophoblastic invasiveness. The pattern of expression of certain oncofetal antigens on normal placenta, hydatidiform mole, and choriocarcinoma may be used to assess trophoblastic differentiation and the relationships among these three gestational tissues.

We have recently studied the localization of murine stage-specific embryonic antigens (SSEA-1 and SSEA-3) and human sperm antigens in normal human placenta, hydatidiform mole and choriocarcinoma in both immunofluorescence and immunoperoxidase assays (Berkowitz et al., 1985a; Berkowitz et al., 1985b). SSEA-1 is expressed on pre-implantation murine trophectoderm and on a variety of human tumors and tumor cell lines (Solter and Knowles, 1978; Fox et al., 1983). SSEA-3 is expressed on the early mouse embryo and on human teratocarcinoma-derived cell lines (Shevinsky et al., 1982). Some sperm antigens persist in the early embryo and antisperm antibodies react with malignant embryonal cell lines (Menge and Fleming, 1978; Anderson et al., 1983). SSEA-3 did not react with any cellular components in normal placentae, hydatidiform moles, or choriocarcinoma cell lines (Berkowitz et al., 1985). SSEA-1 and two antihuman sperm antibodies (MA2 and MA7) reacted with human choriocarcinoma cell lines but not with the trophoblastic cells of hydatidiform moles or normal placentae. Therefore, according to this oncofetal marker system, molar trophoblast is more like normal placental trophoblast than choriocarcinoma.

The expression of certain sperm-embryonic antigens may be useful as markers of trophoblastic differentiation. Because the complete mole bears a double set of identical paternal 23X chromosomes, molar trophoblast may continue to express a higher level of certain sperm-derived antigens than would normal trophoblast. Additional monoclonal antibodies to sperm antigens should be screened against hydatidiform mole and placenta to identify sperm antigens that may be preferentially expressed on molar tissue.

We recently tested an expanded panel of monoclonal antibodies reactive with a variety of human tumor antigens against two human gestational choriocarcinoma cell lines (Table 1). Four of the monoclonal antibodies (115D8, OC125, SSEA-1, 126G5) reacted with both choriocarcinoma cell lines but did not react with a normal human fibroblast cell line (negative control). These four monoclonal antibodies were then tested for reactivity in an indirect immunofluorescence assay against fresh-frozen tissue sections from three normal placentae (10, 20, and 38 wks gestation) and three complete hydatidiform moles (Table 2). The villous trophoblast in both normal placenta and complete mole did not react with 115D8, OC125, or SSEA-1. In contrast, in both normal placenta and complete mole, the villous trophoblast was brightly reactive with 126G5 (a monoclonal antibody to breast cancer). This study provides further evidence that molar villous trophoblast has the same pattern of reactivity with monoclonal antibodies to oncofetal antigens as the villous trophoblast from normal placentae.

Because hydatidiform mole is androgenic in origin and in some cases precedes choriocarcinoma, hydatidiform mole may represent a biologic transition which links teratogenesis and carcinogenesis. Further understanding of the relationships among normal conception, complete hydatidiform mole, and choriocarcinoma may provide important insights into basic mechanisms of teratogenesis and carcinogenesis. Understanding the relationship among these three gestational tissues may also help to explain the unique curability of gestational trophoblastic tumors and this information may hopefully be applied to improved treatment of other tumors.

ACKNOWLEDGEMENTS

Research supported by National Institutes of Health Grant CA32132

Table 2

Reactivity of monoclonal antibodies with normal placenta, hydatidiform mole and
gestational choriocarcinoma*

	Monoclonal Antibodies			
	115D8	OC125	SSEA-1	126G5
Placenta				
10 weeks	–	–	–	+
20 weeks	–	–	–	+
38 weeks	–	–	–	+
Complete mole (3)	–	–	–	+
BeWo cell line	+ membrane	weakly + membrane	+ membrane	strongly + cytoskeletal
Jar cell line	+ membrane	weakly + membrane	+ membrane	strongly + cytoskeletal

*   These four ascites monoclonal antibodies were diluted in phosphate buffered
    saline containing 1% bovine serum albumin at the following dilutions: 115D8-
    1:100, OC125-1:50, SSEA-1-1:50, 126G5-1:100. The indirect immunofluores-
    cence assay was performed as previously described (Berkowitz et al., 1984).

REFERENCES

Anderson, D.J., Adams, P.H., Hamilton, M.S., and Alexander, N.J. (1983) Antisperm
    antibodies in mouse vasectomy sera react with embryonal teratocarcinoma. *J.
    Immunol.* 131, 2908-2912.

Anderson, D.J. and Berkowitz, R.S. (1985) Gamma-interferon enhances expression of
    Class I MHC antigens in the weakly HLA-positive human choriocarcinoma cell
    line BeWo but does not induce MHC expression in the HLA-negative
    choriocarcinoma cell line Jar. *J. Immunol.* 135, 2498-2501.

Bagshawe, K.D. (1969) *Choriocarcinoma*. Edward Arnold, Ltd., London.

Bast, R.C., Jr., Feeney, M., Lazarus, H., Nadler, L.M., Colvin, R.B., and Knapp, R.C. (1981) Reactivity of a monoclonal antibody with human ovarian carcinoma. *J. Clin. Invest.* 68, 1331-1337.

Berkowitz, R.S. and Goldstein, D.P. (1981) Pathogenesis of gestational trophoblastic neoplasms. *Pathobiol. Ann.* 11, 391-411.

Berkowitz, R.S., Hornig-Rohan, J., Martin-Allosco, S., Klein, S., Goldstein, D.P., Bast, R.C., Jr., and DeWolf, W.C. (1981) HLA antigen frequency distribution in patients with gestational choriocarcinoma and their husbands. *Placenta Suppl.* 3, 263-267.

Berkowitz, R.S., Mostoufizadeh, M., Kabawat, S.E., Goldstein, D.P., and Driscoll, S.G. (1982a) Immunopathologic study of the implantation site in molar pregnancy. *Am. J. Obstet. Gynecol.* 144, 925-930.

Berkowitz, R.S., Sandstrom, M., Goldstein, D.P., and Driscoll, S.G. (1982b) 45, X Complete hydatidiform mole. *Gynecol. Oncol.* 14, 279-283.

Berkowitz, R.S., Anderson, D.J., Hunter, N.J., and Goldstein, D.P. (1983a) Distribution of major histocompatibility (HLA) antigens in chorionic villi of molar pregnancy. *Am.J. Obstet. Gynecol.* 146, 221-222.

Berkowitz, R., Kabawat, S., Lazarus, H., Colvin, R., Knapp, R., and Bast, R.C., Jr. (1983b) Comparison of a rabbit heteroantiserum and a murine monoclonal antibody raised against a human epithelial ovarian carcinoma cell line. *Am. J. Obstet. Gynecol.* 146, 607-612.

Berkowitz, R.S., Lahey, S.J., Rodrick, M.L., Rayner, A.A., Goldstein, D.P., and Steele, G., Jr. (1983c) Circulating immune complex levels in patients with molar pregnancy. *Obstet. Gynecol.* 61, 165-168.

Berkowitz, R.S., Hoch, E.J., Goldstein, D.P., and Anderson, D.J. (1984) Histocompatibility antigens (HLA-A,B,C) are not detectable in molar villous fluid. *Gynecol. Oncol.* 19, 74-78.

Berkowitz, R.S., Alberti, O., Jr., Hunter, N.J., Goldstein, D.P., and Anderson, D.J. (1985a) Localization of stage-specific embryonic antigens in hydatidiform mole, normal placenta and gestational choriocarcinoma. *Gynecol. Oncol.* 20, 71-77.

Berkowitz, R.S., Alexander, N.J., Goldstein, D.P., and Anderson, D.J. (1985b) Reactivity of antihuman sperm monoclonal antibodies with normal placenta, hydatidiform mole and gestational choriocarcinoma. *Gynecol. Oncol.* 22, 334-340.

Bernirschke, K. and Driscoll, S.G. (1967) *The Pathology of the Human Placenta.* Springer-Verlag, New York.

Bhan, A.K., Mihm, M.C., Jr., and Dvorak, H.F. (1982) T-cell subsets in allograft rejection: In-situ characterization of T-cell subsets in human skin allografts by the use of monoclonal antibodies. *J. Immunol.* 129, 1578-1583.

Daasch, V.N., Fernsten, P.D., and Metzgar, R.S. (1982) Radioimmune assay studies using a monoclonal antibody to detect an antigen (DU-PAN-2) in the serum and ascites of adenocarcinoma patients. *Proc. Am. Assoc. Can. Res.* (abstract) p. 1051.

Davis, J.R., Surwit, E.A., Garay, J.P., and Fortier, K.J. (1984) Sex assignment in gestational trophoblastic neoplasia. *Am. J. Obstet. Gynecol.* 148, 722-725.

Faulk, W.P. and Temple, A. (1976) Distribution of beta-2 microglobulin and HLA in chorionic villi of human placentae. *Nature* 262, 799-802.

Fisher, R.A., and Lawler, S.D. (1984) Heterozygous complete hydatidiform moles: Do they have a worse prognosis than homozygous complete moles? *Lancet* 2, 51.

Fellous, M., Nir, V., Wallach, D., Merlin, G., Rubenstein, M., and Revel, M. (1982) Interferon dependent induction of mRNA for the major histocompatibility antigens in human fibroblasts. *Proc. Natl. Acad. Sci. USA* 79, 3082-3086.

Fox, N., Damjanov, I., Knowles, B.B., and Solter, D. (1983) Immunohistochemical localization of the mouse stage-specific embryonic antigen I in human tissues and tumors. *Canc. Res.* 43, 669-678.

Goldstein, D.P. and Berkowitz, R.S. (1982) *Gestational Trophoblastic Neoplasms: Clincal Principles of Diagnosis and Management.* W.B. Saunders, Philadelpha, PA.

Heron, I., Hokland, M., and Berg, K. (1978) Enhanced expression of beta 2-microglobulin and HLA antigens on human lymphoid cells by interferon. *Proc. Natl. Acad. Sci. USA* 77, 6215-6219.

Hilkens, J., Baijs, F., Hilgers, J., Hageman, Ph., Calafat, J., Sonnenberg, A., and Van der Valk, M. (1984) Monoclonal antibodies against human milk-fat globule membranes detecting differentiation antigens of the mammary glands and its tumors. *Int. J. Canc.* 34, 197-206.

Jacobs, P.A., Wilson, C.M., Sprenkle, J.A., Rosenshein, N.B., and Migeon, B.R. (1980) Mechanism of origin of complete hydatidiform moles. *Nature (London)* 286, 714-716.

Kabawat, S.E., Mostoufi-Zadeh, M., Berkowitz, R.S., D riscoll, S.G., Goldstein, D.P., and Bhan, A.K. (1985) Implantation site in complete molar pregnancy: A study of immunologically competent cells with monoclonal antibodies. *Am. J. Obstet. Gynecol.* 152, 97-99.

Kajii, T. and Ohama, K. (1977) Androgenetic origin of hydatidiform mole. *Nature (London)* 268, 633-634.

Kato, M., Tanaka, K., and Takeuchi, S. (1982) The nature of trophoblastic disease initiated by transplantation into immunosuppressed animals. *Am. J. Obstet. Gynecol.* 142, 497-505.

Kawata, M., Parnes, J.R., and Herzenberg, L.A. (1984) Transcriptional control of HLA-A,B,C antigen in human placental cytotrophoblast isolated using trophoblast- and HLA-specific monoclonal antibodies and the fluorescence-activated cell sorter. *J. Exp. Med.* 160, 633-651.

Klouda, P.T., Lawler, S.D., and Bagshawe, K.D. (1972) HL-A matings in trophoblastic neoplasia. *Tiss. Antigens* 2, 280-284.

Kufe, D., Inghirami, G., Abe, M., Hayes, D., Justi-Wheeler, H., and Schlom, J. (1984) Differential reactivity of a novel monoclonal antibody (DF3) with human malignant versus benign breast tumors. *Hybridoma* 3, 223-232.

Lahey, S.J., Steele, G., Jr., Berkowitz, R.S., Rodrick, M.L., Ross, D.S., Goldstein, D.P., Zamcheck, N., Wilson, R.E., and Deasy, J.M. (1984a) Identification of material with paternal HLA antigen immunoreactivity from purported circulating immune complexes in patients with gestational trophoblastic neoplasia. *J. Nat. Canc. Inst.* 72, 983-990.

Lahey, S.J., Steele, G., Jr., Rodrick, M.L., Berkowitz, R.S., Goldstein, D.P., Ross, D.S., Ravikumar, T.S., Wilson, R.E., Byrn, R., Thomas, P., and Zamcheck, N. (1984b) Characterization of antigenic components from circulating immune complexes in patients with gestational trophoblastic neoplasia. *Canc.* 53, 1316-1321.

Lawler, S.D., Klouda, P.T., and Bagshawe, K.D. (1971) The HL-A system in trophoblastic neoplasia. *Lancet* 2, 834-837.

Lawler, S.D., Klouda, P.T., and Bagshawe, K.D. (1974) Immunogenicity of molar pregnancies in the HLA system. *Am. J. Obstet. Gynecol.* 120, 857-861.

Lawler, S.D., Pickthall, V.J., Fisher, R.A., Povery, S., Evans, M.W., and Szulman, A.E. (1979) Genetic studies of complete and partial hydatidiform moles. *Lancet* 2, 580.

Lewis, J.L., Jr., and Terasaki, P.I. (1971) HL-A leukocyte antigen studies in women with gestational trophoblastic neoplasms. *Am. J. Obstet. Gynecol.* 111, 547-552.

Li, M.C., Hertz, R., and Spencer, D.B. (1956) Effect of methotrexate therapy on choriocarcinoma and chorioadenoma. *Proc. Soc. Exp. Biol. Med.* 93, 361-366.

Loke, Y.W., Whyte, A., and Davies, S.P. (1980) Differential expression of trophoblast-specific membrane antigens by normal and abnormal human placentae and by neoplasms of trophoblastic and non-trophoblastic origin. *Int. J. Canc.* 25, 459-461.

Magnani, J., Steplewski, Z., Koprowski, H., and Ginsburg, V. (1983) Identification of the gastrointestinal and pancreatic cancer-associated antigen detected by monoclonal antibody 19-9 in the sera of patients as a mucin. *Canc. Res.* 43, 5489-5492.

Menge, A.C. and Fleming, C.H. (1978) Detection of sperm antigens on mouse ova and early embryos. *Develop. Biol.* 63, 111-117.

Miotti, S., Aguanno, S., Canevari, S., Diotti, A., Orlandi, R., Sonnion, S., and Colnaghi, M.I. (1985) Biochemical analysis of human cancer-associated antigens defined by murine monoclonal antibodies. *Canc. Res.* 45, 826-832.

Mittal, K.K., Kachru, R.B., and Brewer, J.I. (1975) The HL-A and ABO antigens in trophoblastic disease. *Tissue Antigens* 6, 57-69.

Morgensen, B., Kissmeyer-Nielsen, F., and Hauge, M. (1969) Histocompatibility antigens on the HL-A locus in gestational choriocarcinoma. *Transplant. Proc.* 1, 76-80.

Pattillo, R.A. and Gey, G.O. (1968) The establishment of a cell-line of human hormone-synthesizing trophoblastic cells in vitro. *Canc. Res.* 28, 1231-1236.

Pattillo, R.A., Ruckert, A., Hussa, R.O., Bernstein, R., and Delfs, E. (1971) The Jar cell line. Continuous human multihormone production and controls. *In Vitro* 6, 398.

Pattillo, R.A., Sasaki, S., Katayama, K.P., Roesler, M., and Mattingly, R.F. (1981) Genesis of 46, XY hydatidiform mole. *Am. J. Obstet. Gynecol.* 141, 104-105.

Saksela, O., Wahlstrom, T., Lehtovirta, P., Seppala, M., and Vaheri, A. (1981) Presence of alpha 2-macroglobulin in normal but not in malignant human syncytiotrophoblasts. *Canc. Res.* 41, 2507-2513.

Sasagawa, M., Kajino, T., Kanazawa, K., and Takeuchi, S. (1984) Immunohisto-chemical study on subpopulations of trophoblasts in normal pregnancy and hydatidiform mole. *Proc. Second World Cong. Trophoblastic Neoplasms*, abstract, p. 13.

Sasaki, S., Katayama, P.K., Roesler, M., Pattillo, R.A., Mattingly, R.F., and Ohkawa, K. (1982) Cytogenetic analysis of choriocarcinoma cell lines. *Acta Obstet. Gynecol. Jpn.* 34, 2253-2256.

Seppala, M., Wahlstrom, T., and Bohn, H. (1979) Circulating levels and tissue localization of placental protein five (PP5) in pregnancy and trophoblastic disease: Absence of PP5 expression in the malignant trophoblast. *Int. J. Canc.* 24, 6-10.

Shaw, A.R.E., Dasgupta, M.K., Kovithavongs, T., Johny, K.V., LeRiche, J.L., Dossetor, J.B., and McPherson, T.A. (1979) Humoral and cellular immunity to paternal antigens in trophoblastic neoplasia. *Int. J. Canc.* 24, 586-593.

Shevinsky, L.H., Knowles, B.B., Damjanov, I., and Solter, D. (1982) Monoclonal antibody to murine embryos define a stage-specific embryonic antigen expressed on mouse embryos and human teratocarcinoma cells. *Cell* 30, 697-705.

Solter, D. and Knowles, B.B. (1978) Monoclonal antibody defining a stage-specific mouse embryonic antigen (SSEA-1). *Proc. Natl. Acad. Sci. USA* 75, 5565-5569.

Sunderland, C.A., Redman, C.W.G., and Stirrat, G.M. (1981) HLA-A,B,C antigens are expressed on non-villous trophoblast of the early human placenta. *J. Immunol.* 127, 2614-2615.

Sunderland, C.A., Redman, C.W.G., and Stirrat, G.M. (1985) Characterization and localization of HLA antigens on hydatidiform mole. *Am. J. Obstet. Gynecol.* 151, 130-135.

Surti, U., Szulman, A.E., and O'Brien, S. (1979) Complete (classic) hydatidiform mole with 46, XY karyotype of paternal origin. *Hum. Genet.* 51, 153-155.

Takeuchi, S. (1980) Immunology of spontaneous abortion. *Am. J. Reprod. Immunol.* 1, 23-28.

Tomoda, Y., Fuma, M., Saiki, N., Ishizuka, N., and Akaza, T. (1976) Immunologic studies in patients with trophoblastic neoplasia. *Am. J. Obstet. Gynecol.* 126, 661-667.

Trowsdale, J., Travers, P., Bodmer, W.F., and Pattillo, R.A. (1980) Expression of HLA-A,B, and C and beta-2 microglobulin antigens in human choriocarcinoma cell lines. *J. Exp. Med.* 152, 11S-17S.

Wake, N., Takagi, N., and Sasaki, M. (1978) Androgenesis as a cause of hydatidiform mole. *J. Natl. Canc. Inst.* 60, 51-53.

Wake, N., Tanaka, K., Chapman, V., Matsui, S., and Sandberg, A.A. (1981) Chromosomes and cellular origin of choriocarcinoma. *Canc. Res.* 41, 3137-3143.

Wake, N., Seki, T., Fujita, H., Okubo, H., Sakai, K., Okuyama, K., Hayashi, H., Shiina, H.J., Sato, H., Kuroda, M., and Ichinoe, K. (1984) Malignant potential of homozygous and heterozygous complete moles. *Canc. Res.* 44, 1226-1230.

Yamashita, K., Wake, N., Araki, T., Ichinoe, K., and Makoto, K. (1979) Human lymphocyte antigen expression in hydatidiform mole: Androgenesis following fertilization by a haploid sperm. *Am. J. Obstet. Gynecol.* 135, 597-600.

Yamashita, K., Ishikawa, M., Shimizu, T., and Kuroda, M. (1981a) HLA-antigens in husband-wife pairs with trophoblastic tumor. *Gyneco. Oncol.* 12, 68-74.

Yamashita, K., Wake, N., Araki, T., Ichinoe, K., and Kuroda, M. (1981b) A further HLA study of hydatidiform moles. *Gynecol. Oncol.* 11, 23-28.

Trophoblast Research 2:139-148, 1987

O-GLYCOSYLATION OF PROTEINS IN THE NORMAL AND NEOPLASTIC TROPHOBLAST

Laurence A. Cole[1]

Department of Internal Medicine
University of Michigan Medical School
Ann Arbor, Michigan 48109, USA

INTRODUCTION

hCG is a glycoprotein hormone composed of two dissimilar subunits, α and β, joined non-covalently. The hormone, in addition to having N-linked sugar units, two on each subunit, contains four O-linked structures, all attached to the COOH-segment of the β subunit (Birken and Canfield, 1977; Cole et al., 1985). Gonadotropin large free α (LFA) is an oversized ($M_r = 24,000$) non-combining form of the glycoprotein hormone common α-subunit ($M_r = 22,000$). LFA can contain a single O-linked sugar unit attached to Thr 39 which may in part account for its size difference with the α- subunit incorporated into glycoprotein hormones (Cole et al , 1984a; Parsons and Pierce, 1984). Both hCG and LFA are secreted by the normal trophoblast in pregnancy and by the neoplastic trophoblast in hydatidiform mole and choriocarcinoma (Hussa, 1981; Cole et al., 1983; Cole et al., 1984b).

The carbohydrate units on proteins are major elements in some of the most crucial biologic events. On secreted molecules, sugar units have major effects on metabolic clearance, and important roles in protease protection and expression of biologic activities (Pen Loh and Gainer, 1980; Rosa et al., 1984). On membrane proteins, sugar units are key components in cell-cell recognition and cell-substratum adhesion (Warren et al., 1978; Damsky et al., 1984). Several laboratories have noted distinct structural changes in sugar units on proteins produced by cancer versus normal tissue, usually accompanied by changes in cellular glycosyltransferase activities (Warren et al., 1978; Funakoshi and Yamashina, 1982). Oligosaccharide structural changes may dramatically effect cellular interactions. On membrane proteins, for instance, whereas oligosaccharide mediated cell-cell and cell-substratum interactions proceed with precision in normal development leading to the formation of functional organs, in neoplasms with carbohydrate structural alterations these processes could fail possibly leading to autonomous growth or tumor tissue (Warren et al., 1978; Damkey et al., 1984). Several hCG-orientated studies have indicated differences in the O-linked sugar units on neoplastic compared to normal trophoblast glycoconjugates. Differences have been suggested by immunoassays of urinary hCG using anitsera sensitive to β-subunit O-linked oligosaccharides (Birken et al., 1981), by lectin binding assays of urinary hCG (Imamura et al., 1984), and by isoelectric

[1]Current Address: Department of Obstetrics and Gynecology, Yale University, 333 Cedar Street, New Haven, Connecticut 06510
Received: 10 October 1985; Accepted: 10 December 1985

focussing studies with the serum hormone (Yazaki et al., 1980). To elucidate the effects of cancer on trophoblast glycoproteins, the structures of the O-linked sugar units on hCG and LFA from both pregnancy and choriocarcinoma tissues were examined.

MATERIALS AND METHODS

Purified urinary hCG (batch CR121) and its isolated α and β subunits (both batch CR123) were kindly provided by the Hormone Distribution Program, NIADDK. All culture fluids were purchased from M.A. Bioproducts, [6-^{3}H]glucosamine and [^{3}H]sodium borohydride from Research Products International, Bio-Gels from Bio-Rad Laboratories, and all other reagents from Sigma.

JAr choriocarcinoma cells were routinely grown in Dulbecco's modified Eagle's medium with 10% fetal bovine serum and subcultured with trypsin-EDTA as previously described; as were explants of trophoblast tissue second trimester pregnancy terminations in similar medium (Cole et al., 1983). JAr cells at 70% dish confluency, and day 5 trophoblast explants were washed in phosphate-buffered saline (PBS), then cultured 24 hr in Eagle's minimal essential medium (5 mM glucose) supplemented with 10% (v/v) dialyzed fetal bovine serum, 10 mM HEPES buffer and [6-^{3}H]glucosamine (GlcN), 45 Ci/mmol, 0.3 Ci/ml. [6-^{3}H]GlcN is incorporated into sialic acid (NeuAc), N-acetylglucosamine (GlcNAc) and N-acetylgalactosamine (GalNAc) residues on glycoproteins. Spent fluids were collected and preserved with 20 mM EDTA, 10 mM iodoacetic acid, 2 mM phenylmethylsulfonyl fluoride and 0.1% soybean trypsin inhibitor.

Radioactive hCG and LFA in spent fluids were purified by immunoaffinity chromatography as follows: IgG isolated from 50 ml H2 rabbit anti β-subunit antisera and from a similar volume H7 rabbit anti LFA antisera were each linked to 50 ml Sepharose, according to the previously described procedures (Cole et al., 1983). With constant mixing, spent fluids were incubated 16 hr at 4°C with anti β-subunit IgG linked to Sepharose (0.02 ml settled gel being used to adsorb 1 microgram hCG). Suspensions were then centrifuged and the supernatants, unbound components containing LFA, removed. Gels were washed five times with 50 volumes PBS in cycles of gel dispersion, centrifugation, and supernatant aspiration. hCG was eluted from gels by a 30 min incubation at room temperature with five volumes 4 M guanidine-HCl, pH 4. Following the addition of myoglobin (to 0.05% w/v) as carrier protein, hormone-containing eluates were desalted on Bio-Gel P6DG columns (equilibrated with 0.1 M ammonium bicarbonate) and then lyophilized. LFA, not bound on anti β-subunit IgG-Sepharose, was mixed with anti LFA IgG-Sepharose (0.05 ml settled gel being used to adsorb 1 microgram LFA) for 16 hr at 4°C. The gel-IgG-LFA was washed, and LFA eluted and desalted as described for anti β-subunit IgG-Sepharose and hCG. hCG and LFA were further purified by gel filtration on 0.9 x 60 cm columns of Bio-Gel P100 (-400 mesh). Samples (0.5 ml) were applied to gel beds in water containing 10% (v/v) glycerol. Gels were washed with 0.1 M ammonium bicarbonate at 2 ml/hr and 1 ml fractions collected. hCG and LFA in fractions were detected by their radioactivities. Peak fractions were pooled and lyophilized.

β-eliminations were carried out using 0.1 N sodium hydroxide and 1.0 M sodium borohydride at 45°C as previously described (Cole et al., 1985). Reactions were terminated by the addition of acetic acid to pH 5. Products (oligosaccharitols plus peptide and glycopepetide fragments) were dried under nitrogen, and then borate removed by multiple evaporations from 50% (v/v) methanol. Samples were taken up in 1 ml of 0.1 M ammonium acetate, pH 4, and applied to 0.8 x 5 cm columns of AG50-X8 cation-exchange resin equilibrated in water (binds peptides and glycopeptides). Resins were washed with 3 ml water to recover the unbound oligosaccharitols. To ensure complete removal of peptide materials, unbound components were adjusted to pH 4 and ion-exchange on AG50-X8 resin repeated. Unbound oligosaccharitols were lyophilized, taken up in 0.5 ml water and then applied to and separated on 1 x 110 cm Bio-Gel P4 (-400 mesh) columns by the procedures previously described (Cole et al., 1984a; Cole et al., 1985). Oligosaccharitol peaks were detected by fraction radioactivity. To determine the relative quantities of peaks, radioactivity in the derivatized GalNAc attachment sugar (N-acetylgalactosaminitol, GalNAc-ol, 1 per structure) was determined. Peak fractions were pooled, dried under nitrogen, and acid hydrolyzed. [^{3}H]galactosaminitol (the hydrolysis product of [^{3}H]GalNAc-ol) was separated from other amino sugars on an HPLC amino acid analysis system and quantitated according to the procedures previously described (Cole et al., 1984a).

To prepare oligosaccharitol standards from O-linked sugar units, 1 mg purified β-subunit from urinary hCG was alkaline-borohydride β-eliminated by the procedures described above. To label the oligosaccharitol products, 25 mCi [^{3}H] sodium borohydride was added to reaction mixtures (procedure of Hull et al., 1984). Following neutralization of the β-elimination products, multiple evaporations from methanol and AG50-X8 cation exchange, the [^{3}H]oligosaccharitol products were separated on Bio-Gel P4. Peaks of radioactivity were identified in the elution volumes, and the following standard structures identified (Cole et al., 1985): 1, NeuAc α2→3 Gal β1→3 (NeuAc α2→3 Gal β1→4 GlcNAc β1→6) GalNAc-ol; 2, NeuAc α2→3 Gal β1→3 (NeuAc α2→6) GalNAc-ol; 3, NeuAc α2→3 Gal β1→3 GalNAc-ol; and 4, NeuAc α2→6 GalNAc-ol. Aliquots of standards 1 and 3 were desialylated by neuraminidase treatment (Cole and Hussa, 1981) to produce standards 5, Gal β1→3 (Gal β1→4 GlcNAc β1→6)GalNAc-ol, and 6, Gal β1→3 GalNAc-ol and 7, GalNAc-ol. Peak fractions were pooled and used as standards.

RESULTS

Second trimester pregnancy trophoblast explants and JAr choriocarcinoma cells were cultured in media containing [6-^{3}H]GlcN, which is incorporated into NeuAc, GlcNAc, and GalNAc residues on glycoproteins. Radioactive hCG and LFA were isolated from spent fluids by immunoaffinity chromatography. As a final purification and as a characterization step, preparations were separated on Bio-Gel P100 columns as shown in Fig. 1. The hCG preparations from the fluids of pregnancy (Fig. 1A) and choriocarcinoma cultured cells (Fig. 1B) predominantly eluted from Bio-Gel P100 in the same volume as hormone standard (NIADDK batch CR121), arrow 2. A small shoulder was observed on both hCG peaks in the position of hCG β-subunit standard, arrow 3. With the absence of an α-subunit peak (arrow 4 is α-subunit standard) which would suggest subunit dissociation, we assumed the shoulder to represent free β-subunit, which is similar to that component incorporated into the hormones and is secreted in small amounts by trophoblastic tissue (Cole et al., 1983; Cole et al., 1984b). Free β would co-purify with hCG on β-subunit IgG-Sepharose

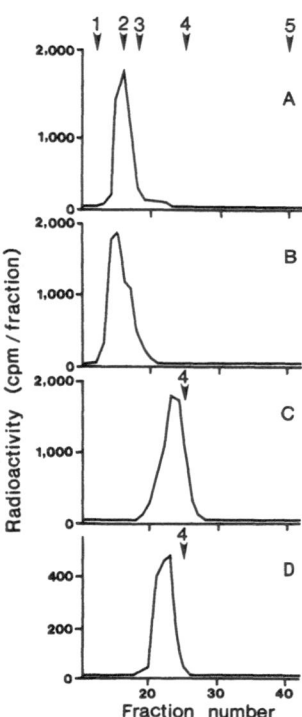

Figure 1. Gel filtration on Bio-Gel P100 of biosynthetically-labeled hCG and LFA from normal trophoblast (panels A and C, respectively), initially purified by immunoaffinity chromatography. Values are radioactivity in 10 microliter aliquots of fractions. The void and total volumes (arrows 1 and 5), and the elution positions of hCG, hCG β subunit and hCG α subunit standards chromatographed on the same column (arrows 2, 3, and 4, respectively) are indicated.

(Cole et al., 1983). LFA preparations from pregnancy (Fig. 1C) and choriocarcinoma cells (Fig. 1D) eluted as single peaks from Bio-Gel P100, in the volume previously reported, just prior to hCG α-subunit standard (Cole et al., 1984a).

Peak hCG and LFA gel filtration fractions were pooled, samples were concentrated and then alkaline-borohydride β-eliminated to release intact O-linked oligosaccharides. In this procedure, the linkage GalNAc residues on the released sugar units are reduced generating sugar units with terminal GalNAc-ol residues or oligosaccharitols. Following neutralization of samples, multiple methanol evaporations to volatilize and remove borate and cation exchange to extract amino acids and glycopeptides, oligosaccharitols were sized on analytical Bio-Gel P4 columns (Figures 2 and 3). Elution volumes were compared to those of standard sugar units of established structures (Cole et al., 1985). To compare the relative quantities of each peak, radioactivity in GalNAc-ol (derivatized linkage sugar, one per structure) was determined. Table 1 shows the distributions of hCG and LFA oligosaccharitols eluting in the positions of specific structural standards ([³H]GalNAc-ol, percentage of total in specific peaks). In the described oligosaccharitol preparations chromatographed on Bio-Gel P4, >90% of the radioactivity was recovered at the elution volumes of the seven sugar unit standards. Considering the absence of other peaks, it was concluded that all hCG and LFA oligosaccharitol structures were included in those measured.

Table 1

Distribution of hCG and LFA sugar units

Structure	Pregnancy		Choriocarcinoma	
	hCG	LFA	hCG	LFA
	%	%	%	%
1. NeuAc α2→3 Gal β1→4 GlcNAc β1→6 Gal NAc-ol NeuAc α2→3 Gal β1→3	14	18	51	55
2. NeuAc α2→6 GalNAc-ol NeuAc α2→3 Gal β1→3	73	71	29	25
3. GalNAc-ol NeuAc α2→3 Gal β1→3	a	3	10	9
4. NeuAc α2→6 GalNAc-ol	5	1	2	<1
5. Gal β1→4 GlcNAc β1→6 GalNAc-ol Gal β1→3	<1	<1	1	2
6. GalNAc-ol Gal β1→3	7	6	7	8
7. GalNAc-ol	a	a	a	a

Biosynthetically-labeled hCG and LFA O-linked sugar units released by β-elimination and sized on Bio-Gel P4. Values are from Figures 2 and 3, and are percentages of total [³H]GalNAc-ol eluting from Bio-Gel P4 in the positions of specific standards.

a Peak not detected

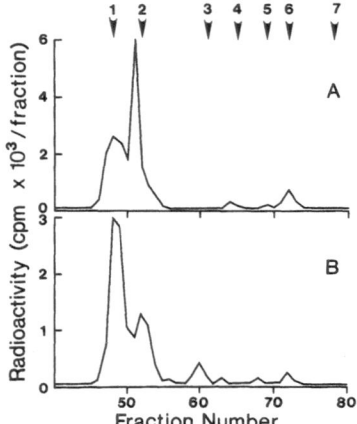

Figure 2. Gel filtration on Bio-Gel P4 of the oligosaccharitols released by the β elimination of hCG from normal trophoblast (panel A) and choriocarcinoma cells (panel B). Oligosaccharitols, containg [³H]GalNAc-ol and on some structures [³H]GlcNAc and [³H]NeuAc, were detected by their radioactivities. Arrows show the elution volumes of labeled oligosaccharitol standards chromatographed on the same column; numbers refer to structures in Table 1.

The results indicate the predominant sugar-unit from normal trophoblast hCG and LFA (73 and 71% of total) to be a tetrasaccharitol structure (Table 1, structure 2), with lesser amounts (14 and 18% of total) of a larger hexasaccharitol structure (structure 1) and di- and tri-saccharitol structure minor components (structures 3-6). By contrast, the hexasaccharitol structure was predominant among the sugar units from the choriocarcinoma hCG and LFA preparations (51 and 55% of total). The choriocarcinoma oligosaccharitols (from hCG and LFA) were 3-4 times more abundant in the hexasaccharitol and 2.5-3 times less abundant in the tetrasaccharitol than those from the normal trophoblast preparations (Table 1). No distinct differences in the distributions of the minor components, di- and tri-saccharitols were observed between the normal and cancer cell line sugar unit preparations (Figures 2 and 3, Table 1).

DISCUSSION

hCG and LFA have been used as model glycoproteins to examine the structures of O-linked oligosaccharides on molecules produced by normal and neoplastic trophoblast cells. As shown in Figures 2A and 3A, the oligosaccharitols released from pregnancy explant LFA and hCG have very similar elution profiles on Bio-Gel P4, and as indicated in Table 1, like distributions of structures. The same is observed with the oligosaccharitols from the choriocarcinoma cell line molecules; oligosaccharitols from LFA and hCG having like elution profiles (Figures 2B and 3B) and distributions of structures (Table 1). The relatively small differences in the hCG and LFA oligosaccharitol distributions may be accounted for by experimental error (estimated at ± 5%), so that the hormone with 4 Ser O-linkage sites on the β subunit

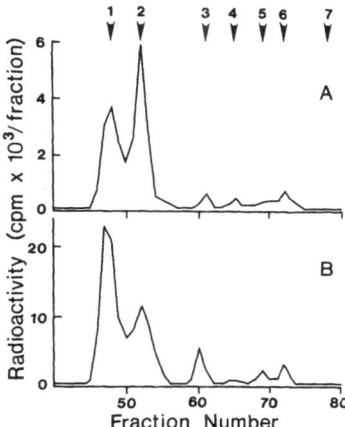

Figure 3. Gel filtration on Bio-Gel P4 of the oligosaccharitols released by the β elimination of LFA from normal trophoblast (panel A) and choriocarcinoma cells (panel B). Oligosaccharitols, containing [3H]GalNAc-ol and on some structures [3H]GlcNAc and [3H]NeuAc, were detected by their radioactivities. Arrows show the elution volumes of labeled oligosaccharitol standards chromatographed on the same column; numbers refer to structures in Table 1.

and LFA with just one at Thr 39 (Parsons and Pierce, 1984; Cole et al., 1985) seemingly have the same heterogeneous mixtures of attached sugar structures, suggesting tissue- rather than peptide-specific glycosylation. Tissue-specific glyco-sylation reflects the availability of glycosyltransferases and other oligosaccharide processing enzymes in a cell type; the structures shown for hCG and LFA, therefore, possibly being representive of those of other trophoblast glycoproteins.

As shown in Table 1, a clear difference is observed in the distribution of sugar units on pregnancy and choriocarcinoma cell line proteins. A hexasaccharide structure, NeuAc α2→3 Gal β1→3 (NeuAc α2→3 Gal β1→4 GlcNAc β1→6) GalNAc, seemingly accounts for just 14% of those on the pregnancy hCG preparation, or one structure on every other molecule of hormone. However, this structure accounts for as much as 51% of those on choriocarcinoma hCG, or two structures on every hormone molecule. The abundance of this structure on these and possibly other choriocarcinoma proteins is interesting, particularly when considering that the only other reports of it, to the best of my knowledge, have been on tumor-derived molecules (Funakoshi and Yamashina, 1982; Nilsson et al., 1982; Hull et al., 1984). The hexasaccharide structure could be a marker of trophoblast malignancy, as other O-linked structures are in different malignancies (Van Beek et al., 1983; Hull et al., 1984). With tissue- rather than peptide-specific glycosylation, the hexasaccharide abundance on choriocarcinoma proteins is indicative of altered glycosyltransferase activity(ies). Changes in glycosyltransferase activities or the abundance of the hexasaccharide itself could effect cellular chemistry or cell-cell interaction (Warren et al., 1978; Damsky et al., 1984), and possibly participate in trophoblast carcinogenesis.

These studies have been carried out and repeated with cultured trophoblast explants and JAr cultured choriocarcinoma cells. That the results are representative of those for glycoproteins produced in vivo, and are not artifacts of tissue culture, is indicated by our recent similar findings on urines from pregnancy and choriocarcinoma subjects (Cole, 1985, abstract). The hexasaccharitol constituted an average of 6% (n = 10) of the released oligosaccharitols released from pregnancy urine hCG, and 50% of that from the urine of two choriocarcinoma subjects.

SUMMARY

hCG and LFA produced by cultured normal and neoplastic cells were used as models to examine O-glycosylation of proteins in the trophoblast. Similar distributions of heterogeneous O-linked oligosaccharide structures were found on both molecules from each tissue, even though hCG has four and LFA just one O-glycosylation site. This suggested tissue - rather than peptide - specific glycosylation, and that the structures found on these molecules could be representative of those on other trophoblast proteins. A hexasaccharide structure was the principal sugar unit on molecules from choriocarcinoma cells, occurring in 3-4 times greater abundance than that on hCG and LFA molecules from pregnancy trophoblast explants (51 and 55% versus 14 and 18% of total). Differences in O-glycosylation of proteins were apparent in neoplastic versus normal cultured trophoblast cells. The hexasaccharide structure on trophoblast glycoconjugates, circulatory hCG or possibly plasma membrane elements, could be a marker of trophoblast malignancy.

ACKNOWLEDGEMENTS

Many thanks to Dr. J. LaFerla of the University of Michigan Medical Center for supplying placentae from pregnancy terminations, and to Dr. R.A. Pattillo of the Medical College of Wisconsin for flasks of JAr choriocarcinoma cells. This investigation was supported in part by Institutional Research Grant No. IN-40Y to the University of Michigan from the American Cancer Society.

REFERENCES

Birken, S. and Canfield, R.E. (1977) Isolation and amino acid sequence of COOH-terminal fragments from the beta subunit of human choriogonadotropin. *J. Biol. Chem.* 252, 5386-5392.

Birken, S.G., Agosto, G., Canfield, R.E., Amr, S., and Nisula, B. (1981) Antisera sensitive to the carbohydrate in the COOH-terminal region of hCG beta: Novel antibodies of clinical significance. *Abstract of the 63rd Annual Meeting of the Endocrine Society*, Cincinnati.

Cole, L.A., Hartle, R.J., Laferla, J.J., and Ruddon, R.W. (1983) Detection of the free beta subunit of human chorionic gonadotropin (HCG) in cultures of normal and malignant trophoblast cells, pregnancy sera, and sera of patients with choriocarcinoma. *Endocrinol.* 113, 1176-1178.

Cole, L.A., Perini, F., Birkin, S., and Ruddon, R.W. (1984a) An oligosaccharide of the O-linked type distinguishes the free from the combined form of hCG alpha subunit. *Biochem. Biophys. Res. Comm.* 122, 1260-1267.

Cole, L.A., Kroll, T.G., Ruddon, R.W., and Hussa, R.O. (1984b) Differential occurrence of free beta and free alpha subunits of human chorionic gonadotropin. *J. Clin. Endocrinol. Metab.* 58, 1200-1202.

Cole, L.A. (1985) Structures of the O-linked oligosaccharides on human chorionic gonadotropin molecules produced by cancer cells. *Abstract of the AAP/ASCI/AFCR Meetings*, Washington, DC.

Cole, L.A., Birkin, S., and Perini, F. (1985) The structures of the serine linked sugar chains on human chorionic gonadotropin. *Biochem. Biophys. Res. Comm.* 126, 333-339.

Damsky, C.H., Knudsen, K.A., and Buck, C.A. (1984) Integral membrane glycoproteins in cell-cell and cell-substratum adhesion. In: *The Biology of Glycoproteins*, (ed.), R.J. Ivatt, New York, Plenum, pp. 1-55.

Funakoshi, I. and Yamashina, I. (1982) Structures of O-glycosidically linked sugar units from plasma membranes of an ascites hepatoma, AH66. *J. Biol. Chem.* 257, 3787.

Hull, S.R., Laine, R.A., Kaizu, T., Rodriguez, I., and Carraway, K.L. (1984) Structures of the O-linked oligosaccharides of the major cell surface sialyl glycoprotein of MAT-B1 and MAT-C1 ascites sublines of the 13762 rat mammary adenocarcinoma. *J. Biol. Chem.* 259, 4866-4877.

Hussa, R.O. (1981) Human chorionic gonadotropin, a clinical marker: review of its biosynthesis. *Ligand. Rev.* 3, 1-43.

Imamura, S., Armstrong, E.G., Birken, S., and Cole, L.A. (1984) Highly specific and sensitive measurements of asialo hCG with lectin-antibody sandwich assays. *Abstracts of the 7th International Congress of Endocrinology*, Quebec City, Canada.

Nilsson, B., DeLuca, S., Lohmander, S., and Hascall, V.C. (1982) Structures of N-linked and O-linked oligosaccharides on proteoglycan monomer isolated from swarm rat chondrosarcoma. *J. Biol. Chem.* 257, 10920-10927.

Parsons, T.F. and Pierce, J.G. (1984) Free α-like material from bovine pituitaries. *J. Biol. Chem.* 259, 2662-2666.

Pen Loh, Y. and Gainer, H. (1980) Evidence that glycosylation of pro-opiocortin and ACTH influences their proteolysis by trypsin and blood proteases. *Mol. Cell. Endocrinol.* 20, 35-44.

Rosa, C., Amr, S., Birken, S., Wehmann, R., and Nisula, B. (1984) Effect of desialyation of human chorionic gonadotropin on its metabolic clearance rate in humans. *J. Clin. Endocrinol. Metab.* 59, 1215-1219.

Van Beek, W.P., Smets, L.A., and Emmelot, P. (1973) Changed surface glycoprotein as a marker of malignancy in human leukaemic cells. *Nature* 253, 457-460.

Warren, L., Buck, C.A., and Tuszynski, G.P. (1978) Glycopeptide changes and malignant transformation: A possible role for carbohydrate in malignant behavior. *Biochem. Biophys. Acta* 516, 97-127.

Yazaki, K., Yazaki, C., Wakabayashi, K., and Igarashi, M. (1980) Isoelectric heterogeneity of human chorionic gonadotropin: Presence of choriocarcinoma specific components. *Am. J. Obstet. Gynecol.* 138, 189-194.

Trophoblast Research 2:149-160, 1987

PLACENTAL PRODUCTION OF IMMUNOREGULATORY FACTORS: TROPHOBLAST IS A SOURCE OF INTERLEUKIN-1

Elliott K. Main, Julie Strizki, and Peter Schochet

Department of Obstetrics and Gynecology
· School of Medicine
University of Pennsylvania
3400 Spruce Street
Philadelphia, Pennsylvania 19104, USA

INTRODUCTION

The placenta is often viewed as an immunologically passive organ either "camouflauged" to avoid immune surveillance or acting as a "sponge" to clear maternal anti-fetal antibodies. We and others have previously isolated low molecular weight lymphokines from human placental explant culture supernatants which would suggest a more active role of the placenta in modulating maternal immune responses (Lala et al., 1984). These include a 70-78 Kilodalton (KD) suppressor of IL-2 dependent cell proliferation (Main et al., 1985), and a 15-17 KD protein which on partial purification (by gel filtration and isoelectric focusing) and by biological activity is identical to adult monocyte-derived Interleukin-1 (IL-1) (Flynn et al., 1982). IL-1 is a key mediator of inflammatory responses leading to lymphocyte activation, systemic metabolic changes, elevated body temperature, local production of collagenase, and stimulation of fibroblast proliferation (Dinarello, 1984; Lachman, 1983). It is now known to be produced by a variety of cell types including monocytes, tissue fixed macrophages, keratinocytes, corneal epithelial cells, astrocytes, and renal mesangial cells. IL-1 is extraordinary potent, with biologic activity demonstrated in the 10^{-10}M concentration range. This high specific activity has facilitated its detection in bioassays but because it is produced in such small quantities, purification and production of monoclonal antibodies has been difficult. We have measured placental IL-1 using both the mouse thymocyte assay and the newer, more sensitive IL-1 dependent tumor line assay. In this report, we address three issues: (1) Can villous tissue from early in gestation produce Interleukin-1? (2) Of the various cell types present in a human term placenta which are capable of producing IL-1? (3) How does the level of IL-1 production by placental cells compare to that produced by human peripheral blood monocytes?

Received: 10 October 1985; Accepted: 15 May 1986

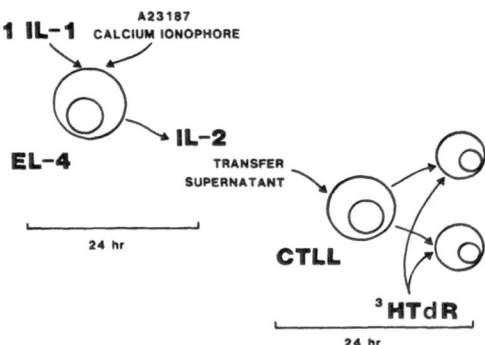

Figure 1. Two/step - two cell line bioassay for Interleukin-1. EL-4 produces IL-2 in an IL-1 dependent manner which can then be measured in a standard CTLL proliferation assay.

MATERIAL AND METHODS

Interleukin-1 Bioassays

(1) Mouse thymocyte assay: thymocytes from juvenile (5-7 wks of age) C3H/HeJ mice were used (Jackson Laboratories, Bar Harbor). These were plated at 1.5×10^6 thymocytes per well in a 96 well flat-bottomed plate (Costar, Cambridge, MA). The thymocytes were co-cultured with a suboptimal dose of PHA-P, 1 µg/ml (Difco, Grand Rapids, MI). Culture media was RPMI 1640 supplemented with 5% fetal bovine serum (FBS), penicillin (100 U/ml), streptomycin (100 µg/ml), glutamine (2 mM), HEPES (20 mM) (designated as PSGH) , and 2-ME (2×10^{-5}M). Supernatants with known and unknown quantities of Interleukin-1 were titrated to achieve a final volume of 0.2 ml per well. The plates were incubated for 72 hr at 37°C in a humidified incubator at 7% CO_2 in air. Before harvesting the cultures were pulsed with 0.5 µCi of (^{3}H)thymidine per culture well. Cells were harvested using a PhD Cell Harveter (Cambridge Technology, Cambridge, MA). Incorporation of (^{3}H)thymidine was measured by standard liquid scintillation counting techniques.

(2) EL4/CTLL assay: This assay takes advantage of the EL4 cell line which produces IL-2 in an IL-1 dependent manner. The IL-2 is then measured in a standard IL-2 assay using the CTLL cell line (Simon et al., 1986). For clarity, the principles of the two-step assay is illustrated in Figure 1. 2×10^5 EL4 cells were cultured per well in RPMI 1640 that contained 5×10^{-7}M calcium ionophore (A23187) (Sigma, St. Louis), 5% FBS and PSGH as noted above. IL-1 containing supernatants were added to each well of the 96 well flat bottom microtiter plate (Costar, Boston, MA) in serial replicate dilutions. Total volume per well was 0.2 ml. The plates were incubated at 37°C for 24 hr in an atmosphere of 7% CO_2 in air at 100% humidity. At that point, 0.1

ml of supernatant was transferred to a new 96 well flat bottom plate. 10^4 CTLL cells (a mouse T-cell tumor line that is exquisitely sensitive to IL-2) were added in 0.1 ml of media per well. The cells were then cultured for an additional 24 hr under the same conditions. Six hr prior to harvesting, wells were pulsed with 0.5 µCi of (^{3}H)thymidine. The cells were harvested and counted as noted above. The standard unit of measurement for IL-1 is the reciprocal of the dilution that gives one half of the maximum stimulation (or half-maximal units).

Preparation of Cytotrophoblast

Cytotrophoblasts were isolated as described by Kliman et al. (1986a,b). This is a modification of the enymzatic dispersion method described by Hall et al. (1977) with the addition of a Percoll gradient centrifugation step. Normal term (37-40 wks of gestation) placentae were obtained immediately following spontaneous vaginal delivery or uncomplicated cesarean section. After removal of membranes, cotelydons were excised and washed thoroughly in 0.9% saline. Soft villus material was excised from underlying connective tissue and blood vessels. Minced tissue was digested in a combination of 0.125% trypsin (Sigma, St. Louis), and 2 mg/ml DNAse (Sigma, St. Louis) at a ratio of 30 gm of tissue/150 ml of media. The tissue was subjected to 2- 3 trypsin-DNA digestions with stirring at 37°C. Following each digestion, supernatant was removed and centrifuged through fetal calf serum. The pellet was resuspended and layered onto a 5-70% Percoll gradient in 5% steps of 3 ml each in a 50 ml conical polystyrene centrifuge tube. The gradient was centrifuged at 1200 x g at room temperature for 20 min. The middle band containing cytotrophoblast (d = 1.045-1.055 g/ml) was removed and washed with Dulbecco's modified Eagle's medium containing 25 mM HEPES and 25 mM glucose (DMEM-HG) and then resuspended in this medium for tissue culture. The area above density of 1.040 encompasses tissue fragments, the area from approximately 1.060-1.075 contained a mixture of lymphocytes, macrophages, and granuocytes. These fractions have been extensively characterized by electron microscopic morphology, immunohistochemistry, steroid production, and behavior in tissue culture (formation of syncytia) by Kliman et al. (1986a,b).

Culture of Placenta Cells

Explant cultures were performed with minced placental cells following extensive washing in either 100 mm Petrie dishes or 6 well plates (Costar, Boston, MA) in a medium of PRMI 1640 with 10% FBS plus PSGH with or without LPS (10 µg/ml final concentration). First trimester villus placental tissue was obtained from tissue from elective pregnancy terminations performed at 7-9 wks from last menstrual period. Under a dissecting microscope villus fronds were teased free in saline. Small pieces, approximately 25 mg wet weight per ml of media were cultured as explants.

Isolated placental cells were dispersed in 6 well plates at 1.0×10^6 cells per ml (2-3 ml/well) in a culture media of DMEM-HG containing 4 mM glutamine, 50 µg/ml gentamycin and 20% fetal calf serum. All cultures were incubated in humified 5% CO_2- 95% air at 37°C.

Isolation of Peripheral Blood Mononuclear Leukocyte Populations

Peripheral blood mononuclear leukocytes (PBL) were obtained by Ficoll-hypaque density gradient centrifugation of peripheral blood from healthy volunteer donors. Adherent and non-adherent fractions of PBL were isolated by adherence to plastic culture flasks as described by Todd and Schlossman (1982). PBL were cultured in T75 flasks pretreated with heat inactivated human serum (Bio Bee, Boston) 1 ml/flask. The incubation lasted for 1 hr at 37°C at which time the supernatant was removed and used as T-cell enriched population. Immunofluorescence analysis using a variety of monocyte markers indicated that there were fewer than 1% monocytes in this fraction. The flask was then rinsed several times. This supernatant included loosely adherent cells and was discarded. Fresh media was placed in the flask and the flask was tapped and observed under an inverted microscope. These adherent cells were greater than 80% positive for monocyte cell surface markers. Both cell populations were cultured in RPMI 1640, PSGH, and 10% heat inactivated human serum.

Monoclonal Antibodies

Monoclonal antibodies directed against T-cell surface antigens, OKT-3 and OKT-8, were purchased from the Ortho Pharmaceutical Laboratories (Raritan, NJ). The monocyte markers included OK-IA and OKM1 (Ortho Pharmaceutical Laboratories, Raritan, NJ), Leu M3 (Becton Dickinson, Sunnyvale, CA), and BRL monocyte 1 (63D3) (Bethesda Research Laboratories, Gaithersburg, MD).

Indirect Immunofluorescence

Cells (2-4×10^5) were incubated for 30 min at 4°C in 0.1 ml staining buffer (PBS with 0.1% bovine serum albumin and 0.02% NaN_3) and 0.005 ml of monoclonal antibody, centrifuged, and washed. The cells were then incubated at 4°C for 30 min with 0.03 ml of a 1/40 dilution of fluorescein isothiocynate-(FITC) labelled goat anti-mouse IgG F(ab')$_2$ fragment (TAGO, Burlingame, CA), The fluorescence intensity of the lymphocytes was measured with an Ortho Spectrum III (Ortho Instruments, Westwood, MA), equipped with a 100 mW Argon Ion laser (488 nM wavelength), and recorded on a pulse-height analyzer. Gating was adjusted so that no viable cells were excluded. The percentage of positive cells was computer calculated from the fluorescent histogram. Five to ten thousand cells were counted for each antibody. A monoclonal mouse antibody directed at rat T-cells with a similar concentration of immunoglobulin (courtesy of Dr. Hiro Kimura) was used as a negative control.

RESULTS

Sensitivity of Interleukin-1 Assays

Figure 2 illustrates standard titration curves of IL-1 containing supernatants in two different IL-1 bioassays. The supernatants are from peripheral blood monocytes cultured at 10^6 per ml, and from villus placental cell cultures, also at 10^6 cells per ml media. Both cell types were treated with LPS (10 μg/ml) for 24 hr. LPS alone has no direct effect in either of these assays. Amount of IL-1 present in these supernatants is quantitated by 1/2 maximal units. Using mouse thymocyte assay,

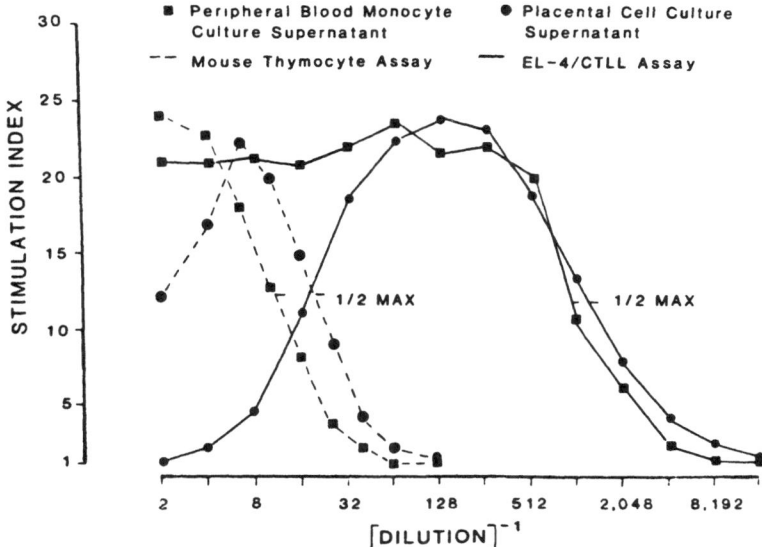

Figure 2. Comparison of sensitivity of IL-1 assays and of comparison of IL-1 production between peripheral blood monocytes and placental cell cultures. See text for details.

peripheral blood monocytes produce 9 half maximal units of IL-1 while placenta explant cultures produce 17 units. The EL4-CTLL assay proved much more sensitive and the same supernatants resulted in 1,020 half maximal units for peripheral blood monocytes and nearly 1,500 units for placental cell culture supernatants. This difference in sensitivity of nearly 2 orders of magnitude persisted in all comparative assays. Therefore, the EL4-CTLL assay was used subsequently. Both assays require that IL-2 not be present in the test supernatant as IL-2 can have direct stimulating activity and hence lead to falsely positive results. The presence of IL-2 was tested for in all of our culture supernatants using the CTLL assay and was found to be negative in each case. Another potential confounding factor is the presence of nonspecific inhibitors in placental culture supernatants. These were definitely present in the crude placental supernatant illustrated in Figure 2 by the delay in reaching maximal stimulation index for several dilutions. Therefore, serial dilutions were performed for all samples. As a standard, commercially available partially purified human monocyte IL-1 (Genzyme, Boston, MA) provided approximately 2,000 half maximal units at 1:10 dilution in the EL4-CTLL assay.

First Trimester Placental Production of Interleukin-1

We asked whether early first trimester villus fronds cultured as explants would produce IL-1. Table 1 illustrates a time course experiment asking whether this tissue would produce IL-1 spontaneously and whether a panel of known IL-1 stimulators would augment its production. Peak production appears to occur by 20 hr in culture

Table 1

IL-1 activity in first trimester villus cultures[a]

Time[b]	Dilution Required for ½ Maximum Stimulation			
	Unstimulated	LPS[c]	CON	PHA-P
5 hr	40	80	—[d]	—
20 hr	110	2400	840	750
29 hr	6	320	350	110
45 hr	5	200	200	20
72 hr	0	200	20	0

a   EL-4/CTLL Assay: Background = 145 cpm, maximum = 1,580 cpm.

b   Media replaced at each time interval, each point performed in triplicate.

c   Mitogen concentrations: LPS 10 µg/ml, CON A 10 µg/ml, PHA-P 2 µg/ml.
    Media with mitogen alone did not stimulate ^{3}H-thymidine incorporation above
    background.

d   Not examined.

with a fall off of IL-1 production after that time. There was significant production of
IL-1 by unstimulated villus cultures which was further augmented by culture with
mitogens. LPS, CON A, and PHA-P, all known stimulators of IL-1, were cultured at
their optimal concentrations (data not shown) with villus tissue. All appeared to have
stimulatory effects, with LPS being the most powerful augmentor.

Immunofluorescence of Trypsin Digest/Percoll Gradient Fraction

To address the question of which cell type(s) in the term placenta are capable of
producing IL-1, we utilized the technique of trypsin-DNAse digestion followed by
density centrifugation using a Percoll gradient perfected in our collaborating
laboratory. We further characterized the cells in the two predominant fractions
(1.045-1.055, "Fraction 2"; and 1.055-1.075, "Fraction 3") by indirect immunofluorescence using monoclonal antibodies against lymphocyte and monocyte cell
surface markers. Table 2 lists the antibodies and their source. In Table 3, we see the
results from two different experiments using these antibodies in indirect
immunofluorescence on an Ortho Spectrum III. None of the lymphocyte-monocyte
surface markers was present in more than 8% of the cytotrophoblast enriched
population and most antibodies were positive on less than 5% of the cells. This is in

Table 2

Antibodies used in immunofluorescence studies

Monoclonal Antibody	Source	Range of Expressions of Antigen
OKT-3	Ortho	Pan T-cells
OKT-8	Ortho	Suppressor T-cells
OK-Ia	Ortho	B cells, many monocytes, activated T-cells
BRL 63D3	BRL	Most monocytes
OK-M1	Ortho	Many monocytes, granulocytes, NK cells
Leu-M3	B-D	Many monocytes
Mouse anti rat T-cell	H. Kimura	Pan rat T-cells (negative control)

contrast to the band immediately below the trophoblasts that contained a mixture of lymphocytes, granulocytes, and monocytes as evidenced both by appearance and staining positively to this panel of antibodies.

As a further control to assess the possibility of enzymatic cleavage of cell surface markers, partially purified T-cells were subjected to the same trypsin-DNAse protocol. Immunofluorescence before and after enzyme exposure was then evaluated (Table 4). There was no loss of expression of any surface marker with this exposure of trypsin.

Using these characterized placental populations of highly enriched trophoblasts (fraction 2) and mixed lymphocytes, granulocytes, and monocytes (fraction 3), we assayed for IL-1 production. As seen in Table 5, the enriched trophoblast population was an excellent producer of IL-1 with LPS stimulation but also had significant production without stimulation. Fraction 3 had but modest production. In comparison, partially purified populations of peripheral blood (PB) T-cells had no IL-1 production and PB-monocytes had activity similar to that of the trophoblast enriched population on a cell to cell basis.

DISCUSSION

We and others have found the placenta to be an unexpectedly rich source of Interleukin-1. This raises important questions regarding (1) the role of IL-1 during maternal-fetal interaction, (2) the placental cellular site of IL-1 production, and (3) the mechanisms of regulation of placental IL-1 production. In the past several years, it has become clear that IL-1 has widespread effects on a variety of cell types and further, that direct lymphocyte activation is a relatively minor one (Dinarello, 1984).

Table 3

Indirect immunofluorescence of percoll fractions[a]

Fraction 2 (Trophoblast Band)	% Positive[b]	
Media	0.6	0.9
OKT-3	4.2	3.7
OKT-8	3.5	2.5
OK-IA	8.2	6.0
BRL 63 D3	2.6	3.1
OK-M1	7.2	4.5
Leu M-3	1.4	1.2
M α R	0.8	0.7
Fraction 3 (Lymphocyte Band)	% Positive	
Media	1.0	1.2
OKT-3	41.8	45.2
OKT-8	11.7	16.8
OK-IA	16.8	20.1
BRL-63 D3	12.1	14.5
OK-M1	23.0	19.8
Leu M3	14.2	13.0
M α R	0.4	0.5

a   Ortho spectrum III cytofluorograph.
b   Data from two different placental digenstions/percoll gradients are
     shown. 5-10,000 cells counted for each antibody.

It is truly a "cytokine" with an apparent key mediating role in chronic inflammation
and in host responses to microbial invasion. Systemic metabolic changes such as
enhanced production of acute phase reactants (e.g., fibrinogen, serum amyloid A) and
elevations in body temperature are IL-1 controlled. These are of particular interest as
gestation in characterized by chronic mild elevation in basal temperature and by
marked liver production of fibrinogen. Furthermore, what was once thought to be

Table 4

Indirect immunofluorescence: trypsin controls

T cells[a]	% Positive[b]
Media	1.9
OKT-3	80.0
OKT-8	18.6
OK-IA	9.4
M α R	1.6
T cells and Trypsin/DNASE[c]	% Positive
Media	0.2
OKT-3	78.1
OKT-8	16.1
OK-IA	8.9
M α R	0.9

[a] T-cells partially purified by removing plastic plate adherent cells.

[b] Ortho spectrum III cytofluorograph: 5-10,000 cells counted for each antibody.

[c] T-cells subjected to same enzyme digestion conditions as trophoblast.

purely a macrophage product has now been shown to be produced by a variety of cells in the nonmonocyte lineage.

In this context, we addressed several issues. Could early first trimester villus trophoblast produce IL-1? What cells type(s) within the placenta could produce IL-1? Our finding of strong production of IL-1 by normal early first trimester villi is of interest for two reasons. First, if IL-1 is important for processes of trophoblast invasion as might be suggested by its ability to induce collagenase and proteolysis, it obviously needs to be present early in gestation. Secondly, while ontogeny of human macrophage function is poorly understood it is probable that should fetal monocytes be present at 5-6 wks from conception, they would be immature.

We utilized a cell separation protocol that highly enriches for cytotrophoblast. The immunofluorscence results presented here indicating that placental monocytes

Table 5

Comparison of IL-1 production from different cell types[a]

Cell Type[b]	Dilution Required for $\frac{1}{2}$ Maximum Stimulation	
	Unstimulated	LPS[c]
Fraction 2 (cytotrophoblast enriched)	160	2200
Fraction 3 (lymphocytes, monocytes)	70	240
PB-T cells[d]	0	0
PB-Monocytes	4	1750

a   EL-4/CTLL Assay: Background = 120 cpm, Maximum = 2,700 cpm Representative of three experiments.

b 10^6 cells/ml cultured for each type.

c   Harvested at 24 hr from LPS pulse (10 µg/ml).

d   Peripheral blood T-cells and monocytes were partially purified by plate adherence.

are but a small (<7%) contaminant of our population confirms the extensive charac-terization of this population done in our department with electron microscopy, cytoplasmic immunohistochemistry, and subsequent production of HCG, HPL, SP-1, and formation of syncytia. While we did not have a monoclonal antibody to cytotrophoblast to use as a positive control (Kawata et al., 1984), using monoclonal antibodies, TROP-1 and TROP-2, characterized a similar population of cells. They also utilized low concentrations of trypsin and DNAse but instead of Percoll gradients they used to a flow cytometer for separation. Over 85% of their intermediate-large cells stained positively for TROP-1 and TROP-2 and 74% of all released cells regardless of size were positive for these antibodies. From the entire population only 8% were considered macrophages. Other groups (Wood, 1980; Flynn et al., 1982; Wilson et al., 1983); using collagenase and higher doses of trypsin identified a high proportion of cells to be macrophages. However, the techniques used for identification were relatively nonspecific. It has now been demonstrated that two of the major methods, surface expression of Fc receptors and cytoplasmic staining for nonspecific esterase can also be positive for trophoblasts (reviewed by Wild, 1983). Therefore, identifying cells as macrophages may not be a straight forward matter. On the other hand, the use of collagenase releases more cells from the connective tissue matrix and hence would lead to a higher proportion of macrophages being released. We also asked whether the use of trypsin in the digest striped the surface markers from the isolated placental cells making them falsely negative for macrophage markers. This is unlikely because the fraction 3 cells did stain positively for the panel of monocyte markers and lymphocytes treated with trypsin/DNAse exactly per the digest protocol did not have a significant change in surface marker expression.

What is the significance of trophoblast production of IL-1? We feel that there is increasing evidence to suggest that trophoblast may have some characteristics of a primitive reticulo-endothelial system. As noted above, several groups have demonstrated that trophoblasts are strongly esterase positive and Fc receptor positive. Production of interleukin-1 may be another example of shared functions. Secondly, "Interleukin-1" is really a misnomer and merely indicates that immunologists were the first to characterize this protein. IL-1 is now understood to be produced by many cell types and have activity in a wide range of systems. A recent review has suggested that IL-1's ability to augment T-cell responses may be among its least important functions (Dinarello, 1984). For the placenta, the IL-1 stimulation of collagenase and the production of prostaglandins may be important for processes of implantation and trophoblast invasion of the decidua as well as maintaining a vasodilated uterine vasculature.

SUMMARY

Interleukin-1 (IL-1) is a key mediator of inflammatory responses and regulator of prostaglandin synthesis. It appears to be a key cytokine controlling a variety of cell interactions. The human placenta has found to be an unexpectedly rich source of IL-1. Using the highly sensitive EL-4/CTLL bioassay, we have demonstrated IL-1 production as early as 5 wks from conception (earliest point tested). A highly enriched cytotrophoblast population was obtained using a recently well characterized trypsin-DNAse/Percoll gradient method. This population produced IL-1 at levels similar to those produced by a standard population of purified peripheral blood macrophages. These findings raise important issues regarding possible roles for IL-1 in the processes of implantation and trophoblast invasion of the uterus as well as suggesting additional capabilities for the already multi-potential trophoblast.

ACKNOWLEDGEMENTS

We wish to thank Drs. Harvey Kliman, Peter Nowell, and Philip Simon for helpful discussions of this project and Ms. Renee Whiting for preparation of this manuscript. Supported by NIH Grant 1-K08-HD-00599-01

REFERENCES

Dinarello, C.A. (1984) Interleukin-1. *Rev. Infect. Dis.* 61, 51-95.

Flynn, A., Finke, J.H., and Hilfiker, M.L. (1982) Placental mononuclear phagocytes as a source of interleukin-1. *Science* 218, 475-477.

Hall, C.St.G., James, T.E., Goodyer, C , Branchand, D., Guyda, H., and Giroud, C.J.P. (1977) Short term tissue culture of human midterm and term placenta: parameters of hormonogenesis. *Steroids* 30, 569-580.

Kawata, M., Parnes, J.R., Herzenberg, L.A. (1984) Transcriptional control of HLA-ABC antigen in human placental cytotrophoblast isolated using trophoblast and HLA specific monoclonal antibodies and the fluorescence-activated cell sorter. *J. Exp. Med.* 160, 633.

Kliman, H.J., Nestler, J.E., Sermasi, E., Sanger, J.M., and Strauss III, J.F. (1986a) Purification, characterization and in vitro differentiation of cytotrophoblasts from human term placentae. *Endocrinol.* 118, 1567-1582.

Kliman, H.J., Feinman, M.A., and Strauss III, J.F. (1986b) Differentiation of human cytotrophoblasts with syncytiotrophoblasts in culture. *Trophoblast. Res.*2:145-155.

Lachman, L.B. (1983) Human interleukin-1: purification and properties. *Fed. Proc.* 42, 2639-2645.

Lala, P.K., Chatterjee-Hasrouni, S., Kearns, M., Montgomery, B., and Colavincenzo, V. (1983) Immunobiology of the feto-maternal interface. *Immunol. Rev.* 75, 87-116.

Lipinski, M., Parks, D.R., Rouse, R.V., and Herzenberg, L.A. (1981) Human trophoblast cell-surface antigens defined by monoclonal antibodies. *Proc. Natl. Acad. Sci. USA* 78, 5147-5150.

Main, E.K., Schochet, P., and Strizki, J. (1985) Characterization of a novel placental suppressor of T-cell proliferation. *Am. J. Reprod. Immunol. Micro.* 7, 130.

Rosenwasser, L.J. and Dinarello, C.A. (1981) Ability of human leukocyte pyrogen to enhance phytohemagglutinin-induced murine thymocyte proliferation. *Cell Immunol.* 63, 134-139.

Rossi, V., Breviario, F., Ghezzi, P., Dejana, E., and Mantovani, A. (1985) Prostacyclin synthesis induced in vascular cells by interleukin-1. *Science* 229, 174-176.

Simon, P.L., Leydon, J.I., and Lee, J.C. (1985) A modified assay for interleukin-1 (IL-1). *J. Immunol. Methods* 84, 85-94.

Todd, R.F. and Schlossman, S.F. (1982) Analysis of antigenic determinants on human monocytes and macrophages. *Blood* 59, 775.

Wild, A.E. (1983) Trophoblast cell surface receptors, In: *Biology of Trophoblast*, Y.W. Loke and A. Whyte, (eds.), Elsevier Science, Amsterdam, pp. 473-487.

Wilson, C.B., Haas, J.E., and Weaver, W.M. (1983) Isolation, purification and characterization of mononuclear phagocytes from human placentas. *J. Immunol. Methods* 56, 305-317.

Wood, G.W. (1980) Mononuclear phagocytes in the human placenta. *Placenta* 1, 113-123.

Trophoblast Research 2:161-171, 1987

REGULATION OF MATERNAL ANTI-PATERNAL IMMUNE RESPONSES IN VITRO BY UTERINE MACROPHAGES

Gary W. Wood, Ossama W. Tawfik, and Joan S. Hunt

Department of Pathology and Oncology
University of Kansas Medical Center
39th Street and Rainbow Blvd.
Kansas City, Kansas 66103, USA

INTRODUCTION

The failure of elements of the maternal immune system to effect rejection of the semiallogeneic fetus remains a major biologic enigma. Currently, the immunologically privileged status of the mammalian embryo is believed to result from 1) failure of expression of histocompatibility antigens by fetal trophoblasts, thus providing an immunologically inert barrier between the mother and the fetus (Simmons and Russell, 1962; Jenkinson and Owen, 1980) and 2) localized immunosuppression in lymph nodes, blood and uterine tissue surrounding the fetus that prevents generation of a cellular immune response (Clark et al., 1984a, 1984b; Hunt et al., 1984b). Two major findings support those hypotheses: 1) trophoblast is poorly or nonstimulatory to lymphocytes (Jenkinson and Billington, 1974; Hunt et al., 1984a) and 2) a cellular immune response to fetal histocompatibility antigens is not generated during pregnancy (Wegmann et al., 1979) although the mother remains generally immunocompetent and antibodies are produced to fetal histocompatibility antigens.

In the current study, we have focused on the second hypothesis by studying the various cell types which exhibit suppressive activity subsequent to their isolation from uterine tissue of pregnant mice. Since the area of greatest contact between maternal and fetal tissues occurs within the uterus, if an effective immune response is to be generated against fetal cells, it is most likely to be initiated within that organ. A high degree of immunosuppressive activity by cells found within the uterus during successful pregnancy would imply that generation of an immune response within the uterine decidua would be difficult. The present study was designed to evaluate the role of decidual macrophages (Wood, 1980; Hunt et al., 1985) in providing an immunosuppressive environment within which the fetus is protected from the maternal immune response.

Received: 10 October 1985; Accepted: 8 July 1986

MATERIALS AND METHODS

Animals

Young adult (6-8 wks old) female BALB/c, male C57B1/6 and male and female Swiss/Webster mice were purchased from Harlan-Sprague Dawley, Inc., and were maintained at the University of Kansas Medical Center Animal Facility. Virgin BALB/c females were mated with C57B1/6 males and female Swiss/Webster with Swiss/Webster males. The onset of pregnancy was determined by the presence of a vaginal plug (day 1). The timed pregnant BALB/c or Swiss females were killed by cervical dislocation between days 14 and 19 of gestation and the uteri were removed to serve as sources of regulator cells. Normal virgin female BALB/c and normal C57B1/6 males served as spleen cell (SC) donors for the mixed lymphocyte reaction (MLR). Normal virgin Swiss/Webster females served as SC donors for phytohemagglutinin (PHA)-induced mitogenesis assays.

Preparation of Cell Suspensions

Uteri were removed and separated from fetal tissue as described previously (Hunt et al., 1984b). Tissue was finely minced, washed, suspended in collagenase (1 mg/ml Type I from Cooper Biomedical) plus Dispase (100 µg/ml, Sigma Chemical Co.) in medium (RPMI 1640 containing penicillin and streptomycin) and agitated in a tissue digestion flask (Wheaton) for 90 min at room temperature. Subsequent to digestion the uterine cells were washed with medium and suspended in medium supplemented with 2 mM glutamine and 10% fetal bovine serum (FBS) for PHA-induced mitogenesis assays or 2.5% type AB pooled human serum and 5×10^{-5}M 2-mercaptoethanol for MLRs. The yield of cells from uterine tissue ranged between 10^7 and 10^8 per gram (wet weight) of tissue, and viability was consistently above 90%.

SC were harvested by sieving intact spleen through stainless steel mesh into medium. SC were washed twice and suspended in the appropriate medium for each assay. SC were used without removal of adherent cells.

Phytohemagglutinin (PHA) Assays

PHA assays were performed in replicates of 4-6 in 96-well flat-bottomed microplates. SC were added to test wells at a concentration of 2×10^6 cells/ml in 0.1 ml. Regulator cells at 1.5×10^6 cells/ml were also added in a volume of 0.1 ml. PHA (Wellcome Laboratories, Beckingham, England) was added at a dilution of 1:20 in 0.02 ml/well. Cultures were incubated at 37°C in 5% CO_2, 95% air, for 64 hr. ^{3}H-methyl-thymidine (1 µCi/well) was added and incubation continued for an additional 8 hr. Cells were harvested and counted in a beta scintillation counter.

Mixed Lymphocyte Cultures (MLR)

BALB/c female responder cells were added to 96-well microplates at a concentration of 5×10^6 SC/ml in medium plus 2.5% type AB pooled human serum and 5×10^{-5}M 2-mercaptoethanol. C57B1/6 stimulator cells were suspended in the same medium at 5×10^6 SC/ml and received 2500 rads ^{61}Co. One-tenth ml of each was added to wells. Uterine cells were similarly irradiated, suspended in the same medium and cells were cocultured for 5 days under the conditions described above. Cultures were harvested as described for PHA assays.

Isolation and Depletion of Macrophages

Macrophages were isolated by plastic adherence or by specific panning and were depleted by those two procedures or by rosette depletion or cytotoxicity. In the first procedure, adherence to plastic dishes, uterine cells were suspended in medium plus 10% FBS and added to 10 cm diameter tissue culture dishes (5×10^6 cells/dish) and incubated for 60 min at 37°C in 5% CO_2, 95% air. Cells adherent to those dishes were greater than 95% Fc receptor positive macrophages (Hunt et al., 1984b). Ahderence to plastic under the stated conditions removed approximately 50% of the macrophages in uterine cell suspensions.

The second method for harvesting or depleting macrophages from uterine cell suspensions, panning, was more effective than was the first. The 10 cm diameter plastic tissue culture dishes were coated with rat IgM monoclonal anti-mouse macrophage reagent, B23.1, (LeBlanc et al., 1982) by diluting the antibody in distilled water, incubating the plates at 4°C overnight, then washing out excess antibody prior to use. Uterine cells were incubated on the plates as described above. This procedure also produced an adherent cell population which was greater than 95% macrophages and approximately 90% of the macrophages were removed from uterine cell suspensions. Adherent cells were recovered from coated dishes by removing nonadherent cells and vigorously washing the monolayer. Adherent cells were exposed to 0.25% trypsin for 2-3 min, trypsin was neutralized with 1 ml of undiluted FBS, cells were scraped from the dish with a plastic scraper and washed with medium.

Macrophages were also depleted from uterine cell suspensions by EA rosette formation using antibody-coated sheep erythrocytes (SRBC) as described previously (Lindsay et al., 1982). SRBC were coated with rabbit IgG anti-SRBC at a subagglutinating concentration. Uterine cells at a concentration of 5×10^6 cells/ml in medium plus 10% FBS were mixed with an equal volume of 5% EA in phosphate buffered saline (PBS) for 5 min at 37°C, then centrifuged at 200 g for 5 min. EA rosetted cells were suspended to the original volume in a 1:25 dilution of goat anti-rabbit IgG in PBS. The suspension was centrifuged at 200 g for 5 min, to enhance agglutination and the cells were resuspended. Rosetted cells were allowed to sediment at unit gravity for 10 min at room temperature. Nonrosetted supernatant cells were collected, recentrifuged at 200 g for 5 min, resuspended gently and resedimented at unit gravity. Supernatant cells were treated with ammonium chloride to lyse SRBC. Those cells were 95% depleted of macrophages as determined by the procedures described above.

In some experiments macrophages were removed from uterine cell suspensions by cytotoxicity using the monoclonal anti-macrophage reagent B23.1, a rat IgM antibody (LeBlanc et al., 1982). Uterine cell suspensions were incubated for 60 min at 4°C with B23.1 diluted 1:50, then rabbit Low-Tox complement (Accurate Scientific, Inc.) diluted 1:20 was added to the mixture and the cells incubated an additional 45 min at 37°C. The resultant cell suspension contained 30-50% dead cells which were removed by centrifugation over Histopaque (specific density 1.077, Sigma). The cytotoxicity protocol has been described in detail (Tawfik et al., in press).

Separation of Uterine Cells by Countercurrent Centrifugation (Elutriation)

Cells were separated using an elutriation rotor (Model JE-6B, Beckman Instruments). Cells were loaded at ambient temperature as described in the rotor instruction manual. Ten ml of sample in RPMI 1640 with 10% FBS were loaded. The medium employed for elutriation was PBS supplemented with 0.5% FBS. The centrifugation was performed at 4°C. The completed system was flushed with 70% ethanol, and all operations were performed using aseptic technique in order to reduce potential contamination. All experiments were performed at a rotor speed of 3260 rpm. Ten fractions were collected in each experiment with flow rates respectively set at 17, 21, 24, 30, 34, 41, 46, 53, 68, and 75 ml/min as described previously (Meistrich et al., 1981).

Quantiation of Macrophages in Cell Suspensions

Two methods were used to quantitate macrophages. Previous studies had demonstrated that macrophage antigen positive and Fcγ receptor positive cell populations in the uterus are the same (Hunt et al., 1985). In the current study, Fcγ receptor positive cells were quantitated by EA rosette formation as described previously (Hunt et al., 1985) and by indirect immunofluorescence using a polyclonal rabbit anti-mouse macrophage serum as described previously (Hunt et al., 1985).

RESULTS

Decidual Cells Suppress Spleen Cell Responses to PHA

Initial studies in our laboratory used PHA-stimulated lymphocyte mitogenesis as the assay to measure immunoregulation by murine uterine cells. We chose that assay because it is simple, reproducible, has a short incubation period and, at least in theory, is relatively difficult to suppress since greater than 90% of all T lymphocytes are stimulated by PHA to divide. Uterine cells from pregnant Swiss/Webster mice were extremely suppressive of PHA-induced mitogenesis. T cell blastogenesis and multiplication were inhibited by greater than 95% whether regulator cells were added at the beginning of cultures (day 0) or on day 3 (n, 5).

Other studies in our laboratory (Wood, 1980; Hunt et al., 1985) had established that macrophages are a major component in the decidual response to pregnancy and that macrophages are the only cells in the pregnant uterus to express receptors for the Fc portion of IgG. Since macrophages have been shown to be immunosuppressive in a variety of assays (Kirchner, 1978; Wood et al., 1983; Goodwin and Cueppens, 1983), we attempted to determine the contribution of macrophages to uterine cell-mediated immunosuppression. Macrophages were removed from cell suspensions by EA rosetting, a procedure that requires expression of Fcγ receptors by macrophages (Lindsey et al., 1982) and results in removal of >90% of the macrophages in uterine cell suspensions. Unfractionated cell suspensions were highly suppressive (3-22% of control values; n, 4). Removal of macrophages provided partial to complete relief of that suppression (32-115% of control values). All increases were significant (P < .05) when unfractionated and macrophage-depleted values in cpm were compared using the paired t test.

Macrophages are known to mediate nonspecific suppressive effects on T lymphocytes through the release of prostaglandin E_2 (Goodwin and Cueppens, 1983). To determine if the suppressive effects of uterine cells were mediated by prostaglandins, indomethacin, a cyclooxygenase inhibitor, was included in the cultures. The suppression of PHA-induced mitogenesis by uterine cells was unaffected by addition of indomethacin (data not shown). When uterine cells were incubated for 24 hr in vitro, and the supernatants of those cultures added to the PHA blastogenesis cultures, mitogenesis was inhibited to a degree comparable to that accomplished by uterine cells (data not shown). We did not test the effect of indomethacin on the generation of those suppressive factors.

Decidual Cells Suppress Mixed Lymphocyte Reactions

The PHA-induced mitogenesis system was useful for preliminary studies, but that assay does not parallel the normal allogeneic situation present during pregnancy. To approximate the conditions operative in vivo between genetically disparate maternal and paternal cells, we utilized mixed lymphocyte reactions between responder BALB/c ($H-2K^dD^d$) SC and irradiated C57Bl/6 ($H-2K^bD^b$) stimulator SC. Uterine cells from BALB/c females (syngeneic to responder) pregnant by C57Bl/6 males were used as regulator cells. Regulator cells were added on the final day of 5-day MLR cultures because regulator cells are effective under those conditions and because that approach avoided the possibility that suppressor cells were activated during the prolonged culture period.

Suspensions of cells from pregnant uteri were highly suppressive of the MLR as shown in Figure 1, Experiment 1, which gives the values obtained in one of two similar experiments. Uterine cells from virgin BALB/c females were slightly suppressive and uterine cells from ovariectomized mice were very poorly suppressive of MLRs. Macrophage percentages in those suspensions were 25% (pregnant), 8-10% (virgin) and 2-3% (ovariectomized). Isolated macrophages were highly suppressive of the MLR but complete adherence reduction of macrophages from uterine cell suspensions did not relieve that suppression (Experiment 2, Figure 1). Experiment 2, Figure 1 shows data collected in a single experiment performed when uterine cells were added at day 4 of culture. Similar results were obtained when cells were added at day 0 (n, 3) and when macrophages were depleted by cytotoxicity using B23.1 anti-mouse macrophage reagent plus complement (n, 3) then added at day 0 or day 4 of the MLR.

The above findings suggested that in addition to macrophages, other uterine cells were capable of inhibiting MLRs. That interpretation was supported by the results obtained when uterine cell suspensions were fractionated by size and density using elutriation centrifugation (Figure 2). Early fractions containing small lymphocyte-like cells that failed to express T or B cell markers were poorly suppressive whereas middle fractions (4-6) which were rich in macrophages were extremely suppressive. The surprising finding from those studies was that large, less dense cells contained in late fractions (7-10) in which macrophage content was minimal were also quite suppressive.

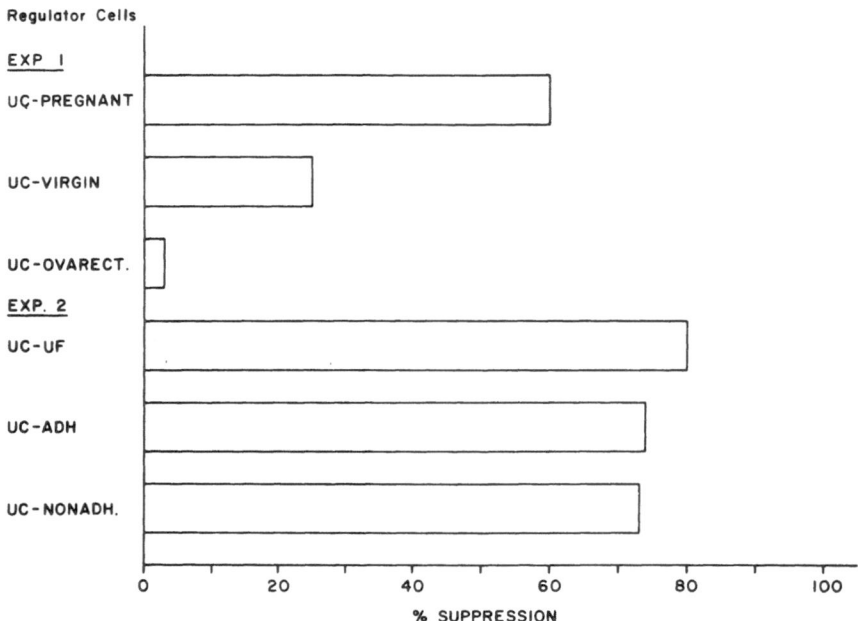

Figure 1. The suppressive capacity of uterine cells (UC) from pregnant, virgin, and ovariectomized BALB/c mice was compared in Experiment 1 and the suppressive capacity of unfractionated (UF), adherent (ADH) and nonadherent (NONADH) cells compared in Experiment 2. Both sets of data were acquired using the mixed lymphocyte reaction (MLR, BALB/c x C57B1/6) and BALB/c uterine cells. 1×10^5 cells/well were added to the MLR at day 4 of a 5 day culture in replicates of 4-6. Adherent and nonadherent cells were acquired by panning, and all cells in Experiment 2 were from allogeneically pregnant mice. Percent suppression was calculated using the following formula. Standard deviations were <10% of the mean for each data point.

$$\% \text{ suppression} = 1 - \left[\frac{\text{cpm (responder x stimulator + regulator)}}{\text{cpm (responder x stimulator)}} \right] \times 100$$

Relief of Decidual Cell-Mediated Immunosuppression by Indomethacin

The data from a representative experiment (n, 10) that tested the ability of supernatants of uterine cells to inhibit MLRs is shown in Figure 3. As had been observed with PHA-induced mitogenesis, supernatants of uterine cells incubated in vitro for 36 hr were highly suppressive. When indomethacin was included in the culture medium during supernatant generation, the suppressive capacity of the supernatants was frequently, but not invariably, abrogated. In approximately 20% of the experiments, only partial relief was observed. Those results suggested that prostaglandins were the major, but probably not the only soluble suppressor factor generated by decidual cells.

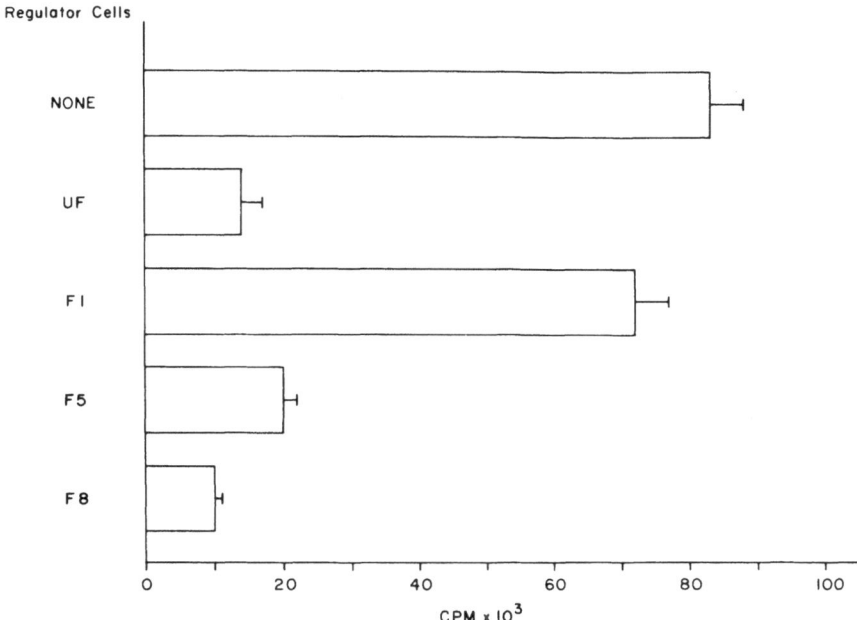

Figure 2. Suppression of the BALB/c x C57Bl/6 MLR by unfractionated (UF) and counterflow elutriator-fractionated uterine cells from allogeneically pregnant mice. The results of adding 1 x 10^5 cells/well at day 4 of a 5 day MLR are shown for fractions, 1, 5, and 8. P values were calculated using the 2 sample t test, comparing test (UF, F1, F5, F8) values with control (none) values (UF: P = 0.00001; F1: P = 0.1305-NS; F5: P = 0.00002; F8: P = 0.00001).

As was observed in the PHA system, when indomethacin was included with decidual cells in the MLR, relief of the suppression of MLRs by decidual cells was variable and incomplete. In six experiments, relief of suppression ranged from none to 50% although total relief was obtained in one experiment. Those results suggested that some suppression in vitro was mediated by short range factors or by direct cell to cell contact.

DISCUSSION

In the present studies we have observed that murine decidual cells and the in vitro generated supernatants of those cells are suppressive to lymphocyte proliferative responses to lectins and to allogeneic cells, thus confirming and expanding the findings of others. Further, we have initiated experiments to 1) identify the types of cells that are responsible for the production of suppressive substances, 2) identify the suppressor molecules, and 3) identify the mode of action of suppressive substances.

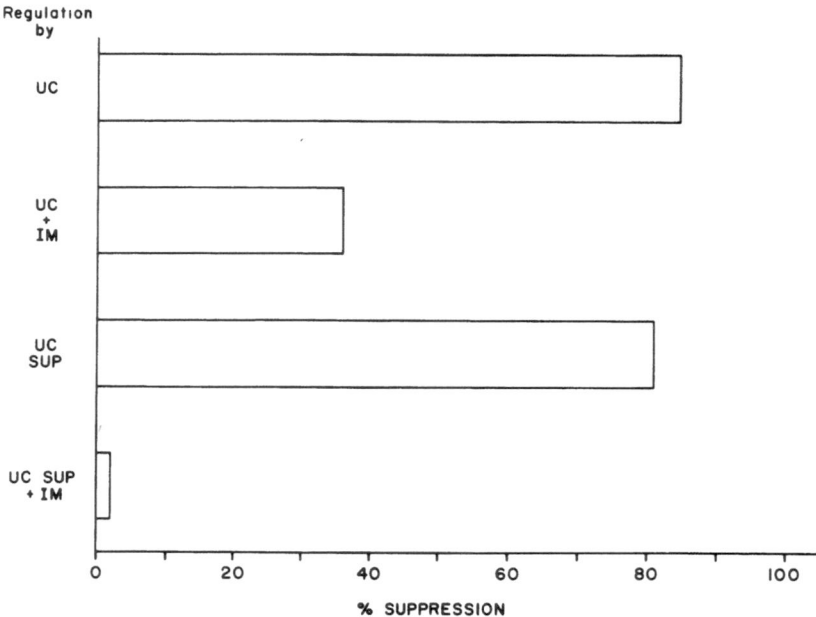

Figure 3. Indomethacin (IM) prevents generation of soluble suppressor factors in supernatants (SUP) of uterine cells (UC), but only partially relieves suppression of mixed lymphocyte reactions (MLRs) by intact UC. Thirty-six hr SUPs of UC incubated with or without 1 µg/well IM were generated and 0.1 ml of SUP/well added to MLRs. 1×10^5 UC or SUP was added to day 0 of the 5 day BALB/c x C57B1/6 MLR. Mean values in cpm ^{3}H-thymidine were used to calculate percent suppression (Figure 1). The standard deviations for 4-6 replicates wells were less than 10% of the mean value.

Fractionation of cells from near-term pregnant uteri by elutriation centri-fugation, which distinguishes cells on the basis of size and density, demonstrated that the murine decidua contains several immunosuppressive cell types. Macrophages were indirectly implicated as a major immunoregulatory decidual cell type by data showing that tissues (virgin uterus, uteri from ovariectomized mice) containing low levels of macrophages yield cells with significantly less ability to suppress MLRs than tissues with high levels of macrophages (pregnant uterus, Hunt et al., 1985). In addition, macrophages purified from decidual cell suspensions by adherence to plastic or by specific panning over a rat anti-mouse macrophage monoclonal antibody, B23.1, were extremely suppressive of maternal anti-paternal MLRs. However, when decidual cells were fractionated by counter-current centrifugation, we found that macrophages were nearly equaled in suppressive ability by large decidual cells, which have recently been identified as the major immunoregulatory cell type in early human decidua (P.K. Lala, personal communication). Some suppression was also mediated by small lymphocyte-like cells in early fractions that may be equivalent to the suppressor cells identified by Clark and coworkers (Clark et al., 1984a, b).

Experiments with supernatants of murine decidual cells documented the suppressive nature of those supernatants on generation of MLRs, and variable inhibition of generation of soluble suppressor factors by the cyclooxygenase pathway inhibitor indomethacin. Characterization studies currently underway in our laboratory suggest that suppressor factors are found primarily in the lipid-rich fractions of decidual cell supernatants rather than in protein-rich fractions. Although suppression has been shown not to be attributable to products of the lipoxygenase pathway, some results, such as failure of indomethacin invariably to prevent suppression of MLRs by decidual cell supernatants, suggest that a second suppressor substance may accompany PGE_2. That possibility is further supported by our finding that suppression of the PHA response by decidual cells could not be inhibited by including indomethacin in the assays. Nevertheless, all avenues of experimentation implicate PGE_2 as a major suppressor factor in decidual cell supernatants. Finally, in preliminary experiments, we have observed that PGE_2 (as identified by radioimmunoassay) is produced by decidual cells from all elutriation fractions, and that the highest levels of PGE_2 on a per cell basis are generated by the macrophage-rich middle fractions. PGE_2 is known to suppress lymphocyte proliferation, but the mode of action of that molecule is still under examination.

Several studies have demonstrated that fetal resorption occurs when pregnant animals are exposed to indomethacin (O'Grady et al., 1972). This observation established the importance of prostaglandin production in the maintenance of pregnancy, but the mechanism(s) by which prostaglandins support the state of pregnancy is not known. The data from our present studies suggest that an early proliferative event in the maternal immune response to fetal antigens may be negated by prostaglandins. Certainly, production of soluble suppressor factors by in vitro cultures of macrophages and other decidual cells was adversely affected by including indomethacin in the culture medium. It is possible that failure of production of prostagladins could lead to termination of pregnancy by immunologic means, but there is as yet no experimental evidence in support of that postulate. The decidual macrophage, which is present in high density in the pregnant uterus and which produces high levels of prostaglandins, may play a central role in prevention of immunologic rejection of the allogeneic fetus throughout pregnancy.

SUMMARY

A major theory to explain survival of the semiallogeneic fetus in a potentially harmful maternal immunologic environment is that local immunosuppression is generated in uterine tissue surrounding the fetus during pregnancy. The current study was designed to study the cellular source of suppressive activity which is associated with pregnancy. Collagenase-digested uteri from pregnant mice yielded cells highly inhibitory to T lymphocyte proliferation following nonspecific mitogen or allogeneic cell stimulation whereas uterine cells from virgin or ovariectomized mice were very slightly or nonsuppressive. Cell suspensions from pregnant uteri contained 25% macrophages, but virgin uteri yielded only 10% and uteri from ovariectomized mice contained only 2-3% macrophages. Purified macrophages were highly suppressive in both assay systems. A second population of large uterine cells were also extremely suppressive of maternal vs. paternal mixed lymphocyte cultures. Both macrophages and the large uterine cells produced soluble factors which were actively suppressive in the absence of intact cells. Soluble factor production was blocked by indomethecin, indicating that suppression may be mediated by prostaglandins.

ACKNOWLEDGEMENTS

Supported by NIH grant #17678HD.

REFERENCES

Clark, D.A., Slapsys, R., Croy, B.A., Krcek, J., and Rossant, J. (1984a) Local active suppression by suppressor cells in the decidua. *Am. J. Reprod. Immunol.* 5, 78-83.

Clark, D.A., Slapsys, R., Croy, B.A., and Rossant, J. (1984b) Immunoregulation of host-versus-graft responses in the uterus. *Immunol. Today* 5, 111-115.

Goodwin, J.S. and Ceuppens, J. (1983) Regulation of the immune response by prostaglandins. *J. Clin. Immunol.* 3, 295-315.

Hunt, J.S., King, C.R., Jr., and Wood, G.W. (1984a) Evaluation of human chorionic trophoblast cells and placental macrophages as stimulators of maternal lymphocyte proliferation in vitro. *J. Reprod. Immunol.* 6, 377-391.

Hunt, J.S., Manning, L.S., and Wood, G.W. (1984b) Macrophages in murine uterus are immunosuppressive. *Cell. Immunol.* 85, 499-510.

Hunt, J.S., Manning, L.S., Mitchell, D., Selanders, J.R., and Wood, G.W. (1985) Localization and characterization of macrophages in murine uterus. *J. Leuko. Biol.* 35, 87-94.

Jenkinson, E.J. and Billington, W.D. (1974) Differential susceptibility of mouse trophoblast and embryonic tissue to immune cell lysis. *Transplant.* 18, 286-294.

Jenkinson, E.J. and Owen, V. (1980) Ontogeny and distribution of major histocompatibility complex (MHC) antigens on mouse placental trophoblast. *J. Reprod. Immunol.* 2, 173-180.

LeBlanc, P.A., Russell, S.W., and Chang, S.M.T. (1982) Mouse mononuclear cell heterogeneity detected by monoclonal antibody. *J. Reticuleondethel. Soc.* 32, 219-228.

Lindsay, J.M., Manning, L., and Wood, G.W. (1982) Exclusive binding of immunoglobulin to Fc receptors on macrophages in 3-methylcholanthrene induced murine tumors. *J. Natl. Canc. Inst.* 69, 1163-1174.

O'Grady, J.P., Caldwell, B.V., Auletta, F.J., and Speroff, L. (1972) The effects of an inhibitor of prostaglandin systhesis (indomethecin) on ovulation, pregnancy, and pseudopregnancy in the rabbit. *Prostaglandins* 1, 97-106.

Simmons, R.L. and Russell, P.S. (1962) The antigenicity of mouse trophoblast. *Ann. NY Acad Sci.* 99, 717-732.

Tawfik, O.W., Hunt, J.S., and Wood. G.W. (1986) Partial characterization of uterine cells responsible for suppression of murine maternal anti-fetal immune response. *J. Reprod. Immunol.*, in press.

Wegmann, T.G., Waters, C.A., Drell, D.W., and Carlson, G.A. (1979) Pregnant mice are not primed but can be primed to fetal alloantigens. *Proc. Nat. Acad. Sci. USA* 76, 2410-2413.

Wood, G.W. (1980) Immunohistologic identification of macrophages in murine placentae, yolk sac membranes and pregnant uteri. *Placenta* 1, 309-317.

Wood, G.W. and Morantz, R.A. (1983) Depressed T lymphocyte function in brain tumor patients: monocytes as suppressor cell. *J. Neuro-Oncol.* 1, 87-94.

Trophoblast Research 2:173-185, 1987

CHARACTERIZATION OF MATERNAL TROPHOBLAST LYMPHOCYTE CROSS-REACTIVE (TLX) IMMUNITY

Donald S. Torry[1,2,4], John A. McIntyre[1,2], W. Page Faulk[2], and Peter R. McConnachie[3]

[1]Department of Medical Microbiology and Immunology
Southern Illinois University School of Medicine

[2,4]Methodist Center for Reproduction and Transplantation Immunology
Methodist Hospital of Indiana
Indianapolis, Indiana

[3]Immuno Transplant Laboratory
Memorial Medical Center
Springfield, Illinois

INTRODUCTION

Women suffering from idiopathic recurrent spontaneous abortions may be grouped as primary (1°) or secondary (2°) aborters (McIntyre and Faulk, 1984; McIntyre et al., 1984). Women classified as 2° aborters have had children, stillbirths or premature births before their abortions began. Unlike 1° aborters, 2° aborters do not share HLA antigens with their spouses and they have high-titred complement dependent cytotoxicity (CDC) and antibody dependent cell-mediated cytotoxicity (ADCC) reactivity against paternal and non-paternal lymphoyctes even when not pregnant (McConnachie and McIntyre, 1984; McIntyre et al., 1984; McIntyre et al., 1986). Although multiparous women are the primary source of HLA typing sera, only 20-60% actually produce demonstrable antibodies and most of these disappear soon after delivery (Amon and Kostyu, 1980). Study of 2° aborters' antibodies on HLA select cell panels before and after absorption with trophoblast membranes has shown patterns of reactivity which indicate specificities for trophoblast lymphocyte cross-reactive (TLX) antigens (McIntyre and Faulk, 1982; McIntyre et al., 1983; McIntyre et al., 1985; Faulk and McIntyre, 1983).

Defined trophoblast membrane preparations remove CDC activity from some but not all 2° aborter sera (McIntyre et al., 1985). This observation coupled with results in the present report which show no correlation between absorption-positive trophoblast and HLA typed target cells suggest TLX represents an important allogeneic antigen in mammalian reproduction (Faulk and McIntyre, 1983; Faulk et al., 1978). IgG antibodies to TLX in 2° aborter sera mediate CDC and ADCC (Torry et al., 1985a; Torry et al., 1985b) and the expression of CDC is under the control of serum inhibitors (Torry et al., 1985c; Torry et al., 1986). These serum inhibitors become

[4]To whom correspondence should be addressed.
Received: 10 October 1985; Accepted: 1 March 1986

activated following serum fractionation, heat treatment (56°C for 30 min) or absorption with heparin covalently linked to Sepharose (Torry et al., 1986). Activation of these inhibitors in vitro by heparin may explain preliminary clinical data which suggest that heparin therapy affords protection for 2° aborters from subsequent miscarriages (McIntyre et al., 1985, 1986).

In this report we provide evidence that 2° aborters produce antibodies to allotypic TLX antigens and we show the IgG subclasses of anti-TLX which are responsible for CDC and ADCC activity in 2° aborter sera. These results support our hypothesis that some antibodies are under the control of serum inhibitors (McIntyre and Faulk, 1985b; Faulk and McIntyre, 1985). We also report characteristics of an inhibitor responsible for altering antipaternal CDC reactions.

MATERIALS AND METHODS

Sera, Plasma, and Lymphocyte Preparation

Blood samples from 18 2° aborting couples and 10 control individuals were collected in vacutainer tubes (Becton-Dickinson, Rutherford, NJ). Plasma and sera were separated within 1 hr and kept at 4°C or frozen (-70°C) until studied. Lymphoyctes were isolated following centrifugation over Lymphocyte Separation Media (Litton Bionetics, Keningston, MD). HLA typing was done by using standardized microlymphocytotoxicity techniques (McIntyre et al., 1984).

Trophoblast Membrane Preparation and Serum Absorptions

Trophoblast membranes were prepared according to the technique of McIntyre et al. (1985). Trophoblast absorptions were done by mixing 40 mg of a trophoblast preparation with 275 µl of maternal serum for 30 min at room temperature: heparin absorptions were performed by mixing 40 mg of heparin-Sepharose CL-6B (Pharmacia, Piscataway, NJ) in 0.5 ml aliquots of 2° aborter sera for 30 min at room temperature; supernatants to be assayed for CDC activity were collected following centrifugation at 12,000 x g for 3 min.

Isolation of IgG and IgG Subclasses

DEAE ion-exchange chromatography was used to isolate gamma globulins according to Garvey et al. (1977). Fractions were monitored at 280 nm and the major protein peaks were pooled, concentrated and dialyzed against phosphate buffed saline (PBS) pH 7.4 (Difco Labortories, Detroit, MI). Immunoglobulin classes were determined by immunodiffusion in 0.5% agarose A (Pharmacia, Sweden) with rabbit antisera to human IgG, IgA, and IgM (Accurate Chemicals, Westburg, NY).

Staphylococcal protein A (SPA) coupled to Sepharose CL-4B (Pharmacia, Sweden) was used to separate IgG3 from IgG1, 2 and 4 (Skvaril, 1976). DEAE fractions (10 ml) were passed through a column (1.9 x 14.5 cm) of SPA with 0.1 M glycine, pH 8.0. IgG1, 2 and 4 were eluted with 0.1 M glycine, pH 3.0, and the fractions were adjusted to pH 7.3 with 0.1 M glycine, pH 9.1, concentrated to 10 ml and dialyzed against PBS, pH 7.4

CDC and ADCC Assays

Complement dependent cytotoxic (CDC) antibodies were determined by two methods: dye exclusion (CDCE) (Amos, 1979) and ^{51}Cr release (CDC^{51}Cr). A modification of the ^{51}Cr release assay orginally described by Rogentine and Plocinik (1979) was used as described by McConnachie et al. (1979). Inhibition studies of 2° aborter antipaternal CDC by monoclonal antibodies to human IgG subclasses (Gibco Laboratories, Grand Island, NY) or inhibitor containing fractions were performed by incubating the reagents with maternal serum for 15 min before adding labeled target cells and complement. Neither the monoclonal antibodies nor the inhibitory fractions themselves were cytotoxic for human lymphocytes in either CDC or ADCC systems. Horse anti-human thymocyte globulin (ATG, Upjohn, Kalamazoo, MI) was used as a positive control serum.

Euglobulin Precipitation and Gel Filtration

Distilled H_2O or 2% boric acid were used to precipitate euglobulins (Garvey et al., 1977). The precipitates were solubilized in 0.1 M Tris saline, pH 8.0, and dialyzed overnight in Tris-saline. The solubilized euglobulins were fractionated through a column (2.6 x 29 cm) of superfine Sephadex G-200 (Pharmacia, Sweden); the fractions were monitored at 280 nm for protein content.

RESULTS

Trophoblast Absorption of 2° Aborter Sera

Absorption with trophoblast differentially removed lymphocytotoxicity from 2° aborters. The results from these experiments can be shown best by presenting the findings from a representative study as follows:

Sera from two 2° aborters (KS and PM) and two chorion preparations (#54 and #91), one able to remove antipaternal CDC activity (absorption positive) and one not altering CDC (absorption negative), were used to determined if a correlation between HLA types of target lymphocytes and chorion sources would explain the absorption capabilities of trophoblast preparations. HLA types of the chorion sources were determined by typing cord blood lymphocytes.

One absorption of 2° aborters KS and PM sera with chorion #54 removed most of their antipaternal CDC activity while chorion #91 produced little effect (Table 1). Retrospective comparisons of HLA types of trophoblast sources and paternal target cells showed chorion #54 to share no HLA antigens with either husband target cells while chorion #91 shared A1 and B8 with the husband of PM and an A1 with the husband of KS.

Unabsorbed KS and PM sera showed reactivity with 51% and 60%, respectively, of a 55 membered HLA select cell panel. Absorption of KS sera with chorion #54 resulted in a 69% loss of cell panel reactivity while absorption with chorion #91 produced no change in panel reactivity (Table 1). Similarly, absorption of PM sera with chorion #54 resulted in a decrease of 73% reactivity while chorion #91 only produced a 30% reduction in cell panel reactivity.

Table 1

Cell panel and antipaternal CDC reactivity of two
2° aborter sera before and after trophoblast absorptions

Absorption with:	Cell Panel Reactivity[a]		Antipaternal Reactivity[b]	
	KS Sera	PM Sera	KS Sera	PM Sera
Nothing	51%	60%	42.6%	56.7%
Chorion #54	20% [-69%][c]	16% [-73%]	8.7% [-80%]	2.1% [-96%]
Chorion #91	51% [0%]	42% [-30%]	40.0 [-6%]	49.8% [-12%]

a. Expressed as % CDCE of 55 member cell panel reactive with respective 2°
    aborter sera.
b. Expressed as % ^{51}Cr release from respective husband target cells.
c. [%] represents % change in reactivity.

Relationships between the HLA type of chorion sources #53 and #91 with
target cells and the absorption potentials of chorion preparations were further studied
by analyzing reactive target cells of the cell panel. Sera from 2° aborter KS was
reactive with 28 out of 55 members (51%) of the cell panel (Table 1). These positive
reactive cells could be divided into two groups; those sharing one or more HLA
antigens with the husband of KS, and those having no HLA antigens in common with
the husband (Table 2). Absorption of KS sera with chorion #54 removed 62% of the
reactivity against cells sharing HLA antigens and 60% of the reactivity against cells
not sharing HLA antigens with the husband (Table 2). Chorion #91 did not alter
panel reactivity. Similar results were obtained with PM sera (data not shown).

Table 2

Effects of trophoblast absorptions on positive reactive panel cells
sharing or not sharing paternal HLA phenotypes*

Absorption with:	Sharing HLA (n = 13)	Not Sharing HLA (n = 15)
Nothing	13	15
Chorion #54	5 [-62%]	6 [-60%]
Chorion #91	14 [+7%]	14 [-7%]

* Expressed as % CDCE of 55 member cell panel.

Table 3

Determination of immunoglobulin class responsible for
2° aborter antipaternal CDC^{51}Cr

Pooled DEAE Fraction #	Ig Class Present	% Specific ^{51}Cr release
1	IgG, IgA	51.5 ± 0.1
2	IgA	9.8 ± 0.3
6	IgA	4.3 ± 0.6
8	IgM	0.8 ± 1.9
9	None	3.1 ± 0.5
Unfractionated Serum	IgG, IgM, IgA	53.0 ± 5.4

The ability of chorion #54 to equally reduce reactivity against 3rd party cells sharing and those not sharing HLA antigens with the husband of KS show that reduction of antipaternal CDC following trophoblast absorptions is not due to the removal of HLA reactive antibodies. Further, the differing absorption capabilities between chorions #54 and #91 support the hypothesis that trophoblast express antigens concomitantly expressed allotypically on peripheral blood lymphocytes (McIntyre and Faulk, 1982).

Immunoglobulin Class and Subclasses Responsible for CDC and ADCC Activity

Ion-exchange chromatography of 2° aborter sera showed antipaternal CDC to be mediated by IgG (Table 3). The polyspecific cell panel reactivity manifested by 2° aborters was also determined to be mediated by IgG: similar results were obtained with other 2° aborter sera. Mouse monoclonal antibodies specific for human IgG 1, 2, 3, or 4 were used to determine which subclasses of IgG mediated 2° aborter antipaternal CDC. Addition of anti-IgG1 resulted in an average of 93.7% inhibition of 2° aborter antipaternal CDC activity (Table 4): anti-IgG3 reduced antipaternal cytotoxicity by 36.4% while anti-IgG2 or 4 were less effective. The subclass specific monoclonals did not inhibit CDC of paternal lymphocytes mediated by ATG, indicating the reduction of CDC mediated by 2° aborter antibodies was not due to anticomplementary effects of the mouse monoclonals.

The IgG subclasses responsible for 2° aborter antipaternal ADCC activity and additional evidence for the subclasses involved in antipaternal CDC^{51}Cr were determined by using SPA chromatography to separate IgG1, 2, and 4 from IgG3. Following passage of DEAE fraction #1 through SPA-Sepharose, 18.8% of the CDC^{51}Cr activity could be recovered in the IgG3 containing effluent, while 90.4% was recovered in the IgG1, 2, and 4 containing eluate (Table 5). The majority of antipaternal activity in the eluate is most likely due to IgG1, since the monoclonal

Table 4

IgG subclass analysis of a representative 2° aborter's
antipaternal CDC[51]Cr reactivity

Added Subclass Reagents	% Specific [51]Cr Release[a]	% Inhibition[b]
None	33.5 ± 12.3	0.0
IgG1	2.1 ± 0.6	93.7
IgG2	30.1 ± 8.7	10.2
IgG3	21.3 ± 8.9	36.4
IgG4	31.0 ± 9.7	7.5

a.   Results represent an average of four experiments run in duplicate, $x \pm$ S.D.

b. $$\% \text{Inhibition} = \frac{\% \text{Specific }^{51}\text{Cr Release of Subclass Treated Sera}}{\% \text{Specific }^{51}\text{Cr Release of Untreated Sera}} \times 100\%$$

antibody results (Table 4) suggest that IgG2 and 4 contribute little CDC[51]Cr activity. Approximately 90% of the ADCC activity in DEAE fraction #1 could be recovered in the eluate while the IgG3 enriched SPA effluent demonstrated no ADCC activity (Table 5). Addition of mouse monoclonal antibodies did not alter antipaternal ADCC activity.

Identification of a Serum Inhibitor to CDC[51]Cr

Although IgG1 and IgG3 appear to be responsible for 2° aborter CDC[51]Cr activity, heat treatment (56°C for 30 min) and heparin absorption of 2° aborter sera abolished antipaternal CDC (Table 6) but the same manipulations performed on IgG containing DEAE fractions caused no loss of CDC[51]Cr activity. Heating or heparin absorptions of subclass enriched SPA fractions also resulted in no loss of activity.

These findings suggest that heat treatment and heparin absorption perform a similar function in serum but not in chromatographically isolated IgG. Incorporation of a wash step following heat treatment restored cytotoxicity to the same serum and plasma, thus, heat treatment does not act directly upon antipaternal antibodies, but seems to act on a serum constituent which inhibits 2° aborter CDC[51]Cr reactivity. In an attempt to localize this inhibitor, the non-IgG containing DEAE fractions were added back to IgG containing DEAE fraction #1; only fraction #8 demonstrated a level of inhibitor analogous with that following heat treatment of sera. Similar results were obtained following fractionation of 4 other 2° aborter sera and normal 3rd party sera.

Table 5

Determination of antipaternal CDC[51]Cr and ADCC activities
in SPA fractions from a 2° aborter[a]

IgG Source	% Specific CDC	% Specific ADCC
DEAE #1 (IgG1, 2, 3, 4)	60.6 ± 4.3 [100%][b]	43.1 ± 1.5 [100%]
SPA Effluent (IgG3)	11.4 ± 3.4 [18.8%]	-2.4 ± 1.0 [0.0%]
SPA Eluate (IgG1, 2, 4)	54.8 ± 3.3 [90.4%]	38.7 ± 3.9 [89.8%]

a. Results represent an average of 3 CDC[51]Cr experiments and 2 ADCC experiments done in duplicate.
b. [%] represents the % recovery of activity from DEAE fraction #1.

Table 6

Effect of heat treatment and heparin absorption on
antipaternal CDC[51]Cr reactivity

Antibody Source	% Specific CDC $\bar{x} \pm SD$	% Inhibition[a]
2° sera	39.4 ± 0.8	–
2° sera HT[b]	2.1 ± 0.5	95.0
2° sera HA[c]	3.9 ± 1.1	90.0
DEAE #1	53.4 ± 2.5	–
DEAE #1 HT	51.0 ± 3.1	4.5
DEAE #1 HA	39.5 ± 2.0	26.0

a.
$$\% \text{ Inhibition} = \frac{\% \text{ specific CDC}^{51}\text{Cr of treated antibody sources}}{\% \text{ specific CDC}^{51}\text{Cr of untreated antibody sources}} \times 100\%$$
b. 56°C for 30 min
c. Heparin absorption

To more accurately characterize the inhibitor, normal 3rd party serum and inhibitor containing DEAE fraction #8 were dialyzed against cold distilled H_2O or 2% boric acid to precipitate euglobulins. The inhibitor was found in the euglobulin fraction of serum, but not in the supernatant fraction from DEAE fraction #8 (Table 7). Similar results have been obtained with all normal 3rd party sera tested, and inhibitor has been precipitated from one additional 2° aborter and one normal 3rd

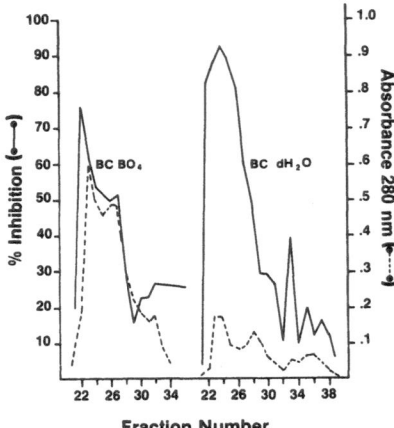

Figure 1. Inhibition of 2° aborter CDC^{51}Cr by gel filtration fractions of dissolved euglobulins precipitated from normal 3rd party female sera. Boric acid or distilled H$_2$O serum precipitates were dissolved in PBS and loaded onto Sephadex G-200 columns. Collected fractions were monitored at 280 nm and assayed for inhibition against husband ^{51}Cr labeled lymphocytes coated with 2° aborter antibodies.

party male source of DEAE fraction #8. To assess the molecular weight of the inhibitor, boric acid and distilled H$_2$O serum precipitates were solubilized in PBS and passed through a Sephadex G-200 superfine column. Inhibitor was found to pass through the column with the void volume (Figure 1), suggesting its molecular weight to be larger than 250,000 daltons.

DISCUSSION

The results of this report provide evidence for the expression of allotypic TLX antigens on trophoblast membranes (McIntyre and Faulk, 1985a; Taylor et al., 1985). Chorion preparations from two separate pregnancies were used to illustrate TLX allotypy by differential absorption of CDC reactivity. Analysis of the HLA phenotypes of husband target cells and chorion sources showed chorion #54 shared none of the husbands' HLA antigens, yet it had a more pronounced effect on antipaternal CDC than did chorion #91 which did share husbands' HLA. Similarly, absorption of 2° aborter sera with chorion #54 resulted in a larger loss of polyspecific cell panel reactivity than that produced with chorion #91. Chorion #54 equally reduced maternal cytotoxicity to 3rd party cells which shared HLA antigens with paternal cells. The lack of correlation between target cell HLA phenotypes and chorion sources suggests that absorption of lymphocytotoxicity with trophoblast is dependent upon the expression on TLX and not HLA allotypes.

DEAE chromatography of 2° aborter sera has shown that antipaternal CDC and ADCC activities are mediated by IgG which appears to cause the polyspecificity of

2° aborter sera. The IgG subclasses responsible for the antipaternal CDC activity in 2° aborter sera were determined using murine anti-human IgG subclass specific monoclonal antibodies. Inhibition assays with these reagents showed that IgG1 and IgG3 are the major CDC mediating antibodies: IgG2 and 4 were shown to contribute little CDC activity. The monoclonal results are supported by SPA column chromatography of 2° aborter IgG enriched DEAE fraction #1 to separate IgG3 from IgG1, 2, and 4. It appears that IgG1 followed by IgG3 represents the majority of IgG mediating antipaternal CDC. Similarly, IgG1, 2, and 4 containing SPA eluates accounted or 90% of 2° aborter antipaternal ADCC activity following fractionation of DEAE fraction #1, and IgG3 containing effluents demonstrated no ADCC. Other investigators have shown inhibition of human ADCC with IgG1 and 3 myeloma proteins and little or no inhibition with IgG2 and IgG4 (Perlman and Perlman, 1970; Spieglberg et al., 1976; Wissloff et al., 1974). Thus, it would appear that IgG1 is the major ADCC-mediating antibody in 2° aborters.

Heating (56°C for 30 min) or solid phase heparin absorptions of 2° aborter sera resulted in a loss of antipaternal $CDC^{51}Cr$, but heating and/or heparin absorption of IgG fractions from 2° aborter sera showed no such loss. These results suggest the involvement of a serum inhibitor not found in the IgG fractions. This is supported by our observations that incorporation of a wash step prior to the addition of complement reverses the inhibitory effects of heating or heparin absorption (Torry et al., 1985c,

Table 7

Euglobulin precipitation of inhibitor from serum and
DEAE fractions

Added Samples[a]	% Inhibition[b] $\bar{x} \pm SD$
Serum[c]	30.0 ± 1.0
Serum BO_4 ppt.	52.8 ± 2.1
Serum dH_2O ppt.	93.9 ± 12.2
DEAE #8[d]	81.3 ± 1.4
DEAE dH_2O ppt.	70.1 ± 16.3
DEAE dH_2O sup.	1.8 ± 5.2

a. All samples were added to husband lymphocytes precoated with a 2° aborter's antibodies which produced >60% ^{51}Cr release.
b. % Inhibition: see Table 6.
c. Normal 3rd party female serum.
d. Represents an inhibitor positive DEAE fraction from a 2° aborter.

1986). In addition, ion-exchange chromatography of sera yielded a CDC^{51}Cr inhibitory fraction. Inhibitor could also be demonstrated in the euglobulin fraction of serum as well as in non-IgG fractions prepared from DEAE chromatography of sera from 2° aborters and normal female and male blood donors. The inhibitor was not effective against CDC reactions mediated by horse ATG; it could be isolated in the void volume of euglobulins fractionated through Sephadex G-200. These results suggest that 2° aborters possess an inactive inhibitor which is activated by heat, heparin absorption or DEAE fractionation. The presence of IgG anti-TLX in the sera of 2° aborters may not be as detrimental to their pregnancy as the lack of functional inhibitors. Normal pregnancy induces the formation of IgG antibodies to trophoblast (Davies, 1985; Davies and Browne, 1985) and normal pregnancy causes the activation of serum inhibitors which are present but inactive in normal sera (Davies and Browne, 1985a, b). We suggest that the presence of IgG anti-TLX antibodies and inactive CDC inhibitors in 2° aborters is associated with recurrent spontaneous abortions.

SUMMARY

Sera from secondary (2°) aborting women contain antibodies that mediate complement dependent cytotoxicity (CDC) and antibody dependent cell-mediated cytotoxicity (ADCC) against both paternal and non-paternal lymphocytes. Antibody activity was shown to be directed to trophoblast lymphocyte cross-reactive (TLX) allotypic antigens which could be removed by absorption with trophoblast membrane preparations from some, but not all pregnancies. Ion-exchange chromatography of 2° aborter sera showed CDC and ADCC antibodies to be IgG. Analyses of the sera with the use of Protein A and IgG subclass specific monoclonal antibodies revealed that 75-90% of antipaternal CDC reactivity was mediated by IgGl and 15-30% was IgG3. Heating (56°C for 30 min) or absorption of 2° aborter sera with heparin-Sepharose resulted in loss of CDC activity. These manipulations had no effect on chromatographically isolated IgG from 2° aborter sera, indicating the presence of non-IgG inhibitors of CDC in sera. Further studies showed these to be euglobulins with molecular weights in excess of 250,000 daltons. These results suggest that recurrent spontaneous abortion is associated with defective interaction between IgG TLX antibodies and their biological inhibitors.

ACKNOWLEDGEMENTS

We thank Dr. Glen Tockstein, Lupe Johnson, Kelly Hartel, Fritz Lower, the personnel of the Renal Transplant Lab at Springfield Memorial Medical Center for their contribution, and Ms. Terry Lynn Bossle for typing the manuscript.

REFERENCES

Amos, D.B. (1979) In: *NIAID Manuel of Tissue Typing Techniques*, (ed.), J.G. Ray, NIH publication No. 80-545, 42.

Amos, D.B. and Kostyu, D.D. (1980) HLA - a central immunological agency of man. In: *Advances in Human Genetics*, Vol. 10.

Davies, M. (1985) Antigenic analysis of immune complex formed in normal human pregnancy. *Clin. Exp. Immunol.* 61, 406-415.

Davies, M. and Browne, C.M. (1985) Anti-trophoblast antibody responses during normal human pregnancy. *J. Reprod. Immunol.* 7, 285-297.

Davies, M. and Browne, C.M. (1985a) Pregnancy associated non-specific immuno-suppression. Kinetics of the generation and identification of the active factors. *Am.J. Reprod. Immunol. Microbiol.* 9, 77-83.

Davies, M. and Browne, C.M. (1985b) Pregnancy associated non-specific immuno-suppression. Mechanism for the activation of the immunosuppressive factiors. *Am. J. Reprod. Immunol. Microbiol.* 9, 84-90

Faulk, W.P. and McIntyre, J.A. (1983) Immunological studies of human trophoblast: Markers, subsets and functions. *Immunol. Rev.* 75, 139-175.

Faulk, W.P., Torry, D.S., and McIntrye, J.A. (1986) Effects of serum versus plasma agglutination of antibody coated indicator cells by human rheumatoid factors. *Clin. Immunol. Immunopath.* (submitted).

Faulk, W.P., Temple, A., Lovins, R., and Smith, N. (1978) Antigens of human trophoblast: a working hypothesis for their role in normal and abnormal pregnancies. *Proc. Natl. Acad. Sci. USA* 75, 1947-1951.

Garvey, J.S., Cremer, N.E., and Sussdorf, D.M. (1977) *Methods in Immunology*, 3rd edition, W.A. Benjamin, Inc., Readings, MA.

McConnachie, P.R., Bloemers, J., Godtferson, D., Leming, G., Riseman, J., Finch, W.T., and Birtch, A.G. (1979) False positive ^{51}Cr release tests during post transplant immunological monitoring due to recipient serum radioactive contaminations. *J. Immunol. Methods* 26, 197-201.

McConnachie, P.R. and McIntyre, J.A. (1984) Maternal antipaternal immunity in couples predisposed to repeated pregnancy losses. *Am. J. Reprod. Immunol.* 5, 145-150.

McIntyre, J.A. and Faulk, W.P. (1982) Allotypic trophoblast-lymphocyte cross-reactive (TLX) cell surface antigens. *Human Immunol.* 4, 27-35.

McIntyre, J.A. and Faulk, W.P. (1984) Histocompatibility and human pregnancy. *Fertil. Steril.* 41, 653-654.

McIntyre, J.A. and Faulk, W.P. (1985) Laboratory and clinical aspects of research in chronic spontaneous abortion. *Diagnost. Immunol.* 3, 163-170.

McIntyre, J.A. and Faulk, W.P. (1985) Antibody responses in secondary aborting women: effects of inhibitors in blood. *Am. J. Reprod. Immunol. Microbiol.* 9, 113-118.

McIntyre, J.A., Faulk, W.P., Verhulst, S.J., and Colliver, J.A. (1983) Human trophoblast-lymphocyte cross-reactive (TLX) antigens define a new alloantigen system. *Science* 222, 1135-1137.

McIntyre, J.A., McConnachie, P.R., Taylor, C., and Faulk, W.P. (1984) Clinical, immunologic and genetic definitions of primary and secondary recurrent spontaneous abortions. *Fertil. Steril.* 42, 849-855.

McIntyre, J.A., McConnachie, P.R., and Faulk, W.P. (1985) Characterization of maternal antipaternal antibodies in secondary aborting women. In: *Contributions to Gynecology and Obstetrics*, (ed.), V. Toder, Karger, Basel. Vol. 14. pp. 131- 137.

McIntyre, J.A., Faulk, W.P., Nichols-Johnson, V.R., and Taylor, C.G. (1986) Immunological testing and immunotherapy in recurrent spontaneous abortion. *Obstet. Gynecol.* 67, 169-175

Perlman, P. and Perlman, M. (1970) Contactual lysis of antibody coated chicken erythrocytes by purified lymphocytes. *Cell Immunol.* 1, 300-315.

Rogentine, G.N. and Plocinik, B.A. (1979) ^{51}Cr release cytotoxicity techniques. *NIAID Manual of Tissue Typing Techniques*, p. 89.

Skvaril, H.L. (1976) The question of specificity in binding human IgG subclasses to protein A-Sepharose. *Immunochem.* 13, 871-872.

Spieglberg, H.L., Perlman, H., and Perlman, P. (1976) Interaction of K lymphocyte with myeloma proteins of different IgG subclasses. *J. Immunol.* 116, 1464-1471.

Taylor, C.G., Faulk, W.P., and McIntyre, J.A. (1985) Prevention of recurrent spontaneous abortions by leucocyte transfusions. *J. Roy. Soc. Med.* 78, 623-627.

Torry, D.S., McConnachie, P.R., and McIntyre, J.A. (1985a) Complement-dependent (CDCE and CDC^{51}Cr) and complement-independent (ADCC) antibodies in secondary aborting women. *Am. J. Reprod. Immunol. Microbiol.* 7, 135A.

Torry, D.S., McIntyre, J.A., and McConnachie, P.R. (1985b) Characterization of immunoglobulin class and subclass responses in secondary aborter sera. *J. Reprod. Immunol.*, in press.

Torry, D.S., McIntyre, J.A., Faulk, W.P., and McConnachie, P.R. (1985c) In vitro control of secondary aborters' antipaternal immunity. In: *Immunology of Reproduction*, Publishing House of the Bulgarian Academy of Sciences, Sophia. (in press).

Torry, D.S., McIntyre, J.A., Faulk, W P., and McConnachie, P.R. (1986) Inhibitors of complement-mediated cytotoxicity in normal and secondary aborter sera. *Am. J. Reprod. Immunol. Microbiol.* 10, 53-57.

Wissloff, F., Michaelson, T.E., and Froland, S.S. (1974) Inhibition of antibody dependent human lymphocyte mediated cytotoxicity by Ig classes, Ig subclasses and Ig fragments. *Scand. J. Immunol.* 3, 29-38.

Trophoblast Research 2:187-197, 1987

THE ASSOCIATION OF MIXED LYMPHOCYTE REACTION (MLR) BLOCKING FACTORS AND MATERNAL ANTIPATERNAL LEUKOCYTOTOXIC ANTIBODIES WITH PREGNANCY OUTCOME IN WOMEN WITH RECURRENT ABORTIONS IMMUNIZED WITH PATERNAL OR THIRD PARTY LEUKOCYTES

Alan E. Beer[1,4], James F. Quebbeman,
A. Enrico Semprini[2], and Zhu Xiaoyu[3]

[1]Department of Obstetrics and Gynecology
University of Michigan
Ann Arbor, Michigan
[2]University of Milano
Milano, Italy
[3]Wuhan, Hubei
People's Republic of China

INTRODUCTION

Our recent investigations have been directed toward analyzing an immunogentic basis for recurrent idiopathic spontaneous abortions in human couples (Beer et al., 1985). We have limited our study to couples who were evaluated medically, genetically, microbiologically, and structurally as previously reported by us and were found to lack any etiology known to be associated with recurrent pregnancy losses (Beer et al., 1983). All couples had an immunogenetic work up as outline in Table 1. The materials and methods utilized in this study have been reported previously (Beer et al., 1981).

One hundred and twenty five couples met our established immunogenetic criteria (Beer et al., 1981), and the female was immunized on two occasions with 50,000,000 paternal lymphocytes intradermally. Of the 125 couples analyzed 41 have delivered a live born child (5 significantly retarded) (Beer et al., 1985b), 24 are currently successfully pregnant, 19 have aborted an additional time and the remainder are attempting to establish another pregnancy. Four of this latter group immunized subsequently with third party HLA incompatible leukocytes are beyond 24 wks of pregnancy.

The purpose of this contribution is to present sequential maternal antipaternal humoral immune response profiles pre- and post-immunization with paternal leukocytes, during pregnancy, at delivery, or at the time of a repeat pregnancy loss. These data are more predictive of subsequent pregnancy outcome than the level of

[4]To whom correspondence should be addressed: L2021-Women's Hospital, University of Michigan, Medical School, Ann Arbor, Michigan 48109
Received: 10 October 1985; Accepted: 10 May 1986

Table 1

Immunogenetic Workup of Couples with Recurrent Abortions

---

1.   HLA, A, B, C, DR tissue typing.

2.   ABO blood group antigen profiles.

3.   Maternal/antipaternal leukocytotoxic antibodies.

4.   Mixed lymphocyte culture responses of maternal to paternal as compared to third party stimulator antigens.

5.   Female serum complement dependent and independent mixed lymphocyte culture blocking factors.

6.   In some selected cases analysis of maternal antibodies to panels of T-cells, B- cells, and monocytes.

---

maternal/antipaternal responsiveness in MLR or the numbers of HLA antigens shared by the couple (Beer et al., 1984).

RATIONALE OF STUDY

For the past 15 years we have questioned if an inappropriate, inadequate, or absent immune response by the mother to paternally derived feto-placental antigens may be related to pregnancy failure and represent a mechanism(s) to ensure the maintenance of major histocompatibility complex polymorphisms (MHC) in humans (Beer et al., 1985). There are selective pressures operating during gestation against conceptuses that are homozygous with their mothers with regard to MHC and non-MHC antigens. These conceptuses are often eliminated either prenatally or early postnatally as a result of immunological responses directed against them (Beer and Billingham, 1977). Recent studies in CBA/J female mice and DBA/2 males, a mating combination that suffers greater than 25% embryonic losses during gestation, are allowing definition of the gene products that incite the appropriate immune responses that protect the fetal/placental unit from resorbtion/rejection (Chaouat et al., 1983).

Immunological studies in couples electing pregnancy terminations when compared with those in gestation matched couples with pregnancies aborting spontaneously have been very informative. In the former group lymphocytes resident in the uterine decidual tissue mantling the conceptus are hyporesponsive/non-cytotoxic in functional assays both to paternal and to third party stimulating alloantigens. In the latter group, lymphocytes from the same decidual compartment are several hundred-fold more cytotoxic/hyperreactive to paternal and third party alloantigens (Beer et al., 1985). Why these immunological features thought necessary for the immunoprotection of the fetal placental unit are not initiated in every pregnancy has been the subject of recent investigations. There is now firm scientific underpinning that intrauterine selection may be one mechanism for maintenance in

Table 2

Mechanisms Operational for the Production of MHC Heterozygotes

1.	Mating preferences (Boyce et al., 1983).
2.	Selective fertilization (Beer et al., 1977; Michie and Anderson, 1971).
3.	Selective abortion (Beer et al., 1985; Gill et al., 1983).
4.	Homozygous lethals (Bennett, 1975; Gill, 1983; Thomas et al., 1985).
5.	Abarrent maternal immune response resulting in lack of immunoprolection required for survival (Palm, 1969, 1970; Hings and Billingham, 1981; Michie and Anderson, 1971; Beer et al., 1984).
6.	Maternal/fetal allelic identity (Beer et al., 1985; McIntyre and Faulk, 1982).

utero of fetuses MHC heterozygous with their mothers or heterozygous with other allotypic or polymorphic antigens. Table 2 lists the mechanisms described for selective production of heterozygous offspring. In this contribution we will focus primarily on mechanism number five, the maternal immune response during pregnancy.

RESULTS

Maternal Anti-Paternal Anti-Leukocytotoxic Antibodies and Their Relationship to Pregnancy Outcome

Not all women develop anti-paternal leukocytotoxic antibodies following a normal pregnancy and their presence is unrelated with pregnancy outcome. It is presumed that the stimulus for their production in the mother are fetal lymphocytes that enter the maternal circulation; however, Class I MHC antigens on the trophoblast may be the responsible stimulus. The incidence of anti-leukocytotoxic antibodies in 206 women with 3 or more spontaneous, consecutive abortions 16% was at initial testing when not pregnant. This is compared to an incidence of 20% in primigravidas delivering an infant at term and the 35 to 64% in multigravid women (Beard et al., 1983; Gill, 1983).

Figure 1 shows the incidence of maternal anti-paternal anti-leukocytotoxic antibodies pre- and post-immuniziation in 41 women who were immunized with paternal leukocytes and delivered a live born child and in 19 women who aborted an additional time post-immunization with paternal leukocytes. Fifty-six percent of immunized women who delivered an infant at term were developed anti-paternal/anti-leukocytotoxic antibody positive. All but 1 of the 19 patients who aborted an additional time post-immunization were positive for the same antibodies at the time of a repeat abortion. This difference is highly significant.

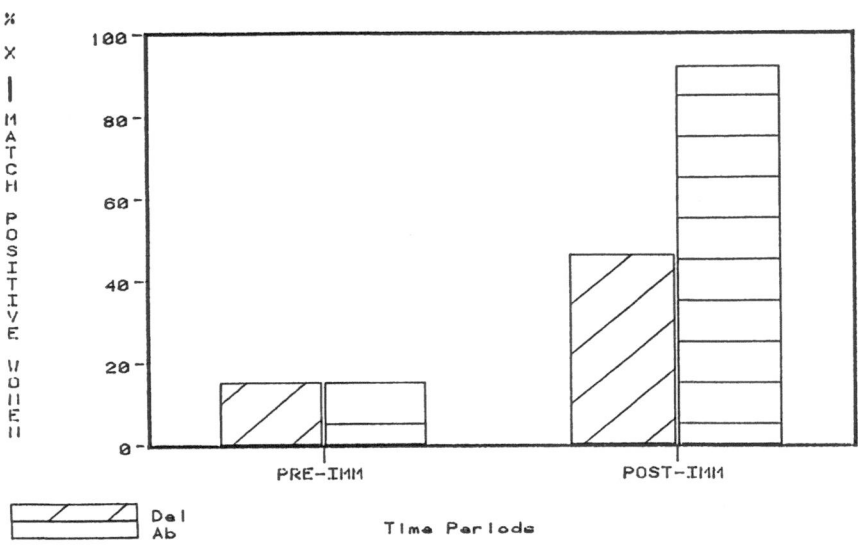

Figure 1. Incidence (%) of maternal anti-paternal anti-leukocytotoxic antibodies in
women with recurrent abortions immunized with paternal leukocytes who delivered a
live born child or aborted an additional time postimmunization.

Mowbray and his colleagues (1985), in a similar study utilizing paternal
leukocyte immunization in women with recurrent abortion excluded any patient with
anti-paternal anti-leukocytotoxic antibodies. They reasoned this to be the expected
response and their presence indicated the mother had responded appropriately to
paternal antigens.   Our data are not in total agreement and highlight that the
presence of maternal anti-paternal anti-leukocytotoxic antibodies in the *absence* of
maternal serum MLR blocking factors in typical of nearly all abortion prone women,
immunized with paternal leukocytes who abort an additional time.

Dynamics of Suppression by Normal Maternal Serum and Heat Inactivated Maternal Serum of MLR Pre- and Post-Immunization: Relationship with Subsequent Pregnancy Outcome

Female serum was tested for the presence or the absence of MLR blocking
activity to paternal and third party stimulating lymphocytes. One half of women at
initial screening showed some serum blocking activity and the others showed serum
factors that significantly enhanced or potentiated MLR reactivity (Figures 2 and 4).
In women who became pregnant and subsequently delivered a live born child
following paternal leukocyte immunization, we found increasing levels of female
serum and heat inactivated serum suppression of MLR.   The female serum
potentiating effects seen in 20 women disappeared post immunization. During

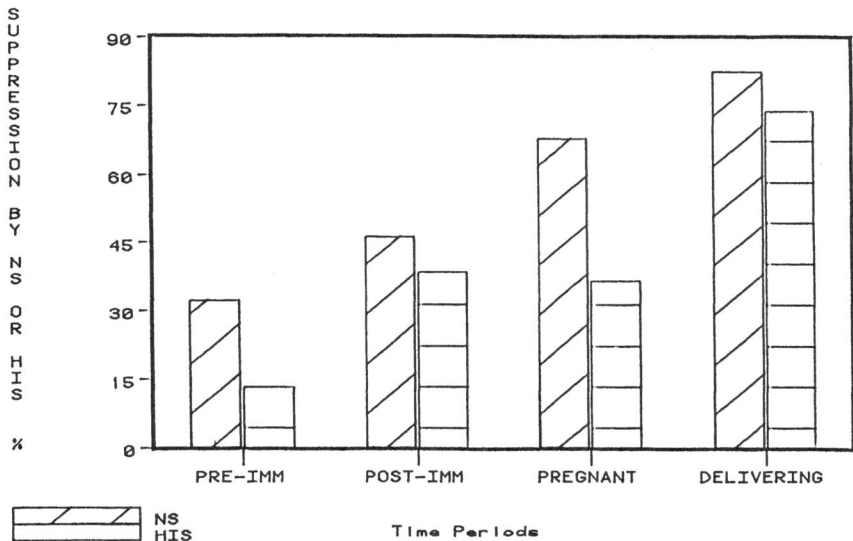

Figure 2. Patterns of humoral suppression by normal female serum (NS) and heat inactivated serum (HIS) of MLR (♂/♀) pre-, post-immunization with paternal leukocytes, during pregnancy, and at delivery of a live born baby.

pregnancy and at the time of delivery all serum tested showed significant MLR blocking activity. The MLR blocking factors did not cross the placenta nor did they appear in the cord sera of the infant. Studies are in progress to determine if the blocking factors can be eluted from the placenta and whether they are IgG antibody as shown by Voisin and Chaouat (1974). Female serum MLR blocking activity of 40% or greater following heat inactivation is seen in all women post immunization with paternal leukocytes who subsequently become pregnant and deliver a live born child.

These same parameters were tested in the 19 couples who experienced a repeat abortion post immunization with paternal leukocytes (Figures 3 and 5). Female serum MLR blocking or potentiating activity was no different in this group when compared with the above group prior to leukocyte immunization. One half demonstrated some MLR blocking activity and the remainder showed MLR potentiating activity. In women that aborted an additional time post immunization, heat inactivated female serum MLR blocking activity of 30% or less was seen. Serum tested at the time of the repeat abortion in those previously showing some female serum MLR suppression did not show any increase in MLR suppressive activity (Figure 3). Ten women who showed serum MLR potentiating activity failed to manifest any MLR blocking activity at the time of a repeat abortion. Their serum continued to potentiate MLR reactivity. We are currently analyzing the nature of the serum factor(s) responsible for suppression and or potentiation of MLR reactivity utilizing antibody separation techniques as well as analyses of non-immunoglobulin substances in serum or plasma that may be involved.

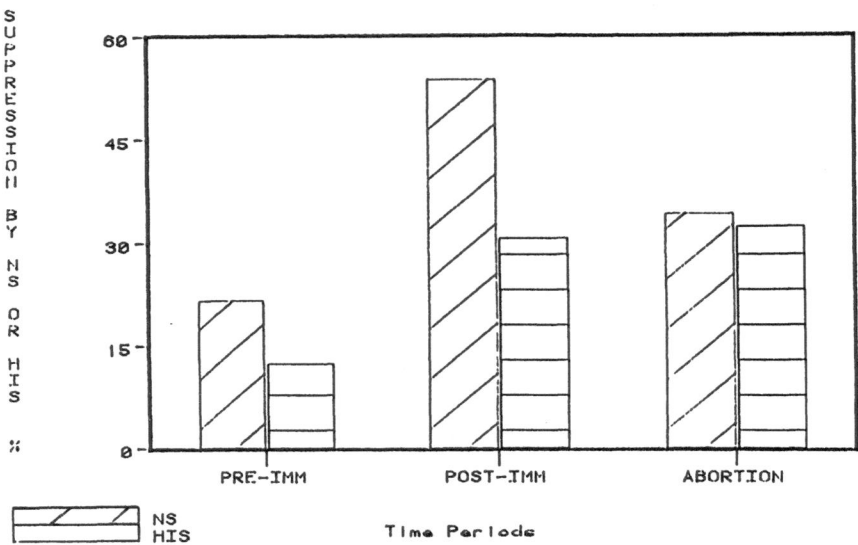

Figure 3. Patterns of humoral suppression by normal female serum (NS) and heat inactivated serum (HIS) of MLR $(\male / \female)$ pre-, post-immunization with paternal leukocytes and after subsequent spontaneous abortion.

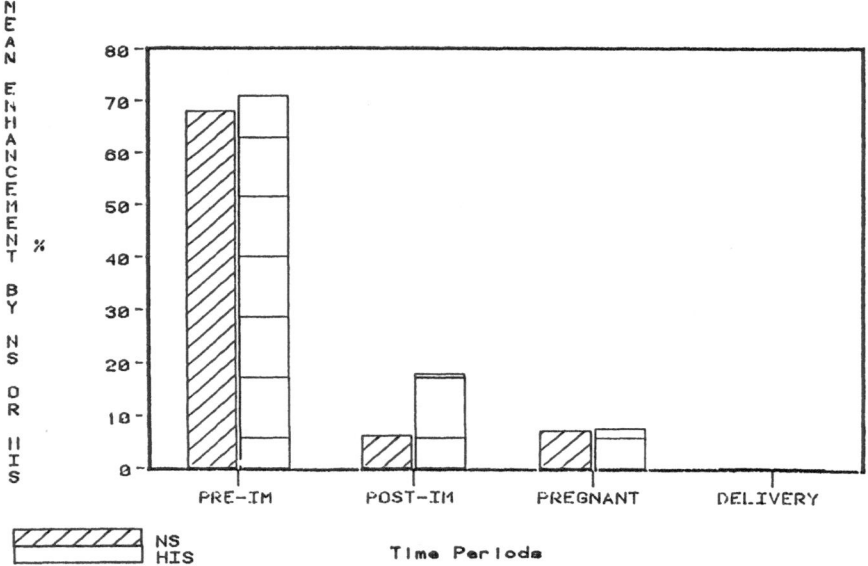

Figure 4. Patterns of humoral enhancement/potentiation by normal female serum (NS) and heat inactivated serum (HIS) of MLR pre- and post-immunization with paternal leukocytes, during pregnancy and at delivery of a live born child.

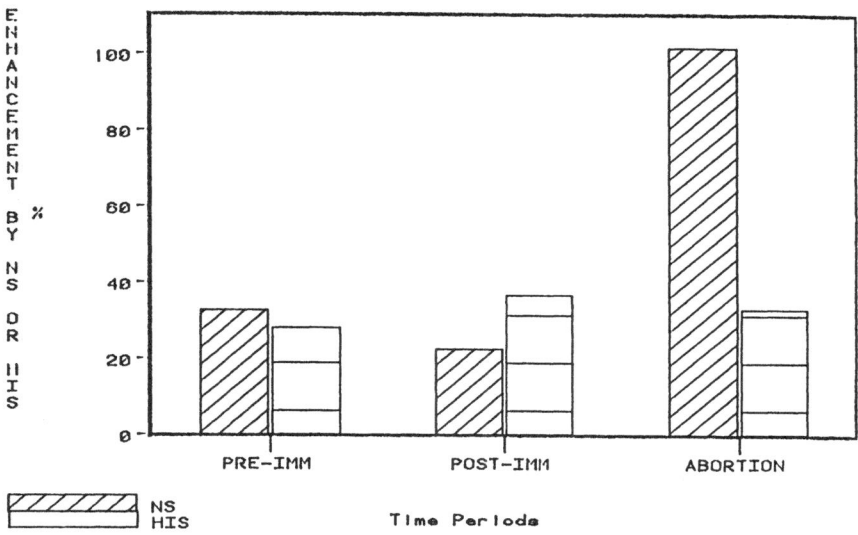

Figure 5. Patterns of humoral enhancement/potentiation by normal female serum (NS) and heat inactivated serum (HIS) of MLR pre- and post-immunization with paternal leukocytes and at the time of a subsequent abortion post immunization.

Blocking Factor: Is the MLR Blocking Factor an Antibody?

In 15 couples we tested if the antibodies in maternal serum showed specificity to panels of nonpaternal T-cells, B-cells and monocytes pre- and post-immunization with paternal leukocytes. Seven women of this group who delivered a child at term developed anti-B-cell cytotoxic antibodies. All others who aborted failed to develop antibodies with B-cell specificity but did develop anti-T-cell antibodies. Studies are in progress to assess if the antibody specificity is to Class II MHC determinants or to a gene product closely aligned (i.e., the TLX complex Faulk et al., 1978 or the growth and reproduction complex (GRC) Gill et al., 1983).

Immunization of Women With Recurrent Abortions With Non-Paternal Third Party Leukocytes

Four of the 19 patients who failed to make MLR blocking factors post-immunization and who aborted an additional time(s) were immunized with HLA incompatible leukocytes from the same female donor. All women developed strong MLR serum blocking activity in MLR to the third party immunizing cells and all 4 developed MLR blocking that was also significantly suppressed reactivity to paternal stimulator cells. Maternal serum also blocked MLR reactivity of 5 additional HLA incompatible individuals tested in all stimulator/responder combination. Repeat immunization was done every 6 wks once pregnancies were established. All four women were beyond 24 wks gestation with normal fetuses by amniocentesis/ultrasound evaluations. One patient was totally compatible with the individual providing the third party immunizing cells at the A, B, and C loci. They shared no antigens in common at the D locus.

DISCUSSION

Recent studies have demonstrated that pretransplantation blood transfusions containing leukocytes have the following effects: 1) significantly increase renal allograft survival (Persijn, 1979; Betuel et al., 1982); 2) generate blocking antibodies directed against T-cell antigens (McLeod et al., 1983; Singal et al., 1983); 3) generate suppressor T-cells (Lenhard et al., 1983); 4) decrease host/donor mixed lymphocyte culture reactivity (Klatzmann et al., 1983); 5) promote the production of autoantiidiotypic antibodies (Sucia-Foca et al., 1983); and 6) transfusions given prior to surgery with patients with colonic tumors are associated with an increased incidence of metastases, aggravated tumor growth, and greater recurrences of the tumors (Woodruff et al., 1983).

Paternal leukocytes therapy in women with recurrent abortion definitely induces, in most women, high levels of serum MLR blocking factors that are associated with successful pregnancy outcome. These blocking factors when present also inhibit MLR reactivity to nonpaternal third party leukocytes in various stimulator/responder combinations. Women who fail to show any changes in immune parameters post- immunization with paternal leukocytes and abort subsequently failed to demonstrate the presence of female serum MLR blocking factors.

The subgroup of 4 of these women who were reimmunized with HLA incompatible nonpaternal third party leukocytes and are now successfully pregnant, deserve comment. Faulk and co-workers (1978) presented a hypothesis for the immunological maintenance of pregnancy. They identified two serologically defined trophoblast antigen groups TA1 and TA2. TA1 is present on trophoblast and certain other cell lines and induces a cytotoxic T-cell response in humans. TA2 is present on the trophoblast, lymphocytes, endothelium, and villous fibroblasts and stimulates B-cells to produce IgG blocking antibody which may protect against abortion/rejection. The latter antigen is called trophoblast lymphocyte crossreacting (TLX) antigen. The TLX system is allotypic (McIntyre and Faulk, 1982). An embryo TLX compatible with its mother would be normal but incapable of stimulating protective blocking antibodies to camouflage the placenta and membranes and would suffer rejection. The four women mentioned who failed to respond to paternal leukocyte immunization and who aborted an additional time(s) may share the same TLX antigens with their spouses. The third party HLA incompatible lymphocytes used to reimmunize these women after paternal leukocyte immunization failed by definition would be TLX incompatible. It is suggested that the TLX genes are located near the D region and are in linkage disequilibrium with certain groups of HLA antigens (Beer et al., 1985). HLA antigen sharing is not increased in these couples; however, analysis of the HLA antigens seen suggests groups that are in linkage disequilibrium (Beer et al., 1985b).

SUMMARY

There is now good evidence that the fetal/placental unit expresses paternally derived antigens including Class I MHC framework alloantigens and most likely TLX antigens. These antigens are presented to the reproductive tract and the maternal immune system in a manner yet unknown, but in a way one that effectively orchestrates an immune response locally and systemically that provides protection of the feto/placental unit from rejection (Beer et al., 1984). The complexities of reproductive immunogenetics are beginning to come into clear focus. There is no simple or singular explanation to explain successful vivaparity. It is very clear that a successful pregnancy evokes in the mother immune responses that have immunological efferent paternal specificity and that the lack of these responses associated with repeat abortions. Why the intracutaneous administration of paternal leukocytes induces the response in most when prior fetuses in utero have been unable to do so is unknown at present. That certain women do not respond following paternal leukocyte therapy but do following immunization with third party HLA incompatible leukocytes suggests the operation of TLX antigenic determinants and immune responses to these in human pregnancy.

REFERENCES

Beard, R.W., Braude, P., Mowbray, J.F., and Underwood, J.L. (1983) Protective antibodies and spontaneous abortion. *Lancet*, Nov. 5, 1090.

Beer, A.E., and Billingham, R.E. (1977) Histocompatibility gene polymorphism and maternal-fetal relationships. *Transplant. Proc.* 9, 1393-1401.

Beer, A.E., Quebbeman, J.F., Ayers, J.W.T., and Haines, R.F. (1981) Major histocompatibility complex antigens, maternal and paternal immune responses, and chornic habitual abortions in humans. *Am. J. Obstet. Gynecol.* 141, 987.

Beer, A.E., Quebbeman, J.F., Semprini, A.E., Smouse, P.E., and Haines, R.F. (1983) Recurrent abortion: Analysis of the roles of parental sharing of histocompatibility antigens and maternal immunological responses to paternal antigens. In: *Reproductive Immunology*, (eds.), S. Isojima and W.D. Billingham, Elsevier Science Publishers, Amsterdam.

Beer, A.E. (1984) How did your mother not reject you? *Ann. Immunol.* (Inst. Pasteur) 135D, 315-318.

Beer, A.E., Quebbeman, J.F., Hamazki, Y., and Semprini, A.E. (1985a) Pregnancy outcome in human couples with recurrent spontaneous abortions: The role(s) of HLA antigen sharing, ABO blood group antigen profiles, female serum MLR blocking factors, antisperm antibodies and immunotherapy. In: *Proceedings of Second Banff Conference on Reproductive Immunology: Immunoregulation of the Maternal-Fetal Interface*, (eds.), T.G. Wegmann and T.J. Gill, III, Banff, Alberta, Canada, in press.

Beer, A.E., Semprini, A.E., Xiaou, Zhu., and Quebbeman, J.F. (1985b) Pregnancy outcome in human couples with recurrent spontaneous abortions: 1) HLA antigen profiles; 2) HLA antigen sharing; 3) Female serum MLR blocking factors, and 4) Paternal leukocyte immunization. *Exp. Clin. Immunogene.* 2(3), 1-12.

Beer, A.E., Xiaoyu, Zhu., Semprini, A.E., and Quebbeman, J.F. (1985c) Pregnancy outcome in human couples with idiopathic recurrent abortions: The role(s) of female serum mixed lymphocyte culture blocking factors, potentiating factors, and local uterine immunity before and after paternal leukocyte immunization. In: *Proceedings of Fogarty International Symposium: Immunological Approaches to Contraception and Promotion of Fertility*, (ed.), P. Talwar, NIH, Bethesda, Maryland, in press.

Bennett, D. (1975) The T locus of the mouse cell. *Cell* 6, 441-454.

Betuel, H., Touraine, J.L., Malik, M.D., Bonnet, M.C., Carrie, J., and Traeger, J. (1982) A policy of deliberate transfusions before kidney transplantation. *Transplant. Proc.* 14, 151-155.

Boyce, E.A., Beauchamp, G.K., and Y amakazi, K. (1983) The sensory perception of genotypic polymorphism of the major histocompatibility complex and other genes: Some physiological and phylogenetic implications. *Hum. Immun.* 6, 177-183.

Chaouat, G., Kiger, N., and Wegmann, T.G. (1983) Vaccination against spontaneous abortion in mice. *J. Reprod. Immunol.* 5, 389-392.

Faulk, W.P., Temple, A., Lovins, R.E., and Smith, N. (1978) Antigens of human trophoblasts: A working hypothesis for their role in normal and abnormal pregnancies. *Proc. Nat. Acad. Sci. USA* 75, 1947-1951.

Gill, T.J. (1983) Immunogenetics of spontaneous abortions in humans. *Transplant.* 35, 1-6.

Gill, T.J. III, Siew, S., and Kunz, H.W. (1983) MHC-linked genes affecting development. *J. Exp. Zool.* 228, 325.

Hings, I.M., and Billingham, R.E. (1981) Splenectomy and sensitization of Fischer female rats favors histoincompatibility of R2 back-cross progeny. *Transplant. Proc.* 13, 1253-1255.

Klatzmann, D., Gluckman, J.C., Foucault, C., Buboust, A., Bensussan, A., Assogba, J., and Berthelot, J.M. (1983) Modification of mixed lymphocyte reactivity after blood transfusions in man. *Transplant. Proc.* 15, 1011-1015.

Lenhard, V., Maassen, G., Grosse-Wilde, H., Wernet, P., and Opelx, G. (1983) Effect on blood transfusions on immunoregulatory mononuclear cells in prospective transplant recipients. *Transplant. Proc.* 15, 1011-1015.

McIntyre, J.A. and Faulk, W.P. (1982) Allotypic trophoblast lymphocyte crossreactive (TLX) cell surface antigens. *Hum. Immunol.* 4, 27.

McLeod, A.M., Mason, R.J., Stewart, K.N., Power, D.A., Shewan, W.G., Edward, N., and Gatto, G.R.D. (1983) Fc-receptor blocking antibodies develop after blood transfusions and correlate with good graft outcome. *Transplant. Proc.* 15, 1019- 1021.

Michie, D., and Anderson, N.F. (1971) A strong selective effect associated with a histocompatibility gene in the rat. *Ann. NY Acad Sci.* 129, 88-93.

Mowbray, J.F., Gibbings, C., Liddell, H., Reginald, P.W., Underwood, J.L., and Beard, R.W. (1985) Controlled trial of treatment of recurrent spontaneous abortion by immunisation with paternal cells. *Lancet* 1, 941-943.

Palm, J. (1969) Association of maternal genotype and excess heterozygosity for Ag-B histocompatibility antigens among male rats. *Transplant. Proc.* 1(1), 82-84.

Palm, J. (1970) Maternal-fetal interactions and histocompatibility antigen polymorphisms. *Transplant. Proc.* 2, 162-173.

Persijn, G.G., Cohen, B., Lansbergen, Q., and van Rood, J.J. (1979) Retrospective and prospective studies on the effect of blood transfusions in renal transplantation in the Netherlands. *Transplant.* 28, 296-401.

Singal, D.P., Fagnilli, L., and Joseph, S. (1983) Blood transfusions induce anti-idiotypic antibodies in renal transplant patients. *Transplant. Proc.* 15, 1005-1008.

Sucia-Foca, N., Reed, E., Rohowsky, C., Kung, P., and King, D.W. (1983) Anti-idiotypic antibodies to anti-HLA receptors induced by pregnancy. *Proc. Natl. Acad. Sci.* Feb., 830.

Thomas, M.L., Harger, J.H., Wagener, D.K., Rabin, B.S., and Gill, T.J., III (1985) Sharing and spontaneous abortion in humans. *Am. J. Obstet. Gynecol.* 151, 1053-1058.

Voisin, G.A., and Chaouat, G. (1974) Demonstration, nature, and properties of maternal antibodies fixed on placenta, and directed against paternal antigens. *J. Reprod. Fertil.* 21 (Suppl.), 89.

Woodruff, M.F.A. and van Rood, J.J. (1983) Possible implications of the effect of blood transfusion on allograft survival. *Lancet* 1, 555-556.

Trophoblast Research 2:199-214, 1987

MONOCLONAL ANTIBODIES TO EQUINE TROPHOBLAST

Douglas F. Antczak[1,3], J. Clare Poleman[1], Laura M. Stenzler[1],
Stephen G. Volsen[2], and W.R. Allen[2]

[1] James A. Baker Institute for Animal Health
New York State College of Veterinary Medicine
Cornell University
Ithaca, New York 14853, USA

[2]Thoroughbred Breeders' Association Equine Fertility Unit
Animal Research Station
307 Huntingdon Road
Cambridge CB3 0JQ United Kingdom

INTRODUCTION

The equine trophoblast exists in two distinct forms. The non-invasive and major component constitutes the outer layer of the allantochorionic villi that interdigitate with the endometrial epithelium. The invasive and minor component consists of the chorionic girdle. This discrete, annulare band of tissue begins to differentiate at approximately day 25 of gestation, and it invades the endometrium between days 36 and 38 to form 10-20 endometrial cups (Allen et al., 1973). The large cells from the fetal endometrial cups secrete the high concentrations of equine Chorionic Gonadotrophin (eCG) that are present in the serum of pregnant mares between days 40 and 120. The cup cells also elicit a maternal cell mediated reaction which hastens the destruction of the cups and their deliverance from the endometrium between days 100 and 140. No such leukocyte reaction occurs at the interface between the endometrial epithelium and the normal trophoblast of the allantochorion (Allen, 1975).

The morphological differences between the invasive and non-invasive components of equine trophoblast, the contrast in the maternal cellular response that those two kinds of trophoblast evoke, and the unique ability of the endometrial cup cells to secrete eCG suggest that the two forms of equine trophoblast might differ in cell surface molecules. This paper describes the use of monoclonal antibody technology to explore this possibility.

[3]Mailing address: Douglas F. Antczak, VMD, Ph.D., James A. Baker Institute of Animal Health, NYS College of Veterinary Medicine, Cornell University, Ithaca, New York 14853, USA
Received: 15 November 1985; Accepted: 30 May 1986

MATERIALS AND METHODS

Preparation of Cells and Tissues

Spleen cells from immunized rats were prepared from whole spleen using cell culture techniques and were washed in Dulbecco's phosphate-buffered saline (PBS) + 2.0% bovine fetal serum (BFS, Gibco Laboratories, Grand Island, NY) as previously described (Antczak et al., 1979).

Peritoneal macrophages were harvested from healthy rats by peritoneal lavage, using the PBS-BFS solution containing 20 IU of preservative-free sodium heparin/ml (Sigma Chemical Co., St. Louis, MO). The cells were washed once, counted in a hemocytometer and kept at 4°C until needed.

Equine lymphocytes were isolated from peripheral blood as described by Antczak et al. (1982).

Endometrial cup tissues were recovered between days 60 and 80 of gestation by hysterotomy performed under general anesthesia. A typical pregnancy yielded 10-15 cups of 0.5-2.0 cm diameter to give at total of 5-8 grams wet weight of tissue. The endometrial cup tissue was dissected free of adjacent maternal endometrium, minced with scissors, and forced through a fine nylon mesh sieve using the rubber-coated plunger of a disposable hypodermic syringe. The material that passed through the sieve was then allowed to settle for 10 min on ice. The debris that settled in the bottom of the tube was discarded and the remaining supernatant was centrifuged at 400 x g for 10 min at 4°C to give a pellet that consisted of individual endometrial cup cells and some contaminating maternal cells. The pellet was washed and centrifuged twice more before it was weighed and either used for immunization or frozen. The supernatant from the first centrifugation was opaque and contained subcellular fragments. This suspension was frozen at -70°C and used for booster immunizations.

Cell Culture Medium and Techniques

Cell culture medium for somatic cell hybridization consisted of Dulbecco's modified Eagle's medium (DMEM; Gibco Laboratories, Grand Island, NY) with a high glucose content. The DMEM was modified by the addition of BFS, to a final concentration of 10% or 20%, plus antibiotics, amino acids, and sodium pyruvate as described by Oi and Herzenberg (1980). The medium was modified further by the addition of either hypoxanthine and thymidine (HT-DMEM) or hypoxanthine, aminopterin, and thymidine (HAT-DMEM), for the selective growth of hybridoma cells (Oi and Herzenberg, 1980). All cell cultures were performed at 37.5°C in a humidified atmosphere of 5% CO_2 in air.

Production of Hybridomas

Female rats of two inbred strains bred and maintained at the Baker Institute were used for immunization. PVG-(A0) rats were used for Fusion 67. The rats were injected with antigen four times. The primary immunization used 10 mg (wet weight) of fresh horse endometrial cup tissue administered subcutaneously in 0.2 ml in Freund's Complete Adjuvant. One year later the rats were boosted on three consecutive days using the intraperitoneal (IP) route. The first boost used 1×10^7 endometrial cup cells in 1.0 ml PBS; the second and third boosts used 0.5 ml of supernatant from washed, centrifuged, endometrial cup cells. The supernatant contained cell fragments. The fusion was performed on the day following the third booster immunization. For Fusion 71, A0 rats were immunized twice with donkey endometrial cup cells given IP in 1.0 ml of PBS. The first injection used fresh tissue (approximately 10 mg wet weight) and the second, which was administered 35 days after first, used the cells from the same preparation that had been stored at - 70°C and thawed just prior to injection. The fusion was performed three days after the booster immunization.

Both fusions used methods described previously by Newman et al. (1984). The HAT-sensitive murine plasmacytoma (SP-2/0) was a gift from Dr. T.J. McKearn of the University of Pennsylvania School of Medicine, and these cells were maintained in DMEM + 10% BFS and 8-azaguanine (15 µg/ml; Sigma Chemical Co., St. Louis, MO). Cells used for the production of hybridomas were harvested during active growth, and the method of Galfre et al. (1977) was used for cell fusion.

Culture supernatant fluids from wells containing hybridomas were tested for specific antibodies in two assays (see below). Selected hybridomas were cloned by limiting-dilution technique in 96-well microculture plates (Costar, Cambridge, MA) on a feeder layer of 1×10^5 irradiated (950R) rat spleen cells per well. Clones producing antibody were expanded by serial passage to large tissue culture containers and preserved in liquid nitrogen. The supernatant fluid from these cultures was harvested as the source of monoclonal antibodies.

The Elisa Assay

A solid phase enzyme-linked immunoassay (ELISA) with horse lymphocytes as antigen (Newman et al., 1984) was used for the initial screening of hybridoma cultures.

Immunohistochemistry

Pieces of endometrial cup and the endometrium-allantochorion interface were recovered from mares and donkeys at day 60-80 of gestation. The tissues were snap frozen in liquid nitrogen and stored at -80°C in OCT imbedding compound (Am. Sci. Products). Cryostat sections (5-7 µm) were processed for immunohistochemical staining as described by Mason et al. (1982). The first stage antibody reagents were the monoclonal antibodies produced in the present study; controls consisted of normal rat serum, serum from the immunized rats, and a rat monoclonal antibody to canine parvovirus (Parrish et al., 1982). The second stage reagent was a peroxidase labeled goat anti-rat Ig (heavy and light chains; Cappel Laboratories, Cochranville, PA). The slides were developed with amino-ethyl-carbamizol (AEC) and counterstained with hematoxylin.

Small (1.0-2.0 cm) pieces of control tissues from a variety of organs (skin, kidney, small intestine, lymph nodes) were obtained from horses at necropsy. These were frozen in liquid nitrogen and processed as described above.

Isotype Analysis

The isotypes of the monoclonal antibodies were determined by agar-gel immunodiffusion using 1% agarose (Marine Colloids, Rockland, ME) gel made in 0.015 M borate buffered solution (pH 8.3). Rabbit antisera specific for rat Ig isotypes (Miles Laboratories Inc., Elkhart, IN) were used to develop the precipitin lines.

RESULTS

Production and Screening of Hybrids

In both fusions, supernatants were screened initially against horse lymphocytes (Fusion 67) or donkey lymphocytes (Fusion 71) in the ELISA assay. Any supernatants which gave a positive result were set aside. The remaining (lymphocyte negative) supernatants were then screened on cryostat sections of horse and donkey endometrial cups and endometrium/allantochorion. Nineteen hybrids that gave positive reactions were cloned, isotyped and frozen in liquid nitrogen (Table 1). Tissue culture supernatants from actively growing cultures were harvested as the source of monoclonal antibodies for further study.

Five hundred microtitre wells containing growing hybrids were screened in the two fusion experiments. About 10% of the wells contained antibodies reactive with lymphocytes. About 80% of the remaining wells with growing hybrids produced rat immunoglobulins, as determined by an ELISA assay (data not shown). About 20% of the Ig positive wells yielded supernatants reactive with placental tissues. The 19 hybrids chosen for cloning were selected to represent several different reaction patterns, many of which were found several times among the large number of positive hybridomas.

Immunohistochemical Characterization

The immunohistochemical reactivity of the 19 monoclonal antibodies was assessed on cryostat sections of equine placental tissues and on various other control tissues.

Four of the antibodies (F67.2, F71.4, F71.5, and F71.6) gave very broad staining patterns on sections of endometrial cups and adjacent tissues, reacting with the large cup cells, the various components of the allantochorion and many cell types in the maternal endometrium. Two others (F67.3 and F67.4) reacted only with scattered cells in the endometrial stroma. These six antibodies were not considered further in this study.

The remaining 13 monoclonal antibodies were assigned to one of four groups on the basis of their reactivity with sections of horse and donkey endometrial cups, adjacent feto-placental, and other non-reproductive control tissues from adult horses.

The largest group contained seven monoclonal antibodes which reacted strongly with endometrial cup cells (F67.1, F71.1, F71.2, F71.3, F71.8, F71.10, and F71.14); they gave six different reaction patterns (Table 2).

Table 1

Hybrid cell lines secreting monoclonal antibodies to antigens of the equine
placenta

Fusion 67		Fusion 71	
Hybrid cell line	Isotype	Hybrid cell line	Isotype
F67.1	IgG2a	F71.1	IgG2a
.2	IgG1	.2	IgG2a
.3	IgM	.3	IgG2a
.4	IgG1	.4	IgM
		.5	IgM
		.6	IgM
		.7	IgG2c
		.8	IgG2a
		.9	IgG2c
		.10	IgG2c
		.11	IgM
		.12	IgM
		.13	IgM
		.14	IgM
		.15	IgG2a

Six of the seven Group I antibodies reacted with one or more other tissues or cell types in addition to endometrial cup cells. F71.8 and F71.10 reacted very strongly with endometrial cup cells and less intensely with the normal, single-celled trophoblast layer of the allantochorion (Figure 1b, Figure 2a). The two antibodies did not stain any other placental tissues. They gave identical reaction patterns on other maternal tissues and both produced a distinctive "honey-comb" pattern on the endometrial cup cells, indicative of strong surface labeling. F71.14 produced a more uniform staining of the endometrial cup cells, and strong labeling of the non-invasive trophoblast (Figure 2b). The three monoclonal antibodies gave weak reactivity with some non-uterine maternal tissues (Table 2), the significance of which is not known.

Table 2

Reactivity patterns of 7 monoclonal antibodies with specificity for endometrial cup cells[1]

Hybrid cell line	Endo-metrial cup cells	Allantochorion		Endometrium			Other adult tissues			
		Trophoblast	Non-trophoblast[2]	Lumenal epi-thelium	Glandular epi-thelium	Inter-stitial tissue	Lympho-cytes	Kid-ney[3]	Intes-tine[3]	Skin[3]
67.1	++	-	-	-	-	-	-	-	-	-
71.1	++	++	-	+	+	-	-	-	-	-
71.2	++	++	-	-	+	+	±[4]	+	++	±
71.3	++	-		++	-	++	-	-	+	++
71.8	++	+	-	-	-	-	-	+	+	-
71.10	++	+	-	-	-	-	-	+	+	-
71.14	++	+	-	-	-	-	-	-	±	-

1   Determined by reactivity on cryostat sections of the tissues indicated using indirect immunoperoxidase labeling.

2   Chorionic mesoderm and allantoic endoderm.

3   The cell types stained by the monoclonal antibodies in these adult tissue sections were not identified.

4   Positive on a small subpopulation of lymphocytes surrounding the endometrial cups, but negative when tested on peripheral blood lymphocytes in ELISA.

F71.1 and F71.2 reacted strongly with endometrial cup cells and the trophoblast layer of the allantochorion, whereas the remaining elements (mesoderm and allantoic endoderm) of the allantochorion were not stained. F71.1 also stained the endometrial epithelium of the endometrial glands. This antibody likewise produced a honey-comb staining pattern on endometrial cup cells. In contrast, F71.2 gave a diffuse, granular staining of endometrial cup cells. It failed to stain the lumenal epithelium of the endometrium, although the epithelial cells of the endometrial glands were labeled (Figure 2c).

F71.3 stained the endometrial cup cells with the typical honey-comb pattern. In addition, it intensely labeled the lumenal epithelium of the endometrium and the elastic layer of maternal arteries. The endometrial gland epithelium was not labeled, however, and neither were any of the cells of the allantochorion, including the trophoblast (Figure 2d).

F67.1 stained only endometrial cup cells, and no other fetal or maternal tissues tested (Figure 2e). Subsequent studies involving comparison with characterized polyclonal sera have demonstrated that this antibody reacts with equine Chorionic Gonadotrophin (Volsen et al., unpublished).

Groups II consisted of three monoclonal antibodies (F71.7, F71.11, and F71.12). All reacted strongly with the trophoblast layer of the allantochorion but not with endometrial cup cells. The antibodies reacted to varying degrees with maternal tissues (Table 3). The reaction pattern of 71.7 on a section of endometrial cup is shown in Figure 2f.

Group III contained two monoclonal antibodies: F71.9 and F71.13. These antibodies gave very restricted reaction patterns on sections of equine placenta, producing an intense and discrete line of staining at the fetal-maternal interface on the endometrium/allantochorion border, but failing to stain endometrial cup cells. F71.13, but not F71.9, stained the allantoic endoderm. Both antibodies weakly labeled the lumen of endometrial glands.

Group IV contained only a single antibody. F71.15 reacted strongly with both placental and maternal tissues recovered from pregnant donkeys, but did not react with equivalent tissues taken from pregnant mares. In addition to this species specificity, the antibody also showed some tissue specificity in failing to label donkey lymphocytes. However, it was broadly reactive, staining both the endometrial cups and allantochorion and most tissues of the maternal endometrium.

DISCUSSION

Monoclonal antibody technology has been used to identify molecules unique to, or largely restricted to, the human placenta (Johnson et al., 1981; Sunderland et al., 1981; also reviewed by Bulmer and Johnson, 1985). This has been especially helpful in identifying the diverse cell types of human trophoblast and in separating fetal from maternal tissue.

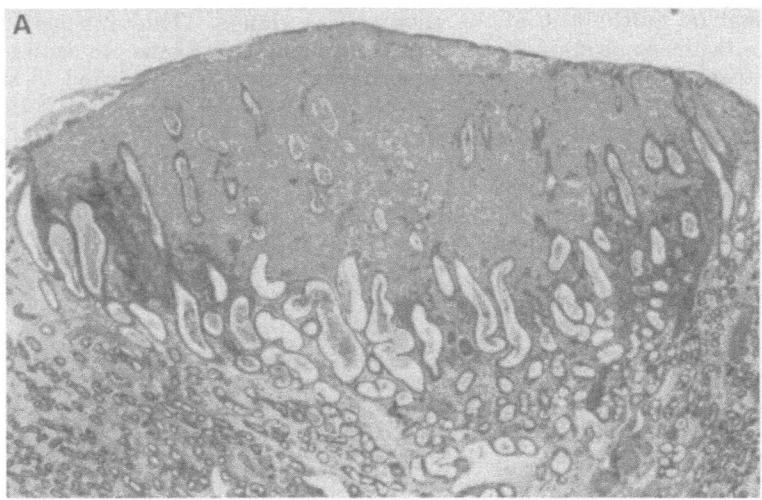

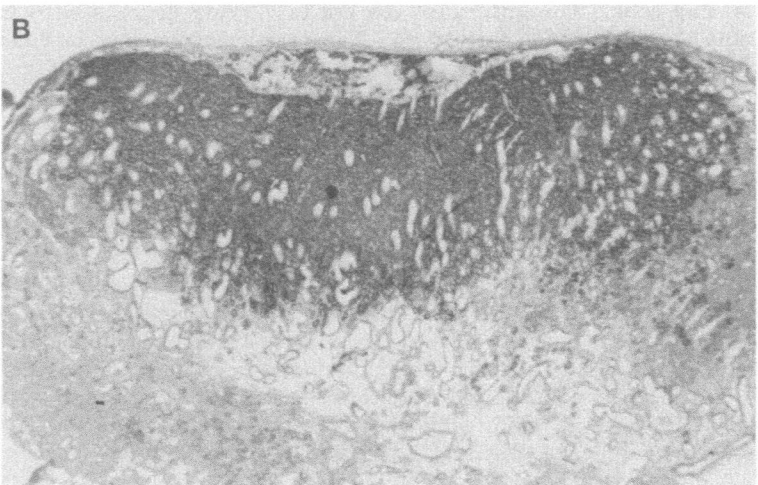

Figure 1.   Photomicrographs of equine endometrial cups.   A) Bouin's fixed endometrial cup from a 60 day pregnancy, 32X magnification, hematoxylin and eosin staining. B) Cryostat section of an endometrial cup from a 60 day horse pregnancy labeled with monoclonal antibody F71.10 using an indirect immunoperoxidase technique developed with Amino-ethyl-carbamizol (AEC), and counterstained with hematoxylin, 16X magnification. The dark staining trophoblast-derived endometrial cup cells are easily distinguishable from the surrounding maternal endometrial tissue.

Table 3

Reactivity patterns of 3 monoclonal antibodies with specificty for non-invasive equine trophoblast[1]

Hybrid cell line	Endometrial cup cells	Allantochorion		Endometrium			Other adult tissues			
		Trophoblast	Non-trophoblast[2]	Lumenal epithelium	Glandular epithelium	Interstitial tissue	Lympho-cytes	Kidney[3]	Intestine[3]	Skin[3]
F71.7	−	++	±[4]	±	±[5]	+	−	±	±	±
F71.11	−	++	−	+	±	+	−	++	++	++
F71.12	−	++	−	+	±	+	−	++	++	++

1 Determined by reactivity on cryostat sections of the tissues indicated using indirect immunoperoxidase labeling.
2 Chorionic mesoderm and allantoic endoderm.
3 The cell types stained by the monoclonal antibodies in these adult tissue sections were not identified.
4 Strongly positive on endoderm, but negative on mesodermal tissue.
5 Strongly positive on the dilated glands within and around the endometrial cups, but negative on other endometrial glands.

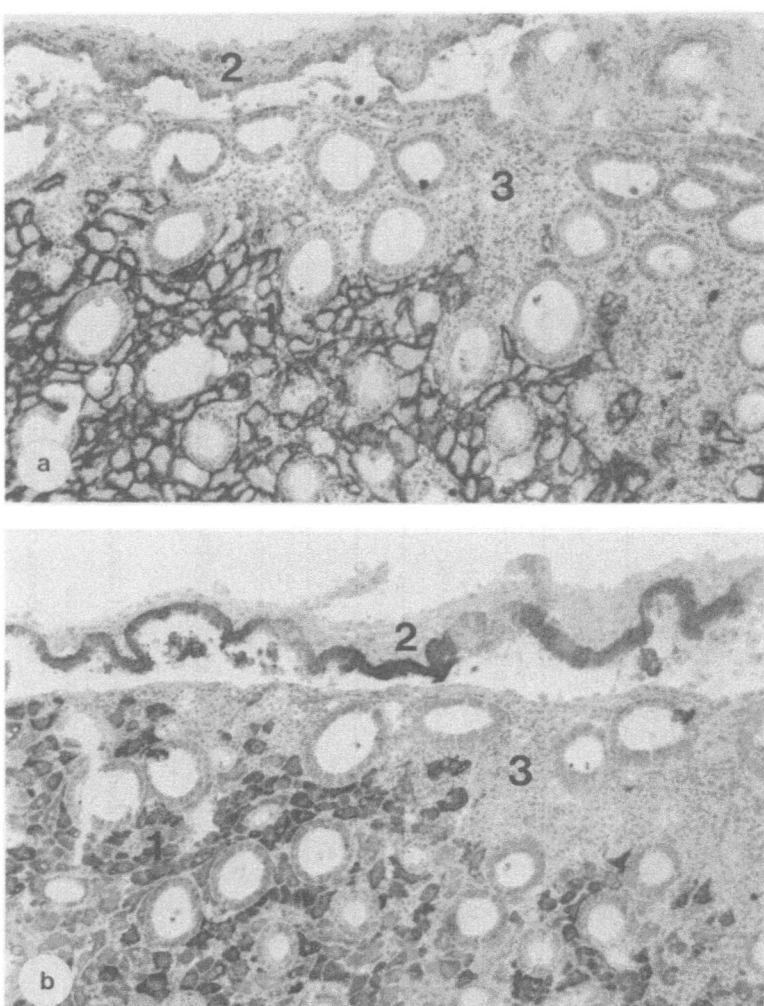

Figure 2. Photomicrographs of tissues of the fetal-maternal interface in equine pregnancy. Cryostat sections of endometrial cup tissue (a,b,e,f) or of the endometrium-allantochorion border (c,d) from 60 day horse pregnancies. The sections were labeled with various monoclonal antibodies using an indirect immunoperoxidase technique, developed with AEC, and counterstained with hematoxylin. 100X magnification. The numbers identify components of placental or maternal tissues: 1) Endometrial cup cells (placental). 2) Allantochorion: Chorion or single-celled non-invasive trophoblast, chorionic mesoderm, and allantoic endoderm. 3) Maternal endometrium.

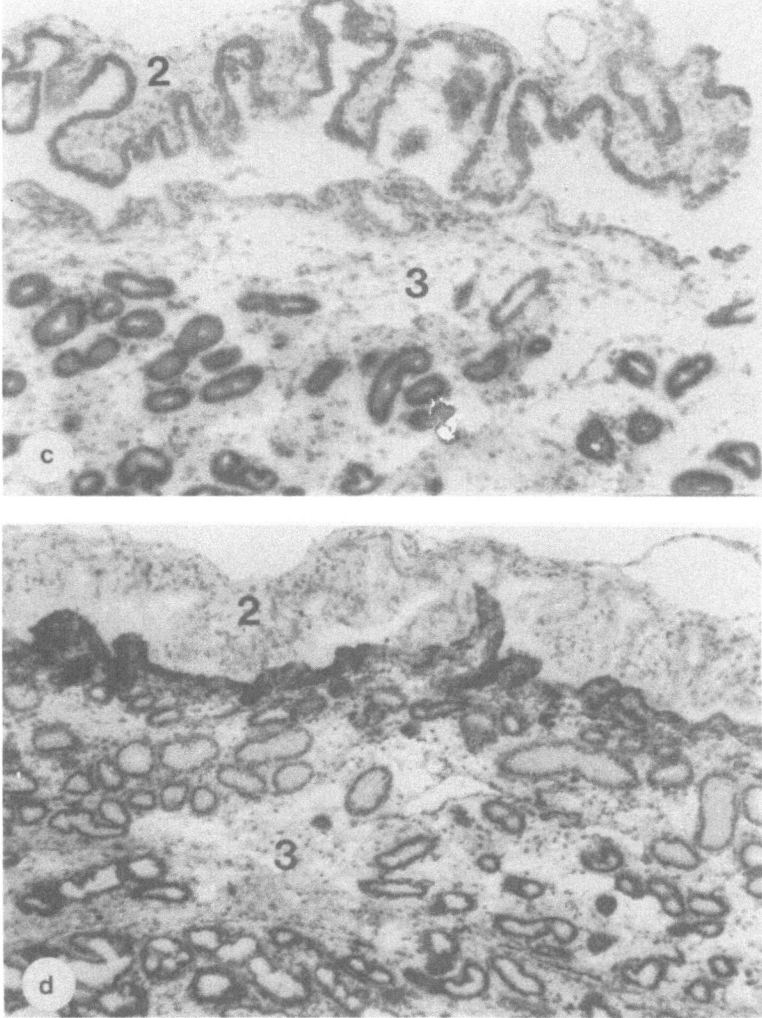

The sections show lumenal epithelium, dilated and/or normal endometrial glands, endometrial stroma, and leukocyte accumulations (only around the endometrial cups). a) Monoclonal antibody (McAb) 71.8, which produced intense honey-comb labeling of endometrial cells and faint labeling of the non-invasive trophoblast. No other cells from the placenta or endometrium were labeled by this antibody; b) McAb F71.14, which produced a more uniform pattern on endometrial cup cells than 71.8, and which labeled the non-invasive trophoblast strongly; c) McAb F71.2, which labeled the non-invasive trophoblast of the allantochorion and the epithelium of the superficial endometrial glands, but not the lumenal epithelium of the endometrium; d) McAb F71.3, which labeled the epithelium of the lumenal endometrium and endometrial glands, but not any of the tissues of the allantochorion;

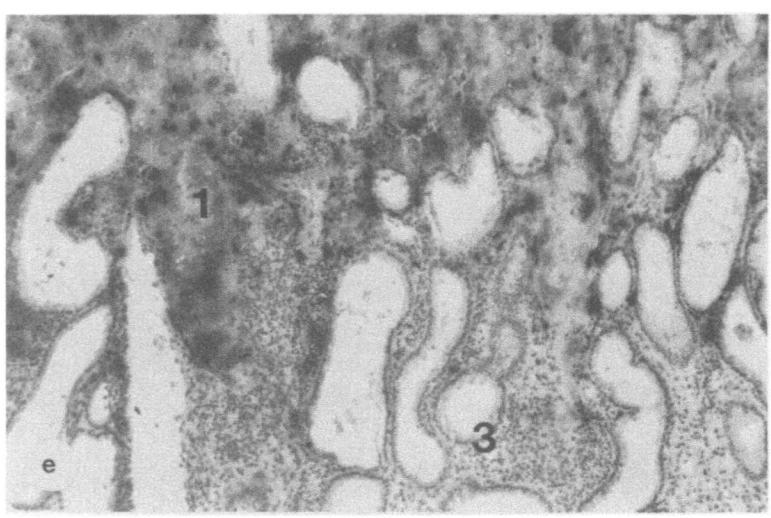

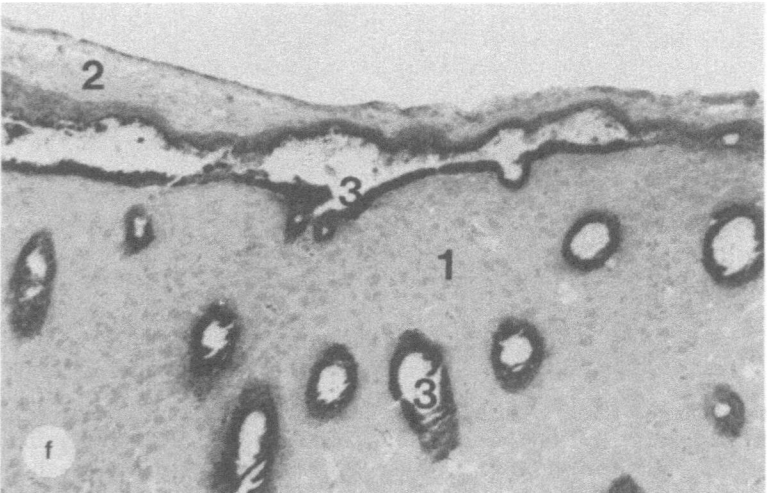

e) McAb F67.1, labeled the endometrial cup cells with uneven intensity and produced faint staining of surrounding tissue; f) McAb F71.7, produced strong labeling of the non-invasive trophoblast, and the epithelium of the endometrium and dilated endometrial glands, but failed to label the endometrial cup cells.

Many of the cell identification problems encountered in the study of hemochorial placentae do not arise in species with epitheliochorial placentae, such as the horse and pig. However, the relative simplicity of the equine placenta provides a unique opportunity for the study of particular aspects of trophoblast structure and function. It is easy to identify fetal and maternal tissues in ordinary histological sections, and the two forms of equine trophoblast, the non-invasive villi of the allantochorion and the invasive endometrial cup cells, can also be readily distinguished. The endometrial cup cells are the sole source of eCG and the maternal mononuclear cell accumulation which surrounds the endometrial cups, but which does not normally occur at the border of the non-invasive trophoblast, suggests that the cup cells are involved in immunological reactions during pregnancy. Thus, the invasive trophoblast appears to determine endocrinological and immunological aspects of pregnancy, while the non-invasive trophoblast is entirely responsible for fetal nutrition. Equine trophoblast can be studied in situ by biopsy section, in cell culture (Allen and Moor, 1972), and in vivo using hormonal and immunological assays (Allen, 1969; Antczak et al., 1984).

The critical importance of the endometrial cups is demonstrated vividly by the extraspecific donkey-in-horse pregnancy model established by embryo transfer. In this one type of equine pregnancy, endometrial cups do not develop and the allantochorion is subject to a severe maternal cellular reaction that results in abortion by day 95 in virtually all cases (Allen, 1982; Allen et al., 1985). The observations that treatment of mares carrying donkey conceptuses with exogenous eCG does not prevent pregnancy loss (Allen, 1982) whereas certain immunological treatments result in a marked increase in the fetal survival to term (Allen et al., 1986), suggest that an immunological function of the endometrial cups may be more important to equine pregnancy than their hormonal production.

The monoclonal antibodies described here will be used in attempts to identify the immunological function of the unique invasive trophoblast cells that comprise the endometrial cups. The screening strategy employed in the present study eliminated monoclonal antibodies that identified molecules expressed by both equine trophoblast and lymphocytes. From the cultures of two cell fusion experiments, 19 hybridomas which labeled cryostate sections of equine placental tissue were cloned. The 19 monclonal antibodies produced a variety of reaction patterns when assayed on sections of equine tissue using an indirect immunoperoxidase staining technique. These patterns provide evidence for a mosaic of molecules expressed by equine trophoblast and a means of distinguishing invasive from non-invasive components of the trophoblast using molecular markers.

Seven antibodies identified six different components of the endometrial cup cells (Table 2, Figures 1 and 2). These included antigens unique to endometrial cups (F67.1), antigens restricted largely to trophoblastic tissue (F71.8, F71.10, and F71.14), and antigens expressed by both placental and maternal tissues (F71.1, F71.2, and F71.3). Antibody F71.3 is of particular interest. Although this antibody stained several tissues of the cryostat sections of the maternal endometrium, it did not label the non-invasive trophoblast of the allantochorion. It thus distinguished non-invasive trophoblast from endometrial cups on the basis of a molecule that the cup cells share with adult tissues. This suggests a differentiation step during endometrial cup development which is not taken by the non-invasive trophoblast. In contrast to this difference between non-invasive trophoblast and endometrial cup cells, however, is the reaction pattern given by the monoclonal antibodies F71.7, F71,11, and F71.12 (Table 3, Figure 2f). These antibodies identified an antigen(s) shared by several adult cell types and the non-invasive trophoblast, but not by the endometrial cups. This suggests that either the endometrial cup cells lose the ability to express this antigen(s), or that the non-invasive trophoblast of the allantochorion develops the capacity to express it. Longitudinal studies of equine non-invasive trophoblast and endometrial cups at different gestational ages should distinguish these possibilities.

The antigens identified by monoclonal antibodies F71.9 and F71.13 may represent secreted extracellular products. The antibodies labeled material in the lumen of endometrial glands and at the interface of the allantochorion and endometrium, but no other tissues.

Only one antibody, F71.15, demonstrated species specificity, reacting with donkey, but not horse, tissues. In light of the close evolutionary relationship between horses and donkeys it is not surprising that most of the rat monoclonal antibodies produced in this study reacted with tissues from both species. F71.15 was broadly reactive with both placental and maternal tissues, and it is therefore not very useful in distinguishing fetal from adult components of the endometrial cup reaction. However, the antibody should be valuable in future studies of interspecies hybrids, and in the particularly interesting extraspecific donkey-in-horse model of pregnancy failure (Kydd et al., 1982; Allen et al., 1985).

SUMMARY

Monoclonal antibodies were raised to isolated placental tissues of horse and donkey (endometrial cup cells and non-invasive trophoblast) using conventional cell hybridization techniques. Supernatants from growing hybrids were screened for lack of reactivity with horse lymphocytes and for positive labeling of cryostat sections of endometrial cup tissue and/or allantochorion using an indirect immunoperoxidase technique. Nineteen cell lines were cloned from two fusion experiments. Thirteen of these antibodies were assigned to one of four groups, based on their reactivity patterns; Group I monoclonal antibodies identified antigens unique to endometrial cup cells or shared by cup cells and fetal or maternal tissues (F67.1, F71.1, F71,2, F71.3, F71.8, F71.10, and F71.14); Group II, antigens shared by the non-invasive trophoblast of the allantochorion and some maternal tissues (F71.7, F71.11, and F71.12); Group III, antigen(s) expressed strongly at the interface of the endometrium and allantochorion (F71.9, F71.13); Group IV, an antigen expressed widely on donkey fetal and maternal tissues, but absent from horse tissue (F71.15). These antibodies are being used to explore the structural and functional characteristics of the invasive and non-invasive forms of trophoblast in equids.

ACKNOWLEDGEMENTS

This work was supported by USPHS grant HD-15799, the Harry M. Zweig Memorial Fund for Equine Research, the Dorothy Russell Havemeyer Foundation, Inc., the Thoroughbred Breeders' Association, and the Horse Race Betting Levy Board of Great Britain.

REFERENCES

Allen, W.R. (1969) Factors influencing pregnant mare serum gonadotrophin production. *Nature* 223, 64-66.

Allen, W.R. and Moor, R.M. (1972) The origin of the equine endometrial cups. I. Production of PMSG by fetal trophoblast cells. *J. Reprod. Fert.* 29, 313-316.

Allen, W.R., Hamilton, D.W., and Moor, R.M. (1973) The origin of the equine endometrial cups. II. Invasion of the endometrium by trophoblast. *Anat. Rec.* 177, 485-502.

Allen, W.R. (1975) Immunological aspects of the equine endometrial cup reaction. In: *Immunology of the Trophoblast*, (eds.), R.G. Edwards, C. Howe, and M.H. Johnson, Cambridge: Cambridge University Press, pp. 217-253.

Allen, W.R. (1982) Immunological aspects of the endometrial cup reaction and the effect of xenogeneic pregnancy in horses and donkeys. *J. Reprod. Fert.*, Suppl. 31, 57-94.

Allen, W.R., Kydd, J., Boyle, M.S., and Antczak, D.F. (1985) Between-species transfer of horse and donkey embryos: a valuable research tool. *Eq. Vet. J.*, Suppl. 3, 53-62.

Allen, W.R., Kydd, J., and Antczak, D.F. (1986) Maternal immunological response to the trophoblast in xenogeneic equine pregnancy. In: *Proc. Second Banff Conference on Reproductive Immunology*, (eds.), T.J. Gill and W. Wegmann, in press.

Antczak, D.F., Brown, D., and Howard, J.C. (1979) Analysis of lymphocytes reactive to histcompatibility antigens. I. A quantitative titration assay for mixed lymphocyte interactions in the rat. *Cell. Immunol.* 43, 304-316.

Antczak, D.F., Bright, S.M., Remick, L.H., and Bauman, B. (1982) Lymphocyte alloantigens of the horse. I. Serologic and genetic studies. *Tissue Antigens* 20, 172-187.

Antczak, D.F., Miller, J.M., and Remick, L.H. (1984) Lymphocyte alloantigens of the horse. II. Antibodies to ELA antigens produced during equine pregnancy. *J. Reprod. Immunol.* 6, 283-297.

Bulmer, J.N. and Johnson, P.M. (1985) Antigen expression by trophoblast populations in the human placenta and their possible immunobiological relevance. *Placenta* 6, 127-140.

Galfre, G., Howe, S.C., Milstein, C., Butcher, G.B., and Howard, J.C. (1977) Antibodies to major histocompatibility antigens produced by hybrid cell lines. *Nature* 266, 550-552.

Johnson, P.M., Cheng, H.M., Molloy, C.M., Stern, C.M.M., and Slade, M.B. (1981) Human trophoblast-specific surface antigens identified using monoclonal antibodies. *Amer. J. Reprod. Immunol.* 1, 246-254.

Kydd, J., Miller, J., Antczak, D.F., and Allen, W.R. (1982) Maternal anti-fetal cytotoxic antibody responses to equids during pregnancy. *J. Reprod. Fertil.,* Suppl. 32, 361-369.

Mason, D.Y., Naiem, M., Abdulaziz, Z., Nash, J.R.G., Gatter, K.C., and Stein, H. (1982) Immunohistological applications of monoclonal antibodies. In: *Monoclonal Antibodies in Clinical Medicine,* (eds.), A.J. McMichael and J.W. Fabre, New York: Academic Press Inc., pp. 585-635.

Newman, M.J., Beegle, K.H., and Antczak, D.F. (1984) Xenogenic monoclonal antibodies to cell surface antigens of equine lympocytes. *Amer. J. Vet. Res.* 45, 626-632.

Oi, V.T. and Herzenberg, L.A. (1980) Immunoglobulin-producing hybrid cell lines. In: *Selected Methods in Cellular Immunology,* (eds.), B.B. Mishell and S.M. Shiigi, San Francisco: WH Freeman Co., pp. 351-372.

Parrish, C.R., Carmichael, L.E., and Antczak, D.F. (1982) Antigenic relationships between canine parvovirus type-2, feline panleukopenia virus, and mink enteritis virus using conventional antisera and monoclonal antibodies. *Arch. Virol.* 72, 267-278.

Sunderland, C.A., Redman, C.W.G., and Stirrat, G.M. (1981) Monoclonal antibodies to human syncytiotrophoblast. *Immunol.* 43, 541-546.

Trophoblast Research 2:215-222, 1987

THE EXPRESSION OF EPITHELIAL ANTIGENS BY TROPHOBLASTS AND TUMOR CELLS USING MONOCLONAL ANTIBODIES TO HUMAN AMNION (GB3 AND GB5)

Chang-Jing G. Yeh and Bae-Li Hsi

INSERM U210
Faculté de Médecine
06034 Nice-Cedex, FRANCE.

INTRODUCTION

The differentiation processes of the amnion and the chorion take place simultaneously with the development of the human embryo (Boyd and Hamilton, 1970). The amnion has been considered to be derived from the trophoblasts which adjoin to the embryonic ectoderm (Bourne, 1962). At term, the amniotic epithelium of reflected and placental amnion lines the amniotic cavity, wraps around the umbilical cord and continues with the fetal skin. By using polyclonal antibodies to human amnion, three groups of antigens in common with some ectodermal tissues were identified (Hsi et al., 1984a). Recently, eight monoclonal antibodies in human amnion have been reported (Hsi and Yeh, 1986). Among them, GB3 reacted with epithelial basement membranes, but the antigen was not among the known ubiquitous components of basement membranes and GB5 recognized the junctional substances of epithelial cells. These antigens are likely to contribute to the cohesion with the subjacent tissues and to the intercellular binding of the epithelium. We report here the expression of these epithelial antigens by trophoblasts in chorion laeve and basal plate, as well as that of breast tumor cells.

MATERIALS AND METHODS

Tissues

Placental tissues from ten normal pregnancies were collected from vaginal deliveries at St. Roch Hospital, Nice, France. Tissues from amniochorions and basal plates were sampled according to Hsi and Yeh (1986). Specimens of amniochorions were sandwiched between two slices of fresh mouse liver (Hsi et al., 1982). Five normal adult skin specimens were collected at the Department of Dermatology, Pasteur Hospital, Nice, France. Biopsies of ten human breast tissues were collected from the Department of Pathology, Centre Antoine Lacassagne, Nice, France. Parallel sections from each specimen were stained with hematoxylin and eosin for examination by pathologists. The tissue blocks were wrapped with aluminum foil and snap-frozen in liquid nitrogen. The frozen tissues were kept at -20°C until use.

Submitted: 10 October 1985; Accepted: 31 July 1986.

Antibodies

GB3 and GB5 were mouse monoclonal antibodies (IgGl) secreted by hybridoma cell lines derived from the fusion of P3-NS1/1-Ag4-1 (NS-1) cells and spleen cells of a mouse immunized with detergent solublized human amnion (Hsi and Yeh, 1986). On reflected and placental amnion, GB3 reacted with amniotic epithelial basement membranes, and GB5 recognized amniotic epithelial cells. Culture medium form NS-1 cells was used as negative control. Fluorescein isothiocyanate (FITC) labeled rabbit anti-mouse immunoglobulin (Ig) was purchased from Dakopatts (Copenhagen, Denmark).

Immunofluorescence Procedure

Frozen sections (4.5 μm) were cut using a cryostat (Bright Instrument, Huntingdon, England). The sections were air-dried and washed in 0.15 M phosphate buffered saline (PBS), pH 7.2 for 10 min. Each of the prewashed tissue sections or fixed cells was incubated with 50 μl supernatant fluid of the culture medium for 2 hr. After washing in PBS for 5 min, the sections were reacted with a 1:40 dilution of 50 μl FITC conjugated rabbit anti-mouse Ig for 20 min. The samples were further washed for 5 min in PBS, then mounted in mounting medium AF1 (Citifluor Ltd., London, England). In some experiments, they were counter stained with propodium iodide as described by Ockleford et al. (1981).

Microscopy

Slides were examined by epi-illumination with the use of Zeiss Universal Microscope fitted with an HBO 50 mercury arc lamp, epifluorescence condenser III RS, and filter sets for FITC and propidium iodide studies (Yeh et al., 1981). Immunofluorescence results were recorded on Kodak Ektachrome 200 films using a Zeiss MC63 camera system.

RESULTS

Amniochorion

In amnion, GB3 reacted with the amniotic basement membrane (Figure 1a), and GB5 recognized the amniotic epithelial cells (Figure 1b). In chorion laeve, some segments of cytotrophoblastic basement membranes and some cytotrophoblasts near the basement membranes reacted with GB3 and GB5 (Figures 1a and 1b). In contrast, most cytotrophoblasts near the maternal decidua and the maternal decidual tissues did not react with these antibodies.

Basal Plate

GB3 reacted with a few cytotrophoblasts, the reactivity of GB3 was not in the cytoplasm, it seemed to be located on the cell membranes or around the trophoblasts (Figure 2). GB5 did not recognize any cytotrophoblasts.

Skin

In epidermis, GB3 reacted only with the epithelial basement membranes (Figure 3a). The basement membranes of blood vessels and nerve bundles were non-

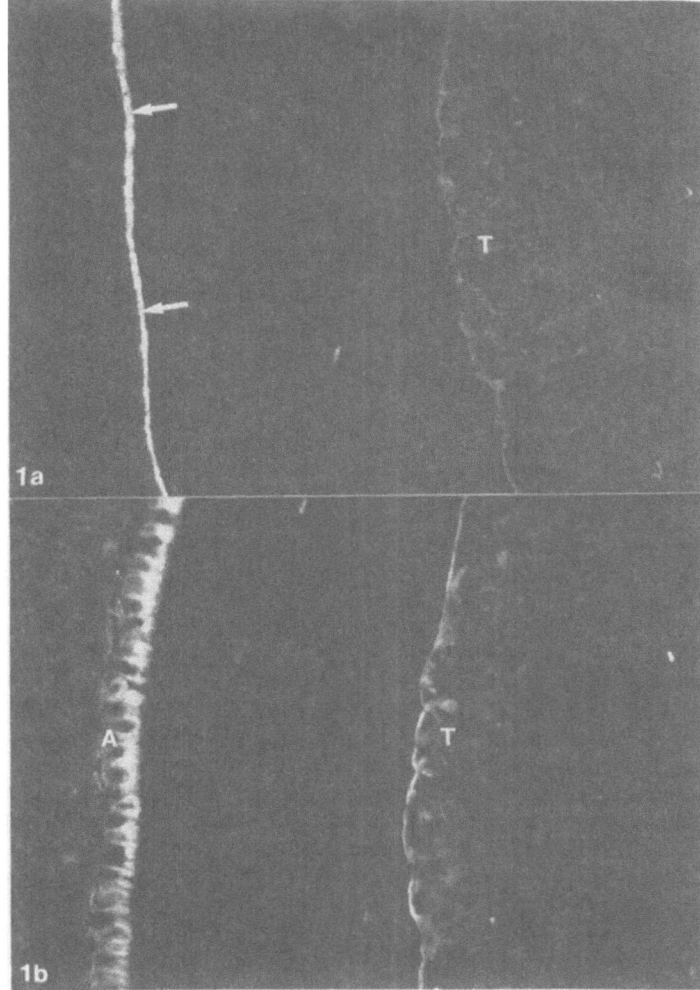

Figure 1. The reactivities of GB3 and GB5 on the amniochorion. X275. a) GB3 reacted with amniotic epithelial basement membrane (arrows), cytotrophoblastic basement membrane and around some cytotrophoblasts (T). b) GB5 recognized amniotic epithelial cells (A) and the basal part of the cytotrophoblastic layer (T). The fetal mesenchyme and maternal decidua were non-reactive with GB3 and GB5.

reactive. GB5 recognized the cells in the stratum granulosum, stratum spinosum and stratum basal, those of stratum corneum were non-reactive (Figure 3b). The reactivities were localized between the epithelial cells. In dermis, GB3 and GB5 reacted with the basement membranes and the intercellular materials of epidermal appendages, respectively.

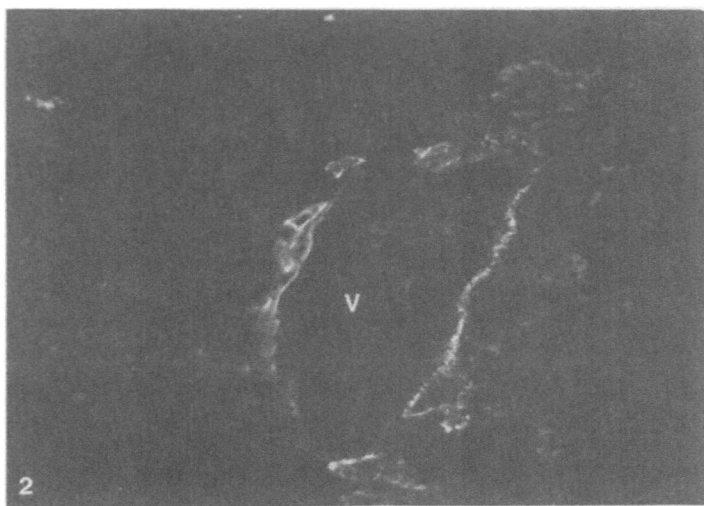

Figure 2. The reactivity of GB3 on the basal plate. X275. GB3 reacted with some cytotrophoblasts surrounding a degenerated chorionic villus (V) in the fibrous tissue of the basal plate.

Breast Tissues

In normal breast tissues, GB3 reacted strongly with the basement membranes of mammary glands, and GB5 recognized the glandular epithelium, the reactivities resided between the epithelium. In intraductal carcinoma specimens, the reactivity of GB3 on the ductal basement membranes reduced to a thin line which encircled the tumor cells or became completely negative (Figure 4a), and that of GB5 also greatly decreased (Figure 4b). Most strikingly, in areas of infiltrated carcinoma, the tumor cells did not react with GB3 or GB5.

DISCUSSION

The results of this study demonstrated that essentially two types of cytotrophoblasts in chorion laeve and basal plate could be distinguished by their expression of the epithelial antigens recognized by GB3 and GB5. Namely, the stationary cytotrophoblasts which were located near the fetal mesenchyme reacted with both antibodies; and the invasive cytotrophoblasts which were situated near the maternal decidua did not. On normal adult skin, GB3 detected the basement membranes of epidermis and epidermal appendages; and GB5 reacted with junctional structures of the epithelium. By immunoelectron microscopic studies, the antigen of GB3 was localized on lamina lucida of basal lamina; and that of GB5 was situated at the desmosome region (unpublished data). These results suggest that these epithelial antigens may play a role in maintaining the integrity of the epithelial tissues. It could be postulated that the cytotrophoblasts near the cytotrophoblastic basement membranes manifest both antigens for binding to the fetal mesenchyme, whereas the de-expression of antigens of GB3 and GB5 in cytotrophoblasts near the maternal

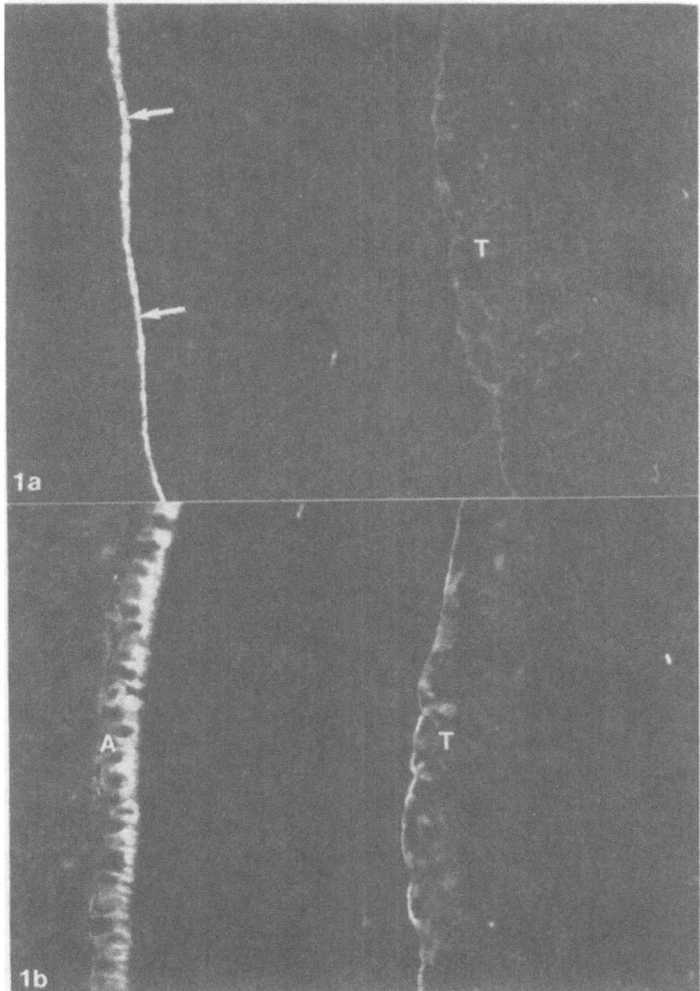

Figure 3. The reactivities of GB3 and GB5 on the skin. X275. Epidermis (E), Dermis (D). a) GB3 reacted only with the epithelial basement membrane. b) GB5 recognized the epithelial cells of the epidermis. The stratum corneum on the surface was non-reactive.

decidua may facilitate the mobility of these trophoblasts. Interestingly, the GB3 and GB5 reactive cytotrophoblasts were usually non-reactive with the monoclonal antibody to HLA-A,B,C common determinant (clone W6/32) (Hsi et al., 1984b). It has been reported that this population of cytotrophoblast reacted with monoclonal antibody to human milk fat globule (clone HMFG-1) (Bulmer and Johnson, 1985). From the pattern of HMFG-1 on normal mammary epithelial cells (Burchell et al., 1983), it seems that GB5 and HMFG-1 detect similar antigens on epithelial cells.

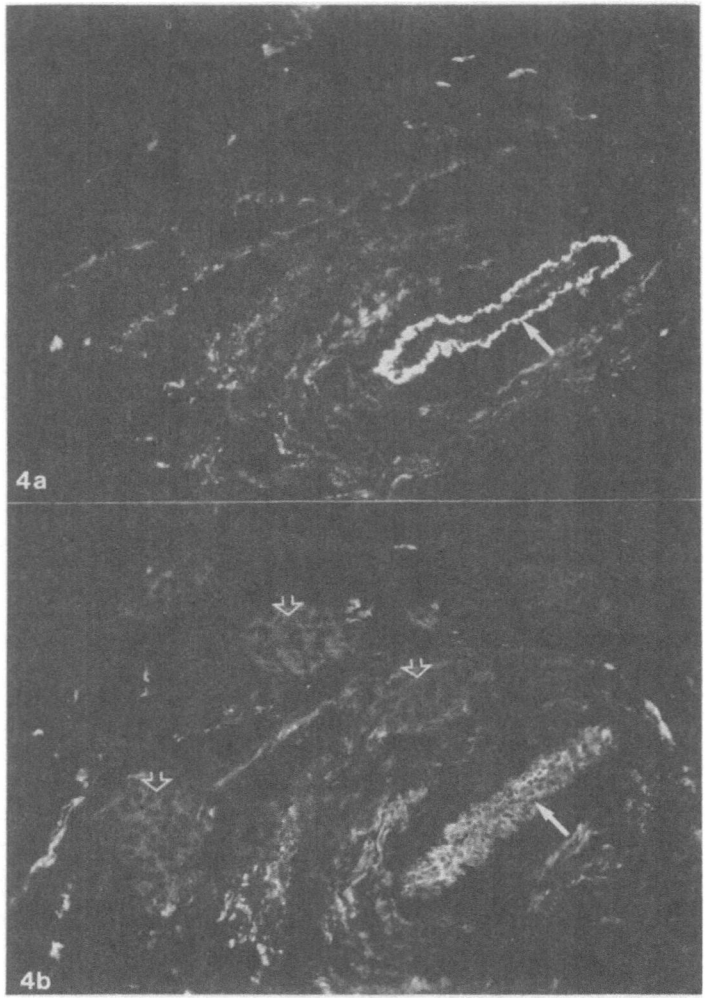

Figure 4. The reactivities of GB3 and GB5 on parallel sections of the breast adenocarcinoma. X275. a) GB3 recognized the glandular epithelial basement membrane (arrow). The surrounding tumor nodules were completely negative. b) GB5 reacted strongly with the glandular epithelium (solid arrow) and weakly with the nodules of carcinoma cells (open arrows). Fibrillar structures in the connective tissues are autofluorescence.

Taken into account the ectodermal origin of breast glandular epithelium, it is not surprising that this epithelium was also recognized by GB3 and GB5. However, for the intraductal carcinoma, the basement membranes and the epithelial cells delineated by GB3 and GB5 greatly diminished. Moreover, when the tumor cells broke away from the breast ducts to become infiltrated carcinoma, their reactivity with GB3 and GB5 were beyond the detectibility by immunofluorescence (Yeh et al., 1986). Thus, these epithelial antigens may serve as limiting factors for tumor metastasis. The reactivity pattern of the trophoblast can thus be extrapolated from that of the tumors to that of the invasive cells which are GB3 and GB5 non-reactive. More experiments are in progress to determine the trigger and the behavior of the growth of these tumor cells, in turn to shed light on the materno-fetal immunobiology.

SUMMARY

GB3 and GB5 are mouse monoclonal antibodies raised against human amnion; GB3 recognized epithelial basement membranes and GB5 detected junctional substances between the epithelial cells. In chorion laeve, the cytotrophoblastic basement membranes and some cytotrophoblasts near the fetal mesenchyme were reactive with both GB3 and GB5. However, the majority of the cytotrophoblasts in basal plate and in chorion laeve which were usually in the proximity of the maternal decidua, were reactive with neither GB3 nor GB5, a situation similar to the metastatic breast adenocarcinomas in which the synthesis of the antigens of GB3 and GB5 discontinued. In conclusion, this study showed that trophoblasts at different anatomical interfaces, can modulate their gene expression of the otherwise normal epithelial antigens which were recognized by GB3 and GB5.

ACKNOWLEDGEMENTS

We thank Mrs. H. Duplay and Mrs. F. Ettore, Centre Antoine Lacassagne, Nice, France for supplying the breast tissues and providing the pathological reports. This research was supported in part by the Federation Nationale des Centres de Lutte Contre le Cancer, Paris and the Association pour la Recherche sur le Cancer, Villejuif, France. The authors are Charges de Recherche of the Centre National de la Recherche Scientifique, Paris, France.

REFERENCES

Bourne, G.L. (1962) *The Human Amnion and Chorion*. London: Lloyd-Luke.

Boyd, J.D. and Hamilton, W.J. (1970) *The Human Placenta*. Cambridge, MA: W. Heffer.

Bulmer, J.N. and Johnson, P.J. (1985) Antigen expression by trophoblast populations in the human placenta and their possible immunobiological relevance. *Placenta* 6, 127-140.

Burchell, J., Helg, D., and Taylor-Papadimitriou, J. (1983) Complexity of expression of antigenic determinants, recognized by monoclonal antibodies HMFG-1 and HMFG-2, in normal and malignant human mammary epithelial cells. *J. Immunol.* 131, 508-513.

Hsi, B.-L., Yeh, C-J.G., and Faulk, W.P. (1982) Human amniochorion: tissue-specific markers, transferrin receptors and histocompatiblity antigens. *Placenta* 3, 1-12.

Hsi, B.-L., Yeh, C.-J.G., and Faulk, W.P. (1984a) Characterization of antibodies to antigens of the human amnion. *Placenta* 5, 513-522.

Hsi, B.-L., Yeh, C-J.G., and Faulk, W.P. (1984b) Class I antigens of the major histocompatibility complex on cytotrophoblast of human chorion leave. *Immunology* 52, 621-629.

Hsi, B-L., and Yeh, C.-J.G. (1986) Monoclonal antibodies to human amnion. *J. Reprod. Immunol.* 9, 11-21.

Ockleford, C.D., Hsi, B.-L., Wakely, J., Badley, R.A., Whyte, A., and Faulk, W.P. (1981) Propidium iodide as a nuclear marker in immunofluorescence. I. Use with tissue and cytoskeleton studies. *J. Immunol. Methods* 43, 261-267.

Yeh, C.-J.G., Hsi, B.-L., and Faulk, W.P. (1981) Propidium iodide as a nuclear marker in immunofluorescence. II. Use with cellular identification and viability studies. *J. Immunol. Methods* 43, 269-275.

Yeh, C.-J.G., Hsi, B.-L., and Ettore, F. (1986) Normal epithelial antigens recognized by GB3 and GB5 are diminished in intraductal and lost in infiltrating human breast carcinoma. *Breast Cancer Res. Treat.*, in press.

Trophoblast Research 2:223-231, 1987

STUDIES OF C1q DEPOSITS IN THE HUMAN PLACENTA USING MONOCLONAL ANTIBODIES TO HUMAN EXTRA-EMBRYONIC TISSUES

Bae-Li Hsi and Chang-Jing G. Yeh

INSERM U210
Faculté de Médecine
06034 Nice-Cedex, France

INTRODUCTION

The presence of the first complement component around some fetal stem vessels in the chorionic villous stroma of term human placentae was first reported by Faulk and Johnson (1977) using the antiserum to C1q. This finding was later extended to the immature placenta (Johnson and Faulk, 1978). Sinha et al. (1984) showed that more anti-C1q positive fetal blood vessels and fibrinoid structures could be seen in the placentae from patients with pre-eclampsia. Although, the pathogenesis of this type of fetal stem vessels was unclear, it has been suggested that they may be mediated by immunological mechanisms (Sinha et al., 1984). Morphologically, two types of lesions of fetal stem vessels have been described, namely, fibromuscular sclerosis and obliterative endarteritis (Fox, 1978). The fibromuscular sclerosis of fetal stem arteries was often associated with villous infarction and the villi which were embedded in the fibrinous plaque. The cause of this lesion was suggested to be the cessation of fetal blood flow in these blood vessels. On the other hand, the fetal arterial obliterative endarteritis was found more often in the placentae of women with pre-eclampsia, essential hypertension or diabetes melitus. It has been proposed that it was the consequence of hypoxia of the placentae (Fox, 1967a).

During our continuous program of screening monoclonal antibodies raised against human extra-embryonic membrane, six different mouse monoclonal antibodies to human extra-embryonic tissues GB9, GB10, GB11, GB37, GB42, and GB45 were found to specifically react with different structures in the placenta. GB11 and GB45 reacted with the fibrinoid structures. GB10 reacted with the connective tissues of all the chorionic villous stroma and GB9 detected the villous stroma which were embedded in the fibrin plaque. GB37 and GB42 reacted with two different types of fetal stem vessels. These antibodies were used as markers for the placental tissues and studied with anti-C1q antiserum using double labeling immunofluorescence on term human placentae. The results showed that the anti-C1q positive chorionic villi were also positive for GB9 and they were always closely associated with the fibrinoid structures. Furthermore, the anti-C1q positive fetal blood vessels were consistently recognized by GB42.

Received: 15 September 1985; Accepted: 31 July 1986

MATERIALS AND METHODS

Tissues

Placental tissues from ten normal pregnancies were collected from vaginal deliveries at St. Roch Hospital, Nice, France. Tissue blocks of 1 cm³ from the central cotyledon of the placenta were wrapped with aluminum foil and snap-frozen in liquid nitrogen. The frozen tissues were kept at -20°C until use.

Antibodies

GB9, GB10, and GB11 were mouse monoclonal antibodies secreted by three hybridoma cell lines derived from the fusion of P3-NS1/1-Ag4-1 (NS-1) cells and the spleen cells of a mouse immunized with detergent solubilized human amnion (Hsi and Yeh, 1986). GB37, GB42, and GB45 are monoclonal antibodies secreted by three hybridoma cell lines from the fusion of NS-1 cells and the spleen cells of a mouse immunized with placental syncytiotrophoblast microvilli. The detail fusion and cloning procedures of GB37, GB42, and GB45 were identical to the procedures as described by Hsi and Yeh (1986). The syncytiotrophoblast microvilli were prepared according Smith et al. (1974). Culture medium from NS-1 cells was used as negative control. The immunoglobulin (Ig) class and subclass of the monoclonal antibodies were determined by Ouchterlony immunodiffusion using mouse monoclonal antibody typing kit (Serotec Ltd., Bicester, Oxon, England). GB9, GB10, and GB37 were IgG1; GB42 and GB45 were IgG3; and GB11 was IgM. Rabbit antiserum to human C1q, tetramethyl rhodamine isothiocyanate (TRITC) conjugated rabbit antiserum to human fibrinogen/fibrin (Fib), TRITC and fluorescein isothiocyanate (FITC) conjugated swine anti-rabbit Ig as well as FITC conjugated rabbit anti-mouse Ig were purchased from Dakopatts (Copenhagen, Denmark).

Immunofluorescence Procedures

Frozen sections (4.5 µm) were cut using a cryostat (Bright Instrument, Huntingdon, England). The sections were air-dried and washed in 0.15 M phosphate buffered saline (PBS), pH 7.2 for 10 min. For monoclonal antibody studies, the sections were incubated with 50 µl supernatant fluid of the culture medium for 2 hr. After washing in PBS for 5 min, sections were reacted with a 1:40 dilution of 50 µl FITC conjugated rabbit anti-mouse Ig for 20 min then washed again for 5 min in PBS. For anti-Fib, after washing in PBS for 10 min, the sections were incubated with TRITC conjugated rabbit anti-Fib at dilutions of 1:100 and 1:200 for 20 min, then washed for 10 min, 2 times each in PBS. For anti-C1q, the sections were incubated with rabbit anti-C1q at dilutions of 1:100 and 1:200 for 20 min, washed 2 times in PBS, then incubated with FITC or TRITC conjugated swine anti-rabbit Ig at 1:40 dilution for 20 min, and finally washed 2 times in PBS. Double labeling experiments were performed using anti-C1q, anti-Fib anitsera and the monoclonal antibodies. The sections were mounted in 40% AF1 mounting medium (Citifluor Ltd., London, England) in PBS. In some experiments, the sections were also stained with propidium iodide as described by Ockleford et al. (1981).

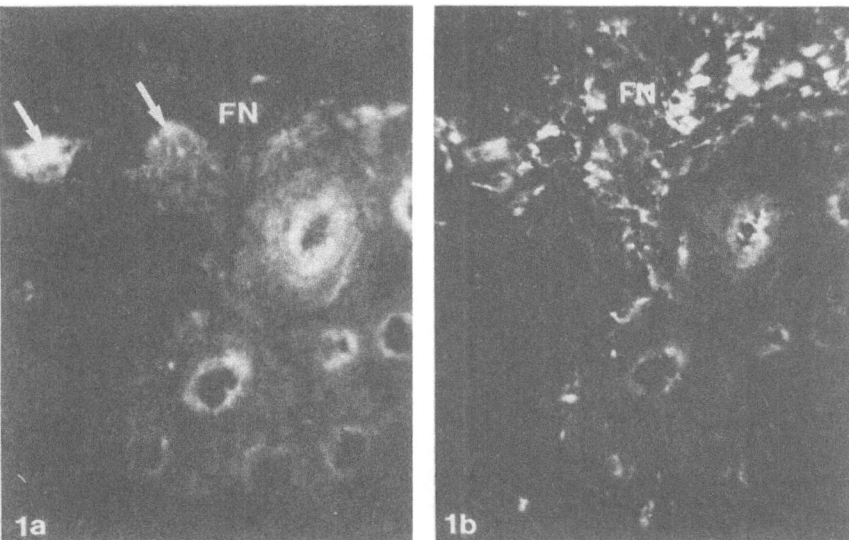

Figure 1. A placental section double labeled with anti-C1q and GB11. (X175)
a) The reactivity of anti-C1q can be identified around 4 of the 8 fetal blood vessels in a
chorionic villus. The reactivity can also be seen in 2 areas (arrows) near the fibrinoid
structures (FN).
b) GB11 reacted with the fibrinoid structures (FN) attached to this villus.

Microscopy

Slides were examined by epi-illumination with the use of Zeiss Universal
Microscope fitted with an HBO 50 mercury arc lamp, epi-fluorescence condenser III
RS and filter sets for FITC, TRITC, and propidium iodide studies as detailed in Yeh et
al. (1981). Immunofluorescence results were recorded on Kodak Ektachrome 200 film
using a Zeiss MC63 camera system.

RESULTS

Chorionic Villous Stroma

Two types of reactivities of anti-C1q antiserum on the chorionic villous stroma
could be distinguished. The first type of reactivity was located around some of the
fetal stem vessels, the reactivity was often extended into the surrounding chorionic
stroma. The second type of reactivity was identified in the chorionic villous stroma
which had no fetal blood vessels and was surrounded by the fibrinoid plaque (Figure
1a). The anti-Fib antiserum showed that large amounts of fibrin could be seen in the
villous stroma which were positive for anti-C1q (Figure 2b). In many conditions, the
clear differentiation of villous stroma from fibrinoid plaque was difficult. However,
the double labeling experiments with GB10 which reacted only with the stroma of the
villi, showed that all of the reactivity of anti-C1q was restricted in the villous stroma
but not in the fibrinoid (Figure 2a). GB9 has similar reactivity on the chorionic

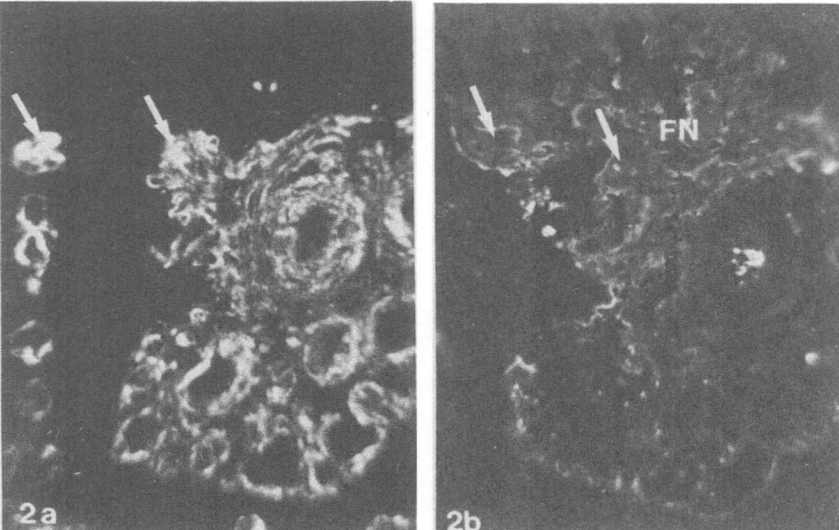

Figure 2. The parallel section of Figure 1 double labeled with GB10 and anti-Fib. (X175)
a) GB10 reacted with the chorionic stroma. Noted that the two anti-C1q positive areas near the fibrinoid structure (arrows) were GB10 positive.
b) Anti-Fib reacted with the fibrinoid structure (FN). Note that the reactivity also extended into the anti-C1q positive chorionic stroma (arrows).

villous stroma as the anti-C1q antiserum. Most of the villi which were anti-C1q positive were also GB9 positive (Figure 3). By double labeling with GB9 and anti-C1q, less than 1% of the villi showed reactivity only with GB9 or only with anti-C1q.

Fetal Stem Vessels

Of all the anti-C1q positive fetal blood vessels, only 50% of them showed constricted or obliterated lumens. GB37 reacted with trophoblastic basement membranes and some fetal blood vessels in the chorionic stem villi (Figure 4). The reactivity on the blood vessels was located on several layers of cells closely surrounding the endothelium. No apparent relationship existed between anti-C1q positive and GB37 positive blood vessels. Only 40% of the blood vessels which were positive with GB37, were also anti-C1q positive, and only 30% of the blood vessels which were anti-C1q positive, were also GB37 positive. On the contrary, there is a distinct relationship between GB42 positive fetal blood vessels and the reactivity of anti-C1q (Figures 5a and 5b). GB42 reacted with the muscular cells surrounding most of the fetal stem vessels. All the GB42 positive blood vessels were also anti-C1q positive.

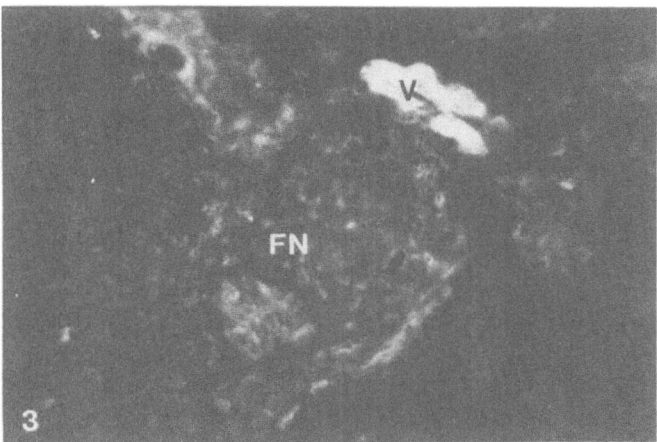

Figure 3. The reactivity of GB9 in the chorionic villi. (X275)
A chorionic villous stroma (V) which was attached to a fibrinoid structure (FN) was reactive with GB9. Other chorionic villi were negative.

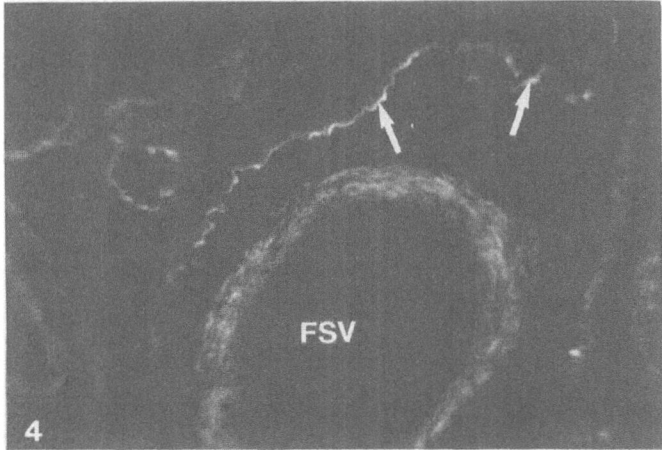

Figure 4. The reactivity of GB37 in the chorionic villi. (X275)
The reactivity can be identified on trophoblastic basement membrane (arrows) and around a fetal stem blood vessel (FSV).

Fibrinoid

Both GB11 and GB45 showed reactivity on the fibrinoid structures surrounding the chorionic villi (Figures 1b and 6). Unlike anti-Fib antiserum, both antibodies did not react with the fibrin in the intervillous space or in the villous stroma. The reactivity of GB11 was localized mostly surrounding the "holes" of the

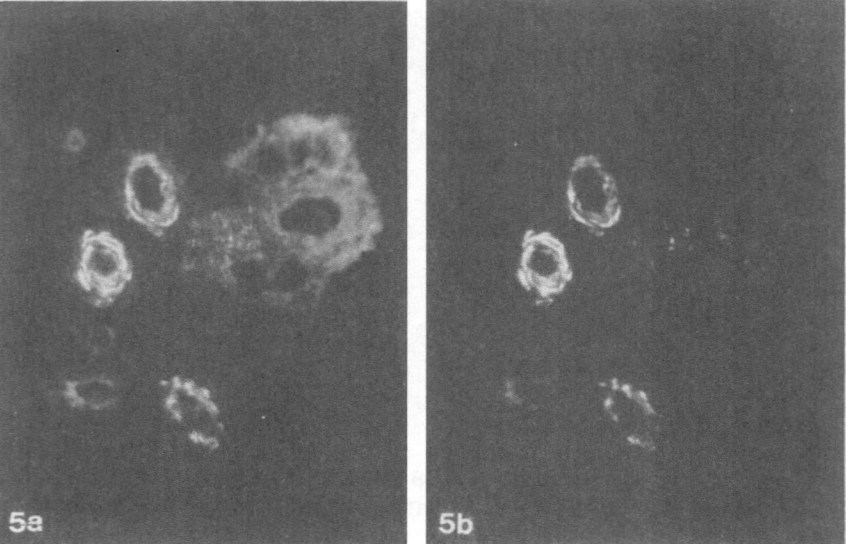

Figure 5. A placental section double labeled with anti-C1q and GB42. (X275)
a) The reactivity of anti-C1q can be identified around three fetal blood vessels and a part of the chorionic villous stroma.
b) The reactivity of GB42 can only be detected around the anti-C1q positive blood vessels.

foamy fibrinoid structures, whereas, GB45 showed a patchy, granular reactivity pattern in the fibrinoid. The double labeling experiments showed that 99% of the chorionic villi which were positive with anti-C1q antiserum, had fibrinoid structures attached to them. In some instances, where the syncytiotrophoblast was still intact, a thin layer of foamy GB11 positive fibrinoid structure could be seen between the trophoblast and the anti-C1q positive villous stroma.

DISCUSSION

The results of this study showed that anti-C1q positive chorionic villi could be identified in the normal term human placentae. They probably belong to the two types of chorionic villi, the stem villi and the mature intermediate villi as described by Kaufmann (1982). The presence of these villi are always closely associated with the fibrinoid structures in the placenta. Two types of anti-C1q reactivity in the chorionic villi could be differentiated, the first type was located around the fetal stem blood vessels which were positive for GB42, and the second type was located in the chorionic stroma which were positive for GB9 and were attached to the fibrinoid. Unlike the previous reports (McCormick et al., 1971; Faulk and Johnson, 1977; Sinha et al., 1984), all of the reactivities of anti-C1q were intravillous, no reactivity could be identified in the fibrinoid. The possible explanation for this is that without using chorionic villous stroma markers such as GB9 and GB10, it is very difficult to identify the villous structures which are embedded in the fibrinoid under the dark field immunofluorescence microscopy.

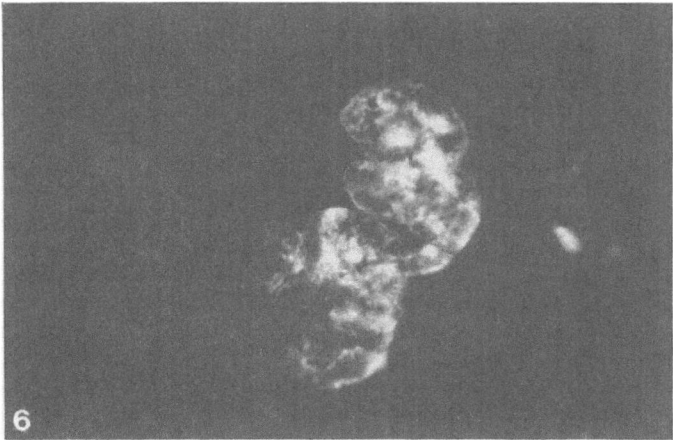

Figure 6. The reactivity of GB45 in the chorionic villi. (X275)
Patchy, granular activity of GB45 can only be seen in the fibrinoid structure.

Although the number of anti-C1q positive fetal stem vessels has been reported to be increased in the case of pre-eclampsia, it is unlikely that these fetal stem vessels are specifically related to the obliterative endarteritis (Fox, 1978). Since anti-C1q positive fetal stem vessels are present in all normal human placentae, while obliterative endarteritis is present only in about 10% of the term placentae from normal pregnancies (Fox, 1978). On the other hand, a close relationship can be drawn between arteries from fibromuscular sclerosis and the anti-C1q positive fetal stem vessels. Both fibromuscular sclerotic arteries and anti-C1q positive fetal stem vessels were shown to be closely related with the fibrinoid structures. It is unclear whether the fibrinoid formation around the chorionic villi is the cause or the result of the changes in fetal stem vessels.

Fox (1967b) considered that perivillous fibrin deposits, which many others call fibrinoid (Boyd and Hamilton, 1970), were a result of the thrombosis of the maternal blood in the intervillous space, a view shared by Moe and Jorgensen (1968) and Moe (1969). However, this hypothesis failed to explain why C1q was present in the chorionic villi which were associated with the perivillous fibrinoid structures. Faulk and Johnson (1977) pointed out that the identification of C1q may suggest the presence of immune complexes resulted from the reaction of maternal anti-allotype antibodies to fetal antigens at the level of placental stem vessels. It is possible that this reaction causes damage to the villous trophoblasts and the subsequent formation of fibrinoid. This hypothesis is supported by the finding that both GB9 which reacted with anti-C1q positive villous stroma and GB11 which reacted with the fibrinoid were found to recognize the intracellular structures of some nonvillous trophoblasts (Hsi and Yeh, 1986). Finally, this study demonstrated the usefulness of this series of GB monoclonal antibodies to extra-embryonic tissues in placental research. They are helpful in identifying and differentiating many different structures in the placentae which would be otherwise impossible by traditional histological means.

SUMMARY

In the normal human placenta, the chorionic villi which were positive for anti-C1q antiserum were studied by double labeling immunofluorescence with six monoclonal antibodies to human extra-embryonic tissues, GB9, GB10, GB11, GB37, GB42, and GB45. GB11 and GB45 reacted with the fibrinoid structures. These two antibodies showed that almost all of the anti-C1q positive villi had fibrinoid attached to them. GB10 reacted with all of the connective tissue stromas of the chorionic villi. This antibody indicated that the reactivity of anti-C1q was always restrictive in the villous stroma, but not in the fibrinoid structures. GB9 reacted with the same chorionic villous stroma which were positive for anti-C1q. GB37 and GB42 recognized two different types of fetal stem vessels. The GB42 positive vessels were also consistently reactive with anti-C1q antiserum. These findings suggest that the fibrinoid information in the placenta may be attributed by an immunologically mediated damage to the fetal blood vessels and their surrounding villous stroma, and this series of GB monoclonal antibodies will be useful tools for studying placental pathology.

REFERENCES

Boyd, J.D. and Hamilton, W.J. (1970) *The Human Placenta*. London: Heffer.

Faulk, W.P. and Johnson, P.M. (1977) Immunological studies of human placentae: identification and distribution of proteins in mature chorionic villi. *Clin. Exp. Immunol.* 27, 365-375.

Fox, H. (1967a) Abnormalities of foetal stem arteries in the human placenta. *J. Obstet. Gynaecol. Brit. Cwlth.* 74, 734-738.

Fox, H. (1967b) Perivillous fibrin deposition in the human placenta. *Am. J. Obstet. Gynecol.* 98, 245-251.

Fox, H. (1978) *Pathology of the Placenta*. London: W.B. Saunders, Co.

Hsi, B.-L. and Yeh, C.-J.G. (1986) Monoclonal antibodies to human amnion. *J. Reprod. Immunol.* 9, 11-21.

Johnson, P.M. and Faulk, W.P. (1978) Immunological studies of human placentae: identification and distribution of proteins in immature chorionic villi. *Immunol.* 34, 1027-1035.

Kaufmann, P. (1982) Development and differentiation of the human placental villous tree. In: *Structural and Functional Organization of the Placenta*, (eds.), P. Kaufmann and B.F. King, London: Karger.

McCormick, J.N., Faulk, W.P., Fox, H., and Fudenberg, H.H. (1971) Immunohistological and elution studies of the human placenta. *J. Exp. Med.* 133, 1-18.

Moe, N. (1969) Deposits of fibrin and plasma proteins in the normal human placenta: an immunofluorescence study. *Acta Pathologica et Microbiologica Scandinavica* 76, 74-88.

Moe, N. and Jorgensen, L. (1968) Fibrin deposits on the syncytium of the normal human placenta: evidence of their thrombogenic origin. *Acta Pathologica et Microbiologica Scandinavica* 72, 519-541.

Ockleford, J.A., Hsi, B.-L., Wakely, J., Badley, R.A., Whyte, A., and Faulk, W.P. (1981) Propidium iodide as a nuclear marker in immunofluorescence. I. Use with tissue and cytoskeleton studies. *J. Immunol. Methods* 43, 261-267.

Sinha, D., Wells, M., and Faulk, W.P. (1984) Immunological studies of human placentae: complement components in pre-eclamptic chorionic villi. *Clin. Exp. Immunol.* 56, 175-184.

Smith, N.C., Brush, M., and Luckett, L. (1974) Preparation of human placental villous surface membranes. *Nature* 252, 302-303.

Yeh, C.-J.G., Hsi, B.-L., and Faulk, W.P. (1981) Propidium iodide as nuclear marker in immunofluorescence. II. Use with cellular identification and viability studies. *J. Immunol. Methods* 43, 269-275.

PLACENTAL PHARMACOLOGY

Trophoblast Research 2:235-263, 1987

P-450 CYTOCHROMES IN THE HUMAN PLACENTA: OXIDATIONS OF XENOBIOTICS AND ENDOGENOUS STEROIDS

- A Review -

Mont R. Juchau, Moses J. Namkung, and Allan E. Rettie

Department of Pharmacology
School of Medicine
SJ-30
University of Washington
Seattle, Washington 98195, USA

INTRODUCTION

The purpose of this article is to provide the reader with a concise view of current concepts concerning the family of P-450 hemoproteins as they exist and function in the human placenta. The subject is discussed with focus on the monooxygenation of small, foreign organic molecules (xenobiotics) in placental tissues and cells, and of the relationships of these monooxygenation reactions to placental P-450-dependent oxidations of endogenous steroids. Several past reviews of placental biotransformation of xenobiotics have appeared in the open literature and references to these are provided for readers more seriously interested in this highly specialized topic. The first review of this research area appeared in 1969 (Juchau and Yaffe, 1969). A very comprehensive and detailed review of the earlier literature appeared in 1973 (Juchau, 1973). A more recent, comprehensive, and readily accessible review appeared in 1980 (Juchau, 1980) and the latest review is in press at the time of this writing (Juchau and Rettie, 1985). Dr. Olvai Pelkonen of Finland has actively researched this area and has also provided relatively recent reviews (Pelkonen, 1980; Pelkonen and Pasanen, 1984). Other reviews appearing between 1969 and 1985 are referenced in those cited above.

The format of this review will be a consideration of twelve current concepts concerning the quantity, number, and functions of placental P-450 cytochromes. It should be emphasized that these concepts vary in the degree of firmness to which they are established.

1.   Spectrally Detectable Cytochrome P-450 in Microsomal Fractions of Human Placental Homogenates Functions Almost Exclusively in the Conversion of Androgens to Estrogens (Aromatase)

The most common method of spectral detection of cytochrome P-450 is the measurement of the ferrous-carbon monoxide (Fe^{2+}-CO) comlex (Omura and Sata, 1964) which, in the difference spectrum yields an absorption maximum at 450 ± 4 nm

Submitted: 10 October 1985; Accepted: 10 June 1986

depending upon the specific isozyme or set of isozymes subjected to measurement. Carbon monoxide binds directly to the heme iron of the reduced holocytochrome. The unusual absorption maximum of this specific Fe^{2+}-CO complex is now generally attributed to the provision by cysteinyl sulfhydryl of thiolate in the fifth liganding position. Certain other chemicals (e.g., stannous fluoride) will also bind to the ferrous iron to yield complexes with absorption maxima near 450 nm but have been utilized far less frequently than carbon monoxide. Basic nitrogenous compounds such as aniline, nicotinamide, various imidazoles, etc. also complex to yield absorption maxima near 430 nm and minima near 390 nm, often referred to as "type II" binding spectra, but these are less useful in distinguishing P-450 from other hemoproteins.

Another commonly used spectral assay that is less sensitive but more discriminating is the substrate binding spectrum. Substrates for P-450 bind to a specific (substrate-binding site) site on the apoprotein part of P-450 molecule. The interaction results in displacement of the endogenous sixth ligand, a shift in equilibrium from low to high spin state and spectral changes with shifts resulting in difference spectra with absorption maxima near 390 nm and minima near 420 nm. The resultant shift in the difference spectrum is also referred to as a "type I" binding spectrum and is, with certain exceptions, produced only by lipophilic chemicals that are substrates for the specific holocytochrome species. The assay cannot be used to quantitate P-450 content except under certain rigid, specified conditions. This is simply because the magnitude of the spectral change depends upon the initial spin state equilibrium of all isozymes in the sample, the degree to which the chemical is capable of altering spin states of each isozyme, and the extinction coefficient of the high-spin hemoprotein(s). Thus the asay is useful primarily for determing *whether* given substrate-specific species may be present in a given preparation and the apparent affinity with which the substrate binds.

Each of these assays has demonstrated that cytochrome P-450 is present in microsomal fractions of human placental homogenates. The dithionite-reduced, carbon monoxide complex provides a quantitative estimate of *total* P-450 isozymes and indicates that amounts of these hemoproteins present in human placental microsomes at term are considerably less than those meaured in rodent hepatic microsomes from untreated animals. When expressed as nmoles P-450/mg microsomal protein, estimates vary but usually range from 1-10%. Thus, although quite variable, quantities of placental microsomal P-450 per unit weight of tissue may be as little as 0.05% of those obtained from the liver. Indeed, some of the earliest reports of attempts to detect cytochrome P-450 in the placenta suggested that P-450 may not be present. Difficulties arose in measurements of placental P-450 by the method of Omura and Sata (1964) because of the tenacious binding of (what appears to be) hemoglobin to placental microsomes (Juchau and Symms, 1972; Symms and Juchau, 1972, 1973, 1974; Hodgson and Juchau, 1977). This phenomenon plagues analyses of P-450 in other extrahepatic tissues as well and remains unexplained in view of the ease with which hemoglobin can be washed free from hepatic microsomes. A more satisfactory method is that described by Greim (1970) in which CO is present in both cuvettes and the reducing agent only in the sample cuvette. The most accurate method in our hands was one which incorporated the use of germanium fluoride, described by Hodgson and Juchau (1977). Despite the analytical difficulties encountered, it is now generally agreed that total quantities of all isozymic placental

P-450s are between 2-3 orders of magnitude less than those measurable in equal quantitites of liver tissue.

Placental cytochrome P-450 is also detectable via analysis of substrate binding (type I) spectra. It is of interest and importance to note that the only chemicals currently known to elicit these spectral changes in human placental microsomes are steroids (Chao and Juchau, 1980). Profound type I spectral changes are produced by androstenedione, testosterone, and other structurally related steroids that are attacked by placental aromatase. All tested steroids that are known to be substrates for the aromatization reaction (including 19-hydroxy-androsteneidone and 19-oxoandrostenedione) have produced these pronounced type I spectral changes (e.g., Zachariah and Juchau, 1977). The only chemicals tested that consistently produce type I spectral changes in placental microsomes but are not aromatase substrates are steroidal estrogens (Chao and Juchau, 1980). However these compounds elicited only very weak spectral changes. The magnitude of the spectral changes produced by androstenedione and other aromatizable androgens are truly remarkable in view of the low concentrations of placental P-450 when assayed as the P-450-CO complex. Ratios of androstenedione-elicted type I spectral maxima (390-420 nm)/P-450-CO complex (450 nm) in human placental microsomes may be higher than for any other known P-450- substrate complex (Chao and Juchau, 1980). These facts already strongly suggest that human placental microsomal P-450 is highly specific for androstenedione and structurally-related chemicals. Also, the fact that only aromatizable chemicals elicit such spectra strongly suggests that the principal function of placental P-450 is to catalyze aromatization reactions.

However, because of the fact that carbon monoxide failed to inhibit placental aromatase under conditions in which low O_2 concentrations (5%) limited the reaction rate (Meigs and Ryan, 1968, 1971; Thompson and Siiteri, 1974; Zachariah et al., 1976; Chakraborty et al., 1972), many investigators were reluctant to believe that placental aromatization was a P-450-dependent reaction. The explanation provided by Thompson and Siiteri (1974) was clearly unsatisfactory on theoretical grounds because carbon monoxide should inhibit the reaction under conditions in which O_2 is rate- limiting -- due to competition of CO for the O_2-binding on the ferroheme.

The experiments leading to the resolution of this enigma now provide a truly remarkable story that provides considerable insight into the functional nature and regulation of the placental P-450(s) which catalyzes the aromatization reaction. In early, preliminary experiments (Symms and Juchau, 1973), we observed that androstenedione appeared capable of displacing carbon monoxide from the sixth liganding position of the reduced hemoprotein. With sodium dithionite as the reducing agent, these observations were difficult to repeat but, with NADPH as the reducing agent, consistent displacement was observed at nanomolar concentrations of androstenedione (Juchau and Zachariah, 1975; Lee et al., 1975). Importantly, androstenedione appeared to displace CO to the extent that a microsomal CO-P-450 complex was no longer discernible. It was subsequently shown that testosterone and certain other aromatizable C-19 androgens (including 19-OH and 19-oxo-androstenedione) could also displace CO from NADPH-reduced placental P-450 but that other steroids did not evoke this effect (Zachariah et al., 1976; Juchau et al., 1976). Most interestingly, 19-norandrostenedione and certain other aromatizable 19-norsteroids could not only prevent the CO-displacing effect of androstenedione but actually facilitated CO-binding to the placental cytochrome. (Aromatization of the 19-norsteroids is readily inhibited by CO). The opposing effects of the C-19 and 19-nor steroids on CO binding could actually be titrated against each other in the same cuvettes, even though the facilitating effects of the latter steroids required concentrations in the μmolar range (Juchau and Zachariah, 1975; Lee et al., 1975; Juchau et al., 1976). With 19-norandrostenedione present in aromatase reaction mixtures at concentrations that permitted significant binding of CO to placental P-450, carbon monoxide was shown to inhibit the aromatization of androstenedione (Lee et al., 1975; Zachariah and Juchau, 1977). More significantly, the inhibition could be prevented by illumination of the reaction mixture with visible light and maximal reversal occurred at a wavelength of 450 nm (Zachariah and Juchau, 1977). These investigations established definitively that human placental microsomal cytochrome P- 450 functions to catalyze the aromatization of androstenedione to estrone, a reaction which represents one of the most important metabolic functions of the human placenta.

It is of interest that ovarian aromatase catalyzes the same reaction but that CO inhibits the ovarian reaction without the above described manipulations (Meigs and Ryan, 1971). This provides evidence for differential modes of regulation of the placental vs. ovarian aromatase systems and suggests the possibility that different P-450 isozymes may be involved. This represents an interesting area for future investigations.

Particularly striking to us was the correlation between the effect of aromatizable steroids on the P-450-CO difference spectrum and capacity of CO to inhibit aromatization of the same steroids. All 19-norsteroids tested (Juchau et al., 1976; Zacharaiah and Juchau, 1977) exhibited at least some capacity to facilitate CO binding. Placental aromatization of these compounds is readily inhibited by CO. All C-19 steroids whose placental aromatization is not normally inhibited by CO (under the usual reaction conditions) exhibited the capacity to displace CO from its heme binding site.

At least one C-19 steroid, 16α-hydroxyandrostenedione, undergoes CO-inhibited aromatization under unadjusted (vide supra) reaction conditions in human placental microsomes (Canick and Ryan, 1975). This compound neither displaced CO from the P- 450-CO complex nor facilitated CO-binding to the ferrous heme. These observations strongly suggest that interactions of steroids with human placental aromatase P-450 are critically dependent on the steroid's chemical structure. Presence of a 19- methyl group is necessary but not sufficient to displace CO from the sixth liganding position. It is significant that the same 19-methyl is attacked by placental aromatase. It would appear that the orientation of the 19-methyl group toward the sixth liganding position is critical for the CO-displacing effect. A question that arises concerns the binding of molecular O_2. If androstenedione can displace CO from the sixth liganding position, why should it not also displace O_2 and thus prevent hydroxylation at C-19? The answer may lie in the difference in configuration of binding of CO vs. O_2 to the ferrous iron. CO binds in a linear configuration (Fe^{2+}- $C = O$) whereas O_2 is bound in a "bent" mode (Fe^{2+}-$o_{\backslash_0}$). This bent mode may allow O_2 to remain bound. Such a scenario would predict that androstenedione should more readily displace CO in the presence of higher O_2 concentrations. Indeed, we have shown (Juchau, 1984) that androstenedione will not displace CO under totally anaerobic conditions. This would also explain the difficulty in displacing CO from dithionite-reduced P-450 because excess dithionite would also remove residual O_2 from the cuvette. Further work will be required to establish these concepts more definitively. A recent, elegant discussion (Frauenfelder and Wolynes, 1985) of modes of interactions of O_2 and CO with hemeproteins provides a number of ideas for rationalizing the interesting effects that we observed.

The capacity of steroids to alter CO-binding to placental aromatase P-450 also provides a useful tool for determining whether this P-450 may catalyse other placental monooxygenation reactions. If so, then androstenedione should tend to prevent or attenuate the capacity of CO to inhibit the reaction by virture of its displacing effect. 19-Norandrostenedione, on the other hand, should enhance the inhibitory effect of CO by viture of its capacity to facilitate CO-binding. We tested these ideas with the placental arylhydrocarbon hydroxylase system (Zachariah et al., 1976; Zachariah and Juchau, 1977) and found that neither steroid significantly affected the capacity of CO to inhibit the reaction. This provided an additional piece of evidence that aryl hydrocarbon hydroxylase P-450 and aromatase P-450 are entirely separate entities and that the spectrally detectable placental microsomal P-450, possibly except for small traces, functions exclusively in steroidal aromatization reactions. Future research could, of course, reveal additional functions but, as of this writing, convincing experimental data supporting additional functions have not been been forthcoming. In fact, aside from the possibility that spectrally detectable placental P-450 could function to catalyze various xenobiotic oxidations, only one additional function has been suggested. On the basis of some observations indicating that certain substances known to bind to *hepatic* P-450s could impede the transfer of O_2 from maternal to fetal blood, it was suggested (Burns and Gurtner, 1973; Gurtner and Burns, 1973) that placental P-450 functions in the O_2 transport process. However, it was not shown that those chemicals bind to placental P-450 (they probably do not) and no additional evidence has emerged to support the idea. In fact, at least one study has provided data that tend to discount the idea (Gilbert et al., 1979).

2.   The Placental P-450(s) That Catalyzes the Aromatase Reaction is Highly Substrate- Specific and Attacks Xenobiotics Minimally if at All.

Much of the above discussion is already supportive of this concept but considerable additional corroborating evidence also exists and will be summarized in this section. The oxidative biotransformation of many (but not all) xenobiotics known to undergo monooxygenation in placental tissues is markedly increased in placentae of smokers (Welch et al., 1969; Juchau, 1980). The very marked increases in xenobiotic monooxygenation appear to occur as the result of exposure of pregnant women to polynuclear aromatic hydrocarbon inducing agents present in the tobacco smoke. At least three laboratories (Conney et al., 1971; Zachariah and Juchau, 1977; Pasanen, 1984) have reported that aromatase activities remain unchanged in the same placentae in which xenobiotic monooxygenase activities are increased by as much as 100-fold. These observations provide convincing evidence that aromatase does not catalyze these xenobiotic oxidations.

Other evidence rests with the stringent structural requirements for elicitation of substrate-binding spectra (Chao and Juchau, 1980), capacity to prevent or facilitate

CO-binding, and considerations of spectral dissocation constants. Androstenedione elicits detectable type I binding spectra when added to human placental microsomes at concentrations as low as 10^{-11}M (Zachariah and Juchau, 1977; Juchau, 1980) with spectral dissociation constants in the nanomolar range. This is quite remarkable when one considers that spectral dissociation constants for xenobiotics in hepatic microsomes are in the millimolar range (usually around 0.1 mM). Only very slight changes in the androstenedione molecule result in profound changes in the spectral dissocation constant. For example, testosterone, which differs from androstenedione only by virtue of a reduced oxygen at carbon 17, exhibits a spectral dissociation constant nearly 2 orders of magnitude higher than that of androstenedione (Symms and Juchau, 1973). 19-Norandrostenedione, which differs only by the absence of the 19-methyl group, displays a spectral dissociation constant nearly 3 orders of magnitude greater than that of androstenedione when direct comparisons are made (Juchau et al., 1976). Millimolar concentrations of classical xenobiotic substrates have thus far failed to elicit consistent spectral changes (Juchau, 1980) in human placental microsomes. Recently, in collaboration with Dr. Frank Bellino, we found that antibodies toward human placental aromatase, while effectively inhibiting the aromatase reaction, had no effect on placental ethoxyphenoxazone deethylase activity (unpublished). Taken together, the above observations provide highly convincing evidence that the human placental aromatase would attack xenobiotic substrates only if the substrate were steroidal and structurally similar to androstenedione. Some of the oral contraceptive progestational steroids may fit into that category (Juchau et al., 1976). However, the extent to which such xenobiotic steroids would undero significant aromatization in vivo remains questionable due to their comparatively (i.e., compared to androstenedione) low affinity for aromatase P-450 (Juchau et al., 1976). Eventually, sufficiently purified preparations of aromatase P-450 will be obtained such that its substrate specificity can be tested directly in reconstituted systems. Some progress is being made but has not reached the point at which such experiments would be meaningful.

3.   P-450(s) Spectrally Detectable in Placental Mitochondrial Fractions Function Only in the Conversion of Cholesterol to Pregnenolone (via side-chain oxidation) and, Likewise, Utilize Xenobiotics as Substrates Minimally if at All.

This concept rests on a base of lesser supportive evidence simply because it has not been tested directly. Early investigations (Juchau and Smuckler, 1973) suggested that human placental aryl hydrocardon hydroxylase was localized principally in the endoplasmic reticulum of trophoblast cells. Activity was also detected in fractions containing predominantly mitochondria and marker enzyme analyses suggested that the hydroxylase activity might be partially distrubted in the mitochondrial membranes. The results howver, did not permit a conclusive statement in this regard, and retrospectively, in view of the considerable research suggesting localization of xenobiotic-biotransforming P-450s in the endoplasmic reticulum of cells of other tissues, it seems more likely that little if any aryl hydrocarbon hydroxylase resides in the mitochondria. Recent research in Pelkonen's laboratory (Pasanen and Pelkonen, 1984) tends to corrobate this viewpoint. All results reported to date indicate that subcellular distribution of aryl hydrocarbon hydroxylase differs substantially from that of the cholesterol side-chain cleavage system. The latter system is clearly closely associated with innner mitochondrial membranes whereas aryl hydrocarbon hydroxylase is predominantly microsomal. If it were verified that only one isozymic

P-450 catalyzed each reaction, this would already constitute fairly good evidence that the side-chain oxidizing P-450 and the xenobiotic-oxidizing P-450 were totally independent entities. The possibility that various isozymic forms may be capable of catalyzing each reaction prevents such a conclusion based on distributional evidence alone. However, the fact that an inverse correlation between cholesterol side chain oxidase and aryl hydrocarbon hydroxylase activities were observed (Juchau et al., 1972; Juchau et al., 1974) argues quite strongly for the independence of the two systems. Studies by Gunasegarem et al. (1982) tend to substantiate these observations. Also, studies with inhibitors (Pasanen and Pelkonen, 1984) support this view.

On the basis of their studies with electron paramagnetic resonance spectroscopy, Simpson and Miller (1978) stated that "it appears likely that all of the P-450 in placental mitochondria is involved in cholesterol side chain cleavage". Some investigators (e.g., Canick and Ryan, 1978) have indicated that human placental mitochondria may also contain an aromatase system localized on the outer mitochondrial membrane (the side chain oxidizing system is localized on the inner membrane). In studies with antibodies to cytochrome c reductase, Thompson and Siiteri (1979) concluded that "inhibition of placental mitochondrial aromatization by anti-NADPH-cytochrome c reductase precludes the involvement of mitochondrial cytochrome P-450". The microsomal (hepatic) antibody could, of course, inhibit a mitochondrial aromatase but, in view of the parallel inhibition reported, this seems less likely. The results of Thompson and Siiter also indicated that specific activities in mitochondrial fractions were approximately 1/8 those measured in microsomal fractions. This ratio would be compatible with microsomal cross- contamination of the mitochondrial fraction. However, Meigs and Moorthy (1984) report that mitochondrial specific activities are approximately 50% of those measured in placental microsomes.

Final conclusions regarding the validity of concept #3 await the isolation, purification, and reconstitution of the various placental P-450 isozymes and a thorough examination of their substrate specificities. Additionally, however, experience with human placental microsomes warns of the possibility that spectrally undetectably P-450 isozymic species with high substrate turnover numbers may also exist in the mitochondria of human placentae.

4.   Placental Xenobiotic-Oxidizing P-450s are Severely Restricted in Quantity and Variety of Isozymes. However, Some are Exceptionally Inducible.

Some of the very early studies dealing with the placental mixed-function oxidation of xenobiotic substrates left the impression that the human placenta was a major drug-oxidizing organ. This was probably due, at least in part, to the methology available at that time for measurements of P-450 dependent monooxygenase activities. It became increasingly evident, however, that the xenobiotic biotrans-forming activity of human placental tissues was extremely limited, with certain exceptions. The most interesting and potentially most important exception was readily observed in measurements of aryl hydrocarbon hydroxylase activities in placentae of smokers. A very striking close correlation between smoking habits and activity of the placental hydroxylase was observed in several laboratories (Welch et al., 1969; Nebert et al., 1969; Juchau, 1971; Pelkonen et al., 1972). In some placentae, activities were comparable to those commonly measured in adult rat livers -- most frequently, these placentae were from women who smoked heavily and who had also been exposed to various other drugs and foreign chemicals. Importantly, in spite of these astounding increases in monooxygenase activity, no increases in placental cytochrome P-450 could be detected when measured as the CO-complex. In addition, no changes in the spectral absorption maximum could be detected. The substrate, benzo(a)pyrene also did not elicit observable type I binding spectra in microsomes of such placentae. Attempts to measures these spectra are difficult because benzo(a)pyrene absorbs at similar wavelengths, but several other substrates that do not absorb at these wavelengths but appear to be biotransformed by the same placental enzyme system, likewise fail to elicit detectable binding spectra.

Attempts to detect increases in placental microsomal protein associated with increases in placental aryl hydrocarbon hydroxylase activities have also failed. In our laboratory, we utilized two-dimensional gel electrophoresis-isoelectric focusing in attempts te detect such increases and were unable to find any evidence of correlative increases (Juchau, 1984). These observations are consistent with the idea expressed above, that the P-450 isozyme(s) catalyzing this reaction is spectrally undetectable. We examined the possibility that this placental hydroxylation reaction might not be catalyzed by a P-450 isozyme (Zachariah and Juchau, 1977). However, the reaction was inhibited by CO. CO-inhibition was reversed by visible light and maximal reversal occurred with monochromatic light at a wavelength near 450 nm. These observations provided conclusive evidence that placental aryl hydrocarbon (benzo(a)pyrene) hydroxylase activity was P-450- dependent. The conclusions that must be drawn from the above observations are: 1. that at least one xenobiotic-oxidizing P-450 is present in placentae of women exposed to the appropriate inducing agent -- presumably polycyclic aromatic hydrocarbons represent the inducers in smokers' placentae, 2. based on studies with the specific inhibitor, α-naphthoflavone (Moilanen and Pelkonen, 1979), it is unlikely that the enzyme exists constitutively, 3. the induced enzyme has a sufficiently high substrate turnover number that a relatively rapid reaction can be catalyzed in the absence of spectrally detectable levels of the isozyme. The *implications* are even broader, suggesting the distinct possiblity that *several* spectrally undetectable P-450 isozymes with relatively high substrate turnover numbers may be present in various tissue preparations. Such a scenario will greatly complicate attempts to purify the various individual P-450 isozymes to homogeneity and to characterize their substrate specificities.

The evidence for a restricted quantity of xenobiotic-oxidizing P-450 in placental tissues has been presented above. Evidence for a restricted *variety* of placental xenobiotic-oxidizing P-450 isozymes is provided by considerations of the many substrates that do not undergo detectable mixed-function oxidation when incubated together with various placental preparations and excess quantities of cofactors (e.g., Juchau and Zachariah, 1975b). Many of the classical "drug substrates" are not detectably attacked by placental P-450s even when highly sensitive methods are utilized for measurements of the appropriate reaction products. This aspect has been extensively discussed in a previous review (Juchau, 1980). Other substrates are attacked only at certain positions on the molecule. Aspects of regio- and stereoselectivity of placental P-450s and implications for the involvement of specific P-450 isozymes are discussed further below (see discussion of concept no. 7). In spite of the limited variety of P-450 isozymes present in human placental tissues, indirect evidence continues to increase for a *multiplicity* of isozymes and it appears that the number of such isozymes will prove to be much larger than previously anticipated. Juchau (1980) summarized the evidence for existence of at least 5 isozymes and research published since the appearance of that review suggest that additional isozyme species may be present. This is also discussed below.

5.   Aromatizing and Cholesterol Side-Chain Oxidizing P-450s are Regulated by Polypeptides and Cyclic AMP but are not Inducible by Xenobiotics.

It has been known for several years that gonadotropins, dibutyryl cyclic AMP, theophylline, and prostaglandins produce increases in aromatase activity (Torminaga and Troen, 1967; Cedard et al., 1970; Alsat and Cedard, 1973; Wolf et al., 1978; Bellino and Hussa, 1978). The susceptibility of cholesterol side-chain oxidizing systems to regulation by polypeptdie hormones and cyclic AMP also has been recognized for several years but, the adrenal system has been studied far more intensively than the placental system. The former is heavily regulated by ACTH while the latter is not known to be influenced by ACTH. A generalized concept that is now rapidly gaining credence (e.g., John et al., 1985; Anderson and Mendelson, 1985; Lin et al., 1985; Dee et al., 1985) is that P-450s functional in catalyzing important, rate- limiting steps in the *synthesis* of steroid hormones are regulated primarily (possibly exclusively in some cases) by polypeptide hormones as a first messenger and by cyclic AMP as a second messenger. The most important overall effect of these messengers is to increase rates of synthesis of the appropriate P-450 isozyme by increasing transcriptional activity of the corresponding gene. The mechanism by which increased transcription occurs, however, is presently unknown. Phosphorylation of an enzyme or other protein critical to the regulation of transcription rates would seem likely and future research will almost certainly focus on this aspect. Phosphorylation of P-450 per se does not appear to be an important regulatory factor.

Steroid-*degrading* and xenobiotic-oxidizing P-450s are generally felt not to be regulated by polypeptide hormones (Ruckpaul et al., 1985; Nebert and Gonzales, 1985) although some notable exceptions exits. For example, sex-specific P-450s in rat livers are known to be regulated by growth hormone or a very similar polypeptide (e.g., see Morgan et al., 1985; Virgo and Vockentanz, 1985). It seems possible that the hepatic xenobiotic-oxidizing P-450's may belong to one of two categories -- those that are constitutive and regulated by polypeptide hormones and/or cyclic AMP and those that are not constitutive but respond to xenobiotic inducing agents. P-450's catalyzing steroid homrone degradation would also belong in both categories whereas P-450's catalyzing steroid hormone synthesis could be placed only in the first category at present. Although possibly simplistic, the search for generalizations in the confusing field of P-450 isozyme regulation would seem to render the idea worthy of a closer examination. All data obtained thus far on human placental cytochrome P- 450 isozymes are compatible with the idea regarding these two categories. Nevertheless, much remains to be learned concerning the specificity of hormonal vs. xenobiotic regulation of individual P-450 isozymes before the idea can be generally accepted by biological scientists. Future testing of the validity of the idea should prove to be highly interesting and informative.

6.   At Least Some Placental Xenobiotic-Oxidizing P-450s are Highly Inducible by "3-Methylcholanthrene-Type" Inducing Agents but not by Most Other Inducing Agents. They do not Respond to Polypeptide Hormones or Cyclic AMP.

Until recently, inducers of xenobiotic-oxidizing P-450s were classified either as "3-methylcholanthrene-type" (MC-type) or as "phenobarbital-type" (PB-type) inducing agents. The classification is now well recognized to be overly simplistic but the terminology continues to persist. The terminology originated as the result of perceived similarities in the effects of numerous (more than 400) inducing agents to be characteristic inducing properties of either MC or PB which were two of the first xenobiotic inducers to be recognized. Subsequent studies have revealed that several chemicals will induce P-450 isozymes not induced (or minimally induced) by either MC or PB. Some of the first of these to be recognized were referred to as "unique" inducers but this term is now fading into disuse because of the necessity to classify many, if not most, other inducing agents as "unique".

In very early investigations of placental xenobiotic biotransformation (Juchau, 1973), it was recognized that the placenta did not respond (or responded very poorly) to the effects of "PB-type" inducers. In contrast, placental response to the "MC- type" inducers was very profound. Later it was recognized that these are properties displayed by most extraheptic tissues, a notable exception being the intestine.

The capacity of cells to respond to the inducing effects of MC-type compounds (e.g., 2,3,7,8-tetrachlorodibenzo-p-dioxin (TCDD), certain polyhalogenated biphenyls, B-naphthoflavone, etc.) appears to represent a somewhat primitive function common to most cells, including those undergoing rapid proliferation. Response to the P-450 inducing effect of phenobarbital, on the other hand, appears to be a highly specialized function, occurring principally in highly differentiated, non-proliferating cell populations.

For ethical reasons, the capacity of the human placenta to respond to individual xenobiotic inducing agents (particularly those regarded as "unique") has received little investigative attention. Studies with placentae of experimental animals and with other extrahepatic tissues are presumably indicative of the capacity of the human placenta to respond to various inducers. Recently, we studied the capacity of rat placentae to respond to a series of commonly studied inducing agents. Benzo(a)pyrene, estradiol-17β (Namkung et al., 1985) and 2-acetylaminofluorene (AAF) (unpublished) were utilized as probe substrates. Inducers studied were ethanol, caffeine, TCDD, β-naphthoflavone, AAF, phenobarbital, 3-methylcholanthrene, isosafrole, pregnenolone-16α-carbonitrite (PCN) and Aroclor 1254 (a mixture of polychlorinated biphenyls also referred to as PCB). For comparative purposes, the maternal livers and fetal livers of the same experimental animals were directly compared with the placentae. The maternal liver was responsive to virtually all of the inducers studied. The fetal liver responded in a more limited fahsion, displaying substrate specificity and regioselectivity. The placenta displayed a very contrasting lack of overall responsiveness to most inducers except those often regarded as "MC-type" inducers (MC, β-naphthoflavone, TCDD, and PCB). Particularly striking was the difference between the fetal liver and the placenta with respect to aromatic hydroxylation of estradiol following exposure to PCN. The fetal liver responded with 20-30 fold increases in rates of catecholestrogen formation whereas the placenta displayed no significant response. Surprisingly, the placenta did exhibit 5-6 fold increases in rates of benzo(a)pyrene hydroxylation (no increase in estradiol or AAF hydroxylation) after exposure to ethanol (Namkung et al., 1985). Whether placentae of other species, particularly humans, would respond similarly to the inducing effects of ethanol or other chemicals now known to exhibit similar inducing properties (e.g., imidazole, 4-methylpyrazole, isoniazid, etc.) will be of considerable interest in future investigations. Also of interest will be a determination of the number of different placental P-450 isozymes undergoing induction in response to MC-type inducers and an exmaination of the responsiveness of placental tissue to yet other inducing agents.

It should be emphasized that there is currently little experimental evidence to support the latter part of concept No. 6. Nevertheless, in view of observations in the liver and other tissues, it seems prudent at present to retain the concept until the results of future research should dictate otherwise.

7.   The Impracticality of Studying Purified Placental Xenobiotic-Oxidizing P-450s Has Led to the Use of Substrate, Inhibitor/Activator, Inducer, and Antibody Probes.

Attempts to purify human placental P-450 isozymes have met with limited success thus far. Partial purification of the microsomal aromatizing P-450 has been achieved (Tseng and Bellino, 1985; Washida et al., 1985) and the mitochondrial cholesterol side-chain oxidizing P-450 has been purified to a reasonably high degere of purity (Simpson and Miller, 1978; Pasanen and Pelkonen, 1981) but in neither case have the purifications achieved homogeneity by rigorous biochemical criteria. Recent progress has been encouraging and future preparations may be expected to exhibit specific concentrations approaching those expected from pure P-450 isozymes.

However, the prospects for purifying placental xenobiotic-oxidizing P-450's appear considerably dimmer. Although several attempts have been made (Symms and Juchau, 1973; Manchester and Jacoby, 1982; Namkung et al., 1983), none have achieved a significant degree of success measurable in terms of purities. In view of considerations of the characteristics of placental xenobiotic-oxidizing P-450s discussed in previous sections, it is not surprising that the standard purification techniques have not been particularly useful. Purification with immunoaffinity columns could prove useful but, to our knowledge, has not been attempted as yet. In view of this lack of success, investigators have begun to utilize various probes to investigate the various xenobiotic-oxidizing P-450 isozymes and allozymes present in human placental tissues. Investgiations in Dr. Gelboin's laboratories (Fujino et al., 1982, 1984; Song et al., 1985) have utilized a monoclonal antibody to the major MC-induced rat hepatic P-450, most commonly designated at P-450$_c$. They studied the 3-hydroxylation of benzo(a)pyrene and the O-deethylation of 7-ethoxycoumarin in human placental microsomes and found that the antibody preferentially inhibited monooxygenase activity in placentae of smokers. Because the antibody did not inhibit ethoxycoumarin O-deethylase activity in placentae of nonsmokers, the investigators suggested that two xenobiotic-oxidizing P-450s are present in human placental microsomes -- one with an epitope common to the major MC-inducible isozyme of rat liver and an additional P-450 that does not interact with a monoclonal antibody to the same rat liver P-450. The former P-450 is increased in placentae of smokers but the latter is unaffected (or minimally affected) by cigarette smoking.

   Identical conclusions can be drawn from studies with substrate probes (Juchau, 1980; Rettie et al., 1985) with the exception that substrate probe studies indicate the presence of at least two and possibly more PAH-inducible, xenobiotic-oxidizing P-450s, one of which is extremely sensitive to induction. Ethoxyphenoxazone appears to be an excellent marker substrate for the highly inducible enzyme when its metabolism is highly correlated with the 6- and 8-hydroxylation of R-warfarin (Rettie et al., 1985). Other human placental P-450s, somewhat less sensitive to increases associated with cigarette smoking and apparently under differential regulation by other factors, would include those which preferentially catalyze the N-hydroxylation of 2- acetylaminofluorene (Juchau et al., 1975) the 4-hydroxylation of estradiol-17β (Juchau et al., 1982a; Namkung et al., 1985) and the hydroxylation of 7,12-dimethylbenz(a)anthracene (Juchau et al., 1978). However, the extent to which these latter three reactions are *commonly* regulated has not been investigated intensively. (It must also be borne in mind that differential regulation is not necessarily proof of different P-450 isozymes because substrate and/or metabolite factors can enter into regulatory processes.)

   The number of xenobiotic substrates attacked by placental P-450s that are significantly biotransformed without induction by MC-type inducers has increased significantly recently. These include N-methylaminobenzoic acid (Zachariah and Juchau, 1975), ethoxycoumarin (Kaelin and Cummings, 1983, 1984), benzyloxyphenoxazone (Rettie et al., 1985), N-methylaniline (Pelkonen et al., 1972) and possibly others (Juchau, 1980). As these reactions occur at extremely low rates in placental tissues and do not appear to be catalyzed by inducible placental P-450 isozymes, they have attracted little attention to date. Whether the apparently constitutive placental P- 450s that attack these substrates may subserve important physiological functions is entirely unknown at present.

   The use of substrate probes for identification and quantification of functional extrahepatic P-450s now appears increasingly promising. Research has shown that many more P-450s exist than previously thought (Nebert and Gonzales, 1985) and that each has characteristic substrate specificities. Of particular value are substrates whose various carbons are attacked specifically by different P-450 isozymes. Warfarin is a particularly valuable probe substrate because various P-450 isozymes attack different positions of the warfarin molecule with a fairly high degree of regiospecificity. In addition, the R and S stereoisomers can be utilized in combination to provide an even sharper tool. Recently we reported metabolic studies with R- and S-warfarin in human placental microsomes (Rettie et al., 1985). We found that microsomes isolated from placentae of smokers preferentially attacked the R-isomer at the 6 and 8 positions while the placenta of nonsmokers preferentially attacked carbon 7 with little if any stereoselectivity. Preferential attack at carbons 6 and 8 of the R-isomer is characteristic of MC-induced P-450s, providing good evidence that the polynuclear aromatic hydrocarbons present in cigarette smoke are responsible for the high aryl hydrocarbon hydroxylase activities measurable in smokers' placentae and that the effect occurs via induction. We also studied a series of phenoxazone ethers in the same placentae. This series represents a highly valuable group of probe substrates because of the availability of a highly sensitive and specific assay which continuously measures the appearance of the intensely fluorescent, common reaction product, resorufin. Each ether is converted to the same metabolite but each displays varying degrees of specificity as substrates for individual P-450 isozymes. Ethoxyphenoxazone, commonly referred to as ethyoxyresorufin, is highly selective for rat hepatic $P-450_c$ and closely related isozymes while the pentoxy and benzyloxy ethers are more selective for isozymes similar to rat hepatic $P-450_b$. Microsomes isolated from smokers' placentae displayed very high ratios of ethoxy/benzyloxy biotransformation whereas placentae of nonsmokers utilized the benzyloxy derivative as the preferred substrate. The results provided further evidence that a P-450 isozyme closely related to rat hepatic $P-450_c$ is the major form in placentae displaying high aryl hydrocarbon hydroxylase activities.

Testosterone has been employed as a probe substrate by several laboratories and it is thus interesting that this molecule is attacked almost exclusively at carbon 19 by human placental P-450s. Lack of attack at other positions suggests that P-450 isozymes capable of catalyzing hydroxylations at those positions are either absent from or present in extremely low quantities (or in functionally inactive forms) in the placenta. It is also of interest that investigations of the capacity of placental tissues to catalyze aniline and acetanilide hydroxylation have yielded only negative results (Juchau, 1980). The p-hydroxylation of aniline is preferentially catalyzed by an ethanol-inducible P-450 in hepatic microsomes (Koop et al., 1985; Coon et al., 1985) and is termed $P-450_{3a}$ in rabbits. Acetanilide p-hydroxylation has served as a marker reaction for a mouse liver isozyme referred to commonly as P_3-450 (Nebert and Gonzales, 1985). P_3-450 is analogous to $P-450_d$ in the rat liver and both are induced by methylcholanthrene but are preferentially induced by isosafrole. Failure to detect aniline or acetanilide p-hydroxylation in human placental microsomes suggests the possible absence of P-450 forms analogous to those present in livers of common experimental animals (e.g., $P-450_d$, $P-450-LM_4$ (rabbits), P_3-450). On the other hand, placentae from women exposed to the appropriate inducing agent may not have been studied. This possibility is more attractive if one considers the strong probability that isozymic forms likely to catalyze these hydroxylation reactions would not exist constitutively in the placenta. Of interest in this regard is the recent finding (Namkung et al., 1985) that chronic ethanol treatment of pregnant rats resulted in increased placental hydroxylation of BaP. It would be of interest to ascertain whether aniline p-hydroxylation, nitrosamine dealkylation, or other monooxygenation reactions were increased in the same placentae.

Another tool available for identification and potential aid in quantitation of functional P-450 isozymes is the use of classical inhibitors and activators. One of the most useful agents discovered to date is 7,8-benzoflavone which acts to activate certain P-450s, inhibit others, and exert no apparent effect on yet others. For example, Huang et al. (1981) found that this chemical would inhibit catalysis of BaP hydroxylation by rabbit liver isozymes LM_2, LM_{3b}, and (particularly) LM_6 but markedly increased the same reaction as catalyzed by isozymes LM_{3c} and LM_4. These studies have been extended by Raucy and Johnson (1985). When added to human liver preparations, increased rates of monooxygenation of BaP, BaP-7,8-diol, zoxazolamine, antipyrine and aflatoxin B_1 are observed but rates of oxidation of 7-ethoxycoumarin, coumarin, and hexobarbital are unaffected. From this viewpoint, it is interesting that 7,8-benzoflavone is a highly potent inhibitor of placental BaP hydroxylation, particularly in smokers' placentae (Pelkonen and Moilanen, 1979; Pelkonen, 1980, 1984). Metyrapone inhibits monooxygenation reactions catalyzed by rat $P-450_b$, $P-450_e$, and $P-450_p$ but increases rates of acetanilide hydroxylation (presumably via an effect on P-450$_d$). Metyrapone is an extremely weak inhibitor of placental xenobiotic monooxygenation reactions studied to date (Pelkonen and Moilanen, 1979; Kaelin and Cummings, 1983). Several other activators and inhibitors with varying degrees of specificity are known. Of particular promise are the suicide inhibitors which, theoretically, should be very selective for particularly isozymes. Interest in this area of research is currently at a high level and it is to be expected that excellent probes will become available.

As yet, individual probes are unsatifactory, but the judicious use of a *combination* of substrate, inducer, antibody, inhibitor, and activator probes can provide a wealth of fairly definitive information concerning the content and functionality of P-450's in various tissues (including the placenta) for which purification procedures are impractical.

Unfortunately, many (including some investigators who have utilized antibodies as probes) either have not recognized the limitations associated with this approach or have failed to communicate such limitations, thereby giving the impression that such experiments are conclusive. Limitations of the use of antibody/immunoassay probes include:

1.   Antibody probes (particularly monoclonal antibody probes) may not distinguish among functionally different P-450 isozymes having the same epitopes (antibody recognition sites).

2.   Antibody probes may not distinguish between functionally active and inactive P-450s. Functionally inactive forms include P-450 isozymes, apoisozymes, chemically inactivated P-450 isozymes, and partially degraded (e.g., proteolytically cleaved) isozymes.

3.   Antibody probes may *differentially* inhibit various functionally active isozymic species. Specific epitope interactions may even fail entirely to inhibit the functional activity of certain isozymes (e.g., see Park et al., 1985).

4.   Varying concentrations of functionally inactive (or differently active) proteins with common epitopes may differentially inhibit the capacity of antibodies to inhibit the activity of functionally active isozymes via competition.

Nevertheless, when used in *combination* with substrate probes, inhibitor/activator probes and inducer probes, the use of antibody probes promises to yield much useful information concerning individual P-450 isozymes and their functions in the placenta as well as other tissues for which isolation/purification is impractical.

8.   Human Placental Catecholestrogen-Forming P-450s Exhibit Similarities and Differences to Placental Xenobiotic-Oxidizing P-450s.

For several years we have been interested in the possibility that the placental xenobiotic-oxidizing P-450s may utilize endogenous chemicals as "natural" substrates. We felt that we may have discovered such a phenomenon when we found a very close correlation between measured rates of hydroxylation of the aromatic ring of estradiol-17β (E_2) and the aromatic ring hydroxylation of BaP in placental microsomes (Chao and Juchau, 1980). The former reaction was studied with the assay developed by Paul and Axelrod (1977) which is highly sensitive and reliable but does not distinguish between 2- and 4-hydroxylation and is also indirect. Both BaP and estrogen hydroxylating reactions were increased in placentae of smokers and both were activated by additions of micromolar quantities of hematin to reaction flasks. However, the magnitude of smoking-associated increases in estrogen hydroxylation were far less than those of BaP hydroxylation. Also, characteristics of hematin activation (discussed under concept 12) differed for the two substrates. It was considered possible that either 2-hydroxylation or 4-hydroxylation of the steroid estrogen might be even more closely correlated with BaP hydroxylation in placental preparations. The tritium-release assays utilized to explore that possibility indicated that the 4-hydroxylation was strongly correlated but that the 2-hydroxylation reaction correlated only very weakly. Since the placental 2- and 4-hydroxylation reactions exhibited other dissimilarities (as measured with the tritium-displacement assays), it was felt possible that the placental P-450(s) catalyzing BaP hydroxylation could preferentially attack E_2 at the 4-carbon and that this might be its normal physiologic function. However, this possibility is clouded by a number of factors. Firstly, the (apparent) magnitude of the increases in estrogen 4-hydroxylation in response to smoking was much less than that commonly seen with BaP as substrate. Also, only poor correlations were observed when the extremes of the regression line were analyzed. A confounder of data interpretation is the now well-known fact that estrogen hydroxylation at the 2-carbon can result in release (via nonenzymatic and enzymatic mechanisms including glutathione conjugation and keto-enol tautomerization) of tritium label at the 4-carbon (Jellinck et al., 1984). This phenomenon, however, should tend to *mask* correlations between estrogen 4-hydroxylation and BaP hydroxylation. Therefore, since reasonable correlations were *still* observed, it is considered likely that the correlation may actually be much *better* than was detectable with the assay systems employed. This seems deserving of further study. At present, BaP hydroxylation and estradiol 4-hydroxylation are the only xenobiotic-endobiotic hydroxylation reactions known to exhibit significant correlations in placental tissues. However, research in this area has been sufficiently scant to preclude extensive speculation. It seems entirely possible that future research would disclose hydroxylation of endobiotic and xenobiotic substrates by a common placental P-450. It is of interest to note that Schwab and Johnson (1985) recently reported that rabbit hepatic P-450-LM1 is the principal estradiol 2-hydroxylase in rabbit liver but that LM-6 would also attack the substrate. Although they detected only 2-hydroxylation, it seems possible that an "LM-6-type" placental cytochrome may preferentially attack the 4-position. This, of course, is entirely speculative.

9.   The Preponderance of Placental Xenobiotic-Oxidizing P-450s is Functionally and Immunologically Similar to $P-450_c$ (Rats), P_1-450 (Mice), and $P-450-LM_6$ (Rabbits).

Thus far "extensive" characterization of individual isolated and purified xenobiotic-oxidizing P-450s has been attempted in only a very few tissues. Currently, the most detailed information is available from studies with livers of rats, rabbits, and mice and lungs of rabbits. The literature describing these investigations has been extremely confusing because of the lack of a standard nomenclature, often without reference to names used by competing laboratories. For this discussion, we will utilize what we perceive to be the most commonly used names for individual P-450 isozymes. Rat liver isozymes are named in accordance with the lower case alphabetized ("Levin") system (P-450$_a$, P-450$_b$, etc.) commonly reported by the group at Hoffman-LaRoche (Ryan et al., 1984; Levin et al., 1985; Astrom and DePierre, 1985). At least 13 forms have been identified and partially characterized in rat liver microsomes. Ten forms (a-j) have been electrophoretically purified by the same group. A form induced by PCN, dexamethasome and macrolides has been purified by Guzelian's research group and is now commonly referred to as P-450$_p$ (Wrighton et al., 1985). Forms inducible by ethanol and clofibrate have been identified but not yet extensively purified. In mice, the nomenclature adopted by Nebert's research group (Tukey and Nebert, 1984; Gonzales et al., 1985) is utilized. Only 3 mouse isozymes have been named thus far. These are P$_3$-450, a major MC- inducible form with the highest tunrover number for hydroxylation of acetanilide, P$_1$-450, a major MC-inducible form with the highest substrate turnover number for hydroxylation of BaP and P$_2$-450, an isosafrole-induced P-450 with the highest turnover number for isosafrole metabolite formation. For rabbits, the nomenclature is that utilized by Coon's research group (Koop et al., 1985; Black and Coon, 1985). These P-450s have been names LM1, LM2, LM3$_a$, LM3$_b$, LM3$_c$, LM4, LM5, and LM6 or simply as isozymes 1, 2, 3a, 3b, 3c, 4, 5, and 6. Rabbit lung P-450s are also named according to the latter nomenclature (Philpot et al., 1985).

Insofar as the placental xenobiotic-oxidizing P-450s are concerned, evidence accumulated thus far indicates strongly that the major form associated with smoking is one that is very similar to rat hepatic P-450$_c$, murine hepatic P$_1$-450 or rabbit isozyme 6. Evidence for this is based on the following observations pertaining to P-450-dependent biotransformation in the placentae of smokers:

1.   Placental monooxygenase activities are markedly increased following exposure to polynuclear aromatic hydrocarbons.

2.   Activities are minimally affected by phenobarbital and other inducers not regarded as "MC-type".

3.   Activities are markedly inhibited by 7,8-benzoflavone but minimally affected by metyrapone (Pelkonen and Moilanen, 1979) or (-) maackiain acetate (Gelboin et al., 1981).

4.   Substrates attacked are those expected also to be attacked by P-450$_c$, P$_1$-450 or isozyme 6.

5.   Antibodies to P-450$_c$ strongly inhibit placental monooxygenase activity in smokers (discussed under concept 8).

6.   Analysis of placental warfarin metabolism (Rettie et al., 1985) on smokers yields the same stereospecifity and regioselectivity patterns exhibited by P-450$_c$. Analyses of the biotransformation of a series of phenoxazone ethers in smokers placentae are also compatible with the concept that a "P-450$_c$-type" cytochrome predominates. This is also discussed more extensively under concept no. 8.

10. Inducibility of Placental Xenobiotic-Oxidizing P-450s Increases with Gestational Age.

In our earliest investigations of the capacity of human placental tissues to biotransform foreign compounds, we utilized placentae from the very early stages of gestation. At that time, a widely believed concept was that the placenta "aged" metabolically as gestation proceeded and that the term placenta was metabolically "senescent". The rationale for use of early placentae was that our best opportunity to detect placental drug biotransformational activity would be to utilize freshly delivered tissues from early therapeutic abortions. Even though we were able to measure the reduction of compounds containing aromatic nitro groups or azo linkages in those tissues (Juchau et al., 1968; Juchau, 1969; Symms and Juchau, 1972), monooxygenase activities were below the limits of detectability. Reports of readily measurable benzo(a)pyrene hydroxylase activity in term placentae of smokers (Welch et al., 1969; Nebert et al., 1969) prompted us to investigate the role of gestational age more carefully. We found (Juchau, 1971) that irrespective of smoking habits, monooxygenase activities were extremely low or undetectable prior to the 12th week of gestation. With advancing gestational age, activities in smokers' placentae increased and maximal activities were measured in placentae delivered at term. These observations were confirmed in a subsequent report from Dr. Pelkonen's laboratory (Pelkonen et al., 1972). Data accumulated since that time have further corroborated the idea that early placentae are less susceptible to the inducing effects of substances present in tobacco smoke. Experimentally validated reasons for the reduced susceptibility have not been forthcoming, but it has been speculated that demands for the synthesis of proteins required for cellular proliferation would tend to compromise the synthesis of more highly differentiated P-450 proteins (Nebert and Gonzales, 1985; Marx, 1985). It should be pointed out that extremely few comparisons between term placentae and early gestation placentae have been made and in each of those comparisons only BaP has been utilized as a substrate for P-450-dependent oxidation. Future investigations may well uncover P-450-dependent xenobiotic oxidations that occur at relatively rapid rates during early gestation. At present, however, this seems relatively unlikely.

11. Placental Phase II Conjugation of Products of P-450-Dependent Oxidations Occurs to Varying Degrees Depending Upon the Oxidation Product and the Specific Conjugation Reaction.

The major xenobiotic conjugating or phase II reactions include glucuronidation, sulfation, gluathione conjugation, acetylation, methylation, and amino acid (principally glycine and glutamine) conjugation. Most studies indicate that unless the xenobiotic is a close structural relative of an endogenous substrate(s), foreign chemicals would be slowly conjugated, if at all, in human placental tissues and cells (Juchau, 1973, 1980, 1984). Exceptions to this generalization have been found (e.g., see Namkung et al., 1977) and possibly one of the most important exceptions pertains to the relatively high levels of placental glutathione conjugating activity that have been observed (Juchau and Namkung, 1974; Polidoro et al., 1980). The placenta appears to contain only one GSH S-transferase isozyme and this reportedly is identical to the major erythrocytic form (Awasthi and Singh, 1984; Satoh et al., 1985; Radulovic and Kulkarni, 1985; Guthenberg and Mannervik, 1981; Alin et al., 1985). By contrast, the liver contains at least 8 individual isozymic forms, each of which has been reasonably well characterized. From this, one would expect that human placental transferases would effectively catalyze glutathione conjugation for only a relatively limited number of substrates when compared with adult hepatic tissues. This expectation is in accord with experimental findings reported by Polidoro et al. (1980) in which they found that 1-chloro-2,4-dinitrobenzene (DCNB) was conjugated at a relatively rapid rate while several other potential substrates were conjugated extremely slowly or undetectably. Interestingly, the rate of human placental glutathione conjugation of DCNB closely approached that measured in the adult human liver and was roughly 50% of that measured in analogous preparations of adult rat liver. Importantly, it has been shown that GSH-transferase activities measured in human placental tissues can be almost entirely separated from activities present in contaminating erythrocytic elements. Studies from our laboratory (Juchau and Namkung, 1974) indicated that cigarette smoking did not affect human placental GSH-transferase activities and these results were confirmed by Manchester and Jacoby (1982) with DCNB as substrate. These observations are in agreement with the concept (Juchau, 1973, 1980, 1984; Juchau and Rettie, 1985) that smoking by pregnant women results in increases in ratios of bioactivation/bioinactivation in human placental tissues and may predispose the developing conceptus to the toxic effects of foreign organic chemicals. To our knowledge, all experimental data reported to date indicate that placental conjugating reactions, including glucuronidation, sulfation, acetylation, methylation, amino acid conjugation as well as glutathione conjugation are not increased in placentae of smokers. We recently failed to detect increases in glucuronidation or sulfation of 3-hydroxy-BaP in explanted preparations of smokers placentae (unpublished) in agreement with this concept. In fact, it appears that the only xenobiotic-biotransforming placental enzymes affected by smoking are those of the P-450 family. Nevertheless, the extent to which placental P-450 induction by xenobiotic chemicals may influence reproductive processes is highly speculative at present and much further research will be required to elucidate the importance of placental P-450 induction. Other aspects of placental conjugation reactions have been reviewed very recently (Juchau, 1984; Juchau and Rettie, 1985).

12.   Placental Tissue Contain a Relatively Large Complement of Non-functional Apo-P-450. When Combined with Hematin, the Holocytochrome(s) Functions in the Catalysis of Specific Monooxygenations.

During investigations designed to characterize aryl hydrocarbon hydroxylase activity in aortic tissues (Bond et al., 1979), it was discovered that additions of micromolar quantities of hematin to reaction vessels containing aortic homogenates resulted in profound (25-30 fold) increases in observed monooxygenase activities. Subsequent investigations (Omiecinski et al., 1978, 1980a,b,c, 1982; Bond et al., 1980; Chao et al., 1981) demonstrated that hematin, when added to reaction flasks containing any one of several different extrahepatic tissue preparations, could produce up to 100-fold increases in rates of cytochrome P-450-dependent monooxygenation reactions with benzo(a)pyrene as substrate. Other tested substrates (e.g., acetanilide, 7,12-dimethylbenz(a)anthracene, estradiol-17β, certain other polycyclic aromatic hydrcarbons, etc.) generally yielded somewhat lesser increases. With the exception of fetal rabbit liver, hepatic preparations of several species did not respond to the activating effects of added hematin. A body of accumulated evidence has indicated that the mechanism whereby the hematin-mediated increases were produced was via a combination of the iron protoporphyrin with a species (or possibly multiple species) of free apocytochrome P-450 present in extrahepatic tissues. Apparently, a product(s) of the monooxygenation reaction interacts with the apocytochrome to produce an allosteric or conformational change leading to an increased affinity of the apoprotein for its heme prosthetic group, thereby resulting in increased levels of the holocytochrome. Early evidence for the mechanism has been summarized in a review (Juchau et al., 1982b). Investigations with placental tissues from humans, monkeys, rabbits, rats, and mice (Namkung et al., 1983a) showed that the phenomenon is also applicable to placentae from several species. Interestingly, the placenta of DBA/2 mice, a strain that is resistant to the inducing effects of the "MC-type" inducing agents, responded only minimally to the effects of hematin while placentae of C57BL/6 mice exhibited a 4.2-fold increase. However, this lack of response in the former strain remains unexplained as various other extrahepatic tissues from the same strain responded with sizable increases in activity (Namkung et al., 1983b). The placentae of rabbits and monkeys exhibited the greatest responses (BaP as substrate) but hematin also produced respectable increases in various preparations of human placentae. Preparations included tissue minces, 9000 g supernatant fractions, washed microsomal fractions and solubilized, partially purified microsomal preparations.

Recently we discovered that for most extrahepatic tissues, a cytosolic factor(s) markedly increases the capacity of hematin to increase rates of P-450-dependent monooxygenation in microsomal prepartions (Dean et al., 1984). Human placental cytosolic fractions, however, were only minimally effective, for reasons not yet elucidated. For isolated rat lung microsomes, additions of micromolar quantities of hematin to reaction vessels resulted in 5-7 fold increases in rates of hydroxylation of BaP. Lung cytosol, added together with hematin, elicited 25-30 fold increases. By comparison, placental cytsol produced only approximately 8-10 fold increases when added to rat lung microsomes and hematin-dependent increases in human placental microsomes were only minimally affected by additions of placental cytosol to placental microsomes (unpublished). The cytsol of human placentae appears to lack a heat-labile protein factor capable of interacting with hematin to produce profound increases in monooxygenase activities. It is clear that much remains to be learned concerning these highly interesting phenomena, particularly concerning the mechanistic aspects. The apparent complexity of this mode of monooxygenation regulation suggests that very specialized and important basic functions are subserved, but the nature of such functions remains to be explored in future investigations. It is of interest that reactions involving both synthesis (Bellino and Hussa, 1985) and degradation (Namkung et al., 1983a) of estrogens are markedly affected by hematin in placental tissues. Perhaps these observations will lead to an elucidation of the physiologic significance of this interesting phenomenon.

REFERENCES

Alin, P., Mannervik, B., and Jornvall, H. (1985) Structural evidence for three different types of gluathione transferase in human tissues. *FEBS Lett.* 182, 319-322.

Alsat, E. and Cedard, L. (1973) The stimulating action of the prostaglandins on the production of estrogens by the human placenta. *Prostaglandins* 3, 145-153.

Anderson, C.M. and Mendelson, C.R. (1985) Regulation of steroidogenesis in rat leydig cells in culture: Effect of human choronic gonadotropin and dibutyryl cyclic AMP on the synthesis of cholesterol side-chain cleavage cytochrome P-450 and adrenodoxin. *Arch. Biochem. Biophys.* 238, 378-387.

Astrom, A. and DePierre, J.W. (1985) Metabolism of 2-acetylaminofluorene by eight different forms of cytochrome P-450 isolated from rat liver. *Carcinogenesis.* 6, 113-120.

Awasthi, Y.C. and Singh, S.V. (1984) Purification and characterization of a new form of glutathione-S-transferase from human erythrocytes. *Biochem. Biophys. Res. Comm.* 1265, 1053-1060.

Bellino, F.L. and Hussa, R.O. (1978) Trophoblast estrogen synthetase stimulation by dibutyryl cyclic AMP and theophylline: Increase in cytochrome P-450 content. *Biochem. Biophys Res. Comm.* 85, 1588-595.

Bellino, F.L. and Hussa, R.O. (1985) Estrogen synthetase stimulation by hemin in human choriocarcinoma cell culture. *Biochem. Biophys. Res. Comm.* 127, 232-238.

Black, S.D. and Coon, M.J. (1985) Studies on the identity of the heme-binding cysteinyl residue in rabbit liver microsomal cytochrome P-450 isozyme 2. *Biochem. Biophys. Res. Comm.* 128, 82-89.

Bond, J.A., Omiecinski, C.J., and Juchau, M.R. (1979) Kinetics, activation and induction of aortic monooxygenases: Anlysis of the mixed-function oxidation of benzo(a)pyrene with high-pressure liquid chromatography. *Biochem. Pharmacol.* 28, 305-312.

Bond, J.A., Yang, H.L., Majesky, M.W., Benditt, E.P., and Juchau, M.R. (1980) Metabolism of benzo(a)pyrene and 7,12-dimethylbenz(a)anthracene in chicken aortas: Monooxygenation, bioactivation to mutagens and covalent binding to DNA in vitro. *Toxicol. Appl. Pharmacol.* 52, 323-335.

Burns, B. and Gurtner, G.H. (1973) A specific carrier for oxygen and carbon monoxide in the lung and placenta. *Drug Metab. Disp.* 1, 374-387.

Canick, J.A. and Ryan, K.J. (1978) Properties of the aromatase system associated with the mitochondrial fraction of human placenta. *Steroids* 32, 499-509.

Cedard, L., Alsat, E., Urtasun, M., and Varnagot, J. (1970) Studies on the mode of action of luteinizing hormone and chronic gonadotropin on estrogenic biosynthesis and glycogenolysis by human placental perfused in vitro. *Steroids* 16, 361-375.

Chakraborty, J., Hopkins, R., and Parke, D.V. (1972) Inhibition studies on the aromatization of androst-4-ene-3,17-dione by human placental microsomal preparations. *Biochem. J.* 130, 19P.

Chao, S.J. and Juchau, M.R. (1980) Interaction of endogenous and exogenous estrogenic compounds with human placental microsomal cytochrome P-450 (P-450$_{hpm}$). *J. Ster. Biochem.* 13, 127-133.

Chao, S.T., Omiecinski, C.J., Namkung, M.J., Nelson, S.D., Dvorchick, B.H., and Juchau, M.R. (1981) Catechol estrogen formation in placental and fetal tissues of humans, macaques, rats and rabbits. *Dev. Pharmacol. Ther.* 2, 1-17.

Conney, A.H., Welch, R., Kuntzman, R., Poland, A., Poppers, P.J., Finster, M., Wolfe, J.A., Munro-Faure, A.D., Peck, A.W., Bye, A., Chang, R., and Jacobson, M. (1971) Effects of environmental chemicals on the metabolism of drugs, carcinogens and normal body constituents in man. *Ann. NY Acad. Sci.* 179, 155-167.

Dean, P.A., Namkung, M.J., and Juchau, M.R. (1984) Potentiation of hematin-mediated increases in pulmonary monooxygenase activity cytosolic protein. *Toxicologist* 4, 388.

Dee, A., Carlson, G., Smith, C., Masters, B.S., and Waterman, M.R. (1985) Regulation of synthesis and activity of bovine adrenocortical NADPH-cytochrome P-450 reductase by ACTH. *Biochem. Biophys. Res. Comm.* 128, 650-656.

Frauenfelder, H. and Wolynes, P.G. (1985) Rate theories and puzzles of hemeprotein kinetics. *Sci.* 229, 337-345.

Fujino, T., Parks, S.S., West, D., and Gelboin, H.V. (1982) Phenotyping of cytochromes P-450 in human tissues with monoclonal antibodies. *Proc. Natl. Acad. Sci. USA* 79, 3682-3686.

Fujino, T., Gottleib, K., Manchester, D.K., Park, S.S., West, D., Gurtoo, H.L., Tarone, R.E., and Gelboin, H.V. (1984) Monoclonal antibody phenotyping of interindividual differences in cytochrome P-450-dependent reactions of single and twin human placenta. *Can. Res.* 44, 3916-3923.

Gelboin, H.V., West, D., Gozukara, E., Natori, S., Nagao, M., and Sugimura, T. (1981) Maackiain acetate specifically inhibits different forms of aryl hydrocarbon hydroxylase in rat and man. *Nature* 291, 659-661.

Gilbert, R.D., Cummings, L.A., Juchau, M.R., and Longo, L.D. (1979) Placental diffusing capacity and fetal development in exercising or hypoxic guinea pigs. *J. Appl. Physiol.* 46, 828-835.

Gonzales, F.J., Kimura, S., and Nebert, D.W. (1985) Comparison of the flanking regions and introns of the mouse 2,3,7,8-tetrachlorodibenzo-p-dioxin-inducible cytochrome P$_1$-450 and P$_3$-450 genes. *J. Biol. Chem.* 260, 5040-5049.

Greim, H. (1970) Synthesesteigerung und Abbauhemmung bei der Vermehrung der mikrosomalen Cytochrome P-450 und b$_5$ durch Phenobarbital. *Naunyn-Schmeideberg's Archiv. fur Pharmakologie* 266, 261-275.

Gunasegarem, R., Peh, K.L., Loganath, A., Karim, S.M.M., and Ratnam, S.J. (1982) Inhibition of second trimester human placental cholesterol C-20, 22-desmolase activity by soluble constituents of cigarette smoke. *IRCS Med. Sci.* 10, 322-323.

Gurtner, G.H. and Burns, B. (1973) The role of cytochrome P-450 of placenta in facilitated oxygen diffusion. *Drug Metab. Disp.* 1, 368-374.

Guthenberg, C. and Mannervik, B. (1981) Glutathione S-transferase (Transferase π) from erythrocytes. *Biochim. Biophys. Acta* 662, 255-261.

Hodgson, E. and Juchau, M.R. (1977) Ligand binding to human placental cytochrome P-450: Interaction of steroids and heme-binding ligands. *J. Ster. Biochem.* 8, 669-675.

Huang, M., Johnson, E.F., Muller-Eberhard, U., Koop, D.R., Coon, M.J., and Conney, A.H. (1981) Specificity in the activation and inhibition by flavonoids of benzo(a)pyrene hydroxylation by cytochrome P-450 isozymes from rabbit liver microsomes. *J. Biol. Chem.* 256, 10897-10901.

Jaiswal, A.K., Gonzales, F.J., and Nebert, D.W. (1985) Human dioxin-inducible cytochrome P_1-450: Complementary DNA and amino acid sequence. *Sci.* 228, 80-82.

Jellinck, P.H., Hahn, E.F., Norton, B.I., and Fishman, J. (1984) Catechol estrogen formation and metabolism in brain tissue: Comparison of tritium release from different positions in ring A of the steroid. *Endocrinol.* 115, 1850-1856.

John, M.E., John, M.C., Simpson, E.R., and Waterman, M.R. (1985) Regulation of cytochrome P-450$_{11B}$ gene expression by adrenocorticotropin. *J. Biol. Chem.* 260, 5710-5767.

Juchau, M.R. (1969) Studies on the reduction of aromatic nitro groups in human and rodent placental homogenates. *J. Pharmacol. Exp. Ther.* 165, 1-8.

Juchau, M.R. (1971) Human placental hydroxylation of 3,4-benzopyrene during early gestation and at term. *Toxicol. Appl. Pharmacol.* 18, 665-675.

Juchau, M.R. (1973) Placental metabolism in relation to toxicology. *CRC Crit. Rev. Toxicol.* 2, 125-159.

Juchau, M.R. (1980) Drug biotransformation in the placenta. *Pharmacol. Ther.* 8, 501-524.

Juchau, M.R. (1984) Placental drug metabolism with respect to transplacental carcinogenesis. In: *Comparative Perinatal Carcinogenesis*, (ed.), H.M. Schuller, CRC Press, Inc., Boca Raton, Florida, pp. 117-137.

Juchau, M.R., Krasner, J., and Yaffe, S.J. (1968) Studies on reduction of azo linkages in human placental homogenates. *Biochem. Pharmacol.* 17, 1969-1979.

Juchau, M.R. and Yaffe, S.J. (1969) Biotransformations of drug substrates in placental homogenates. In: *The Foeto-Placental Unit*, (eds.), A. Pecile and C. Finzi, Amsterdam, Excerpta Medica Foundation, pp. 269-270.

Juchau, M.R. and Symms, K.G. (1972) Aniline hydroxylation in the human placenta -- mechanistic aspects. *Biochem. Pharmacol.* 21, 2053-2065.

Juchau, M.R., Lee, Q.H., and Blake, P.H. (1972) Inverse correlation between aryl hydrocarbon hydroxylase activity and conversion of cholesterol to pregnenolone in human placentas at term. *Life Sci. II: Biochem. Gen. Molec. Biol.* 11, 949- 956.

Juchau, M.R. and Smuckler, E.A. (1973) Subcellular localization of human placental aryl hydrocarbon hydroxylase. *Toxicol. Appl. Pharmacol.* 126, 163-180.

Juchau, M.R., Zachariah, P.K., Colson, J., Symmns, K.G., Krasner, J., and Yaffe, S.J. (1974) Studies on human placental carbon monoxide-binding cytochromes. *Drug Metab. Disp.* 2, 79-86.

Juchau, M.R. and Namkung, M.J. (1974) Studies on the biotransformation of naphthalene-1,2-oxide in fetal and placental tissues of humans and monkeys. *Drug Metab. Disp.* 2, 380-386.

Juchau, M.R. and Zachariah, P.K. (1975a) Displacement of carbon monoxide from placental cytochrome P-450 by steroids: Antagonistic effects of androstenedione and 19-norandrostenedione. *Biochem. Biophys. Res. Comm.* 65, 1026-1032.

Juchau, M.R. and Zachariah, P.K. (1975b) Comparative studies on the oxidation and reduction of drug substrates in human placental versus rat hepatic microsomes. *Biochem. Pharmacol.* 24, 227-233.

Juchau, M.R., Namkung, M.J., Berry, D.L., and Zachariah, P.K. (1975) Oxidative biotransformation of 2-acetylaminofluorene in fetal and placental tissues of humans and monkeys: Correlations with aryl hydrocarbon hydroxylase activities. *Drug Metab. Disp.* 3, 494-502.

Juchau, M.R., Mirkin, D.L., and Zachariah, P.K. (1976) Interactions of various 19-norsteroids with human placental microsomal cytochrome P-450 (P-450$_{hpm}$). *Chem. Bio. Interactions.* 15, 337-347.

Juchau, M.R., Namkung, M.J., Jones, A.H., and DiGiovanni, J. (1978) Biotransformation and bioactivation of 7,12-dimethylbenz(a)anthracene in human fetal and placental tissues. *Drug Metab. Disp.* 6, 273-281.

Juchau, M.R., Namkung, M.J., and Chao, S.T. (1982) Monoxygenase induction in the human placenta: Interrelationships among position-specific hydroxylation of 17β-estradiol and benzo(a)pyrene. *Drug Metab. Disp.* 10, 220-225.

Juchau, M.R., Omiecinski, C.J., and Namkung, M.J. (1982) Hematin-mediated increases in P-450-dependent monooxygenation reactions: An apparent short-term regulatory mechanism. In: *Microsomes, Drug Oxidations and Drug Toxicities*, (eds.), R. Sato and R. Kato), Wiley-Interscience, New York, pp. 371-381.·

Juchau, M.R. and Rettie, A.E. (1985) The metabolic role of the placenta. In: *Principles of Drug and Chemical Action in Pregnancy*, (eds.), S.E. Fabro and A.R. Scialli, Marcel Dekker, Inc., New York, in press.

Kaelin, A.C. and Cummings, A.J. (1983) A study of monooxygenase activity in human placental homogenates: In vitro behavior towards a number of substrates and inhibitors. *Biochem. Pharmacol.* 32, 2421-2426.

Kaelin, A.C. and Cummings, A.J. (1984) Differential inhibition of human placental monooxygenase activity: Evidence for multiple forms of 7-ethoxycoumarin O-deethylase. *Biochem. Pharmacol.* 33, 505-508.

Koop, D.R., Crump, B.L., Nordblom, G.D., and Coon, M.J. (1985) Immunochemical evidence for induction of the alcohol-oxidizing cytochrome P-450 of rabbit liver microsomes by diverse agents: ethanol, imidazole, trichloroerthylene, acetone, pyrazole and isoniazid. *Proc. Natl. Acad. Sci. USA* 82, 4065-4069.

Lee, Q.P., Zachariah, P.K., and Juchau, M.R. (1975) Differential inhibition of androst-4-en-3,17-dione aromatization by carbon monoxide in the presence of estr-4-en-3,17-dione. *Steroids* 26, 571-579.

Levin, W., Thomas, P.E., Reik, L.M., Ryan, D.E., Bandiera, S., Haniu, M., and Shively, J.E. (1985) Immunochemical and structural characterization of rat hepatic cytochromes P-450. In: *Microsomes and Drug Oxidations*, (eds.), A.R. Boobis et al., Taylor and Francis, London, pp. 13-23.

Lin, T. (1985) Mechanism of action of gonadotropin-releasing hormone stimulated Leydig cell steroidogenesis III. The role of arachidonic acid and calcium/phospholipid dependent protein kinase. *Life Sci.* 1255-1264.

Manchester, D.K. and Jacoby, E.H. (1982a) Resolution and reconstitution of human placental monooxygenase activity responsive to maternal cigarette smoking. *Dev. Pharmacol. Ther.* 5, 162-172.

Manchester, D.K. and Jacoby, E.H. (1982b) Glutathione S-transferase activities in placentas from smoking and nonsmoking women. *Xenobiotica* 12, 543-547.

Marx, J.L. (1985) The cytochrome P-450's and their genes. *Sci.* 228, 975-976.

Meigs, R.A. and Moothry, K.B. (1984) The support of steroid aromatization by mitochondrial metabolic activities of the placenta. *J. Ster. Biochem.* 20, 883-886.

Meigs, R.A. and Ryan, K.J. (1968) Cytochrome P-450 and steroid synthesis in the human placenta. *Biochem. Biophys. Acta* 165, 476-483.

Meigs, R.A. and Ryan, K.J. (1971) Enzymatic aromatization of steroids. I. Effects of oxygen and carbon monoxide on the intermediate steps of estrogen biosynthesis. *J. Biol. Chem.* 239, 2370-2377.

Morgan, E.T., MacGeoch, C., and Gustafsson, J. (1985) Sexual difference of cytochrome P-450 in rat liver: Evidence for a constitutive isozyme on the male-specific 16α-hydroxylase. *Mol. Pharmacol.* 27, 471-479.

Namkung, M.J., Zachariah, P.K., and Juchau, M.R. (1977) O-sulfonation of N-hydroxy-2-fluorenylacetamide and 7-hydroxy-N-2-fluorenylacetamide in fetal and placental tissues of humans and guinea pigs. *Drug. Metab. Disp.* 5, 288-294.

Namkung, M.J., Chao, S.T., and Juchau, M.R. (1983) Placental monooxygenation: Characteristics and partial purification of a hematin-activated human placental monooxygenase. *Drug Metab. Disp.* 11, 10-15.

Namkung, M.J., Faustman-Watts, E.M., and Juchau, M.R. (1983) Hematin-mediated increases in benzo(a)pyrene monooxygenation in maternal, fetal and placental tissues of inducible and non-inducible mouse strains. *Dev. Pharmacol. Ther.* 6, 199-207.

Namkung, M.J., Porubek, D.H., Nelson, S.D., and Juchau, M.R. (1985) Regulation of aromatic oxidation of estradiol-17β in maternal hepatic, fetal hepatic and placental tissues: Comparative effects of a series of inducing agents. *J. Ster. Biochem.* 22, 563-568.

Nebert, D.W., Winker, J., and Gelboin, H.V. (1969) Aryl hydrocarbon hydroxylase activity in human placenta from cigarette smoking and nonsmoking women. *Canc. Res.* 29, 1763-1769.

Nebert, D.W. and Gonzales, F.J. (1985) Cytochrome P-450 gene expression and regulation. *Trends Pharmacol. Sci.* 6, 160-164.

Omiecinski, C.J., Bond, J.A., and Juchau, M.R. (1978) Stimulation by hematin of monooxygenase activity in extra-hepatic tissues from rats, rabbits and chickens. *Biochem. Biophys. Res. Comm.* 83, 1004-1011.

Omiecinski, C.J., Chao, S.T., and Juchau, M.R. (1980) Modulation of monooxygenase activities by hematin and 7,8-benzoflavone in fetal tissues of rats, rabbits, and humans. *Dev. Pharmacol. Ther.* 1, 90-100.

Omiecinski, C.J., Namkung, M.J., and Juchau, M.R. (1980) Mechanistic aspects of the hematin-mediated increases in brain monooxygenase activities. *Molec. Pharmacol.* 17, 225-232.

Omiecinski, C.J. and Juchau, M.R. (1980) Hematin-mediated increases in the monooxygenation of polynuclear aromatic hydrocarbons. *Proc. West. Pharmacol. Soc.* 23, 9-13.

Omiecinski, C.J., Namkung, M.J., and Juchau, M.R. (1982) Substrate and position specifity of hematin-activated monooxygenation reactions. *Biochem. Pharmacol.* 30, 2837-2845.

Omura, T. and Sato, R. (1964) The carbon monoxide binding pigment of liver microsomes. I. Evidence for its hemoprotein structure. *J. Biol. Chem.* 239, 2370- 2377.

Park, S.S., Miller, H., Guengerich, P.F., and Gelboin, H.V. (1985) Monoclonal antibodies to liver microsomal pregnenolone-16a-carbonitrile induced cytochrome P-450E of rat liver. *Pharmacologist* 27, 114.

Pasanen, M. (1984) Human placental aromatase activity: Use of a C-18 reversed phase cartridge for separation of tritiated water or steroid metabolites in placentas from both smoking and nonsmoking mothers in vitro. *Biol. Res. Preg.* 5, 1-6.

Pasanen, M. and Pelkonen, O. (1981) Solubilization and purification of human placental cytochromes P-450. *Biochem. Biophys. Res. Comm.* 103, 1310-1317.

Pasanen, M. and Pelkonen, O. (1984) Cholesterol side-chain cleavage activity in human placenta and bovine adrenals: A one-step method for separation of pregnenolone formed in vitro. *Steroids* 43, 517-527.

Paul, S.M., Axelrod, J. and Diliberto, E.J. (1977) Catechol estrogen-forming enzyme of the brain: Demonstration of a cytochrome P-450 monooxygenase. *Endocrinol.* 101, 1604-1610.

Pelkonen, O. (1980) Environmental influences on human foetal and placental xenobiotic metabolism. *Eur. J. Clin. Pharmacol.* 18, 17-24.

Pelkonen, O. (1984) Xenobiotic metabolism in the maternal-placental-fetal unit: Implications for fetal toxicity. *Dev. Pharmacol. Ther.* 7, 11-17.

Pelkonen, O., Jouppila, P., and Karki, N.T. (1972) Effect of maternal cigarette smoking on 3,4-benzpyrene and N-methylaniline metabolism in human fetal liver and placenta. *Toxicol. Appl. Pharmacol.* 23, 399-406.

Pelkonen, O. and Moilanen, M.L. (1979) The specificity and multiplicity of human placental xenobiotic-metabolizing monooxygenase systems studied by potential substrates, inhibitors and gel electrophoresis. *Med. Biol.* 57, 306-312.

Pelkonen, O. and Pasanen, M. (1984) Enzymology and regulation of xenobiotic and steroid metabolism in the placenta. *Biochem. Soc. Trans.* 12, 42-44.

Philpot, R.M., Domin, B.A., Devereux, T.R., Harris, C., Anderson, M.W., Fouts, J.R, and Bend, J.R. (1985) Cytochrome P-450-dependent monooxygenase systems of the lung: Relationships to pulmonary toxicity. In: *Microsomes and Drug Oxidations*, (eds.), A.R. Boobis et al., Taylor and Francis, London, pp. 248-255.

Polidoro, G., Dillio, C., Del Boccio, G., Zulli, P., and Federici, G. (1980) Glutathione S-transferase activity in human placenta. *Biochem. Pharmacol.* 29, 1677-681.

Radulovic, L.L. and Kulkarni, A.P. (1985) A rapid, novel high performance liquid chromatography method for the purification of glutathione S-transferase: An application to the human placental enzyme. *Biochem. Biophys. Res. Comm.* 128, 75-81.

Raucy, J.L. and Johnson, E.F. (1985) Variations among untreated rabbits in benzo(α)pyrene metabolism and its modulation by 7,8-benzoflavone. *Molec. Pharmacol.* 27, 296-301.

Rettie, A.E., Heimark, L., Mayer, R.T., Burke, M.D., Trager, W.F., and Juchau, M.R. (1985) Stereoselective and regioselective hydroxylation of warfarin and selective O-dealkylation of phenoxazone ethers in human placenta. *Biochem. Biophys. Res. Comm.* 126, 1012-1022.

Ruckpaul, K., Rein, H., and Blanck, J. (1985) Regulation mechanisms of the endo-plasmic cytochrome P-450 system of the liver. *Biomed. Biochem. Acta* 44, 351-380.

Ryan, D.E., Dixon, R., Evans, R.H., Ramanthan, L., Thomas, P.E., Wood, A.W., and Levin, W. (1984) Rat hepatic cytochrome P-450 isozymes specifity for the metabolism of the steroid sulfate, 5α-androstane-3α,17β-diol-3,17-disulfate. *Arch. Biochem. Biophys.* 233, 636-642.

Satoh, K., Kitahara, A., Soma, Y., Inaba, Y., Hatayama, I., and Sato, K. (1985) Purification, induction and distribution of placental glutathione transferase: A new marker enzyme for preneoplastic cells in the rat chemical hepatocarcinogenesis. *Proc. Natl. Acad. Sci. USA* 52, 3964-3968.

Schwab, G.E. and Johnson, E.F. (1985) Variation in hepatic microsomal P-450[1] concentration among untreated rabbits alters the efficiency of estradiol hydroxylation. *Arch. Biochem. Biophys.* 237, 17-26.

Simpson, E.R. and Miller, D.A. (1978) Cholesterol side-chain cleavage, cytochrome P-450 and iron sulfur protein in human placental mitochondria. *Arch. Biochem. Biophys.* 190, 800-808.

Song, B., Gelboin, H.V., Park, S.S., Tsokos, G.C., and Friedman, F.K. (1985) Monoclonal antibody-directed radioimmunoassay detects cytochrome P-450 in human placenta and lymphocytes. *Sci.* 228, 490-492.

Symms, K.G. and Juchau, M.R. (1972) Further studies on the catalysis of nitro group reduction in tissue-free systems. *Proc. West. Pharmacol. Soc.* 15, 156-161.

Symms, K.G. and Juchau, M.R. (1973) Stablization, solubilization, partial purification and some properties of cytochrome P-450 present in CaCl$_2$-precipitated human placental microsomes. *Life Sci.* 13, 1221-1230.

Symms, K.G. and Juchau, M.R. (1974) The aniline hydroxylase and nitro-reductase activities of partially purified cytochromes P-450 and P-420 and cytochrome b$_5$ solubilized from rabbit hepatic microsomes. *Drug Metab. Disp.* 2, 194-201.

Thompson, E.A., Jr. and Siiteri, P.K. (1974) The involvement of human placental microsomal cytochrome P-450 in aromatization. *J. Biol. Chem.* 249, 5373-5378.

Thompson, E.A., Jr. and Siiteri, P.K. (1979) Subcellular distribution of aromatase in human placenta and ovary. *Horm. Res.* 11, 179-185.

Torminaga, T. and Toren, P. (1967) Stimulation of aromatization in human placenta by human placental lactogen. *J. Clin. Invest.* 46, 1124A.

Tseng, L. and Bellino, F.L. (1983) Inhibition of aromatase and NADPH cytochrome c reductase activities in human endometrium by the human plcental NADPH cytochrome c reductase antiserum. *J. Ster. Biochem.* 22, 555-558.

Tukey, R.H. and Nebert, D.W. (1984) Regulation of mouse cytochrome P$_3$-450 by the Ah receptor. Studies with a P$_3$-450 cDNA clone. *Biochem.* 23, 6003-6008.

Virgo, B.B. and Vockentanz, B.M. (1985) Feminization of the hepatic monooxygenases by growth hormone is mimicked by puromycin and correlates with a decrease in male-type cytochrome P-450. *Biochem. Biophys. Res. Comm.* 128, 683-688.

Washida, N., Matsui, S., and Osawa, Y. (1985) Preparation of monoclonal antibody against human placental aromatase II cytochrome P-450. *Fed. Proc.* 44, 861.

Welch, R.M., Harrison, Y.E., Gomni, B.W., Poppers, P.J., Finster, M., and Conney, A.H. (1969) Stimulating effects of cigarette smoking on the hydroxylation of 3,4-benzpyrene and the N-demthylation of 3-methyl-4-monomethyl-aminoazobenzene by enzymes in human placentas. *Clin. Pharmacol. Ther.* 10, 100-115.

Wrighton, S.A., Maurel, P., Schultz, E.G., Watkins, P.B., Young, B., and Guzelian, P.S. (1985) Identification of the cytochrome P-450 induced by macrolide antibiotics in rat liver as the glucocorticoid-responsive cytochrome P-450$_p$. *Biochem.* 24, 2171-2179.

Wolf, A.S., Musch, K., Speidel, W., Strecker, J.R., and Lauritzen, C. (1978) Steroid metabolism in the perfused human placenta. I. Metabolism of dehydro-epiandrosterone sulfate. *Acta Endocrinol. (Kb)* 87, 181-191.

Zachariah, P.K. and Juchau, M.R. (1975) Interactions of steroids with human placental cytochrome P-450 in the presence of carbon moxoxide. *Life Sci.* 16, 1689-1693.

Zachariah, P.K., Lee, Q.P., Symms, K.G., and Juchau, M.R. (1976) Further studies on the properties of human placental microsomal cytochrome P-450. *Biochem. Pharm.* 25, 793-800.

Zachariah, P.K. and Juchau, M.R. (1977) Inhibition of human placental mixed-function oxidations with carbon monoxide: Reversal with monochromatic light. *J. Ster. Biochem.* 8, 221-228.

Trophoblast Research 2:265-277, 1987

HUMAN PLACENTAL CYTOCHROME P-450: MICROSOMAL PREPARATION AND PURIFICATION

Charles W. Fisher and John J. Moore

Department of Pediatrics
Case Western Reserve University
Cleveland Metropolitan General Hospital
3395 Scranton Road
Cleveland, Ohio 44109

INTRODUCTION

The human placenta contains several microsomal oxidases important in the oxidation of steroids and drugs (Pelkonen and Pasanen, 1984; Pelkonen, 1980; Namkung et al., 1983). An important function of placental cytochrome P-450 is the aromatization of steroids (Meigs and Ryan, 1968). This aromatase activity is induced by the cAMP analog, dibutyryl cAMP, and theophylline indicating a possible control by kinase (Bellino and Hussa, 1978). Further study of the regulation of this and other cytochrome P-450 activities requires the purification of the terminal oxidase, cytochrome P-450, and the associated cytochrome c reductase. Methods of purifying cytochrome P-450 (Pasanen and Pelkonen, 1981; Pelkonen and Pasanen, 1982) and cytochrome c reductase (Bellino, 1982; Yasukochi and Masters, 1976) from human placenta have been reported. However, due to the low specific content of cytochrome P-450 (0.05-0.15 nmol/mg) in human placental microsomes only small amounts of cytochrome P-450 or the reductase can be obtained from a single placenta.

In an attempt to increase the amounts of placental cytochrome P-450 and the reductase obtained in purification we utilized two methods of bulk microsomal isolation which do not require the extensive use of an ultracentrifuge: precipitating microsomes with polyethylene glycol 6000 (PEG) (Van Der Hoeven, 1981) or with calcium chloride (Kamath and Narayan, 1972). These methods considerably shorten the time required to prepare microsomes and allow the collection of large quantities of placental microsomes.

We were unable to purify cytochrome P-450 to homogeneity from placental microsomes prepared by ultracentrifugation or bulk-isolation, however cytochrome c (P-450) reductase was effectively purified using bulk-isolated microsomes.

Received: 18 November 1985; Accepted: 10 June 1986

METHODS

Microsomes

Placentae were trimmed free of cords and membranes, and washed with cold (4°C) saline (150 mM KCl, 50 mM K phosphate, pH 7.7, 0.1 mM EDTA) to remove clots. The placentae were minced, rinsed with more saline to remove blood, and then ground in a meat grinder. Ground placentae were rinsed, drained, and either processed immediately or frozen at -70°C. This procedure removed the majority of the blood. The ground placentae were thoroughly homogenized in 2 volumes of saline with a Tekmar tissumizer and strained through gauze. A postmitochondrial supernatant was prepared by centrifugation at 12,000 x g for 15 min.

Microsomes were prepared by each of three different methods from the postmitochondrial supernatant.

A.   A 50% solution of polyethlyene glycol 6000 (PEG) was added to 5% (w/v), mixed, and the mixture centrifuged at 12,000 x g for 15 min (Van Der Hoeven, 1983).

B.   Calcium chloride ($CaCl_2$) was added to 10 mM and the mixture centrifuged at 12,000 x g for 15 min (Kamath and Narayan, 1972).

C.   Centrifugation at 100,000 x g for 60 min.

Microsomes were resuspended in 50 mM K phosphate, pH 7.7, 0.1 mM EDTA, 20% glycerol and recentrifuged (12,000 x g for 15 min. (A,B) or 100,000 x g for 60 min (C)).

Protein and Enzyme Assays

Protein was measured by the method of Lowry et al. (1951) using bovine serum albumin as a standard. To compensate for the interference of PEG or detergents in the protein samples 1% SDS was included in the Lowry reagent (Dulley and Grieve, 1975).

Cytochrome P-450 was assayed by the method of Matsubara et al. (1976) in order to compensate for the presence of hemoglobin in the preparations.

Cytochrome c reductase was assayed at 22°C as described by Yasukochi and Masters (1976).

SDS Electrophoresis

Proteins were electrophoresed in 9% acrylamide gels according to the method of Laemmli (1970) using molecular weight markers of 30,000 (carbonic anhydrase), 45,000 (ovalbumin), 68,000 (bovine serum albumin), 92,000 (phosphorylase b), and 116,000 (β-galactosidase).

Purification - Cytochrome P-450

Cytochrome P-450 purification was attempted using the method of Pelkonen and Pasanen (1982) using the microsomes prepared by methods A-C. Microsomes were suspended to a concentration of 10 mg/ml of protein and solubilized by addition of Emulgen 913 to 0.1% (v/v) and Na cholate to 0.5% (w/v). The mixture was stirred on ice for 60 min. and then centrifuged for 60 min. at 100,000 x g. The solubilized supernatant was removed and diluted with 1 volume of K phosphate buffer, pH 6.8, containing 20% glycerol and 0.1% Na cholate and then made 0.5 M NaCl. The supernatant was applied to a 2.5 x 20 cm phenyl Sepharose column equilibrated with 10 mM K phosphate, pH 6.8, 20% glycerol, 0.5 M NaCl, 0.05% Emuglen 913, and 0.1% Na cholate. The column was washed with equilibrating buffer until the OD_{415} nm fell below 0.010 and was eluted with 10 mM K phosphate, pH 7.4, 60% glycerol, 1% Emulgen 913 and 0.1% Na cholate. The cytochrome P-450 containing fractions were pooled and diluted 5-fold with DEAE equilibrating buffer (10 mM K phosphate, pH 7.4, 20% glycerol, 0.05% Emulgen 913, and 0.1% cholate). The material was loaded onto a 2.5 x 20 cm DEAE Sephadex column and washed with DEAE equilibrating buffer until the OD_{415} nm fell below 0.010. Cytochrome P-450 was eluted with a 500 ml 0-0.5 m NaCl gradient.

When using the microsomes isolated by PEG precipitation for cytochrome P-450 purification it was necessary to strip the solution of PEG in order to allow binding of the material to the phenyl Sepharose column. To achieve this, the solubilized supernatant was added to 1/4 volume calcium phosphate gel and eluted with 1/4 volume 300 mM K phosphate, 20% glycerol, 0.1 mM EDTA. The elutant was applied to the phenyl Sepharose column.

Cytochrome c Reductase

Purification of cytochrome c reductase was accomplished by solubilizing a 10 mg/ml solution of microsomes with 0.5% Na deoxycholate and 1% Emulgen 913 for 30 min. The solubilized microsomes were centrifuged for 60 min. at 100,000 x g. The supernatant was batch absorbed with 1/4 volume DEAE cellulose (Yasukochi et al., 1977) and washed with 250 ml 25 mM Tris-HCl, pH 7.6, 0.1 M KCl, and 20% glycerol. The reductase was eluted with 250 ml of the same buffer containing 0.35 M KCl (Yasukochi et al., 1977). The eluted reductase was diluted 1 x with 20% glycerol and applied to a 10 ml 2',5'-ADP-Sepharose column washed with 100 ml of 10 mM K phosphate, pH 7.4, 20% glycerol, and 0.1% Emulgen 913. The reductase was eluted with 10 mM 2'-AMP in the same buffer neutralized to pH 7.4 (~50 ml).

RESULTS

Recovery

Recovery and purification activities for both the cytochrome P-450 and cytochrome c reductase for the three methods of microsomal membrane preparation are shown in Table 1. When compared with the standard method of centrifugation, the use of $CaCl_2$ resulted in microsomes with lower yields of protein, cytochrome c reductase and cytochrome P-450. PEG-prepared microsomes had greater recovery of

Table 1

Comparison of recovery and specific activity of cytochrome c reductase and P-450
with three methods of microsome preparation

RECOVERY	A. PEG	B. CaCl$_2$	C. Centrifuged
Protein (mg/ml supernatant)	1.47 ± 0.67 (111%)	0.932 ± 0.051 (70.0%)	1.33 ± 0.64 (100%)
Cytochrome P-450 (nmol/ml supernatant)	0.0646 ± 0.056 (59.8%)	0.0389 ± 0.0569 (36.0%)	0.108 ± 0.099 (100%)
Cytochrome c reductase (OD/min/ml supernatant)	0.327 ± 0.130 (80.7%)	0.159 ± 0.136 (39.3%)	0.405 ± 0.395 (100%)
SPECIFIC ACTIVITY			
Cytochrome c reductase (OD/min/mg protein)	0.181 ± 0.100 (72.4%)	0.136 ± 0.096 (54.4%)	0.250 ± 0.197 (100%)
Cytochrome P-450 (nmol/mg protein)	0.047 ± 0.018 (58.8%)	0.058 ± 0.029 (72.5%)	0.080 ± 0.029 (100%)

The number in parentheses indicates the percent of the values obtained for
centrifuged microsomes.

No significant differences at the 0.05 level were detected by paired t-tests.

total protein but lower specific activity for both cytochrome c reductase and
cytochrome P-450 than the centrifuged microsomes.

Using PEG to prepare microsomes we were able to process 4 liters of
homogenized placentae within 2 hr. Isolating microsomes from a similar volume of
homogenate using ultracentrifugation required nearly 24 hr.

Solubilization of microsomes usually results in a two-fold increase in total
cytochrome c reductase activity (Bellino, 1982). Each of the preparations tested here
revealed similar activation after solubilization with detergent (Table 2).

Electrophoresis

In addition to the specific enzyme activities discussed above, the general
protein distribution in the three placental membrane preparations was examined by
SDS-polyacrylamide gel electrophoresis. As demonstrated in Figure 1, major
differences in the relative quantities of microsomal protein were apparent in the three
preparations. The CaCl$_2$ prepared microsomes differed most from the others.

Table 2

Cytochrome c reductase solubilization by 1% Emulgen and 0.5% Na deoxycholate

	% Stimulation of reductase*	Fold increase in specific activity**
PEG	194	3.63
CaCl$_2$	204	8.04
Centrifuged	176	2.36

*   Stimulation of reductase = 100 x (units/ml solubilized microsomal supernatant ÷ units/ml microsomes)

**   Fold increase in specific activity = (specific activity solubilized microsomal supernatant ÷ specific activity of micosomes)

      Solubilization with detergent resulted in the release of different proteins from the three microsomal preparations (Figure 2).

Purification

      We could only partially purify cytochrome P-450 using microsomes prepared by any of the methods. Electrophoresis of the final products of the cytochrome P-450 purification from PEG and centrifuged microsomal preparations revealed similar proteins, notably two proteins in the 50,000-55,000 molecular weight range with other minor protein bands (Figure 3).

      Cytochrome c reductase was purified without difficulty with PEG and CaCl$_2$ microsomes (Table 3). On SDS polyacrylamide gels, a band of 73,000 molecular weight was seen with a minor band at 68,000 apparently representing the proteolytically cleaved form of the enzyme (data not shown).

DISCUSSION

      Previous studies have emphasized the similarity of microsomes prepared by alternate techniques using eletrophoretic and enzymatic analysis (Van Der Hoeven, 1981; Kamath and Narayan, 1972; Parkinson and Safe, 1979). We have observed several differences in placental microsomes prepared by these methods. It should be noted that these techniques were developed for hepatic tissues and that their use for placental tissue has not been optimized.

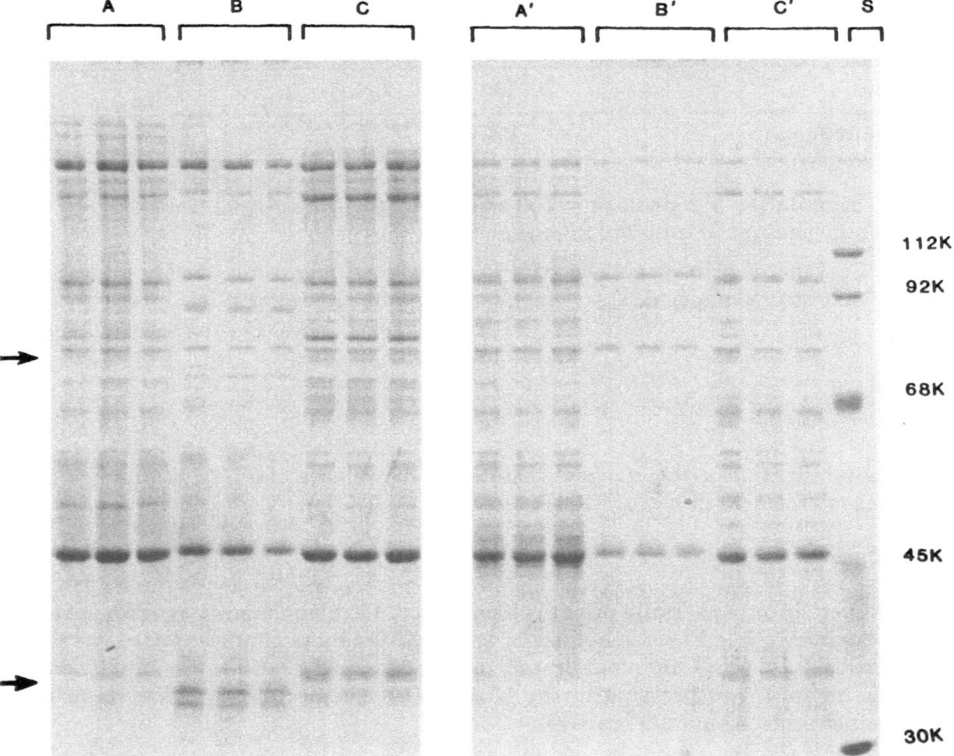

Figure 1. Comparison of the three methods of microsome isolation by SDS-polyacrylamide electrophoresis. The preparations were run in triplicate on two separate placentae. (S) SDS-standards 112K, 92K, 68K, 45K, and 30K (see text) (A and A') centrifuged microsomes, (B and B') CaCl$_2$ prepared microsomes, (C and C') PEG-prepared microsomes. Arrows indicate proteins which exhibit differences in concentration between sample preparations techniques.

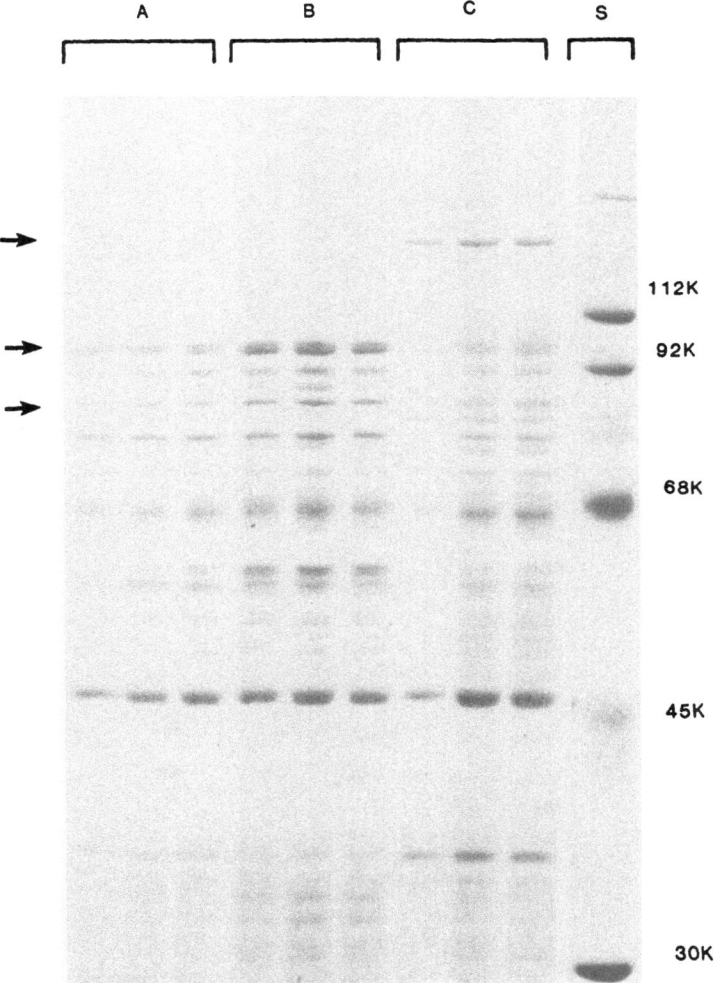

Figure 2. Comparison of solubilization of the three microsomal preparations by SDS-polyacrylamide electrophoresis. The preparations were run in triplicate. Solubilization was performed with 0.5% Na deoxycholate and 1% Emulgen 913 for 30 min at 4°C. (S) SDS-standards 112K, 92K, 68K, 45K, and 30K, (see text) (A) centrifuged microsomes, (B) $CaCl_2$ prepared microsomes, (C) PEG-prepared microsomes. Arrows indicate proteins which exhibit differences in concentration between sample preparation techniques.

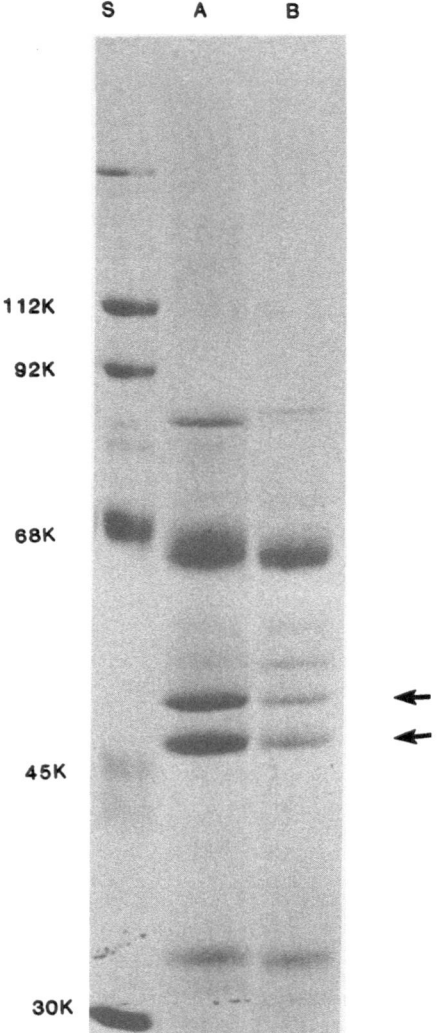

Figure 3. SDS-polyacrylamide electrophoresis of the final products of cytochrome P-450 purification. (S) SDS-standards 112K, 92K, 68K, 45K, and 30K, (see text) (A) centrifuged microsomes, (B) PEG microsomes. Arrows indicate proteins which exhibit molecular weight in the range of known forms of cytochrome P-450.

Table 3

Purification tables for cytochrome P-450 and cytochrome c reductase from the human placenta

Cytochrome P-450

	nmol	nmol/mg	Fold purification	% Recovery
PEG Preparation				
Microsomes	1944	0.148–0.090	1	100
Solubilized supernatant	1289	— — — —	–	66
Calcium phosphate gel	274	0.163	1.1	14
Phenyl sepharose	43.8	0.189	1.28	2.3
DEAE eluted	13.3	0.133	0.98	0.7
CaCl$_2$ Preparation				
Microsomes	133	0.048	1	100
Solubilized supernatant	145	0.100	2.08	109
Phenyl sepharose	20.9	0.333	6.94	16
DEAE wash	6.86	0.431	8.98	5.2
DEAE eluted	N.D.	— — — —	— — — —	— — —
Centrifuged Preparation				
Microsomes	726	0.168	1	100
Solubilized supernatant	621	0.169	1	86
Phenyl sepharose	79.9	0.431	2.56	11
DEAE eluted	10.1	0.693	4.13	1.4

N.D. = not detected

Table 4

Cytochrome C Reductase

	Units*	Units/mg	Fold purification	% Recovery
CaCl₂ Preparation				
Microsomes	44.0	0.237	1	100
Solubilized supernatant	129.6	3.58	15.1	295
0.3 M KCl DEAE	60.8	0.746	3.15	138
ADP-column	13.6	193	312.6	31
PEG Preparation				
Microsomes	291.2	0.243	1	100
Solubilized supernatant	896	1.49	6.13	308
0.3 M KCl DEAE	128.3	4.41	18.15	44
ADP-column	73.1	255	1051	25

N.D. - not detected

* Cytochrome c reductase
  Units in ΔOD_{550}/min

In spite of the observed differences, we were able to purify cytochrome c reductase to homogeneity from PEG or from $CaCl_2$ isolated microsomes as determined by SDS-polyacrylamide electrophoresis. The 73,000 MW protein and a 68,000 molecular weight proteolytic cleavage product which has been described by other investigators (Bellino, 1982) was isolated without detectable contaminating proteins.

Purification of the cytochrome P-450 activity has been more difficult. Although Pasenen and Pelkonen (1981), and others (Barbieri et al., 1983) have claimed to have purified cytochrome P-450 to homogeneity using hydrophobic and ion exchange chromatography, others have been unsuccessful using similiar methodology. Lewandowski and Hodgson (1985) recently reported only partial purification of two forms of cytochrome P-450 using a procedure similar to Pasanen and Pelkonen (1981). In addition, Lewandowski and Hogdson reported two major proteins of 54,000 and 52,000 molecular weight instead of a single protein in the 50,000 molecular weight range reported by Pasanen and Pelkonen (1981). Bellino (1985) also isolated two major proteins of 55,000 and 50,000 molecular weight in an active aromatase fraction using similar methodology.

Our final preparations from the cytochrome P-450 purification contained two major proteins in the 50,000-55,000 molecular weight range. These preparations had low specific content and no aromatase activity when reconstituted with phosphatidyl choline, cytochrome c reductase, and ^3H-androsteinedione. It is possible that the purification method has yielded a hemeless apoprotein form of cytochrome P-450. The ability of heme to reconstitute spectral and enzymatic activity of these products is under study.

SUMMARY

Microsomes prepared by the described methods of bulk-isolation appear to have different protein composition and vary in recovery of microsomal enzymes and chromatographic behavior.

PEG was the best of the two bulk-isolation methods for recovery of microsomal enzymes. PEG-isolated microsomes offered a convenient alternative for cytochrome c reductase purification.

Final products prepared from PEG and centrifuged microsomes in the attempted cytochrome P-450 purification had similar electrophoretic patterns.

REFERENCES

Barbieri, R.L., Petro, Z., Canick, J.A., and Ryan, K.J. (1983) Aromatization of norethindrone to ethinyl estradiol by human placental microsomes. *J. Clin. Endocrinol. Metabol.* 57, 299-303.

Bellino, F.L. (1982) Estrogen synthetase. Demonstration that the high molecular weight form of cytochrome c reductase from human placental microsomes is required for andorgen aromatization. *J. Steroid. Biochem.* 17, 261-270.

Bellino, F.L. (1985) Personal communication.

Bellino, F.L. and Hussa, R.O. (1978) Trophoblast estrogen synthetase stimulation by dibutyryl cyclic AMP and theophylline. Increase in cytochrome P-450 content. *Biochem. Biophys. Res. Comm.* 103, 1310-1317.

Dulley, J.R. and Grieve, P.W. (1975) A simple technique for eliminating interference by detergents in the Lowry method of protein determination. *Anal. Biochem.* 64, 136-141.

Kamath, S.A. and Narayan, K.A. (1972) Interaction of Ca^{+2} with endoplasmic reticulum of rat liver: A standardized procedure for isolation of rat liver microsomes. *Anal. Biochem.* 48, 53-61.

Laemmli, U.K. (1970) Cleavage of structural proteins during the assembly of the head of bateriophage T_4. *Nature* 227, 680-685.

Lewandowski, M.M. and Hodgson, E. (1985) Purification and reconstitution of the cytochrome P-450 monooxygenase system from the human placenta. *Toxicologist* 5, 647.

Lowry, O.H., Rosenbrough, N.J., Farr, A.L., and Randall, R.J. (1951) Protein measurement with the folin phenol reagent. *J. Biol. Chem.* 193, 265-275.

Matsubara, T., Koike, M., Touchi, A., Tochino, Y., and Sugeno, K. (1976) Quantitative determination of cytochrome P-450 in rat liver homogenate. *Anal. Biochem.* 75, 596-603.

Meigs, R.A. and Ryan, K.J. (1968) Cytochrome P-450 and steroid biosynthesis in the human placenta. *Biochim. Biophys. Acta* 165, 476-482.

Namkung, M.J., Chao, S.T., and Mont, R.J. (1983) Placental mono-oxygenation characteristics and partial purification of a hematin-activated human placental mono-oxygenase. *Drug Metabol. Disp.* 165, 476-482.

Parkinson, A. and Safe, S. (1979) The detection of enzyme induction by rat liver microsomes prepared by isoelectric precipitation. *J. Pharm. Pharmacol.* 31, 444-447.

Pasanen, M. and Pelkonen, O. (1981) Solubilization and partial purification of human placental cytochromes P-450. *Biochem. Biophys. Res. Comm.* 103, 1310-1317.

Pelkonen, O. (1980) Environmental influences on human foetal and placental xenobiotic metabolism. *Eur. J. Clin. Pharmacol.* 18, 17-24.

Pelkonen, O. and Pasanen, M. (1982) Purification of human placental cytochromes P-450. In: *Cytochrome P-450 Biochemistry, Biophysics and Environmental Implications*, (eds.), E. Hietanene, M. Laitinen, and O. Hanninen, pp. 449-452.

Pelkonen, O. and Pasanen, M. (1984) Enzymology and regulation of xenobiotic and steroid metabolism in placenta. *Biochem. Soc. Trans.* 12, 42-44.

Van Der Hoeven, T.A. (1981) Isolation of hepatic microsomes by polyethylene glycol 6000 fractionation of the postmitochondrial fraction. *Anal. Biochem.* 115, 398-402.

Yasukochi, Y. and Masters, B.S.S. (1976) Some properties of a detergent-solubilized NADPH-cytochrome c (P-450) reductase purified by biospecific affinity chromatography. *J. Biol Chem.* 251, 5337-5344.

Yasukocki, Y., Peterson, J.A., and Masters, B.S.S. (1977) NADPH-cytochrome c (P-450) reductase: Spectrophotometric and stopped-flow kinetic studies on the formation of reduced flavoprotein intermediates. *J. Biol. Chem.* 254, 7097-7104.

Trophoblast Research 2:279-287, 1987

PARTIAL PURIFICATION OF THE OPIOID RECEPTOR FROM HUMAN PLACENTA

Mahmoud S. Ahmed[1] and Anna G. Cavinato[2]

Department of Biochemistry, College of Medicine
University of Tennesee Center for the Health Sciences
Memphis, Tennessee 38163

INTRODUCTION

The human placental villus tissue contains membrane bound etorphine binding sites (Valette et al., 1980). These sites were shown to have binding properties similar to those of the Kappa opioid receptor subtypes. We solubilized these receptors using the detergent 3-[(3-cholamidopropyl)-dimethylammonio]-1-propane sulfonate (CHAPS) (Ahmed et al., 1981). The binding activity was of a truly soluble protein-detergent macromolecular complex migrating on a Sepharose CL-6B column between two soluble proteins. The apparent Stokes radius of the complex was 70 Å (Ahmed, 1983). The solubilized receptors were shown to be all or in part a protein with an (SH) group at or close to the binding site. This group was found to be necessary for binding activity (Ahmed, 1983).

Several laboratories have reported on the solubilization and partial purification of the opioid receptors from rodent brains (Bidlack et al., 1981; Cho et al., 1985) and tissue culture cell lines (Simonds et al., 1985). The most widely used technique was that of affinity chromatography using an opiod alkaloid derivatized to a matrix.

The opioid receptors from human placenta, Kappa subtype, have low affinity for most of the currently available alkaloids used in affinity chromatography columns (Valette et al., 1980; Ahmed, 1983). Therefore we were unable to utilize any of them for placental receptor purification.

However, we were able to utilize the commercially available glutathione-2-pyridyl disulfide derivative of Sepharose 4B to achieve partial purification. The derivatized Sepharose forms a disulfide bond with the opiod receptors and other (SH) containing proteins. We report here on the use of this covalent chromatography technique followed by preparative gel electrophoresis for the partial purification of the opioid receptors from human placenta.

[1]To whom all correspondence should be addressed: Department of Obstetrics/Gynecology, University of Missouri-Kansas City, School of Medicine, Kansas City, Missouri 64108-2792

[2]Current address: Chemistry Department, Memphis State University, Memphis, Tennessee 38163
Received: 10 October 1985; Accepted: 1 May 1985

MATERIALS AND METHODS

Materials

Radioactive etorphine at a specific activity of 30-50 Ci/mmole was purchased from Amersham or was a gift from The National Institute of Drug Abuse. Dinitrofluorobenzene, 2,4-[3-5-^{3}H]- at a specific activity of 20.4 Ci/mmole was purchased from New England Nuclear. The detergent 3-[(3-cholamidopropyl)-dimethylammonio]-1-propane sulfonate (CHAPS), tris(hydroxymethyl)amino-methane and the activated thiol-Sepharose 4B were products of Sigma Chemical Company. All chemicals used for preparative or analytical gel electrophoresis and for protein estimation were from Bio-Rad. Sephadex G-25 columns used in binding assays of the solubilized receptors were purchased from Isolab, Inc., Akron, Ohio. The microconcentrators "centricon-10" and the ultrafiltration membranes YM5 are products of Amicon Corporation. All placentae were obtained from full term pregnancies with patient consent from the delivery room of the Women's Hospital in Memphis, Tennessee according to an approved protocol.

Methods

The preparation of homogenates of membrane bound opioid receptors from placental villus tissue, and the solubilization of these receptors using the detergent CHAPS was carried out as described earlier (Ahmed et al., 1981; Ahmed, 1983). Binding activity of the membrane bound and solubilized receptors were tested using a radioreceptor assay with etorphine as the ligand (Ahmed et al., 1981).

Covalent Chromatography

Activated thiol-Sepharose 4B/(0.5 g) were suspended in 5 ml of distilled water and poured into a 1.5 x 8 cm column to a final volume of 2 ml. The gel was then washed with 100 ml of 3 mM CHAPS in 50 mM Tris-HCl (pH 7.4). The solubilized receptor was concentrated using an Amicon YM5 membrane. The concentrate, 7-8 ml, was applied to the thiol-Sepharose column and left for 40 min at 4°C, with occasional agitation. This ensured the disulfide bond formation between the receptor SH group and the glutathione 2-pyridyl disulfide of the column. The column was then washed with 80 ml of 3 mM CHAPS in Tris buffer (pH 7.4). The receptor was eluted off the column by treatment with cysteine thus forming a resin-cysteine disulfide and a reduced receptor. This was achieved using a step gradient of 0.25, 1 and 5 mM cysteine in 50 mM Tris (pH 7.4), containing 3 mM CHAPS. All fractions were collected and assayed for ^{3}H-etorphine specific binding.

Preparative Gel-Electrophoresis

Preparative polyacrylamide gel-electrophoresis was performed on a SE-6101 vertical slab gel-electrophoresis unit (Hoefer Scientific Instruments). The basic operation of the system was performed according to the gel system of Ornstein (1964) and Davis (1964) modified by Furlong et al. (1973). However, further modification was done in our laboratory. We substituted the detergent SDS, which causes receptor denaturation, with CHAPS. Molecular weight standards were run in the presence of

CHAPS to calibrate their elution times. The slab gels (3 x 120 mm) consisted of a 12% running gel and a 6% stacking gel.

The fraction, eluted with 0.25 mM cysteine of the thiol-Sepharose column, and containing highest etorphine binding capacity, was prepared for electrophoresis. Glycerol, 25% (w/w) was added to the receptor preparation together with the following Tris-HCl (pH 7.4), CHAPS and bromophenol blue to final concentrations of 0.1 M, 0.7 mM and 0.002% (v/v), respectively. The preparation was kept at 4°C for 1-2 hr with occasional agitation to ensure homogeneity of the preparation. The sample was applied to the top of the stacking gel and the electrophoresis unit placed in a cold room at 4°C. Elution of the sample was carried out at constant voltage of 70V until it entered the running gel. The voltage was then changed to 160V during the time the sample migrated through the running gel. Fractions of 5 ml were collected at a rate of 12 ml/hr.

The molecular weight markers phosphorylase B, 925,000; bovine serum albumin, 66,200; ovalbumin, 45,000; and carbonic anhydrase, 31,000, were separated under the same conditions to calibrate their elution times. Accordingly, fractions for the markers and/or the receptor preparation were pooled at the end of the electrophoretic separation based on their retention time/molecular weight. The pooled fractions from the receptor preparation were concentrated using YM5 ultramembranes, and assayed for protein content by the method of Nishio et al., 1982 and opiod binding capacity as described earlier (Ahmed et al., 1981).

Analytical Slab Gel Electrophoresis

The method used is essentially that described by Laemmli (1970). However, a fundamental modification was introduced in our laboratory. The nondenaturing detergent CHAPS replaced SDS. This allowed us to determine the migration of the molecular weight markers and the opioid receptor fractions under conditions similar to those for the preparative slab gel electrophoresis. Visualization of the gel was done by staining overnight with 0.2% Coomassie blue/10% (vol/vol) acetic acid/25% (vol/vol) methanol and diffusion destained in 10% acetic acid, 5% methanol. Purified receptor fractions were visualized by silver staining according to the method of Morrissey (1981).

Protein Estimation

Protein was measured by the method of Bradford (1976). However, as we achieved purification of the receptor the detection limits of the method was exceeded. The more sensitive method utilizing ^{3}H-dinitrofluorobenzene was used (Nishio et al., 1982). Bovine serum albumin was used as a standard in both methods.

RESULTS

Covalent Chromatography

The solubilized opioid receptor contains an (SH) group necessary for alkaloid binding (Ahmed, 1983). This group was allowed to react with a glutathione-2-pyridyl disulfide derivative of Sepharose 4B by loading the receptor solution on a column

Table 1

Purification of the opioid receptors
using activated thiol-sepharose 4B column

Fraction No.	Eluate Fraction	[3H]-etorphine Specifically Bound (fmoles/mg protein)
1	load	158
2	wash	112
3	wash	70
4	wash	15
5	wash	7
6	wash	0
7	0.25 mM cyst.	26154
8	1 mM cyst.	4580
9	5 mM cyst.	2400

filled with the gel. The column was washed and the receptor eluted with a step gradient of cysteine as described in Methods. All fractions were assayed for their protein content and specific etorphine binding capacity. The highest etorphine specific binding is in the fraction eluted with 0.25 mM cysteine (Table 1). This fraction contained 60% of the total binding activity loaded on the column. The specific binding of etorphine was 2.6×10^{-11} moles per mg protein. This represents approximately 800 fold purification over the membrane bound receptor. Etorphine binding in this fraction was saturable and stereospecific and required the presence of CHAPS.

Sodium dodecyl sulfate slab gel electrophoresis, (Laemmli, 1970) was used to determine the protein composition of this fraction. Figures 1 and 2 are of one slab gel preparation. The gel was divided in two parts prior to staining with silver. This allowed longer times of staining necessary to enhance visualization of the very small amounts of protein present in the fraction containing the opioid receptor (Figure 2). Figure 1 contains fractions with higher protein concentrations and its staining required substantially shorter times. Figure 1, lane 1 showed the molecular weight standards used: phophorylase B (92,500), bovine serum albumin (66,200) ovalbumin (45,000) and carbonic anhydrase (31,000). Lanes 2, 3, and 4 represent the solubilized receptor, prior to covalent chromatography, from three different experiments A, B, and C.

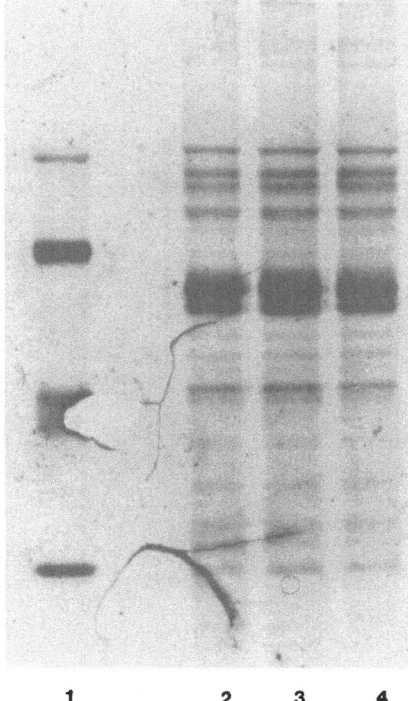

Figure 1. SDS gel electrophoresis of solubilized opiod receptors from human placental villus tissue. Silver stained gel lane 1 contains molecular weight markers as follows (top to bottom) 92.5, 66.2, 45, and 31K daltons. Lanes 2-4 represent the solubilized receptors for 3 different experiments, A, B, and C.

The gel pattern in Figure 2 represents that of the 0.25 mM cysteine eluted fraction of the covalent chromatography column, i.e., the opioid receptor containing fraction under reducing and non-reducing conditions from different experiments as follows: lane 5 represents this fraction separated under reducing conditions from experiment A. Lane 6 is that of experiment B under reducing conditions while lane 7 is of the same experiment A but under non-reducing conditions. Lane 8 represents the molecular weight markers similar to those in Lane 1. It is apparent that there is no difference between the pattern observed under reducing conditions when compared to that under non-reducing conditions.

The pattern in Lane 5, representing the 0.25 mM cysteine eluted fraction, shows two major protein bands corresponding to the approximate molecular weights of 76 and 65 K daltons and three minor bands corresponding to the approximate molecular weights 83, 81, and 62 K daltons.

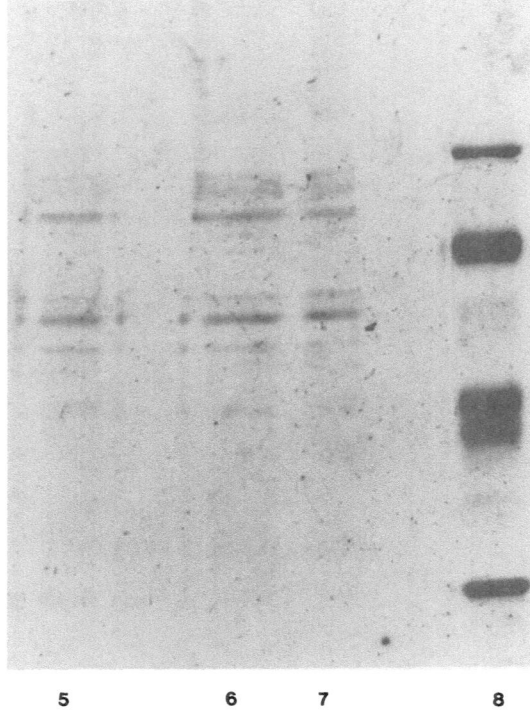

5   6   7   8

Figure 2. This represents the second half of the gel in Figure 1. This part was allowed to be stained for a longer time to enhance visualization. Lanes 5-7 represent the opiod receptor containing fraction eluted off the thiol-Sepharose column with 0.25 mM cysteine. Lanes 5 and 6 are from experiments A and B separated on the gel electrophoresis under reducing conditions. Lane 7 is of experiment A under non-reducing conditions. Lane 8 is identical to Lane 1 of Figure 1.

Preparative Polyacrylamide Gel Electrophoresis

The receptor containing fraction from the covalent chromatography column was concentrated using centricon-10 microconcentrator tubes. The concentrate was applied to the gel and fractions collected as described in Methods. Each 10 fractions, corresponding to a molecular weight range of approximately 10,000 daltons were pooled, concentrated, and assayed for their protein content and opioid binding capacity. The fraction containing highest opioid binding activity eluted after 30 hr (Fraction 8, Table 2), and corresponded to the molecular weight range of 62-72K daltons. Table 2 illustrates the binding activity of the pooled fractions arranged in order of their elution off the running gel, i.e., increasing molecular weight. Fraction 8 represents 40% of the activity applied. The specific etorphine binding of this fraction was 6.9×10^{-11} moles/mg P.

This activity represents a 2000-fold purification over the membrane bound receptor. This partially purified opioid receptor was shown to bind etorphine

Table 2

Purification of the opioid receptors using preparative gel
electrophoresis

Pooled Fraction	[3H]-etorphine Specifically Bound (fmoles/mg protein)
1	--
2	7792
3	--
4	--
5	1000
6	--
7	1960
8	69166
9	6000
10	1635
11	2410

The pooled fractions are arranged in order of their elution off
the running gel, i.e., increasing molecular weight.

specifically and saturably. A scatchard plot of the data revealed a Kd of 0.5 nM and a
B-max of 70 pmoles. Table 3 represents a summary of the purification achieved for
the placental opioid receptors.

DISCUSSION

The opioid receptors from human placenta show binding properties similar to
those of the Kappa subtype (Valette et al., 1980) and retain their properties after
solubization with the zwitterionic detergent CHAPS (Ahmed et al., 1981; Ahmed,
1983).

Several reports have appeared on the purification of opiod receptors from
rodents brain (Bidlack et al., 1981; Cho et al., 1985) as well as tissue culture cell lines
(Simonds et al., 1985) utilizing affinity chromatography. We were able to partially
purify the opioid receptor from human placenta utilizing a covalent chromatography
column followed by preparative gel electrophoresis.

Table 3

Summary of the purification achieved for the
opioid receptors from human placental villus tissue

Fraction	[^{3}H]-etorphine Specifically Bound (fmoles/mg protein)	Recovery of Specific Binding %
membrane	33	
solubilized	95	35
eluate from covalent chromat	26154	12
eluate from prep-gel-electrophoresis	69166	3

We achieved an 800-fold purification using covalent chromatography with thiol-sepharose 4B. The fraction containing etorphine binding at 26.1 pmole/mg P showed 5 protein bands on SDS polyacrylamide gel electrophoresis. These bands ranged in molecular weights between 62 and 83K daltons. The opioid binding activity associated with this fraction is therefore either one or more of these protein bands. However, the fact that the binding activity could be associated with another protein present in amounts beyond the detection limits of silver stain cannot be excluded.

The purified receptor from the preparative gel electrophoresis binds approximately 7 pmole of etorphine per mg of protein. The elution time of this fraction corresponds to that of molecular weight markers ranging between 62-72K daltons. This is in the same molecular weight range reported by others for purified opioid receptors (Simonds et al., 1985). Assuming a molecular weight of 68,000, the theoretical binding of etorphine should be 14,700 pmole per mg protein. It is obvious that further purification is needed before receptor homogeneity is achieved. Currently we are using a lectin column for that purpose.

Our partially purified receptor binds etorphine saturably and specifically with a Kd of 0.5 nM and B-max of 70 pmoles.

SUMMARY

The human placental villus tissue contains opioid receptors with properties similar to that of the Kappa subtype. Solubilization and partial purification of these receptors using covalent chromatography and preparative gel electrophoresis has been achieved. The partially purified protein binds etorphine saturably and specifically. Its apparent molecular weight is in the range of 62,000-72,000 dalton. It has a Kd of 0.5 nM and B-max of approximately 70 pmole per mg protein. This protein fraction represents over 2000-fold purification over the membrane bound receptors. Further purification using a lectin column is underway.

ACKNOWLEDGEMENTS

This work has been supported by National Science Foundation grant #83-17212 to M.S.A.

REFERENCES

Ahmed, M.S., Byrne, W.L., and Klee, W.A. (1981) Solubilization of opiate receptors from human placenta. *Placenta* 3, 115-121.

Ahmed, M.S. (1983) Characterization of solubilized opiate receptors from human placenta. *Memb. Biochem.* 5, 35-47.

Bidlack, J.M., Abood, L.G., Osei-Gyimah, P., and Archer, S. (1981) Purification of the opiate receptor from rat brain. *Proc. Natl. Acad. Sci.* 78, 636-639.

Bradford, M. (1976) A rapid and sensitive method for the quantitation of microgram quantities of protein utilizing the principle of protein-dye binding. *Anal. Biochem.* 72, 248-254.

Cho, T.M., Ge, B.L., and Loh, H.H. (1985) Isolation and purification of morphine receptor by affinity chromatography. *Life Sci.* 36, 1075-1085.

David, B.J. (1964) Disc electrophoresis II. Method and application to human serum proteins. *Ann. NY Acad. Sci.* 121, 404-427.

Furlong, C., Cirakoglu, C., Willis, R.C., and Souty, P.A. (1973) A simple preparative polyacrylamide disc gel electrophoresis apparatus: Purification of three bronchial-chain amino acrol binding proteins from Escherichia Coli. *Anal. Biochem.* 51, 297-311.

Laemmli, U.K. (1970) Cleavage of structural proteins during the assembly of the head of bacteriophage T4. *Nature* 227, 680-685.

Morrissey, J.H. (1981) Silver stain for proteins in polyacrylamide gels: A method procedure with enhanced uniform sensitivity. *Anal. Biochem.* 117, 307-310.

Nishio, K. and Kawakawi, M. (1982) Protein assays in very dilute solutions. *Anal. Biochem.* 726. 239-241.

Orstein, L. (1964) Disc electrophoresis I. Background and theory. *Ann. NY Acad. Sci.* 121, 321-349.

Simonds, W.F., Burke, T.R., Rice, K.C., Jacobson, A.E., and Klee, W.A. (1985) Purification of the opiate receptor of NG108-15 neuroblastoma-glioma hybrid cells. *Proc. Natl. Acad. Sci.* 82, 4974-4978.

Valette, A., Reme, J.M., Pontonnier, G., and Cros, J. (1980) Specific binding for opiate-like drugs in the placenta. *Biochem. Pharmacol.* 29, 2657-2661.

Trophoblast Research 2:289-304, 1987

REGIONAL AND DIFFERENTIAL SENSITIVITY OF UMBILICO-PLACENTAL VASCULATURE TO 5-HYDROXYTRYPTAMINE, NICOTINE, AND ETHYL ALCOHOL

B. V. Rama Sastry and L. K. Owens

Department of Pharmacology
Vanderbilt University School of Medicine
Nashville, Tennessee 37232, USA

INTRODUCTION

The fetal blood concentration of 5-hydroxytryptamine (5-HT) increases before birth (Jones and Rowsell, 1973). 5-HT decreases perfusion flow through umbilico-placental vasculature in isolated human placenta (Gautieri and Ciuchta, 1962). Nicotine is known to release 5-HT from nervous tissues (Goodman and Weiss, 1973; Balfour, 1973; Hery et al., 1977). Among tobacco smokers, nicotine enters fetal circulation from maternal blood. Therefore, a question arises as to whether nictoine releases 5-HT which decreases fetal blood flow through placenta and contributes to fetal hypoxia and growth retardation in tobacco smokers.

Ethyl alcohol causes a marked collapse of umbilical vasculature in primates, rhesus (Macaca mulatta), and cynomolgus (Macaca fasciularis) monkeys (Mukherjee and Hodgen, 1982). It also caused such collapse in isolated umbilical arteries of human placenta (Sastry and Owens, 1983, 1984). The consequence of such collapse of the umbilical arteries is decreased blood flow through placenta and fetal hypoxia. The combined effect of maternal tobacco smoking and alcoholism will be highly hazardous to umbilical and placental vasculature. Therefore, one of the primary objectives of this investigation is to evaluate combined effects of 5-HT, nicotine and ethyl alcohol on the vascular resistance of umbilico-placental vasculature and assess the total damage, if any, to the placental vasculature due to a combination of maternal smoking and alcoholism, acute or chronic.

Perfusion of isolated human placentae in vitro through the umbilical artery has indicated that placental vasculature is sensitive to a variety of compounds with vasoactive properties (von Euler, 1938; Elliasson and Astrom, 1955; Ciuchta and Gautieri, 1964; Gautieri and Mancini, 1967). This preparation provides useful information on the perfusion flow and overall resistance of the system. However, this preparation does not distinguish the sites of vascular resistance and permit quantifying the actions of drugs. Strips of umbilical arteries and veins have been examined for drug sensitivity, but these studies reflect only the sensitivity of cord vessels per se and not of the placental vasculature (Somlyo et al., 1965; Eltherington

Received: 10 October 1985; Accepted: 10 June 1986

et al., 1969; Hillier and Karim, 1968; Altura et al., 1972; Park et al., 1972; Fiscus and Dyer, 1982). Different regions of the placental vascular tree exhibit different sensitivities to vasoactive peptides (Tulenko, 1979). Since the small arteries or arterioles of the placental vasculature are the most probable sites of placental vasculture resistance, the present experiments were undertaken to directly compare sensitivities of resistance of vessels to 5-HT and nicotine with those of the larger arteries supplying the placenta.

The three regions selected for the present study are: (a) umbilical cord, (b) chorionic plate, and (c) villus stem arteries. According to the present investigations, chorionic arteries are the most sensitive segments of the placental vasculature to 5-HT. The vasoconstrictive effects of 5-HT and nicotine on chorionic arteries were enhanced by pharmacological concentrations (43-100 mM) of ethyl alcohol.

METHODS

Materials

5-Hydroxytryptamine-creatinine sulfate complex was supplied by Sigma Chemical Co. (St. Louis, MO). Nicotine was obtained from Eastman (Rochester, NY); absolute ethyl alcohol, U.S.P. (200 proof) was purchased from Aaper Alcohol and Chemical Co. (Shelbyville, KY). All blocking drugs were of U.S.P. grade and were obtained from commercial sources. Potassium chloride, A.R. and other salts of analytical grade were purchased from Mallinckrodt Chemical Works (St. Louis, MO).

Preparation and Measurement of Contractile Responses of the Vascular Segments

Human placentae were obtained immediately after either vaginal or Cesarean section delivery at Vanderbilt University Hospital. Each placenta was washed to remove fetal blood by cannulating the two umbilical arteries with polyethylene tubing and flushing the placenta with ice cold 0.9% NaCl solution containing 0.5 U/ml beef lung heparin (Upjohn, Kalamazoo, MI) until the effluent from the umbilical vein was clear of blood. Several blood free cotyledons were used as the source of placental vasculature.

Umbilical, chorionic, and villus stem arteries are localized as shown in Figure 1. Several lengths of these arteries were dissected, helical strips were prepared by the general procedure described by Furchgott and Badrakom (1953) and placed in Krebs Ringer bicarbonate solution (KRB, pH 7.4) of the following composition: NaCl, 119 mM; KCl, 5 mM; $CaCl_2$, 2.5 mM; $MgSO_4$, 1.2 mM; KH_2PO_4, 1.2 mM; $NaHCO_3$, 25 mM; dextrose, 11 mM.

Helical strips of arterial muscles, 2.0 to 2.5 cm long, were tied at both ends and mounted in a 25 ml organ bath (Phipps and Bird, Inc., Richmond, VA) containing KRB solution. One end of the muscle strip was fastened onto a holder at the bottom of the bath, and the other end was fastened to an F-50 Microdisplacement Myograph Transducer (Narco Biosystems, Houston, TX) to record isometric contractions. The bath was maintained at 37°C. A gas mixture consisting of 95% oxygen and 5% carbon dioxide was bubbled through the bathing solution at all times. The muscle was

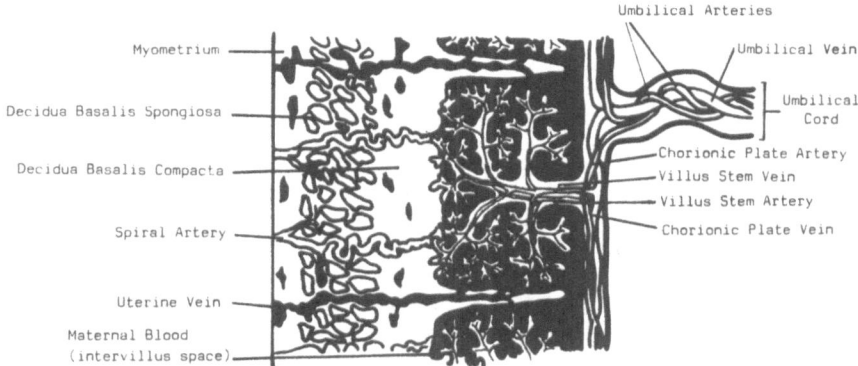

Figure 1. A schematic drawing showing circulation to a single lobe of human placenta. This figure is redrawn from the anatomical details shown by Netter (1967) and Tulenko (1979) in their figures showing circulation to the placenta. Umbilical arteries, chorionic arteries, and villus stem arteries were dissected from the sites shown in the figure, and helical strips were prepared. Contraction of the helical strips corresponds to constriction of the vessels.

allowed to rest 30 min after mounting. Then 0.5-1.0 g tension was placed on the muscle. The muscle was allowed to incubate for about 1 hr. Then the muscle was "normalized" by inducing contractions with two doses of 5-HT (2×10^{-7} M). The muscle was washed and allowed to incubate for an additional period of 1 hr before the comparisons were made. During this period, the bathing fluid was changed frequently. Subsequently, the muscle gave consistent responses with repeated challenges with a fixed dose of 5-HT (800 mg). The contractions were recorded on a DMP-4A Physiograph (Narco Biosystems, Houston, TX).

The following time sequence was followed for construction of the dose response curves for each agonist before and after treatment of the helical strips with ethyl alcohol. After the helical strip was "normalized" and allowed to incubate for 1 hr, it was used to construct the first cumulative dose response curve for the agonist. After each addition of the agonist dose (in 0.05 ml of KRB), a period of 5 min was allowed for the full development of the contraction before the next addition of the agonist dose. About 5-6 doses were used for each dose response curve. After constructing the first dose response curve, the strip was washed and allowed to incubate in the KRB solution for about 30 min. When the muscle relaxed to the original base line, a dose of ethyl alcohol in KRB was added to the bath. A second cumulative dose response curve was constructed for the agonist starting 3-5 min after the addition of ethyl alcohol. The maximum height of the contraction (contraction height) with each dose was measured in mm. The microdisplacement transducer was calibrated with a standard weight (1 g), so that the contraction heights could be expressed as mm or Δ tension. Δ tension developed by each dose of the agonist was expressed as a percentage of the tension developed by the maximal dose of the agonist used.

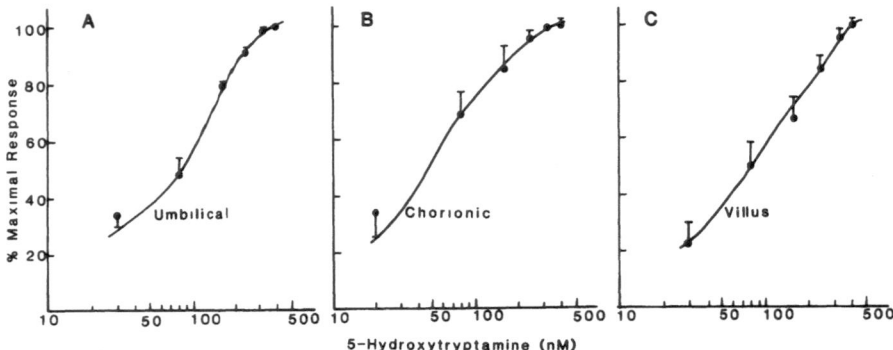

Figure 2. Dose response curves of 5-hydroxytryptamine (5-HT) for the contractile response of helical strips of human umbilical (A), chorionic (B), and villus stem (C) arteries. Each point is a mean $\pm$ standard error from 4 to 6 values.

The isolated umbilical, chorionic, and villus stem arteries gave consistent responses even after 24 hr of storage in a physiological medium (Krebs bicarbonate buffer, pH 7.4, 4°C). However, freshly prepared arterial preparations were used in all experiments.

Statistics

Results are expressed as means and standard errors. The significance of the difference between mean values was calculated by Student's t test. A result was considered significant if its p value was < 0.05.

RESULTS

Dose-Response Curves for 5-HT and Nicotine

The dose response curves for 5-HT induced contractions of helical strips of umbilical, chorionic and villus stem arteries are shown in Figure 2. Chorionic vessels (EC_{50}, 47 ± 5 nM) were more sensitive to 5-HT than umbilical (EC_{50}, 80 ± 7 nM) or villus stem (EC_{50}, 80 ± 2 nM) arteries. Observations by Tuleno (1979) also indicated that chorionic arteries were more sensitive to 5-HT than umbilical arteries. In all cases, maximal contraction height was found at about 400 nM 5-HT. The effect of 5-HT was easily reversible by washing the tissue with KRB solution. The maximal response to 5-HT was antagonized in the presence of methylsergide (40 nM) by about 64-70% in chorionic (IC50, 35.1 ± 1.2 nM) and villus stem (IC50, 35.2 ± 1.3) arteries.

Nicotine caused contractions of the strips of umbilical arteries (EC_{50}, 2.4 ± 0.4 mM), chorionic arteries (EC_{50}, 3.3 ± 0.5 mM), and villus stem vessels (EC_{50}, 2.4 ± 0.4 mM) (Figure 3). There were no significant differences in the responsiveness of the three types of vessels to nicotine. The dose response curves on three types of vessels

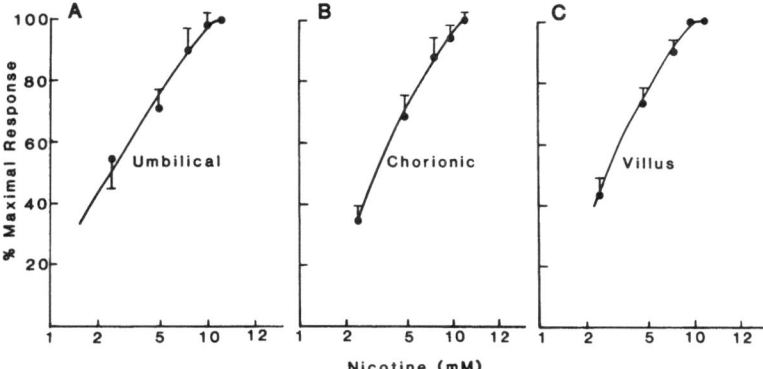

Figure 3. Dose response curves of nicotine for the contractile responses of helical strips of human umbilical (A), chorionic (B), and villus stem (C) arteries. Each point is a mean $\pm$ standard error from 4 to 6 values.

were parallel and the effects of nicotine were reversed by washing the vessels with KRB solution. The maximal contraction height was achieved at 11 mM nicotine with all three types of vessels. The effects of nicotine were partially antagonized by hexamethonium (I50, 13.8 $\pm$ 0.7 µM), d-tubocurarine (I50, 80.7 $\pm$ 5.5 µM), atropine (I50, 0.10 $\pm$ 0.1 µM), methylsergide (I50, 50.0 $\pm$ 2.5 µM), and verapamil (I50, 76.9 $\pm$ 3.8 µM). Of all antagonists, atropine was the most potent to block nicotine induced responses by about 140-800 times depending upon the antagonist.

Dose Response Curves for Potassium Chloride

Depolarizing concentrations of KCl induced contractions in chorionic and umbilical arterial strips (Figure 4). Chorionic vessels (EC_{50}, 19 $\pm$ 2 mM) were about twice as sensitive as umbilical vessels (EC_{50}, 30 $\pm$ 2 mM). In each case, the maximal contraction developed in about 10 min. These contractions were due to influx of Ca^{++} ions into the smooth muscle cells because the contraction heights to KCl were reduced by about 85% in Ca^{++} free medium. Similar observations were made in other arterial tissues (Robinson and Sastry, 1976).

Effects of Ethyl Alcohol on the Contractile Properties of Umbilical and Chorionic Vessels

Ethyl alcohol (0.1 - 2.0% v/v, 17.2 - 344 mM) influenced the different segments of placental vasculature in different ways. Umbilical arterial strips exhibited minimal contractions or became flacid and lost their contractile properties in the presence of ethyl alcohol.

Ethyl alcohol (0.5 - 2.0%, 43-344 mM) caused contractions of chorionic arterial strips. The maximum contraction height (Δ tension, 1 g) developed in about 10 min.

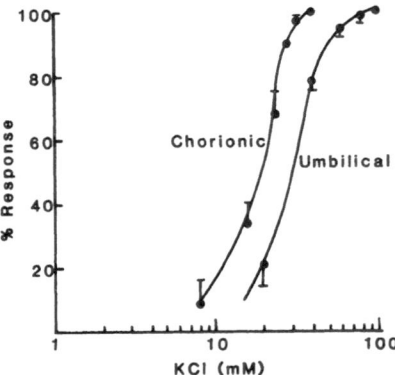

Figure 4. Dose response curves of KCl on the contractile responses of strips of human umbilical vessels and chorionic vessels. Each point is a mean ± standard error from 4 to 6 values.

Influence of Ethyl Alcohol on the Contractile Responses of Chorionic and Umbilical Arterial Strips to 5-HT

The dose response curves of 5-HT on the chorionic and umbilical arterial strips without and with ethyl alcohol (0.5%, v/v) in the bath are shown in Figure 5. In chorionic arterial strips, the contraction heights were enhanced by ethyl alcohol at all concentrations of 5-HT (Figure 5A). As the concentration of 5-HT increased, the enhancement of the contraction height by ethyl alcohol increased. For example, at 20 nM 5-HT, the increase in the contraction height was only 20%, while it was 40% at 150 nM 5-HT. Further, this increase was gradual with increasing concentrations of 5-HT. This means that ethyl alcohol induced an upward shift of the dose response curve to 5-HT in chorionic arterial strips.

The effect of ethyl alcohol on the 5-HT induced contractions in umbilical arterial strips was quite opposite of what was observed in chorionic arterial strips. Ethyl alcohol (0.5%, v/v) decreased the contraction heights (20-40%) of 5-HT at all doses. It induced a downward shift of the dose response curve (Figure 5B).

The enhancing effects of ethyl alcohol on 5-HT induced contractions in chorionic arterial strips were dependent upon its concentration. For example, the maximum enhancement of the contraction height in the chorionic strips by 2% (v/v) ethyl alcohol was about 220% (Figure 6A) while it was only 40% at 0.5% (v/v) ethyl alcohol. Further, the maximum enhancement with 2% (v/v) ethyl alcohol was obtained at a lower dose of 5-HT (40 nM) than the dose of 5-HT (150 nM) when maximum enhancement was observed with 0.5% (v/v) ethyl alcohol.

The depressing or blocking effects of ethyl alcohol on 5-HT induced contractions in umbilical arterial strips were also dependent upon the concentration of ethyl alcohol (Figure 6B). For example, the maximal reduction in the contraction height in

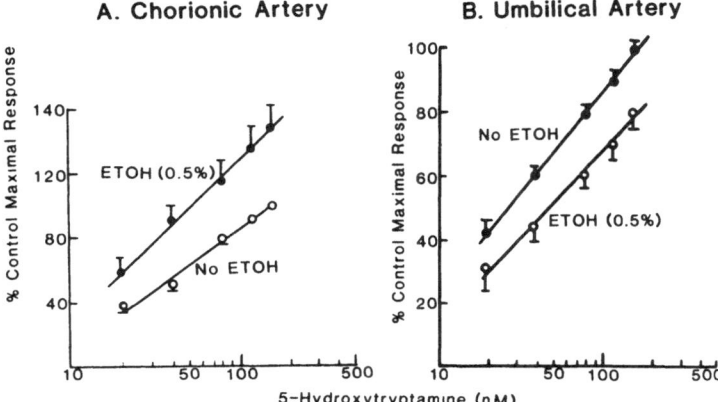

Figure 5. Influence of a low concentration (0.5%, v/v) of ethyl alcohol (ETOH) on the contractile responses of helical strips of chorionic (A) and umbilical (B) arteries of human placentae to 5-hydroxytryptamine (5-HT). Each point is a mean from 8 to 9 values. Vertical lines represent standard errors. The tension developed for each concentration of 5-HT is expressed as a percentage of maximal Δ tension developed by 5-HT in the absence of ETOH.

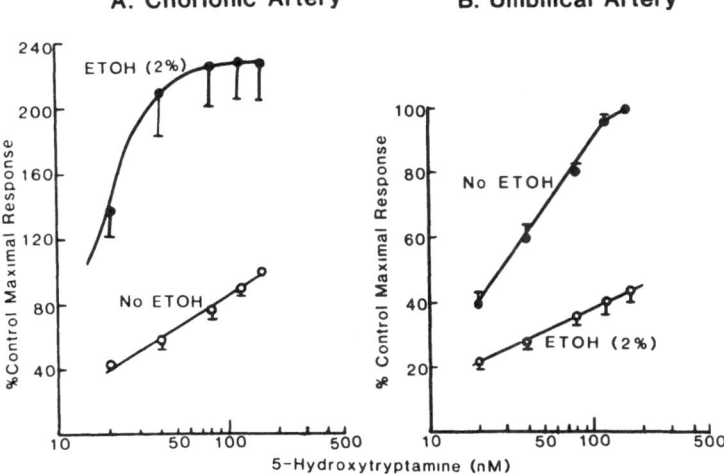

Figure 6. Influence of a high concentration (2%, v/v) of ethyl alcohol (ETOH) on the contractile response of helical strips of chorionic (A) and umbilical (B) arteries of human placentae to 5-hydroxytryptamine (5-HT). Each point is a mean of 10 to 11 values. Vertical lines represent standard errors. The Δ tension developed for each concentration of 5-HT is expressed as a percentage of maximal Δ tension developed by 5-HT in the absence of ETOH.

the umbilical strips by 2% (v/v) ethyl alcohol was about 60%, whereas it was only 20% at 0.5% (v/v) ethyl alcohol. The maximal reduction was observed at 150 nM 5-HT at all concentrations of ethyl alcohol (0.5 - 2.0%, v/v).

Effects of Ethyl Alcohol on the Contractile Responses of Chorionic and Umbilical Arterial Strips to Nicotine

Ethyl alcohol (0.5%, v/v) increased contraction heights induced by nicotine both in the chorionic and umbilical arterial strips (Figure 7). The degree of the enhancement of the contraction height remained constant at about 40% of control in the chorionic arterial strips and at about 20% of control in the umbilical strips. There were upward shifts of the dose response curves at all doses of nicotine.

The degree of enhancement of the contraction heights induced by nicotine was dependent on the concentration of ethyl alcohol. For example, ethyl alcohol at 2% (v/v) increased the contraction heights by about 80% at all doses of nicotine both in chorionic and umbilical arterial strips. There were upward shifts of the dose response curves at 2% (v/v) ethyl alcohol (Figure 8).

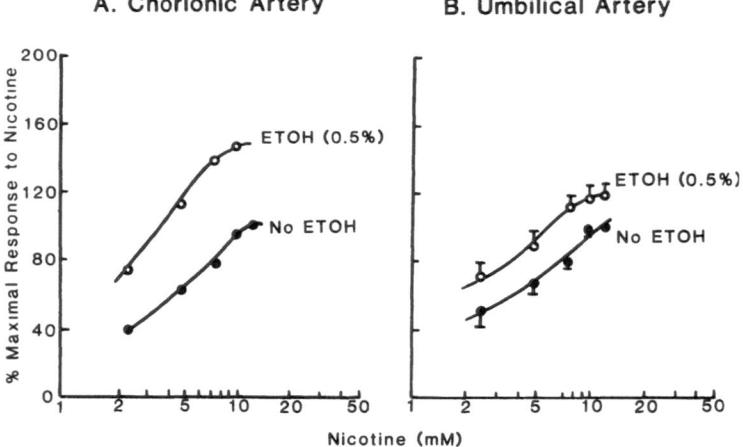

Figure 7. Influence of a low concentration (0.5%, v/v) of ethyl alcohol on the contractile responses of helical strips of human chorionic (A) and umbilical (B) arteries to nicotine. Each point is a mean from 6 to 8 values. Vertical lines represent standard errors. The Δ tension developed for each concentration of nicotine is expressed as a percentage of maximal Δ tension developed by nicotine in the absence of ETOH.

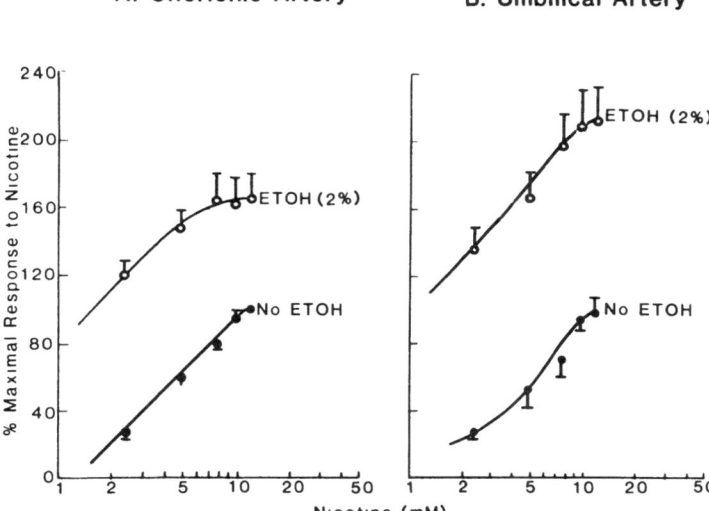

Figure 8. Influence of a high concentration (2.0%, v/v) of ethyl alcohol (ETOH) on the contractile respones of helical strips of human chorionic (A) and umbilical arteries (B) to nicotine. Each point is a mean from 6 to 8 values. Vertical lines indicate standard errors. The Δ tension is expressed as a percentage of maximal Δ tension developed by nicotine in the absence of ETOH.

Reversibility of the Effects of Ethyl Alcohol on the Chorionic and Umbilical Vessels

In order to evaluate whether ethyl alcohol brings about permanent changes in the umbilical and chorionic strips, a series of experiments were conducted as shown in Figure 9. The same group of chorionic and umbilical strips were challenged in sequence with (a) 5-HT, (b) ethyl alcohol, (c) ethyl alcohol plus 5-HT, and (d) 5-HT (Figures 9A and B). Similar experiments were also conducted using nicotine (Figures 9C and D). The strips were washed with KRB buffer after each challenge. The effects of 5-HT and ethyl alcohol were not additive in chorionic or umbilical vessels (Figures 9A and B). The effects of 5-HT were partially antagonized by ethyl alcohol. The effect of ethyl alcohol was additive to the effect of nicotine on chorionic vessels, whereas ethyl alcohol potentiated the effect of nicotine on umbilical vessels (Figures 9C and D). When the strips were washed with the buffer, they recovered completely and gave the same responses.

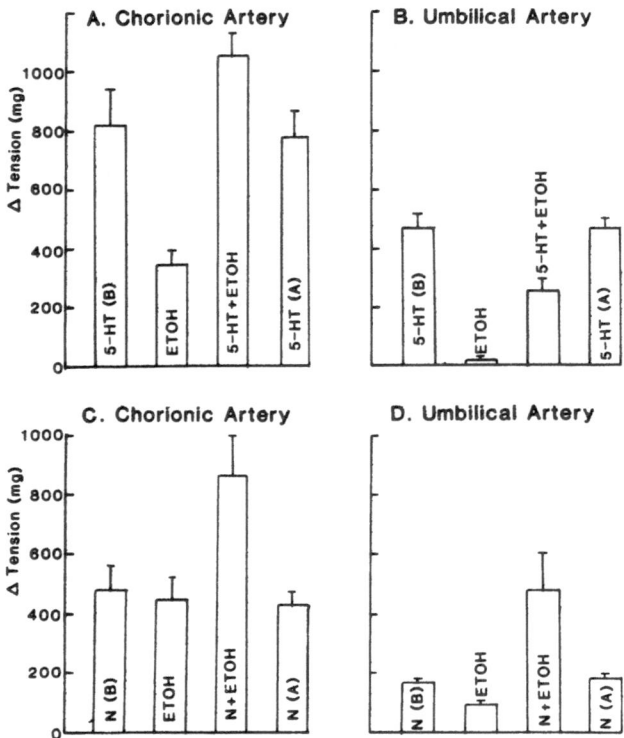

Figure 9. Contractile response of helical strips of human chorionic (A) and umbilical (B) arteries to 5-hydroxytryptamine (5-HT, B) at the beginning of the experiment; and ethyl alcohol (ETOH) only (1%, v/v), a combination of 5-HT and ETOH (5-HT + ETOH) and 5-HT at the end of the experiment (5-HT, A). Contractile response of helical strips of human chorionic (C) and umbilical (D) arteries to nicotine at the beginning of the experiment [N (B)]; and ethyl alcohol (ETOH) only (1%, v/v), a combination of nicotine and ETOH (N + ETOH) and nicotine [N (A)] at the end of the experiment. Each point is a mean ± standard error from 10 experiments. The tissue was washed between each challenge. The tissue gave the same consistent responses to 5-HT or nicotine before and after challenges with ETOH or a combination of ETOH and 5-HT of nicotine.

DISCUSSION

Consideration should be given to various factors to evaluate (a) the differential effects of 5-HT and ethyl alcohol in chorionic and umbilical arterial strips and (b) similar effects of nicotine and ethyl alcohol both in umbilical and chorionic arteries. These include: (a) extracellular Ca^{++} and intracellular Ca^{++} stores necesary for contraction of chorionic and umbilical arteries (Robinson and Sastry, 1976; Janis and Triggle, 1983), (b) alterations in membrane fluidity induced by ethyl alcohol, and (c) the effects of alterations in membrane fluidity on Ca^{++} movements, Ca^{++} stores, 5-HT receptors, and nicotine sites.

Two theories have been proposed for the effects of ethyl alcohol on the membranes: (a) the fluidization hypothesis (Harris and Schroeder, 1981; Hitzemann et al., 1984) and (b) the selective perturbation hypothesis (Richards et al., 1978; Harris and Hitzemann, 1981). Both hypotheses are based upon the effect of ethyl alcohol to fluidize membranes. According to the former, pharmacological concentrations (20-100 mM) increase membrane fluidity, and this effect is greatest in the hydrophobic core and less pronounced on the surface. According to the latter hypothesis, ethyl alcohol selectively disorders certain lipid domains by partitioning into liquid crystalline phases of the membrane. Ethyl alcohol facilitates the Ca^{++} entry into the somatic nerve terminal, possibly by fluidizing the nerve terminal membrane and increases the quantal release of acetylcholine (Van der Kloot, 1978). The effects of different pharmacological treatments upon the microviscosity, lipid structural order parameter and calcium levels of several membrane systems were measured using fluorescence polarization of diphenylhexatriene (DPH) as described in earlier publications (Jaiswal et al., 1983; Landon et al., 1984). Increase in membrane fluidity may increase membrane permeability to calcium.

The microviscosity and lipid structural order parameter of the microsomal membranes of the human chorionic and umbilical arteries were determined by Sastry and Naukam (1985, unpublished observations) using fluorescence polarization. Ethyl alcohol (43-344 mM) decreased the microviscosity of the umbilical and chorionic arterial microsomes in a concentration dependent manner. About a 32% decrease in microviscosity of both types of microsomes was observed at 2% ethyl alcohol. These observations suggest that ethyl alcohol induced membrane fluidization in umbilical and chorionic arterial strips, and may facilitate the uptake of Ca^{++}.

Ethyl alcohol (43-344 mM) increased contraction heights to 5-HT in chorionic arterial strips. This increase was not observed in the Ca^{++} free medium. The contraction height was lower than that of the control by about 27%. These observations suggest that ethyl alcohol fluidizes the chorionic arterial membranes, facilitates increased Ca^{++} uptake and thereby enhances the contraction heights induced by 5-HT. This is further supported by the following observations: (a) ethyl alcohol alone caused contraction of chorionic strips in a medium containing Ca^{++}, (b) chorionic arterial strips were twice as sensitive to potassium depolarization as umbilical arterial strips, and (c) potassium depolarization increased Ca^{++} influx in several cell types.

Ethyl alcohol induced fluidity in the plasma membrane may decrease the number of 5-HT receptor sites (as has been observed in other tissues, Heron et al., 1980; Gould et al., 1985). In chorionic strips, the contraction height observed with 5-HT in the presence of ethyl alcohol was not equal to the additive heights observed when the muscle strip was challenged with 5-HT or ethyl alcohol alone. It was less than the additive height. This indicates that ethyl alcohol affects 5-HT receptor sites and the release of Ca^{++} from receptor-linked Ca^{++} stores. However, ethyl alcohol induced Ca^{++} influx should have satisfied the Ca^{++} requirements for contraction of chorionic arteries.

Umbilical arterial strips were less sensitive then chorionic strips for potassium induced depolarization. Ethyl alcohol alone did not cause contractions of umbilical strips. Therefore, 5-HT induced contractions of umbilical arterial strips should use receptor-linked intracellular Ca^{++} stores for contraction. These will be adversely effected by ethyl alcohol due to the decreased number of receptor sites. In the present investigation, ethyl alcohol decreased the contraction heights of umbilical arterial strips to 5-HT. There was a parallel downward shift of the dose response curve indicating that Ca^{++} release from the 5-HT receptor linked Ca^{++} stores were effected by ethyl alcohol.

Nicotine in mM concentrations caused contraction of both chorionic and umbilical arteries. These contractions were blocked by low doses of atropine (0.1 µM). They were also blocked by several other antagonists at µM concentrations which were higher than that of atropine. Acetylcholine caused contraction of the strips only in very high concentrations (mM). Evidence for the existence of cholinergic sites which may be classified as muscarinic receptors in placental vasculature were not convincing (Sastry and Sadavongvivad, 1979). There are muscarinic receptors in endothelium of several vascular tissues (Furchgott, 1983; Horst and Sastry, 1985). There may be such muscarinic receptors in the endothelium of placental vasculature. Acetylcholine may activate these muscarinic receptors and cause relaxation (Sastry and Horst, 1975, unpublished observations). Therefore, these receptors should not be the site of action for nicotine to cause contraction of the umbilical or chorionic arterial strips.

Both nicotine and atropine are tertiary amines and cross cell membranes. Nicotine may enter the smooth muscle cells and dislodge many units of Ca^{++} from membrane stores. Atropine may also enter the cytoplasm of the smooth muscle cells and interfere with nicotinic sites in the cell. Such actions have been attributed to nicotine in other systems (Douglas and Rubin, 1961; Thorpe and Seeman, 1972). Ethyl alcohol may increase permeability of nicotine into smooth muscle cells to release more bound Ca^{++} and enhance contraction height. It increased contraction heights to nicotine both in chorionic and umbilical strips. There was a parallel upward shift of the dose response curve to nicotine in the presence of ethyl alcohol both in chorionic and umbilical strips. However, these investigations do not exclude the possibility that nicotine does not release an endogenous agent (e.g., prostaglandins, Karim, 1967; Fiscus and Dyer, 1982) which produces similar effects in both chorionic and umbilical vessels.

In view of the potent actions of 5-HT on the uterine and placental vasculature, it has been implicated in a number of physiological and pathophysiological conditions related to pregnancy and uterine hemodynamics which include pre-eclampsia, abortion, and parturition (Robson and Sullivan, 1968; Berman et al., 1978; Clark et al., 1980). According to present investigations, ethyl alcohol enhances the vasoconstrictive effects of both 5-HT and nicotine. Therefore, maternal smoking combined with alcoholism may have hazardous effects on the blood flow through umbilico-placental vasculature and may contribute to the above complications of pregnancy.

SUMMARY

Perfusion of human placenta in vitro does not permit distinguishing its sites of vascular resistance. Therefore, helically cut strips of (a) umbilical, (b) chorionic, and (c) villus stem arteries were examined for their sensitivity to 5-HT, nicotine, and ethyl alcohol (ETOH) in Kreb's bicarbonate buffer (pH 7.4, 37°C). In these preparations, increased vascular resistance is indicated by increased tension or contraction of the vascular strips. These experiments gave the following results: (a) 5-HT was very potent for causing contractions of umbilical, chorionic, and villus stem (EC_{50}, 84, 47, and 80 nM, respectively) arterial strips; (b) nicotine induced contractions in umbilical, chorionic and villus stem arterial strips (EC_{50}, 2.4, 3.3, and 2.4 mM, respectively); (c) ETOH (0.5 to 2.0%, v/v) had differential effects on umbilical and chorionic arterial segments. Its effect on umbilical arterial strips was minimal contractions or flacidity. 5-HT was less effective in the umbilical arterial strips pretreated with ETOH. About 50% of the contraction height induced by 5-HT, was reduced when the umbilical strips were pretreated with ETOH. However, ETOH potentiated the contractile responses of umbilical strips to nicotine. (d) ETOH, by itself, caused contractions of chorionic and villus stem arterial strips and potentiated the effects of 5-HT and nicotine by about 115-160% depending upon the concentration of ETOH and the nature of the strip. It also enhanced the effects of KCl on these strips. These observations suggest that Ca^{++} movements required for nicotine induced contractions may be facilitated in all vessels whereas Ca^{++} movements from its sources for 5-HT induced contractions are partially hindered in umbilical arteries.

ACKNOWLEDGEMENTS

This investigation was supported by grants from The Council for Tobacco Research, Inc., USA and USPHS-NIH grants ES-03172 and HD-10607.

REFERENCES

Altura, B.M., Malrya, D., Reich, C.F., and Orkin, L. (1972) Effect of vasoactive agents on isolated human umbilical arteries and veins. *Am. J. Physiol.* 222, 345-355.

Balfour, D.J.K. (1973) Effects of nicotine on the uptake and retention of ^{14}C-noradrenaline and ^{14}C-5-hydroxytryptamine by rat brain homogenates. *Eur. J. Pharmacol.* 23, 19-26.

Berman, W., Jr., Goodlin, R.C., Heymann, M.A., and Rudolph, A.M. (1978) Effects of pharmacologic agents on umbilical blood flow in fetal lambs in utero. *Biol. Neonate* 33, 225-235.

Ciuchta, H.P. and Gautieri, R.F. (1964) Effect of certain drugs on perfused human placenta. III. Sympathomimetics, acetylcholine, and histamine. *J. Pharmaceutical Sci.* 53, 184-188.

Clark, K.E., Mills, E.G., Otte, T.E., and Stys, S.J. (1980) Effect of serotonin on uterine blood flow in pregnant and non-pregnant sheep. *Life Sci.* 27, 2655-2661.

Douglas, W.W. and Rubin, R.P. (1961) Mechanism of nicotinic action at the adrenal medulla: Calcium as a link in stimulus-secretion coupling. *Nature* 192, 1087-1089.

Eliasson, R. and Astrom, A. (1955) Pharmacological studies on perfused human placenta. *Acta Pharmacol. Toxicol.* 11, 254-264.

Eltherington, L.G., Staff, J., Hughes, T., and Melman, K.L. (1968) Constriction of human umbilical arteries: Interaction between oxygen and bradykinin. *Circul. Res.* 22, 749-752.

Fiscus, R.R. and Dyer, C.C. (1982) Effects of indomethacin on contractility of isolated human umbilical artery. *Pharmacol.* 24, 328-336.

Furchgott, R.F. (1983) Role of endothelium in responses of vascular smooth muscle. *Circul. Res.* 53, 557-573.

Furchgott, R.F. and Badrakam, S. (1953) Reactions of strips of rabbit aorta to epinephrine, isopropylarterenol, sodium nitrite and other drugs. *J. Pharmacol. Exp. Ther.* 108, 129-143.

Gautieri, R.F. and Ciuchta, H.P. (1962) Effect of certain drugs on perfused human placenta. I. Narcotic analgesice, serotonin, and relaxin. *J. Pharmaceutical Sci.* 51, 55-58.

Gautieri, R.F. and Mancini, R.T. (1967) Effect of certain drugs on perfused human placenta. VII. Serotonin versus angiotensin II. *J. Pharmaceutical Sci.* 56, 296-297.

Goodman, R.F. and Weiss, G.B. (1973) Alteration of 5-hydroxytryptamine-^{14}C efflux by nicotine in rat brain area slices. *Neuropharmacol.* 12, 955-965.

Gould, R.J. and Ginsberg, B.H. (1985) Membrane fluidity and membrane function. In: *Membrane Fluidity in Biology: Disease Processes*, (eds.), R.C. Aloia and J.M. Boggs, Academic Press, Inc., Orlando, Florida, volume 3, pp. 258-281.

Harris, R.A. and Hitzemann, R.J. (1980) Membrane fluidity and alcohol actions. In: *Currents in Alcoholism*, (ed.), M. Galanther, Grune and Stratton, New York, pp. 379-404.

Harris, R.A. and Schroeder, F. (1981) Ethanol and physical properties of brain membranes: Fluorescence studies. *Mol. Pharmacol.* 20, 128-137.

Heron, D.S., Shinitzky, M., Hershkowitz, M., and Samuel, D. (1980) Lipid fluidity markedly modulates the binding of serotonin to mouse brain membranes. *Proc. Natl. Acad. Sci. USA* 77, 7463-7467.

Hery, F., Bourgoin, S., Hamon, M., Ternaux, J.P., and Glowinski, J. (1977) Control of the release of newly synthesized ^{3}H-hydroxytryptamine by nicotinic and muscarinic receptors in rat hypothalamic slices. *Nauyn-Schmiedeberg's Arch. Pharmacol.* 296, 91-97.

Hillier, K. and Karim, S.M.M. (1968) Effects of prostaglandins E_1, E_2, $F_1\alpha$, $F_2\alpha$ on isolated human umbilical and placental blood vessels. *J. Obstet. Gynaecol. Br. Cmwlth.* 75, 667-673.

Hitzemann, R.J., Harris, R.A., and Loh, H.H. (1984) Synaptic membrane fluidity and function. In: *Physiology and Membrane Fluidity*, (ed.), M. Shinitzky, CRC Press, Inc., Boca Raton, Florida, volume II, pp. 109-126.

Horst, M. and Sastry, B.V.R. (1985) Interaction of furan analogs of muscarine at endothelial muscarinic receptors of the rat aorta. *Fed. Proc.* 44, 1112.

Jaiswal, R.K., Landon, E.J., and Sastry, B.V.R. (1983) Methylation of phospholipids in microsomes of the rat aorta. *Biochim. Biophys. Acta* 735, 367-379.

Janis, R.A. and Triggle, D.J. (1983) New developments in Ca^{++} channel antagonists. *J. Med. Chem.* 26, 775-785.

Jones, J.B. and Rowsell, A. (1973) Fetal 5-hydroxytryptamine levels in later pregnancy. *J. Obstet. Gynecol.* 80, 687-689.

Karim, S.M.M. (1967) The identification of prostaglandins in human umbilical cord. *Br. J. Pharmacol. Chemother.* 29, 230-237.

Landon, E.J., Jaiswal, R.K., Naukam, R.J., and Sastry, B.V.R. (1984) Effects of calcium channel blocking agents on membrane microviscosity and calcium in the liver of the carbon tetrachloride treated rat. *Biochem. Pharmacol.* 33, 3553-3560.

Mukherjee, A.B. and Hodgen, G.D. (1982) Maternal ethanol exposure induces transient impairment of umbilical circulation and fetal hypoxia in monkeys. *Sci.* 218, 700-702.

Netter, F.H. (1967) Pregnancy and its diseases. In: *The Ciba Collection of Medical Illustrations: Reproductive System*, (ed.), E. Oppenheimer, Ciba Medical Education Division, Summit, NJ, volume 2, p. 219.

Park, M.K., Rishor, C., and Dyer, D.C. (1972) Vasoactive action of prostaglandins and serotonin on isolated human umbilical arteries and veins. *Can. J. Physiol. Pharmacol.* 50, 393-399.

Richards, C.D., Martin, K., Gregory, S., Keightley, C.A., Hesketh, T.R., Smith, G.A., Warren, G.B., and Metcalfe, J.C. (1978) Degenerate perturbations of protein structure as the mechanism of anesthetic action. *Nature (London)* 276, 775-776.

Robinson, C.P. and Sastry, B.V.R. (1976) The influence of mecamylamine on contractions induced by different agonists and on the role of calcium ions in the isolated rabbit aorta. *J. Pharmacol. Exp. Ther.* 197, 57-65.

Robson, J.M. and Sullivan, M. (1968) Effect of 5-hydroxytryptamine on maintenance of pregnancy, congenital abnormalities, and the development of toxemia. *Adv. Pharmacol.* 6, 187-189.

Sastry, B.V.R. and Owens, L.K. (1983) Regional sensitivity of human umbilicoplacental vasculature of 5-hydroxytryptamine and alcohol. *Pharmacologist* 25, 139.

Sastry, B.V.R. and Owens, L.K. (1984) Differential effects of nicotine and ethyl alcohol on the contractile responses of umbilical and chorionic arteries of human placenta. *Fed. Proc.* 43, 350.

Sastry, B.V.R. and Sadavongvivad, C. (1979) Cholinergic systems in non-nervous tissues. *Pharmacol. Rev.* 30, 65-132.

Somlyo, A.V., Woo, C.Y., and Somlyo, A.P. (1965) Responses of nerve free vessels to vasoactive amines and polypeptides. *Am. J. Physiol.* 208, 748-753.

Thorpe, W.R. and Seeman, R. (1972) On the mechanism of the nicotine-induced contracture of skeletal muscle. *Can. J. Physiol. Pharmacol.* 50, 920-923.

Tulenko, T.N. (1979) Regional sensitivity to vasoactive polypeptides in the human umbilicoplacental vasculature. *Am. J. Obstet. Gynecol.* 135, 629-636.

Van der Kloot, W. (1978) Calcium and neuromuscular transmission. In: *Calcium and Drug Action*, (ed.), G.B. Weiss, Plenum Press, New York, London, pp. 261-288.

von Euler, U.S. (1938) Action of adrenaline, acetylcholine and other substances on nerve-free vessels (human placenta). *J. Physiol. (London)* 93, 129-143.

Trophoblast Research 2:305-314, 1987

PLACENTAL DISTRIBUTION AND THE EFFECT OF ANTIHYPERTENSIVE DRUGS ON MONOAMINE OXIDASE AND CATHECHOL-o-METHYL TRANSFERASE ACTIVITY AT TERM

Eytan R. Barnea[1], A.H. DeCherney,
and Frederick Naftolin

Department of Obstetrics and Gynecology
Yale University School of Medicine
333 Cedar Street
New Haven, Connecticut, 06510 USA

INTRODUCTION

Pregnancy associated with hypertension frequently requires antihypertensive medications, which among other effects should improve the uteroplacental circulation. The mechanism and site of action of the commonly used antihypertensive drugs (AHD) used in pregnancy in the circulation has been previously investigated (Lunell et al., 1981; Johanson et al., 1980, Freed et al., 1978; Moskowitz and Cohn, 1980). Unfortunately, knowledge concerning mechanism of action of this group of AHD on the placenta and the uterus is limited. It has been suggested that their effect among others may be mediated via a direct action on the placental blood vessel wall. Moreover, beta adrenergic receptors were found in the brush border membrane of the placenta where some AHD could exert their action (Moore and Whitsett, 1981; Karlsson et al., 1984; Maigaard et al., 1984). Other effects might be exerted on metabolizing enzymes for placental catecholamine and serotonin, monoamine oxidase (MAO) and catechol-o-methyl transferase (COMT). The activities of these metabolizing enzymes have been shown previously to be lower in placentae of patients with hypertension (Barnea et al., 1983, 1986) where a high concentration of norepinephrine was found (DeMaria, 1966). In addition, alpha methyl DOPA and hydralazine can interact with COMT and MAO, respectively in other tissues (Gordonsmith et al., 1982; Lyles et al., 1983).

Using a placental explant system we investigated the effect of different AHD used during pregnancy on placental COMT and MAO activities. Such information could be useful for better directing AHD therapy in the pregnant woman.

[1]To whom correspondence should be addressed: Department of Obstetrics/Gynecology A, Rambam Hospital, Technion Medical School, Haifa, Israel
Submitted: 15 September 1986; Accepted: 30 September 1986

MATERIALS AND METHODS

Chemicals

Tryptamine Bisuccinate (53.9 Ci/mmole TB) and 2 hydroxyestrone - 3H (SA 50 Ci/mmole, 20HE1) were purchased from New England Nuclear (MA). Alpha methyl DOPA, hydralazine, $MgSO_4$, verapamil and propranolol were purchased from Sigma (MO). All other chemicals were of high analytical quality. Dulbecco's modified Eagle media (DMEM) was purchased from Dulbecco Co. (NY). Unlabeled S-adenosyl methionine (SAM), and tropolone were purchased from Sigma (MO). Clorgyline was obtained as a gift from May Baker Laboratories (Dagenham, UK). A (0.1 M) phosphate buffer was prepared by adding Na_2HPO_4 to distilled water and titrating it to pH 7.8 with Na_2HPO_4. Sucrose buffer was prepared by adding TRIS 1.2 g/l (10 mM), EDTA 0.56 gm/l (1.5 mM) and 0.25 M sucrose to distilled water and titrating it to pH 7.4 by concentrated HCl.

Placental Dissection

For regional distribution experiments, two placentae were utilized and they were obtained following elective term, not in labor, cesarean section. The placenta was placed on ice. Using sharp dissection, the placenta was divided into four equal quadrants and further cut into the maternal surface close to the endometrium and fetal surface, the part of the placenta close to the fetal membranes. Several fragments were taken from each quadrant and then were extensively rinsed to remove all blood. Tissue fragments, were homogenized in 4 volumes of cold TRIS-EDTA 0.25 M sucrose buffer with a blender using three strokes of 30 sec duration. The homogenate obtained was used as a source for MAO. The homogenate was then centrifuged at 800 x g for 10 min saving the supernatant followed by 105,000 g spin for 60 min. The supernatant obtained following the last centrifugation was used as a source for COMT estimation.

Placental Explants

Placental tissues were obtained from four healthy subjects following elective cesarean section, not in labor, at term. Small placental fragments were dissected from the meaty layer in a sterile fashion (300 mg) avoiding calcified regions and these fragments were placed in a large excess of 0.9% NaCl solution to remove all blood. Subsequently, tissues were rinsed again in DMEM containing 0.5% antibacterial solution composed of 10000 U/ml penicillin, 100 µg/ml streptomycin and 10 µg/ml fungizone. Incubations were conducted by placing a single fragment of 300 mg placenta per dish (wet weight) in 2 ml MEM media containing 0.5% bovine serum albumin with or without test compounds at 37°C in a gas mixture of 95% air and 5% CO_2 for 6 or 24 hrs. Three concentrations of AHD were used, the estimated therapeutic dose was calculated based upon known plasma concentrations of the drug. The highest dose was considered to be therapeutic. This concentration was estimated by calculating the dose of AHD usually employed per kg body weight. The concentration of $MgSO_4$ was based upon published therapeutic plasma concentrations (Sibai et al., 1984). Following incubation, dishes were placed on ice, the media was discarded, and placental tissue was further processed by homogenizing it in four volumes of 0.25 sucrose buffer with a blender using three strokes of 30 sec. The homogenate obtained was used as the source for MAO. The homogenate was then centrifuged at 800 x g for 10 min saving the supernatant, which was then centrifuged at 105,000 x g for 60 min. The supernatant obtained following the last centrifugation was used as a source for COMT exposure.

MAO Assay

MAO activity was measured using the radioenzymatic assay previously described (Parvez et al., 1973; Barnea et al., 1986). Briefly, 1.5 µM of ^{14}C tryptamine bisuccinate, phosphate buffer (pH 7.8), and 100 µl of placental homogenate (0.05 - 0.2 mg protein/ml) in a total volume of 350 µl were incubated for 10 min at 37°C in a water bath. Following incubation, samples were placed immediately in ice, and the reaction was stopped by adding 0.4 ml of 2 N HCl. Incubates were extracted twice with 3 ml toluene (recovery was 95%). Ten ml of POP-POPOP scintillation fluid were added, and ^{14}C samples were analyzed at 52% efficiency in a Beckman 9000 scintillation counter.

Reaction was linear for 0-15 min, at a protein concentration of 0.05 - 0.5 mg/ml. Inter- and intra-assay variability was less than 10%. Placentae boiled for 30 min at 100°C served as blank; these blanks had an activity which was 1% of the activity of normal tissue. Incubations in the presence of 10^{-3} M clorgyline, a specific MAO A blocker, inhibited completely the enzyme activity. Blank values obtained in this manner were lower than the heat inactivated samples and were used as blanks in the assay.

COMT Assay

COMT activity of placental supernatant fraction was measured by a radioenzymatic assay (Bates et al., 1979). In the placenta, the affinity of 2 hydroxyestrone was 100-fold higher than that of catecholamines (Barnea et al., 1983). Consequently, this method was appropriate for this assay. Briefly, incubation mixture consisted of 40 µM of (3H 2OHE$_1$), 6 mM MgCl$_2$, SAM (1 mM) and supernatant in a total volume of 0.95 ml. Incubations were performed out at 37°C in a water bath for 60 min. At the end of incubation, samples were placed on ice to cool, and the reaction was stopped by adding a 0.5 M borate solution at pH 10.5. Samples were extracted with 5 ml toluene POP-POPOP and were analyzed in a scintillation counter with 52% efficiency. The enzyme activity was estimated by the amount of substrate metabolized to 2- methoxyestrone, which was the sole product formed as identified by high performance liquid chromatography. Product recovery was 95%. Incubates with added 1 mM tropolone (a specific COMT inhibitor) were used as blanks. Blank values were 3% of total activity. Reactions were linear for 60 min at 0.5 - 5 mg/ml. Placental protein concentration was measured according to the method of Lowry et al. (1953).

Statistical Analysis

Statistical significance of differences between mean values was analyzed by one way ANOVA and Student's t test. Data were expressed as mean $\pm$ standard error of the mean from 2-3 individual experiments. In each experiment, 3-6 individual measurements were made for the three doses of each drug tested, at each time interval.

RESULTS

Regional Distribution of COMT and MAO Activity

The regional distribution of COMT and MAO activity was studied (Table 1). The activity of COMT in the region close to the maternal surface was 33-fold higher than the activity found in fetal surface. In contrast, MAO activity in the two regions was similar.

COMT Activity in Placental Explants

COMT activity following 6 hr incubation with alpha methyl DOPA was significantly reduced ($F = 5.12$), ($p < 0.01$). Propranolol and $MgSO_4$ did not show any significant effect on COMT activity (Figure 1). There was a significant dose-dependent increase of COMT activity during incubation with hydralazine and verapamil for 6 hr ($F = 6.3$, $F = 8.05$) both, respectively ($p < 0.001$), (Figure 2). At 24 hr, however, COMT activity was suppressed when incubated with hydralazine ($F = 5.1$), ($p < 0.01$). Verapamil caused a slight increase in COMT activity which was not significant, ($p = .07$), (Figure 3). The other three drugs, alpha methyl DOPA (5.01 x 10^{-5} - 5.01 x 10^{-7} M) propranolol (10^{-6} - 10^{-8} M) and $MgSO_4$ (1.58 x 10^{-4} - 1.58 x 10^{-6}) had no effect on enzyme activity at the 24 hr time interval.

MAO Activity in Placental Explants

At 6 hr, verapamil induced a significant decrease in MAO activity, ($F = 10.1$), ($p < 0.01$), (Figure 4). None of the other drugs tested (in doses similar to that used for COMT estimations) caused any significant changes in MAO activity at 6 hr, and none of the five drugs had a significant effect at 24 hr.

Table 1

MAO and COMT activity in different regions of the human placenta at term

	MAO activity*	COMT activity**
Maternal surface N = 8	165 ± 23	2435 ± 120***
Fetal surface N = 8	187 ± 22	80 ± 38

MAO activity * = pmol/min/mg protein
COMT activity ** = pmol/h/mg protein
*** = P < 0.0001 from maternal surface

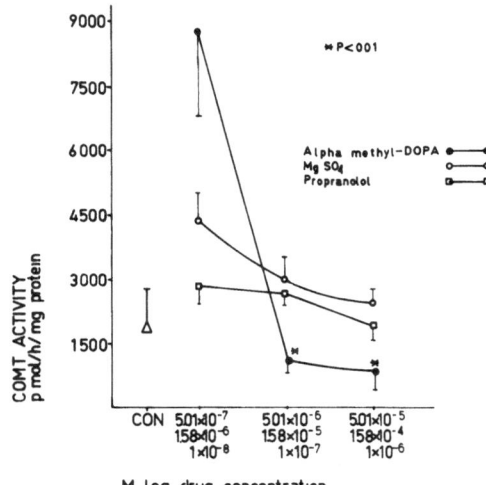

Figure 1. COMT activity (mean ± SEM) in placental explants following exposure to hydralazine and verapamil. Both compounds increased the enzyme activity after 6 hr of incubation. CON = controls.

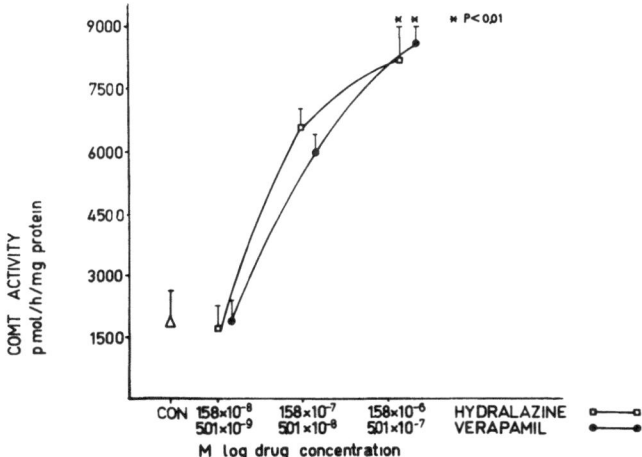

Figure 2. COMT activity (mean ± SEM) in placental explants following exposure to various AHDs. Only alpha methyl DOPA had a significant suppressive effect on the enzyme activity after 6 hr of incubation. CON = controls.

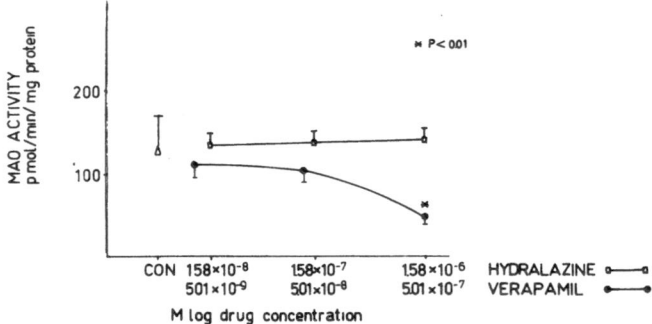

Figure 3. COMT activity (mean ± SE) in placental explants following exposure to hydralazine and verapamil for 24 hr. Hydralazine significantly reduced the enzyme activity,.while the effect of verapamil was borderline stimulatory (P = 0.065). CON = controls.

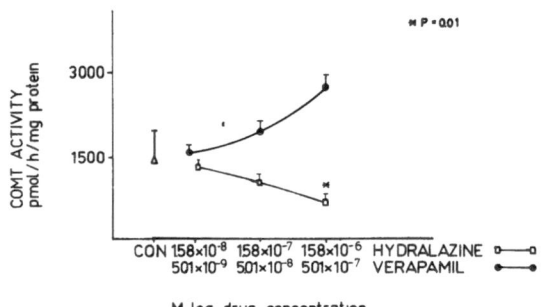

Figure 4. MAO activity (mean ± SE) in placental explants following exposure to AHDs. Only verapamil had a significant suppressive effect on the enzyme activity.

DISCUSSION

Present results suggest that COMT activity is affected by estimated therapeutic concentrations of verapamil, alpha methyl DOPA and hydralazine at 6 hr incubation. Alpha methyl DOPA probably mediated its effect through interaction with the enzyme active site. This is supported by the fact that similar inhibitory effects of alpha methyl DOPA in incubation with semi-purified placental cytosol COMT preparation were noted (Barnea et al., 1983). Similarly, this drug was shown to have a high affinity for COMT in different rat tissues (Gordonsmith et al., 1983). Following 24 hr incubation, however, this inhibitory effect was not observed, probably due to the metabolism of the drug. Indeed, it has been shown that therapeutic concentrations of alpha methyl DOPA did not have a toxic effect on trophoblastic cultures (Lueck and Aladjem, 1982). At 6 hr, COMT activity was significantly increased in the presence of hydralazine. Such an effect of hydralazine on COMT activity has not been previously reported. The lower enzyme activity at 24 hr might be related to the toxic effect of this drug following a longer time of incubation. Lueck and Aladjem (1982) demonstrated that therapeutic concentrations of hydralazine produced a 100% loss in viability of trophoblast cells following prolonged cultures.

COMT activity is suppressed in the presence of high concentration of calcium (Weinshilbum and Raymond, 1976). Verapamil which is a calcium ion channel-blocker raised COMT activity following 6 and 24 hr incubations. This stimulatory effect of verapamil was not found in incubations made with $MgSO_4$. This difference suggests that adequate levels of Mg^{++} are present in trophoblastic cells to permit normal COMT activity. The presence of adequate Mg^{++} concentrations is also supported by the finding that varying the concentration of Mg^{++} (0-12 mM) in the incubation with the untreated semi-purified placental cytosol COMT preparation did not change significantly the obtained enzyme activity (Barnea et al., 1983). Unlike Ca^{++} which is a COMT inhibitor, Mg^{++} is an essential cofactor for this enzyme. The cause for decreased MAO activity in the presence of verapamil at 6 hr incubation is unknown. It is of interest, however, that COMT activity was simultaneously rising in the placental tissue. Similarly, Nandakamuran et al. (1983) found that placental MAO inhibition induced by pargyline caused a significant increase in COMT activity and consequently, increased the concentration of the methylated metabolites of catecholamines. Thus, in the placenta these two drugs can increase COMT activity while decreasing MAO activity by an unknown mechanism. Since it was demonstrated that approximately 50% of COMT is membrane bound, it raises the possibility of interactions between the two enzymes (Jeffery and Roth, 1984).

The drugs tested in these experiments are commonly used for the treatment of hypertension associated with pregnancy. The calcium channel-blocker, nifedipine, increases placental blood flow by antagonizing the effects of calcium and serotonin (Maigaard et al., 1984). Jouppila et al. (1985) have shown that dihydralazine administered intravenously did not change the placental circulation but did increase the umbilical vein flow in hypertensive pregnant patients. Of the five drugs tested, $MgSO_4$ and propranolol did not have any effect on placental COMT and MAO activity. Hydralazine and verapamil had a stimulatory effect on placenta COMT activity. Such stimulation may aid in the inactivation of the excess catecholamines present in the uteroplacental bed in hypertensive pregnancies (O'Shaugnessey and Zuspan, 1982).

There is a very high COMT activity in the placental region close to the maternal circulation compared with the fetal region (Table 1). High activity of this enzyme was found also in the decidua vera and endometrium (Casey and MacDonald, 1983). Perhaps, these high tissue levels of enzyme may serve the same function - inactivating circulating catecholamines which reach these tissues via the maternal circulation. In contrast, the MAO activity was similar in both maternal and fetal regions of the placenta.

In conclusion, AHD used in pregnancy, e.g., alpha methyl DOPA, hydralazine and verapamil, appear to affect preferentially placental COMT activity. The possible therapeutic significance of these findings remain to be established by further studies.

SUMMARY

The effect of therapeutic concentrations of antihypertensive drugs (AHD) used in pregnancy on catechol-o-methyl transferase (COMT) and monoamine oxidase (MAO) activities in term placental explants were studied. The enzyme activities were measured in tissue fractions using radioenzymatic techniques. The COMT activity in the maternal surface of the placenta was 33-fold higher than that on the fetal side. In contrast, MAO activity in both regions was similar. After 6 hr of incubation, COMT activity increased significantly following exposure to verapamil and hydralazine, while exposure to alpha methyl DOPA caused a significant suppression of the enzyme activity. At 24 hr, exposure to hydralazine significantly suppressed COMT activity. By 6 hr MAO activity was significantly suppressed by verapamil. $MgSO_4$ and propranolol had no effect on the activities of COMT and MAO at 6 and 24 hr. These results suggest that COMT and MAO activities in the placenta can be modulated by AHD used in pregnancy and therefore, these drugs might affect local catechol metabolism.

REFERENCES

Barnea, E.R., MacLusky, N.J., and Naftolin, F. (1983) Catecholamine metabolism in the human term placenta. *Soc. Gyn. Invest.* (Abstract), 283.

Barnea, E.R., MacLusky, N.J., and Naftolin, F. (1986) Monoamine oxidase activity in the human placenta at term. *Am. J. Perinat.* 3, 219-224.

Bates, G.W., Edman, C.D., Porter, J.C., Johnston, J.M., and MacDonald, P.C. (1979) An assay for human erythrocyte catechol-o-methyl transferase activity using a catechol estrogen as the substrate. *Clin. Chem. Acta* 94, 63-69.

Casey, M.L. and MacDonald, P.C. (1983) Characterization of catechol-o-methyltransferase activity in human uterine decidua vera tissue. *Am. J. Obstet. Gynecol.* 145, 453-457.

DeMaria, F.J. and See, H.Y.C. (1966) Role of the placenta in pre-eclampsia. *Am. J. Obstet. Gynecol.* 77, 412-416.

Freed, C.R., Murphy, R.C., and Quintero, E. (1978) Blood pressure reduction and hypothalamic a-methyl-DOPA metabolism after long term a-methyl-DOPA infusions. *Clin. Res.* 26, 100A.

Gordonsmith, R.H., Raxworthy, M.J., and Gulliver, P.A. (1982) Substrate stereospecificity and selectivity of catechol-o-methyl transferase for DOPA, DOPA derivatives, and a-substituted catecholamines. *J. Pharmacol.* 31, 433-439.

Jeffery, D.R. and Roth, J.A. (1984) Characterization of membrane-bound and soluble catechol-o-methyl transferase from human frontal cortex. *J. Neurochem.* 42, 826-832.

Johanson, B.B., Auer, L.M., and Trummer, U.G. (1980) Pial vascular reaction to intravenous dehydralazine in the cat. *Stroke* 11, 369-374.

Jouppila, P., Kirkinen, P., Koivula, A., and Ylikorkala, O. (1985) Effects of dehydralazine infusion on the fetoplacental blood flow and maternal prostanoids. *Obstet. Gynecol.* 65, 115-118.

Karlsson, K., Lundren, Y., and Ljungblad, U. (1984) The acute effect of a non-selective β-adrenergic blocking agent in hypertensive pregnant rats. *Acta Obstet. Gynecol. Scand.* (Suppl.) 118, 81-85.

Lowry, O.H., Rosebrough, N., Farr, A.L., and Randall, R.J. (1951) Protein measurement with the pholin phenol reagent. *Biol. Chem.* 193, 265-269.

Lueck, J. and Aladjem, S. (1982) Effect of therapeutic levels of magnesium sulphate, methyl dopa, hydralazine and phenobarbitone on normal human trophoblast in vitro. *Placenta* 3, 39-44.

Lunell, N.O., Hjendahl, P., and Fredholm, B.B. (1981) Circulatory and metabolic effects of a combined and -β-adrenoreceptor blocker (labetalol) in hypertension of pregnancy. *Br. J. Pharmacol.* 15, 345-349.

Lyles, A.G., Garcia-Rodriguez, J., and Callingham, B.A. (1983) Inhibitory actions of hydralazine upon monoamine oxidizing enzymes in the rat. *Biochem.* 32, 2515-2520.

Maigaard, S., Forman, A., and Andersson, K.E. (1984) Effects of nifedipine on human placental arteries. *Gynecol. Obstet. Invest.* 18, 217-221.

Moore, J. and Whitsett, J. (1981) The beta-adrenergic receptor in human placenta: Receptor subtype analysis and partial characterization of the solubilized receptor. *Placenta Suppl.* 3, 103-114

Moskowitz, R. and Cohn, T. (1980) Hemodynamic effects of oxdralazine and hydralazine in hypertension. *Clin. Pharmacol. Ther.* 27, 773-777.

Nandakumaran, M., Gardey, C., Challier, J.C., and Oliver, G. (1983) Placental monoamine oxidase content and inhibition: effect of enzyme inhibition on maternofetal transfer of noradrenaline in the human placenta in vitro. *Placenta* 4, 57-62.

Parvez, H. and Parvez, S. (1973) Micro radioisotopic determination of enzymes catechol-o-methyltransferase, phenyletyhanolamine-n-methyl-transferase and monoamine oxidase in a single concentration of tissue homogenate. *Clin. Chem. Acta* 40, 85-92.

O'Shaugnessy, R.W. and Zuspan, F.P. (1982) Uterine catecholamines in normal and hypertensive human pregnancy. *Clin. Exp. Hypertens.* 2 (Abstract), 183.

Sibai, B.M., Graham, J.M. and McCubbin, J.H. (1984) A comparison of intravenous and intramuscular magnesium sulphate regimens in preeclampsia. *Am. J. Obstet. Gynecol.* 149, 128-150.

Weinshilbum, R.M. and Raymond, F.A. (1976) Calcium inhibition of rat catechol-o-methyl transferase. *Biochem. Pharmacol.* 25, 573-577.

Trophoblast Research 2:315-327, 1987

MATERNOFETAL POTENTIAL DIFFERENCES: STUDIES USING THE IN VITRO PIG PLACENTA

B.S. Ward[1,4], W.M.O. Moore[1], C.P. Sibley[3],
J.D. Glazier[2], and R.D.H. Boyd[2]

Departments of Obstetrics and Gynaecology[1]
Child Health[2] and Physiology[3]
University of Manchester
Saint Mary's Hospital
Manchester, M13 OJH, United Kingdom

INTRODUCTION

A potential difference (pd) between maternal and fetal circulations created by active transport of electrolytes by the placenta would be important not only for the actively transported species themselves, but also for the transfer of passively distributed charged molecules. In several species, it is possible to measure a pd between catheters placed in maternal and fetal vessels. It is possible that this maternofetal pd is a result of active electrolyte transport by the placenta. However, the recorded pd's have a wide range, for example in the sheep it is 34 mV, fetus negative relative to mother (Weedon et al., 1978), whereas in the rat it is 14 mV, fetus positive (Mellor, 1969). Considerable doubt is shed on the placental basis of this pd by at least two lines of evidence. Firstly, there appears to be no relationship between the magnitude and polarity of the pd and the type of placenta, e.g., the guinea pig which has a relatively permeable (Hedley and Bradbury, 1980) haemochorial placenta demonstrates a pd of 18 mV, fetus negative (Mellor, 1969) more similar to that of the sheep, which has a relatively impermeable (Boyd et al., 1976) epitheliochorial type of placenta than to the term human (O mV, Mellor et al., 1969). Further evidence against the placenta being the site of pd generation comes from Binder et al. (1978) and Thornburg et al. (1979) who found in the guinea pig and sheep, respectively, that the steady-state concentration ratio between maternal and fetal plasmas of exogenous ions injected into the mother was consistent with a near zero transplacental pd. They suggest that the maternofetal pd is generated at a site outside of the placental exchange area.

However, the diffuse membranous placenta of the pig, consisting of the chorioallantoic membrane and intimately associated uterine epithelium (Brambel, 1933), can be shown to generate a pd in vitro when mounted in a Ussing chamber (Crawford and McCance, 1960; Michael et al., 1985). This tissue therefore provides a useful system with which to study the transplacental pd directly. We report here some results of our earlier and recent investigations into the electrical activity

[4]To whom corespondence should be addressed.
Received: 10 October 1985; Accepted: 18 June 1986

generated in vitro by the pig placenta and in particular the effect of catecholamines on pd.

MATERIALS AND METHODS

Large White/Landrace sows of 95-105 days gestation (term 115 days) were initially anesthetized with a halothane:nitrous oxide/oxygen gas mixture via a face mask, and, after endotrachael intubation with a cuffed tube, anesthesia was maintained with this mixture. Sows were ventilated using a Cape ventilator while maternal blood pH, pCO_2, and pO_2 were monitored.

The abdomen was opened and the fetuses palpated in order to identify one of average size in the litter; once identified this conceptus and its surrounding uterus was clamped off and excised. The excised region was opened and a piece of uterus and attached placenta removed and placed in a Ringer solution at room temperature. Six portions of full thickness of placenta were then stripped of the underlying myometrium by gentle pulling (Figure 1); the necrotic tips and main vessels were avoided. These placental portions were mounted as vertical membranes between Ussing-type perspex conical half-chambers (Michael et al., 1985). The area exposed to the Ringer solution used as bathing medium was 8 cm^2. Each half chamber was filled with 25 ml of this Ringer which had the following composition (mM): Na^+:143, K^+:5.9, Ca^{2+}:2.5, Mg^{2+}:1.2, Cl^-:127.8, HCO_3^-:25, PO_4^{3-}:1.2 and glucose:11 (final pH = 7.4). The Ringer solution bathing the tissue was gassed and stirred with a mixture of 95% oxygen and 5% carbon dioxide by a gas lift circulatory system and maintained at 39°C using water jackets. The interval of time between excising the conceptus and mounting the sixth membrane was between 20 to 25 min.

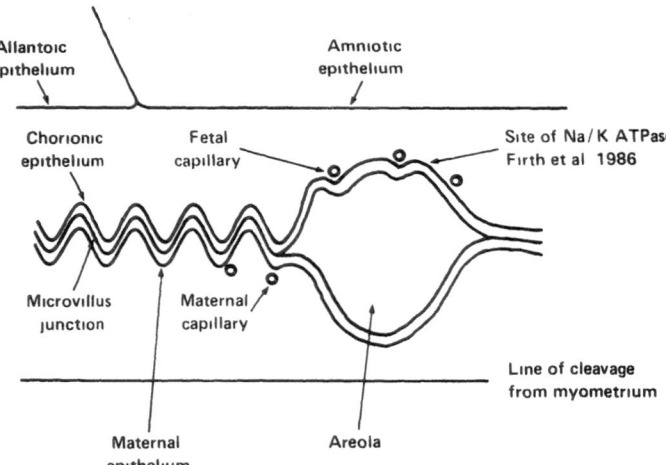

Figure 1. Diagrammatic representation of the fine structure of the fetal membranes as mounted. (After Crombie, 1972).

The transmembrane potential difference (pd) was measured through narrow-tipped PVC bridges placed approximately 1 mm from either side of the membrane and in close proximity to its center. These bridges contained saturated KCl in 3% agar and were connected to a high impedence volt meter (model 1055, Datron Electronics Limited, Norwich, U.K.) via matched calomel electrodes. The asymmetry in the circuit, which was corrected, was never greater than 1.0 mV before or after an experiment. For measurement of short circuit current (scc), a second similar pair of bridges was placed at the apex of each cone and connected to silver/silver chloride electrodes via a saturated KCl solution. Using these electrodes a current could be passed through the membrane from an external source so that the pd across the tissue could be maintained equal to zero. This current was defined as the scc. Fluid resistance between electrode tips was measured in the absence of placenta and allowed for in the scc measurements.

Tissue resistance (R) was determined either by clamping the tissue to various potentials using known currents or by similar application of Ohm's law to the scc and pd readings.

In a normal experiment, pieces of placental tissue were mounted as described above and after a suitable equilibration period were clamped to zero potential and the following protocols followed:

Effect of Epinephrine on pd and scc

Scc and pd (by briefly open circuiting the chambers) were recorded at 10 min intervals until reasonably steady (usually 90 min) and then epinephrine (final concentration 10^{-5}M, Sigma Chemical Company Limited, Poole, UK) was added to fetal compartment of experimental chambers and, at the same time, an equal volume of Ringer's solution was added to the maternal compartment. An equal volume of Ringer's solution was also added to the fetal and maternal compartments of control chambers.

Effect of Adrenergic Agonists and Antagonists on pd

After clamping for 90 min the appropriate adrenergic agonist was added to the fetal compartment of the chambers in an increasing cumulative dose. After each dose of drug was added, no further addition was made until (a) it was apparent that there was no increase in pd, or (b) a rise in pd in response to drug stopped, or (c) the rate of the rise in pd in response to drug had markedly declined. The pd, immediately before addition of the next dose, was taken as the maximum value for the concentration in the chamber. In this series of experiments, the Ringer's solution contained ascorbic acid (5.7×10^{-4}M), to prevent oxidation of the drugs which were: (-) isoprenaline hydrochloride (β-agonist), (-) epinephrine, (-) norepinephrine hydrochloride and L-phenylephrine hydrochloride (α-agonist; all obtained from Sigma Chemical Company Limited, Poole, UK). Stock solutions of these drugs were made up on the day of the experiment using 0.1 M HCl.

In a further set of experiments the pd dose response to epinephrine, measured as above, was observed in the presence of propranolol ("Inderal"; Imperial Chemical

Industries plc, Macclesfield, UK) (final concentration $10^{-7}M$) added to the fetal compartment 30 min after clamping.

Sodium Replacement Studies

After 30 min clamping, the Ringer's solution in the experimental group of chambers was replaced by a Ringer's solution in which Na^+ had been replaced by choline (143 mM), simultaneously in the control chambers the Ringer's solution was replaced by fresh Na^+ containing Ringer's solution; this procedure was then repeated 30 min later. After a further 20 min epinephrine (in appropriate Ringer's solution) was added to the fetal side of all chambers (final concentration $10^{-5}M$). Finally 20-25 min later the Ringer's solution in all chambers was replaced by the standard Na^+ containing Ringer's solution. Pd and scc were recorded at 5 min intervals throughout the experiment.

Data Analysis

The effect of the adrenergic agonists and antagonists on pd was analyzed by calculating the response caused by each concentration of drug relative to the maximally effective dose (designated 100%) in each piece of tissue. Response, $f(x)$, as a function of drug concentration, x, was fitted by non-linear least squares regression to the equation $f(x) = aK\ x/(1 + Kx)$ using sub-routine MINUITS from the CERN library in an interactive program. The drug concentration for 50% maximum response (EC_{50}) were obtained from the dose response curves.

In the sodium replacement studies, the pd values obtained after starting change of Ringer's solution were normalized by expressing these values as a quotient of the pd value immediately prior to the first solution change. These normalized values obtained for the control and experimental groups were compared using the unpaired Student's t test.

RESULTS

Control Values

Table 1 shows the mean values at the start of voltage clamping (t_0) and 60 min later (t_{60}). The pd and scc were always found to be initially fetal side positive with respect to maternal side, although an occasional piece of tissue declined through OmV and became fetal side negative by 60 min. The pd, scc, and R were all significantly lower at t_{60} than at t_0. The site of excision of the placenta from the uterus (i.e. amniotic or allantoic, Figure 1) appeared to make no significant difference to the recorded electrical activity and for consistency amniotic pieces were generally used.

Effect of Epinephrine ($10^{-5}M$) on the Fetal Side

Table 2 shows the mean data for the effect of fetal side epinephrine. There was a significant increase in pd and scc by 10 min after addition and a small further increase during the second 10 min after addition. By contrast in control experiments, there was no change in scc after Ringer's solution addition, and there was a significant decrease in pd (Table 2).

Table 1

Electrical parameters in the pig placenta

	t_0	t_{60}
pd (mV)	6.3 ± 0.3	4.8 ± 0.4**
scc (μA.cm^{-2})	8.6 ± 0.6	7.5 ± 0.8*
R (Ω.cm^2)	824 ± 40	696 ± 24**

Control pd, scc and resistance (R) values fetal side positive with respect to maternal side, t_0 and t_{60} refer to time after start of voltage clamping. Mean ± se is shown, n = 67 (number of placental pieces, taken from 18 different placentae, from 9 animals).

** = $P < 0.001$, * = $P < 0.05$, significantly different from t_0 by paired Student's t test.

Table 2

Effect of epinephrine on electrical parameters

Addition to Fetal Side of Chamber		Before Addition	10 min after Addition	20 min after Addition
Epinephrine (10^{-5}M)	pd (mV)	4.4 ± 0.5	5.0 ± 0.5**	5.4 ± 0.5**
	scc (μA.cm^{-2})	8.3 ± 1.3	9.9 ± 1.3**	10.6 ± 1.3**
Ringer	pd (mV)	3.5 ± 0.8	3.3 ± 0.8°	3.1 ± 0.8°
	scc (μA.cm^{-2})	6.3 ± 1.4	6.0 ± 1.4	5.9 ± 1.5

Effect of epinephrine or an equal volume of Ringer on pd and scc. Mean ± se is shown, for epinephrine group n = 30 (pieces of placenta taken from 16 placentae from 9 animals), for control (Ringer) group n = 14 (pieces of placenta taken from 12 placentae from 6 animals).

** = $P < 0.001$, significantly higher than before addition; ° = $P < 0.05$, significantly lower than before addition (paired Student's t test).

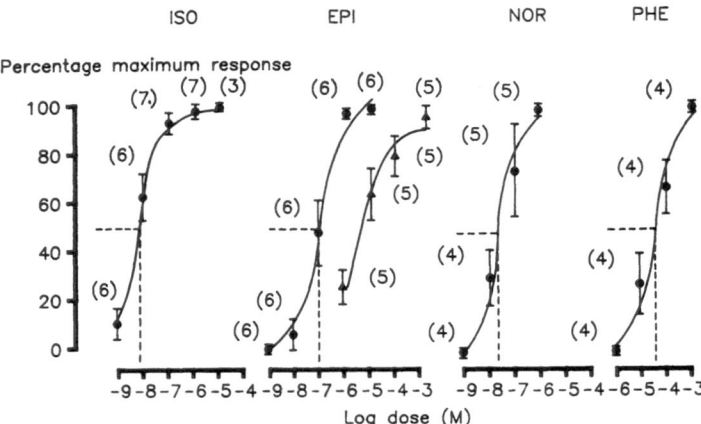

Figure 2. Dose response curves for a series of adrenergic agonists on pd. ISO = isoprenaline, EPI = epinephrine, NOR = norepinephrine, PHE = phenylephrine. The curve for epinephrine in the presence of propranolol (10^{-7}M) is that shown by (Δ) on the EPI panel. Values shown are mean $\pm$ se with n (number of placental pieces) in brackets.

Effect of Adrenergic Agonists and Antagonists

Figure 2 shows the dose response curves for the effect of a series of adrenergic agonists on pd. The EC_{50} values for the series were isoprenaline; 6.2 x 10^{-9}M, norepinephrine; 2.7 x 10^{-8}M, epinephrine; 1.1 x 10^{-7}M and phenylephrine; 3.7 x 10^{-5}M.

There was no significant difference between the maximum pd responses caused by each of the drugs, and these pds were all significantly different from changes in control (no drug) chambers studied simultaneously (Figure 3). When propranolol (10^{-7}M) was added to the fetal side of chambers prior to epinephrine addition, there was a shift in the pd dose response curve (Figure 2) so that the EC_{50} for the effect of epinephrine in the presence of propranolol was 3.3 x 10^{-6}M. The maximum effect on pd in the presence of propranolol (1.3 $\pm$ 0.4 mV; mean $\pm$ se, n = 5) was not significantly different from epinephrine alone.

Effect of Na+ Replacement

The time course of the pd's generated by two individual pieces of placenta are shown in Figure 4. For the first 30 min there was no difference between the experimental and control pieces in terms of bathing Ringer's solution, and both showed a slowly declining pd (cf. Table 1). At this point, the Ringer's solution bathing the experimental piece of tissue was replaced by one which contained choline instead of Na+ and the control tissue Ringer's solution was replaced by fresh normal Ringer's solution. There was an initial decline in the pd generated by both pieces, perhaps due to temperature changes, but then in the control the pd began to become constant whereas in the experimental piece there was a continued decrease in the pd. After a similar second change of Ringer's solution the control piece of placenta showed overall very little effect on pd, whereas the experimental piece showed a further decline. Adding epinephrine (10^{-5}M) to the fetal side of both pieces of placenta was followed by a sharp rise of pd in control but only a very small increase of the pd generated by the Na^+ free, experimental piece. Finally the Na^+ free Ringer's solution bathing the experimental piece of tissue was replaced by normal, Na^+ containing Ringer's solution and this change was followed by a marked rise in pd to a value similar to that of the control. These single experiment results were reflected in the mean values for the effect of Na^+ replacement on pd which was shown in Figure 5. At the start of the experiment before any replacement of Na^+ there was no significant difference in pd between the two groups ("Start" in Figure 5). After the first replacement of the Ringer's solution there was a significant difference between the experimental (Na^+ free) and control groups ("First Change" in Figure 5).

After the second round of Na^+ replacement, however, there was no significant difference between the two groups as control pd also declined (cf. Table 1); the measured Na^+ content in the Ringer's solution from the experimental chamber at this point was always less than 6 mM. After addition of epinephrine (10^{-5}M) to the fetal side of both chambers, there was a significant increase in the pd in control

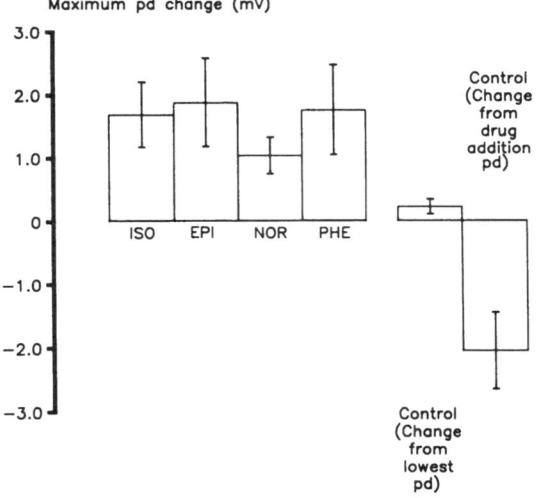

Figure 3. Maximum change in pd caused by the adrenergic agonists. Values are mean ± se; isoprenaline (ISO), n = 7, epinephrine (EPI) n = 6; norepinephrine (NOR), n = 5; phenylephrine (PHE), n = 4. Also shown is the change in pd in control chambers (n = 7) calculated either as the maximum increase from the lowest value attained during the course of an experiment or as the maximum change from the time drug was added to the experimental chambers (negative value indicates a decrease).

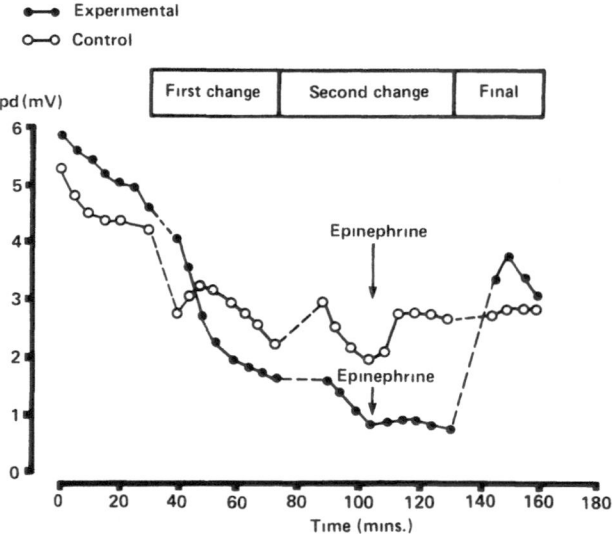

Figure 4. Time course of pd generated by two individual pieces of placenta in their separate Ussing chambers. Initially both pieces were bathed by Ringer's solution of identical composition and then at the start of the "First Change" block the control chamber Ringer was replaced by fresh normal Ringer and the experimental chamber Ringer was replaced by Na+ free Ringer. This was repeated at the start of the "Second Change" block and then epinephrine ($10^{-5}M$) was added to the fetal side of both chambers. At the start of the "Final" block the control chamber Ringer was replaced again by fresh normal Ringer and the experimental chamber Na+ free Ringer was now replaced by the normal, Na+ containing Ringer.

chambers ($P < 0.01$; paired Student's t test) but no significant effect ($P > 0.05$) was noted in the Na+ free chambers, so that the control and experimental chambers were once more significantly different ("Add epinephrine" Figure 5). Finally return of Na+ to the experimental group was followed by an increase in the pd to control levels ("Final", Figure 5). Similar results were found with scc.

DISCUSSION

Crawford and McCance (1960) were the first to make use of the opportunity provided by the membranous nature of the pig placenta to study transplacental ion movements, scc and pd under controlled conditions. In vivo maternofetal ion exchange is presumably predominantly between the maternal and fetal capillaries either side of the chorioallantoic and uterine epithelial cell layers. In vitro the membranes as mounted in the Ussing chamber also include an additional allantoic or amniotic cell layer, depending on the region of excision from the uterus, (Figure 1). It is possible that these nonplacental layers might be generating the pd especially as they enclose compartments of different ion composition from each other and from fetal

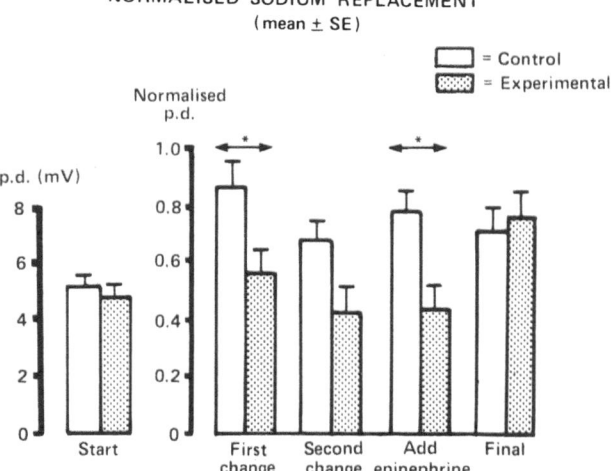

Figure 5. Effect of Na+ replacement by choline on pd. The mean values immediately before the beginning of Ringer changes are shown as "Start"; the values after Ringer changes were normalized to these values. "First Change" is the value 30 min after the first Ringer's solution change (experimental group now Na+ free), "Second Change" is the value 20 min after a second Ringer change, add epinephrine is the value 20 min after addition of epinephrine (10^{-5}M) to the fetal side and "Final" is the value 20 min after the Ringer's solution in all chambers was replaced by normal Na+ containing Ringer. Statistical comparisons shown are experimental (n = 8) v. control (n = 7) by unpaired Student's t test, * = P < 0.05.

plasma (Goldstein et al., 1980). However, this explanation seems unlikely as neither allantoic nor amniotic cell types possess features normally associated with transporting epithelia, such as large number of mitochondria or highly folded basolateral membranes (personal observations). The trophoblast and uterine epithelial cell layers do, on the other hand, possess these features. They are particularly prominent in the trophoblast cells of the areolar regions which have also Na+ - K+-ATPase localized by histochemical techniques. This ATPase is absent from either the allantoic or amniotic cell layers (Firth et al., 1986).

Different values and signs for pds across the pig placenta have been reported (Crawford and McCance, 1960; Bazer et al., 1981). The reason for this variation is uncertain. Differences in the gestational age of pigs used are unlikely to be responsible (Michael et al., 1985; personal observation). Another possibility is a decline in vitro from an in vivo catecholamine stimulated state which may be different in degree depending on the conditions under which placenta is collected.

We found that the effect of fetal side epinephrine in vitro on pd and scc though small was highly reproducible. The order of potency of the series of adrenergic agonists suggests the involvement of β-adrenergic receptors (Lands et al.,1967) on the fetal side of the pig placenta as does the antagonist action of the β-adrenergic blocker, propranolol. Presumably the β-adrenergic receptors are acting via cyclic AMP; in keeping with this hypothesis, dibutyryl cyclic AMP stimulates pd and scc in vitro (Boyd et al., 1984). The action of epinephrine on pd and scc appears to be the result of a stimulation of active Na^+ transport towards the fetal side of the pig placenta as in the absence of Na^+ from the bathing Ringer's solution, there was no significant response to the catecholamines. This effect is in agreement with the measured radioactive Na^+ fluxes across the pig placenta (Sibley et al., 1986) in which fetal side epinephrine stimulated net Na^+ flux towards the fetal side; the magnitude of the stimulated net Na^+ flux was not significantly different from the simultaneously measured scc. This active Na^+ transport presumably involves Na^+ - K^+ - ATPase in the membranes of the pig placenta, and as previously mentioned, this enzyme has been localized histochemically specifically to the fetal areolae within the pig placenta (Firth et al., 1986). Although all cells must have some Na^+ - K^+ - ATPase activity for regulation of cell volume, etc., it is only transporting epithelia which have a sufficiently high activity of this enzyme for it to be histochemically detectable (Firth, 1983). Further evidence for the involvement of Na^+ - K^+ - ATPase in the generation of pd and scc comes from the inhibitory action of fetal side ouabain (Sibley et al., 1985). However, in nonstimulated conditions there is evidence of electrogenic transport of other electrolytes as, for example, the removal of Na^+ from the bathing Ringer's solution to below 6 mM did not completely abolish the pd. This lack of Na^+ effect is another area which deserves further investigation.

The presence of β-adrenergic receptors has been reported in placentae from a number of different species including sheep, rat, rabbit, guinea pig, and human (Padbury et al., 1981; Moore and Whitsett, 1982); and catecholamines have been reported to be associated with a number of effects on placental metabolism (Ginsberg and Jeacock, 1968; Belleville et al., 1978; Caritis et al., 1983). However, to our knowledge there is no previous report of an effect on placental transfer of any molecule. The relevance of our in vitro observations on the effect of epinephrine to placental electrolyte transfer in vivo is of course open to question, although it is interesting to note that Macdonald et al. (1984) have reported concentrations of norepinephrine and epinephrine in fetal pig plasma after fetal hypoxia similar to our EC_{50} values.

In conclusions, the in vitro pig placenta generates an easily measurable pd which may be stimulated by fetal side catecholamines via β-adrenergic receptors and which may be related to the Na^+ - K^+ - ATPase in the areolar region. This system should make it possible to investigate directly the relationship between the transplacental pd and the in vivo maternofetal pd using chronically catheterized fetal and maternal vessels (Silver, 1980).

SUMMARY

When mounted in vitro, pieces of pig placenta from anesthetized animals generated a potential difference (pd) and short circuit current (scc) oriented with fetal side positive. After a short (20-30 min) equilibration period the mean values for these parameters were 6.3 ± 0.3 mV and 8.6 ± 0.6 μ A.cm^{-2}, respectively (mean $\pm$ se, n = 67). A series of catecholamines added to the fetal side caused an increase in pd in a dose-dependent fashion with the order of potency isoprenaline > norepinephrine > epinephrine > > phenylephrine. The maximum increase in pd caused by epinephrine was 1.9 ± 0.7 mV (n = 6). Prior addition of the β-antagonist, propranolol (10^{-7}M), to the fetal side caused a 30-fold increase in the EC_{50} for the effect of epinephrine on pd. Replacement of Na^+ by choline in the Ringer's solution bathing the tissue reduced pd but did not abolish it. In the absence of sodium, epinephrine had no effect on pd. These results suggest that the electrical activity of the pig placenta is not completely accountable for by the active transfer of Na^+ although it is concluded that in vitro β-adrenergic receptors may mediate stimulation of the electrical activity by increasing a sodium transfer towards the fetus.

ACKNOWLEDGEMENTS

We are grateful to the Wellcome Trust for grant support, to Dr. W.G. Bardsley and Dr. M. Hollingworth for their helpful advice, and to Mr. C. Tomlin and Mr. J. Cranley for invaluable assistance.

REFERENCES

Bazer, F.W., Goldstein, M.H., and Barron, D.H. (1981) Water and electrolyte transport by pig chorioallantois. In: *Fertilization and Embryonic Development In Vitro*, (eds.), L. Mastroianni and J.D. Biggers, Plenum Press, New York, pp. 299-321.

Belleville, F., Lasbennes, A., Nabet, P., and Paysant, P. (1978) HSC-HCG regulation in cultured placenta. *Acta Endocrinol.* 88, 169-181.

Binder, N.D., Faber, J.J., and Thornburg, K.L. (1978) The transplacental potential difference as distinguished from the maternal-fetal potential difference of the guinea pig. *J. Physiol.* 282, 561-570.

Boyd, R.D.H., Haworth, C., Stacey, T.E., and Ward, R.H.T. (1976) Permeability of the sheep placenta to unmetabolized polar non-electrolytes. *J. Physiol.* 256, 617-634.

Boyd, R.D.H., Glazier, J.D., Moore, W.M.O., Sibley, C.P., and Ward, B.S. (1984) Effect of dibutyryl cAMP (DBcAMP) on short circuit current (Iscc) and sodium flux across mini-pig placenta. *J. Physiol.* 349, 43P.

Brambel, C.E. (1933) Allantochorionic differentiations of the pig studied morphologically and histochemically. *Am. J. Anat.* 52, 397-459.

Caritis, S.N., Hirsch, R.P., and Zeleznik, A.J. (1983) Adrenergic stimulation of placental progesterone production. *J. Clin. Endocrinol. Metab.* 56, 969-972.

Crawford, J.D. and McCance, R.A. (1960) Sodium transport by the chorioallantoic membrane of the pig. *J. Physiol.* 151, 458-471.

Crombie, P.R. (1972) *The morphology and ultrastructure of the pig placenta throughout pregnancy*. Ph.D. Thesis, University of Cambridge.

Firth, J.A. (1983) Microscopic analysis of electrolyte secretion. In: *Progress in Anatomy*, (eds.), V. Navaratnam and R.J. Harrison, Cambridge University Press, Cambridge, pp. 33-55.

Firth, J.A., Sibley, C.P., and Ward, B.S. (1986) Histochemical localization of phosphatases in the pig placenta. II. Potassium-dependent and potassium-independent p-nitrophenyl phosphatases at high pH: relation to sodium-potassium-dependent adenosine triphosphatase. *Placenta* 7, 27-35.

Ginsberg, J. and Jeacock, M.K. (1968) Effect of epinephrine on placental carbohydrate metabolism. *Am. J. Obstet. Gynecol.* 100, 357-365.

Goldstein, M.H., Bazer, F.W., and Barron, D.H. (1980) Characterization of changes in volume, osmolarity and electrolyte composition of porcine fetal fluids during gestation. *Biol. Reprod.* 22, 1168-1180.

Hedley, R. and Bradbury, M.W.B. (1980) Transport of polar nonelectrolytes across intact and perfused guinea pig placenta. *Placenta* 1, 277-285.

Lands, A.M., Arnold, A., McAuliff, J.P., Luduena, F.P., and Brown, T.G. (1967) Differentiation of receptor systems activated by sympathomimetic amines. *Nature* 214, 597-598.

Macdonald, A.A., Colenbrander, B., Versteeg, D.H.G., Heilhecker, A., and Wensing, C.J.G. (1984) Catecholamines in fetal pig plasma and the response to acute hypoxia and chronic fetal decapitation. *Roux. Arch. Dev. Biol.* 193, 19-23.

Mellor, D.J. (1969) Potential differences between mother and fetus at different gestational ages in the rat, rabbit and guinea pig. *J. Physiol.* 204, 395-405.

Mellor, D.J., Cockburn, F., Lees, M.M., and Blagdon, A. (1969) Distribution of ions and electrical potential differences between mother and fetus in the human at term. *J. Obstet. Gynaecol. Br. Cmwlth.* 76, 993-998.

Michael, K., Ward, B.S., and Moore, W.M.O. (1985) In vitro permeability of the pig placenta in the last third of gestation. *Biol. Neonate* 47, 170-178.

Moore, J.J. and Whitsett, J.A. (1982) The β-adrenergic receptor in mammalian placenta: species differences and ontogeny. *Placenta* 3, 257-268.

Padbury, J.F., Hobel, C.J., Diakomanolis, E.S., Lam, R.W., and Fisher, D.A. (1981) Ontogenesis of beta-adrenergic receptors in the ovine placenta. *Am. J. Obstet. Gynecol.* 139, 459-464.

Sibley, C.P., Ward, B.S., Glazier, J.D., and Boyd, R.D.H. (1985) Placental electrolyte transfer with particular reference to sodium. In: *The Physiological Development of the Fetus and Newborn*, (eds.), C.T. Jones and P.W. Nathanielsz, Academic Press, London, pp. 520-526.

Sibley, C.P., Ward, B.S., Glazier, J.D., Moore, W.M.O., and Boyd, R.D.H. (1986) Electrical activity and sodium transfer across the in vitro pig placenta. *Am. J. Physiol.* 250, R474-R484.

Silver, M. (1980) Intravascular catheterization and other chronic fetal preparations in the mare and sow. In: *Animal Models in Fetal Medicine (1)*, (ed.), P.W. Nathanielsz, Elsevier/North Holland Biomedical Press, Amsterdam, pp. 107-132.

Thornburg, K.L., Binder, N.D., and Faber, J.J. (1979) Distribution of ionic sulfate, lithium and bromide across the sheep placenta. *Am. J. Physiol.* 236, C58-C65.

Weedon, A.P., Stacey, T.E., Canning, J.F., Ward, R.H.T., and Boyd, R.D.H. (1980) Materno-fetal electrical potential difference in conscious sheep: effect of fetal death or acidosis. *Am. J. Obstet. Gynecol.* 138, 422-428.

Trophoblast Research 2:329-342, 1987

HUMAN PLACENTAL Ca^{2+}-ATPases: TARGETS FOR ORGANOCHLORINE PESTICIDES?

Arun P. Kulkarni, Kimberly A. Treinen and Lorelle L. Bestervelt

Toxicology Program
Department of Environmental and Industrial Health
School of Public Health
The University of Michigan
Ann Arbor, Michigan 48109-2029 USA

INTRODUCTION

Millions of tons of p,p'-DDT and other organochlorine (OC) pesticides have been used worldwide in the past thirty years. Although the use of certain OC pesticides has been discontinued in several countries including the United States, they are still the pesticides of choice in developing countries. It is noteworthy that the recognized deleterious effects of OC pesticides on human health will NOT disappear due to this ban, since, for example, DDT residues in the ecosystem are expected to linger well beyond the beginning of the next century (Matsumura, 1975). OC pesticides are highly lipophilic compounds and are avidly stored in internal tissues, notably the fat depots. Several epidemiological surveys have documented a positive correlation between the amount of OC residues in maternal or cord blood, placenta, and fetal tissues and an increased incidence of spontaneous abortion (O'Leary et al., 1970a, Saxena et al., 1980), missed abortion (Bercovici et al., 1983), fetal prematurity (O'Leary et al., 1970b, 1972; D'Ecrole et al., 1976; Saxena et al., 1980) induction of early labor (Polishuk et al., 1977; Saxena et al., 1980, 1981), premature delivery, (Wassermann et al., 1982; Siddiqui and Saxena, 1985) or stillbirths (Curley et al., 1969; Saxena et al., 1983; Siddiqui and Saxena, 1985). Many forms of covert toxicity of OC pesticides which often times are expressed later in life (morphological, physiological, or neuro-behavioral) are suspected in surviving babies but this documentation is lacking at present. However, a significant increase in neonatal mortality has been well documented in rats (Fitzhugh and Nelson, 1947; Clement and Okey, 1974). Persistent stimulatory and behavioral effects are also known to occur in rats following prenatal exposure to dieldrin (Olsen et al., 1980). Although several hypotheses have been put forth, the underlying biochemical mechanism(s) responsible for these undesirable reproductive effects has not yet been established.

In view of its recognized regulatory role in numerous physiological and biochemical processes, calcium is considered an important constituent of mammalian

Received: 10 October 1985; Accepted: 20 June 1986

cells (Campbell, 1983). Therefore, maintenance of normal calcium homeostasis in placental cells is an absolute requirement for cell viability. Any subtle change in calcium homeostasis is expected to result in a variety of functional impairments. Similarly, calcium is integral for fetal development and survival as it serves many essential functions (Varner et al., 1983). It has been repeatedly shown that maternal hypercalcemia does not result in a hypercalcemic fetus possibly due to regulation of calcium transport at the placental level. However, maternal hypocalcemia does result in a hypocalcemic fetus. It has also been demonstrated that premature infants have a high incidence of hypocalcemia (Bruck and Weintraub, 1955; Gittleman et al., 1956; Rosli and Fanconi, 1973).

Calcium accretion by the fetus depends upon the placental supply. Calcium has been shown to be transplacentally transported against a concentration gradient, where it is in higher concentrations in fetal blood than that of the maternal circulation (Delivoria-Papadopoulos et al., 1967). The mechanism responsible for this transport is as yet unknown, but it has been suggested that intracellular and transcellular calcium levels may be regulated by a calcium-stimulated adenosine triphosphatase (ATPase) (Whitsett and Tsang, 1980). This calcium "pump" is localized in placental mitochondria, microsomes, and brush border membranes and may be involved in the maintenance of low intracellular free calcium concentration in syncytial cells despite the millimolar extracellular calcium concentration. Of the different subcellular membranes, brush border membranes exhibit the highest specific activity of the enzyme (Whitsett and Lessard, 1978; Whitsett and Tsang, 1980). Although a calcium-stimulated ATPase has been identified in human (Miller and Berndt, 1973; Whitsett and Lessard, 1978; Whitsett and Tsang, 1980) and guinea pig (Shami and Radde, 1971) placenta, this enzyme was studied using millimolar concentrations of calcium. Thus, the reported activities resemble the low affinity, non-specific (Mg^{2+} or Ca^{2+}) ATPases found in most tissues. In this paper, we discuss the presence and some properties of high affinity calcium-stimulated ATPase in brush border membranes of human term placenta.

Numerous studies have documented the susceptibility of different ATPases to OC pesticides in various animal tissues (Cutkomp et al., 1982; Doherty, 1984). These reports indicate the degree of inhibition depends upon the type of ATPase, tissue, animal species, and the chemical involved. Inhibition of Ca^{2+}-ATPase has been proposed as a mechanism of OC pesticide neurotoxicity (Cutkomp et al., 1982; Doherty, 1984) and the same may be applicable to other tissues. However, work on avian reproductive tissues has shown the inhibitory effects upon this enzyme to be unpredictable since striking species-specific differences exist. Thus, the Ca^{2+}-ATPase in the egg shell gland preparations of Pekin duck was found to be inhibited by p,p'-DDE while that of chicken was not (Miller et al., 1976). These results raise a question regarding the response of human placental Ca^{2+}-ATPase to OC pesticides. If OC pesticides can disturb the cellular calcium homeostasis, then the accompanying various biochemical events are expected to cause placental dysfunction that would affect the fetal viability unfavorably. Present study deals with certain facets of this hypothesis. From the data presented in this paper, it is clear that these chemicals, at biologically attainable levels, can indeed significantly inhibit placental brush border Ca^{2+}-ATPase.

MATERIALS AND METHODS

Materials

ATP (disodium salt), EGTA (Ethyleneglycol-bis-(β-amino ethyl ether)-N,N,N',N'- tetracetic acid), CDTA (trans-1,2-diaminocyclohexane-N,N,N'N'-tetraacetic acid), and sodium orthovanadate were purchased from Sigma Chemical Co. (St. Louis, MO). Trifluoperazine was a gift from Smith, Kline and French. Purified calmodulin was kindly provided by Dr. M. Welch, University of Michigan. The following OC pesticides ($\geq$95% purity) were obtained commercially: p,p'-DDT [2,2-bis(p-chlorophenyl)-1,1,1-trichloroethane] p,p'-DDD [2,2-bis(p-chlorophenyl)-1,1-dichloroethane]; p,p'-DDE [2,2-bis(p-chlorophenyl)-1,1-dicholoroethylene]; p,p'-DDA [2,2-bis(p- chlorophenyl)acetic acid]; p,p'-DDOH [2,2-bis(p-chlorophenyl)-ethanol]; methoxychlor; aldrin; dieldrin; mirex; chlordane; heptachlor; endrin; kepone and lindane.

Preparation of Brush Border Membranes

The procedure followed for isolation of human placental brush border membrane was essentially that outlined by Booth et al. (1980). Upon delivery, full term placentae were immediately placed on ice and processed within 1 hr. One hundred grams of lobular villous tissue were removed and washed several times in 50 mM CaCl$_2$ to remove as much blood contamination as possible. The tissue was then manually teased and stirred in 0.15 mM NaCl with 1 mM EGTA for 1 hr at 5°C. After filtration through gauze, the preparation was centrifuged at 800 x g for 10 min and the resulting supernatant centrifuged at 10,000 x g for 1 hr to obtain microvillous brush border membranes (microvilli). Suspensions of microvilli (approximately 1 mg/ml) were stored frozen (-20°C) for up to 2 wks in buffer (10 mM mannitol, 2 mM Tris- HCl, 1 mM EGTA, pH 7.1) with no significant decrease in enzyme activity. The microvilli (1 ml) were then washed in 30 ml storage buffer containing no EGTA before use to remove EGTA from the enzyme preparation. Protein determination was by the Lowry method using bovine serum albumin as the standard (Lowry et al., 1951). High alkaline phosphatase activity (Bowers and McComb, 1966) and electron microscopy indicated the preparation to primarily consist of human placental microvillous brush border membranes.

Calcium-Stimulated ATPase Assay

The calcium-stimulated ATPase activity was measured by monitoring the release of inorganic phosphorous (Pi) according to the method of Ames (1966). The standard assay medium employed to measure high affinity Ca^{2+}-ATPase activity contained 30 mM Tris-HCl, pH 7.4, 30-100 µg brush border membrane protein, 200 µM EGTA, and 26- 153 µM total CaCl$_2$ (0.01 µM to 0.20 µM free calcium) in a final volume of 1 ml. After preincubation for 10 min at 37°C, the reaction was initiated by the addition of 2 mM disodium-ATP, and terminated at 30 min by the addition of 3 ml sodium dodecyl sulfate (2%). Calcium-stimulated ATPase activity was determined by subtracting values obtained in the absence of calcium from those in its presence. The low affinity Ca^{2+}-ATPase activity was measured similarly in the absence of EGTA using reaction medium containing 1.0 mM CaCl$_2$ and Tris buffer, pH 8.2. The activity is expressed as µmoles or nmoles Pi released/min/mg protein. Under these conditions, the enzyme activity measured was linear with respect to the time and protein concentrations used. In the inhibition studies, the assays were performed in the presence of indicated concentration of test chemical in 10 μl of acetone. Incubation media containing acetone served as controls.

Calculation of Free Ca^{2+} Concentrations

The desired free Ca^{2+} concentrations were obtained by EGTA buffering and were calculated using a computer program kindly provided by Dr. Luciano Soldati, Swiss Federal Institute of Technology (ETH), Zurich, Switzerland. This program was adapted from the program of Fabiato and Fabiato (1978) and accounts for all complexes involving magnesium, calcium, EGTA, and ATP.

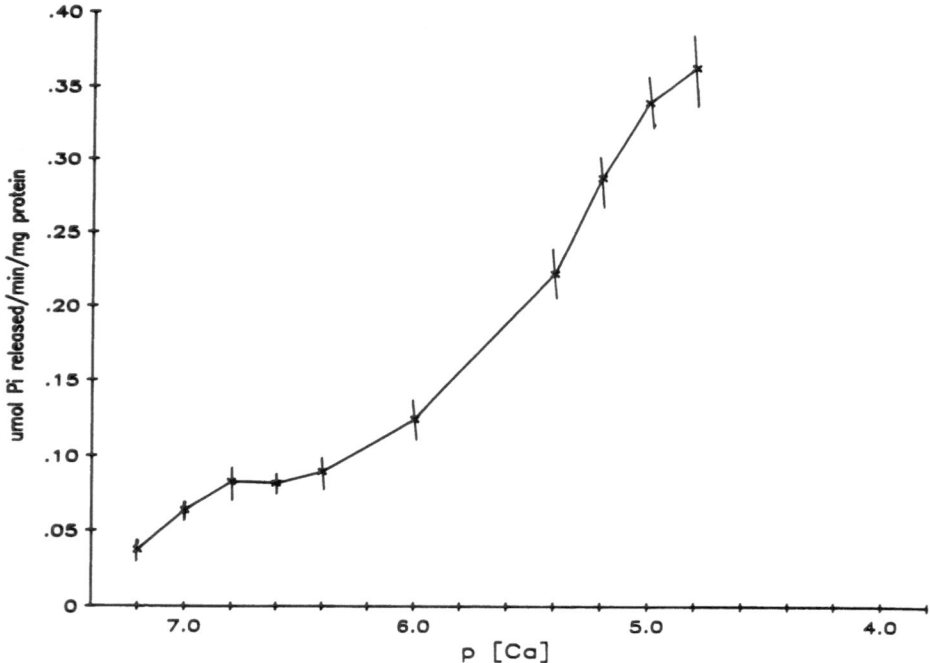

Figure 1. Dependence of calcium-stimulated ATPase activity on free calcium concentration. The log of the free Ca^{2+} concentration is plotted on the abscissa and calcium-stimulated ATPase activity is shown on the ordinate. Experimental conditions were as described in "Experimental Procedures" with free Ca^{2+} concentrations ranging between 66.0 nM to 15.0 μM. Each point represents the mean ± S.E. of three separate membrane preparations.

RESULTS

Identification of Ca^{2+}-ATPase Activities

The dependence of ATPase activity on calcium concentration revealed 2 separate ATPase activities (Figure 1). A high affinity component was active below 0.38 µM free Ca^{2+}, whereas the low affinity enzyme became evident at calcium concentrations greater than 1 µM free Ca^{2+}. This low affinity ATPase has an apparent $K_{0.5}$ for free calcium of 99.7 $\pm$ 22.1 µM and a Vmax of 1.54 $\pm$ 0.17 µmol/min/mg protein at its pH optimum of 8.2, as determined by Lineweaver-Burk analysis of the kinetic data (not shown). Due to this low calcium affinity and a relatively high turnover rate, it appears to represent a non-specific divalent cation ATPase reported in several tissues (Pershadsingh and McDonald, 1980; Verma and Penniston, 1981; Debetto and Cantley, 1984), which often times makes the characterization of the high affinity Ca^{2+}-ATPases more difficult. The high affinity Ca^{2+}-ATPase from human placenta was found to have an apparent $K_{0.5}$ for free calcium of 18.3 $\pm$ 3.7 nM and a Vmax of 233.0 $\pm$ 30.0 nmol/min/mg protein at pH 7.4. Since this enzyme is active at submicromolar calcium concentrations, it is likely to play a role in calcium regulation of the human placenta.

Table 1

Effect of calcium/EGTA and calcium/CDTA buffer systems on high affinity Ca^{2+}-ATPase activity

Ca^{2+}-ATPase activity was assayed in standard incubation medium containing 500 µM CDTA alone, 500 µM CDTA with 487 µM CaCl$_2$, 200 µM EGTA alone, or 200 µM EGTA with 153 µM CaCl$_2$ to maintain free Ca^{2+} concentration at 0 or .20 µM, respectively. Data are expressed as specific activity $\pm$ S.E. of three membrane preparations.

Conditions	Specific Activity*
EGTA alone	62 $\pm$ 6
EGTA + .2 µM free Ca^{2+}	206 $\pm$ 33
CDTA alone	30 $\pm$ 5
CDTA + .2 µM free Ca^{2+}	98 $\pm$ 24

*   nmoles Pi released/min/mg protein. Assays were performed as described in the Materials and Methods.

With permission from Treinen and Kulkarni (1986).

Characteristics of High Affinity Calcium ATPase

Magnesium Requirement:

The high affinity Ca^{2+}-ATPase activity was routinely assayed without exogenously added magnesium. To determine if endogenous magnesium (estimated to be 6-10 µM by atomic absorption spectroscopy) is necessary for the activity of this enzyme, a Ca^{2+}- CDTA buffering system was used instead of Ca^{2+}-EGTA. EGTA is quite specific for Ca^{2+} binding while CDTA binds both Mg^{2+} and Ca^{2+} with equal affinity. Therefore, when the desired calcium concentration of 0.20 µM was maintained with CDTA in these experiments, the endogenous magnesium present was chelated and the enzyme activity was compared to that of the Ca^{2+}-EGTA system. Under these conditions, placental brush border Ca^{2+}-ATPase activity was decreased in the Ca^{2+}-CDTA system, but was not abolished (Table 1). Thus, it appears that although this enzyme may be stimulated by micromolar concentrations of magnesium, the presence of this cation is not essential to observe basal enzyme activity.

Effect of Calmodulin on the Ca^{2+}-ATPase

To date, several plasma membranes Ca^{2+}-ATPases have been shown to be stimulated by the calcium binding protein calmodulin, although liver (Lotersztajn et al., 1981), corpus luteum (Verma and Penniston, 1981), neutrophil (Ochs and Reed, 1984), and murine erythroleukemia cell (Debetto and Cantley, 1984) plasma membrane enzymes are exceptions. It has been suggested, however, the endogenous calmodulin may not be totally removed by repeated EGTA washings of these tissues, thus preventing stimulation of activity by exogenous calmodulin addition. With respect to placental high affinity Ca^{2+}-ATPase, no enzyme stimulation was seen upon the addition of 5 or 10 µg calmodulin at varying free Ca^{2+} concentrations. Since the brush border membrane preparations contained approximately 1.0 µg calmodulin/mg protein after storage in 1 mM EGTA buffer, it is likely that this amount of endogenous calmodulin may be sufficient to keep the enzyme in a maximally stimulated state, thus preventing stimulation by exogenous calmodulin.

Effect of Trifluoperazine on the Ca^{2+}-ATPase

Trifluoperazine is a phenothiazine drug that is known to bind calmodulin in a calcium-dependent manner (Levin and Weiss, 1977), and has been used to determine calmodulin regulation of red cell Ca^{2+}-ATPase (Gietzen et al., 1982). The placental enzyme was found to be quite sensitive to trifluoperazine, with almost total enzyme inhibtion occurring at 250 µM. This suggests the possibility of the compound's interaction with the tightly bound endogenous calmodulin present in the membrane preparations as the mechanism of enzyme inhibition. However, trifluoperazine has been shown to inhibit Ca^{2+}-ATPases through calmodulin-independent mechanisms (Caroni and Carafoli, 1981; Iwasa et al., 1983; Debetto and Cantley, 1984). Therefore, though it remains a possibility that calmodulin may play a regulatory role with respect to this enzyme, inhibition by trifluoperazine alone does not appear to be a sufficient criterion to establish the issue of calmodulin-dependency of placental Ca^{2+}-ATPase. Further studies with purified enzyme are needed to resolve this issue.

Table 2

In vitro inhibition of human placental low affinity Ca^{2+}-ATPase by DDT analogs

Inhibitor (10 µM)	Specific Activity µmoles Pi/min/mg protein	Per cent Inhibition
None (control)	1.18 ± 0.25	–
p,p'-DDT	0.51 ± 0.13*	57.8 ± 2.0
p,p'-DDD	0.46 ± 0.15*	61.5 ± 5.5
p,p'-DDE	0.41 ± 0.12*	67.5 ± 3.2
p,p'-DDA	1.02 ± 0.23	14.0 ± 3.6
p,p'-DDOH	1.00 ± 0.27	19.0 ± 5.3
Methoxychlor	0.69 ± 0.21*	43.8 ± 4.9

The specific activity values represent the mean ± S.E. (n = 4). Control incubations were performed in the presence of 10 µl acetone. See Materials and Methods for further details.

*   Significant at p ≤ 0.01

Effect of Vanadate on the Ca^{2+}-ATPase

Micromolar concentrations of vanadate have been shown to inhibit Ca^{2+}-ATPases from different tissues (Caroni and Carafoli, 1981; Niggli et al., 1981; Debetto and Cantley, 1984). Again, this inhibition appears to be tissue specific. With respect to placental Ca^{2+}-ATPase, no inhibition of enzyme activity was seen with up to 100 µM vanadate at pH 7.4 suggesting that the placental Ca^{2+}-ATPase most resembles that of liver (Iwasa et al., 1983).

Ca^{2+}-ATPases as Targets for Organochlorine (OC) Pesticides

Several OC pesticides were tested at 10 µM concentration as potential inhibitors of low and high affinity Ca^{2+}-ATPase activities of human placenta. The data obtained for the low affinity enzyme are given in Table 2. The highly lipophilic compounds p,p'-DDT, and p,p'-DDE caused a marked inhibition of enzyme activity while the water soluble metabolites of p,p'-DDT such as p,p'-DDA and p,p'-DDOH caused little or no inhibition. The potency of methoxychlor, a relatively biodegradable analog of p,p'-DDT, was found to be intermediate. An apparent I$_{50}$ value of 5.68 ± 1.27 x 10^{-6} M (n = 4) for p,p'-DDT was noted. Under the experimental conditions employed, the inhibitory effect of this compound was not of a competitive nature with respect to ATP.

Table 3

In vitro inhibition of high affinity calcium-stimulated ATPase from
human term placenta by organochlorine (OC) pesticides

OC Inhibitor (10⁻⁵M)	Relative Activity Mean ± S.E.
Control	100
p,p'-DDT	49 ± 2.9*
p,p'-DDD	56 ± 7.7*
p,p'-DDE	42 ± 2.0*
p,p'-DDA	81 ± 2.5
p,p'-DDOH	86 ± 2.5
Methoxychlor	66 ± 3.5*
Aldrin	60 ± 2.6*
Dieldrin	68 ± 6.1*
Mirex	65 ± 4.7*
Chlordane	67 ± 3.4*
Heptachlor	67 ± 4.8*
Endrin	62 ± 10.3*
Kepone	59 ± 7.2*
Lindane	61 ± 9.8*

The specific activities in control experiments ranged from 105 to 332 nmoles
Pi released /min/mg protein.

Assays were performed as described in Methods.

*   Significant at $p \leq 0.05$ (n = 5).

OC insecticides were also tested under standard assay conditions to evaluate
their ability to inhibit placental high affinity Ca^{2+}-ATPase activity (Table 3). At
equimolar concentrations, the parent compound p,p'-DDT and its metabolite p,p'-
DDE, which is stored in human fat, were the most potent inhibitors. p,p'-DDD was
slightly less effective while the urinary metabolites of p,p'-DDT namely p,p'-DDA and
p,p'- DDOH were without any effect, again possibly due to their hydrophilic nature.

All other OC pesticides tested caused statistically significant inhibition. The degree of inhibition varied between 30% and 40% depending upon the insecticide employed.

The inhibition data gathered thus far suggest that the magnitude of inhibition depends upon the chemical used. Of the OC pesticides examined, p,p'-DDT and its major lipophilic metabolite p,p'-DDE were found to be the most potent inhibitors of the enzyme, whereas the water soluble metabolites were essentially ineffective. All the other OC pesticides caused less but statistically significant inhibition. These results suggest that the OC pesticides tested in this study inhibit the Ca^{2+}-ATPases of the human placenta in vitro. Inhibition of these enzymes may interfere with placental calcium homeostasis in vivo and thereby pose a potential danger to the human fetus.

DISCUSSION

The biphasic activation of the ATPase with respect to the free calcium concentration is a common feature of several other Ca^{2+}-ATPase preparations implicated in cellular Ca^{2+} homeostasis (Pershadsingh and McDonald, 1980; Verma and Penniston, 1981; Debetto and Cantley, 1984). To date, only the low affinity calcium-stimulated ATPase has been described in human placenta (Miller and Berndt, 1973; Whitsett and Tsang, 1980). Whether low and high affinity Ca^{2+}-ATPases are two separate proteins, or further activation of a single ATPase is not known. It is of interest, however, that the properties described for this high affinity calcium-stimulated ATPase are similar to those of other Ca^{2+}-ATPases known to be involved in the active transport of calcium, particularly that of liver (Carafoli, 1984).

Briefly, the enzyme described here has a high affinity for calcium (apparent $K_{0.5}$ for calcium equal to 18.3 ± 3.7 nM), is unaffected by vanadate and is inhibited by low concentrations of trifluorperazine. While a direct stimulation of this enzyme by exogenous calmodulin could not be demonstrated, it is possible that sufficient endogenous calmodulin remains after membrane preparation to keep the enzyme maximally stimulated. Uncompetitive inhibition of enzyme activity by trifluoperazine suggests a role for calmodulin with respect to calcium regulation of this enzyme. However, since trifluoperazine has been shown to directly inhibit Ca^{2+}-ATPases in a manner independent of calmodulin inhibition, unequivocal affirmation of calmodulin in regulation of this enzyme was not possible under the experimental conditions employed.

We feel that this high affinity Ca^{2+}-ATPase may be the enzyme primarily responsible for regulating intracellular calcium concentration by pumping Ca^{2+} ions out of the placental cells (Whitsett and Tsang, 1980). These authors reported the $K_{0.5}$ for free calcium of ^{45}Ca uptake in a similar membrane preparation to be 50 ± 6 nM, with inhibition of this process occurring at free calcium concentrations greater than 1 mM. These findings appear to rule out the low affinity Ca^{2+}-ATPase as the enzyme responsible for calcium uptake in these membranes. The high affinity Ca^{2+}-ATPase described here, however, has an affinity for free calcium in the range used to determine placental brush border membrane calcium uptake. Experiments are in progress to determine if this suggested link between high affinity Ca^{2+}-ATPase and calcium uptake in the human placenta exists.

At present, very little is known about the underlying biochemical mechanisms responsible for transplacental toxicity of OC insecticides in humans. Many acknowledge that no single biochemical event can fully explain the observed reproductive failures. Rather it is believed that cumulative disturbances in several mechanisms which are essential for the maintenance of normal pregnancy renders the fetal environment unfit to support normal growth and development of the conceptus. Based on animal data, the current major hypotheses explaining OC fetotoxicity in humans include the induction of maternal liver enzymes, estrogenicity (Kupfer, 1975), and induction of prostaglandin biosynthesis (Rogers et al., 1976). However, due to the lack of human data, validity of these postulates remains debatable.

In this paper, we have presented evidence showing a significant inhibition of human placental low and high affinity Ca^{2+}-ATPases by micromolar concentrations of OC insecticides. In a recent study, Siddiqui and Saxena (1985) examined 1 g samples of human placenta for OC pesticides content and reported the presence of 6 to 12 µg of p,p'-DDT, 3 to 5 µg of p,p'-DDD and 7 to 15 µg of p,p'-DDE. Many investigators have reported similar data for these and other OC insecticides (Saxena et al., 1980, 1981, 1983; Eckenhausen et al., 1981; Bercovici et al., 1983; and others). In general, a close association of high OC pesticide content with unfavorable pregnancy outcome was noted in each study. In view of these reports and the fact that significant inhibition of human placental Ca^{2+}-ATPases can be observed at biologically attainable levels of OC insecticides, we propose that disturbed calcium homeostasis resulting in compromised placental function may be involved in the OC fetotoxicity in humans. Further studies are needed to demonstrate a direct correlation between the OC content of the placenta, a decline in Ca^{2+}-ATPase activities, and unfavorable pregnancy outcomes to establish the proposed hypothesis.

SUMMARY

ATPase activity which is stimulated by nanomolar concentrations of Ca^{2+} was identified in human placental microvillous brush border membranes. The high affinity enzyme has an apparent $K_{0.5}$ for free Ca^{2+} of 18.3 $\pm$ 3.7 nM and a Vmax of 233.0 $\pm$ 30.0 nmol/min/mg protein. Studies using CDTA show this enzyme to require submicromolar concentrations of Mg^{2+} for maximal activity, but appears to have a low basal activity in the absence of this cation. The high affinity Ca^{2+}-ATPase was unaffected by up to 100 µM vanadate, but was extremely sensitive to trifluoperazine inhibition ($I_{50} < 50$ µM). It was not found to be stimulated by the addition of up to 10 µg calmodulin, but this lack of effect may be related to the endogenous calmodulin content of the membrane preparation. A low affinity, non-specific divalent cation ATPase was also identified in this membrane preparation. In contrast to the high affinity enzyme, it has an apparent $K_{0.5}$ for calcium of 99.7 $\pm$ 22.1 µM, and a Vmax of 1.54 $\pm$ 0.17 µmol/min/mg protein.

The characteristics of high affinity placental Ca^{2+}-ATPase are similar to those of other Ca^{2+}-ATPases known to transport and regulate intracellular calcium concentrations in other tissues. By analogy, we propose that the high affinity Ca^{2+}-ATPase described here plays an important role in cellular calcium homeostasis in the human placenta.

Both low and high affinity Ca^{2+}-ATPases activities of human placental brush border membranes were found to be inhibited by micromolar concentrations of OC insecticides. These in vitro data suggest that disturbed placental calcium homeostasis in vivo may be one of the mechanisms responsible for OC insecticide fetotoxicity in humans.

ACKNOWLEDGEMENTS

We wish to express our gratitude to the personnel of St. Joseph Mercy Hospital, Ann Arbor, Michigan for their cooperation in providing placentae. This work was supported, in part by grants from the Horace Rackham School of Graduate Studies, The University of Michigan, and the Charles A. Lindbergh fund.

REFERENCES

Ames, B. (1966) Assay of inorganic phosphate, total phosphate and phosphatases. *Methods Enzymol.* 8, 115-118.

Bercovici, B., Wassermann, M., Cucos, S., Ron, M., Wassermann, D., and Pines, A. (1983) Serum levels of polycholorinated biphenyls and some organochlorine insecticides in women with recent and former missed abortions. *Environ. Res.* 30, 169-174.

Booth, A.G., Olaniyan, R.O., and Vanderpuye, O.A. (1980) An improved method for the preparation of human placental syncytiotrophoblast microvilli. *Placenta* 1, 327-336.

Bowers, G.N. and McComb, R.B. (1966) A continuous spectrophotometric method for measuring the activity of serum alkaline phosphatase. *Clin. Chem.* 12, 70-89.

Bruck, E. and Weintraub, D.H. (1955) Serum calcium and phosphorus in premature and full term infants. *Amer. J. Dis. Child.* 90, 653-668.

Campbell, A.K. (ed.), (1983) In: *Intracellular Calcium: Its Universal Role as Regulator*, John Wiley and Sons Ltd., New York, p. 9.

Carafoli, E. (1984) Calmodulin-sensitive calcium-pumping ATPase of plasma membranes: isolation, reconstitution, and regulation. *Fed. Proc.* 43, 3005-3010.

Caroni, P. and Carafoli, E. (1981) The Ca^{2+}-pumping ATPase of heart sarcolemma: characterization, calmodulin dependence, and partial purification. *J. Biol. Chem.* 256, 3263-3270.

Clement, J.G. and Okey, A.B. (1974) Reproduction in female rats born to DDT-treated parents. *Bull. Environ. Contam. Toxicol.* 12, 373-377.

Curley, A., Copeland, M.F., and Kimbrough, R.D. (1969) Chlorinated hydrocarbons in organs of stillborns and blood of newborn babies. *Arch. Environ. Hlth.* 19, 628-632.

340   **Kulkarni et al.**

Cutkomp, L.K., Koch, R.B., and Desaiah, D. (1982) Inhibition of ATPases by chlorinated hydrocarbons. In: *Insecticide Mode of Action*, (ed.), J.L. Coats, Academic Press, New York, pp. 45-69.

Debetto, P. and Cantley, L. (1984) Characterization of a Ca^{2+}-stimulated Mg^{2+}-dependent adenosine triphosphatase in friend murine erythroleukemia cell plasma membranes. *J. Biol. Chem.* 259, 13824-13831.

D'Ecrole, A.J., Arthur, R.D., Cain, J.D., and Barrentine, B.F. (1976) Insecticide exposure of mothers and newborns in a rural agricultural area. *Ped.* 57, 869-874.

Delivoria-Papadopoulos, M., Battaglia, F.C., Bruns, P.D., and Meschia, G. (1967) Total, protein-bound and ultrafilterable calcium in maternal and fetal plasmas. *Amer. J. Physiol.* 213, 363-366.

Doherty, J.D. (1984) Insecticides affecting ion transport. In: *Differential Toxicities of Insecticides and Halogenated Aromatics*, (ed.), F. Matsumura, Pergamon Press, New York, pp. 423-452.

Eckenhausen, F.W., Bennett, D., Beynon, K.I., and Elgar, K.E. (1981) Organochlorine pesticide concentrations in perinatal samples from mothers and babies. *Arch. Environ. Hlth.* 36, 81-92.

Fabiato, A. and Fabiato, F. (1979) Calculator programs for computing the composition of the solutions containing multiple metals and ligands used for experiments in skinned muscle cells. *J. Physiol. Paris* 75, 463-505.

Fitzhugh O.G. and Nelson, A.A. (1947) The chronic oral toxicity of DDT (2,2-bis(p-chlorophenyl)-1,1,1-trichloroethane. *J. Pharmacol. Exp. Ther.* 89, 1830-1836.

Gietzen, K., Sadorf, I., and Bader, H. (1982) A model for the regulation of the calmodulin-dependent enzymes: erythrocyte Ca^{2+}-transport ATPase and brain phosphodiesterase by activators and inhibitors. *Biochem. J.* 207, 541-548.

Gittleman, I.F., Pincus, J.B., Schmerzler, E., and Saito, M. (1956) Hypocalcemia occurring on the first day of life in mature and premature infants. *Ped.* 18, 721-728.

Iwasa, T., Iwasa, Y., and Krishnaraj, R. (1983) Comparison of high affinity Ca^{2+}-ATPase and low affinity Ca^{2+}-ATPase in rat liver plasma membranes. *Arch. Int. Pharmacodyn.* 264, 40-58.

Kupfer, D. (1975) The effects of pesticides and related compounds on steroid metabolism and function. *CRC Crit. Rev. Toxicol.* 4, 83-124.

Levin, R.M. and Weiss, B. (1977) Binding of trifluoperazine to the calcium-dependent activator of cyclic nucleotide phosphodiesterase. *Molec. Pharmacol.* 13, 690-697.

Lotersztajn, S., Hanoune, J., and Pecker, F. (1981) High affinity calcium-stimulated magnesium-dependent ATPase in rat liver plasma membranes. *J. Biol. Chem.* 256, 11209-11215.

Lowry, O.H., Rosebrough, N.J., Farr, A.L., and Randall, R.J. (1951) Protein measurement with the folin phenol reagent. *J. Biol. Chem.* 193, 265-275.

Matsumura, F. (1975) In: *Toxicology of Pesticides*, p. 460, Plenum Press, New York.

Miller, D.S., Kinter, W.B., and Peakall, D.B. (1976) Enzymatic basis for DDE-induced eggshell thinning in a sensitive bird. *Nature* 259, 122-124.

Miller, R.K. and Berndt, W.O. (1973) Evidence for Mg^{2+}-dependent Na$^+$+K$^+$ activated ATPase and Ca^{2+}-ATPase in the human placenta. *Proc. Soc. Exp. Biol. Med.* 143, 118-122.

Niggli, V., Adunyah, E.S., Penniston, J.T., and Carafoli, E. (1981) Purified (Ca^{2+}-Mg^{2+})-ATPase of the erythrocyte membrane. *J. Biol. Chem.* 256, 395-401.

Ochs, D. and Reed, P.W. (1984) Ca^{2+}-stimulated, Mg^{2+}-dependent ATPase activity in neutrophil plasma membrane vesicles. *J. Biol. Chem.* 259, 102-106.

O'Leary, J.A., Davies, J.E., and Feldman, M. (1970a) Spontaneous abortion and human pesticide residues of DDT and DDE. *Am. J. Obstet. Gynecol.* 108, 1291-1292.

O'Leary, J.A., Davies, J.E., Edmundson, W.R., and Feldman, M. (1970b) Correlation of prematurity and DDE levels in fetal whole blood. *Am. J. Obstet. Gynecol.* 106, 939.

O'Leary, J.A., Davies, D.E., Edmundson, W.F., and Feldman, M. (1972) Correlation of prematurity and DDE levels in fetal whole blood. In: *Epidemiology of DDT*, Futura Publishing Company, New York, pp. 55-56.

Olson, K.L., Boush, G.M., and Matsumura, F. (1980) Pre- and post-natal exposure to dieldrin: Persistent stimulatory and behavorial effects. *Pest. Biochem. Physiol.* 13, 20-33.

Pershadsingh, H.A. and McDonald, J.M. (1980) A high affinity calcium-stimulated magnesium-dependent adenosine triphosphatase in rat adipocyte plasma membranes. *J. Biol. Chem.* 255, 4087-4093.

Polishuk, Z.W., Wassermann, D., Wassermann, M., Cucos, S., and Ron, M. (1977) Organochlorine compounds in mother and fetus during labor. *Environ. Res.* 13, 278-284.

Rogers, A., Mellors, A., and Safe, S. (1976) Lysosomal membrane labilization by DDT, DDE and polychlorinated biphenyls (PCB). *Res. Commun. Chem. Path. Pharmacol.* 13, 341-344.

Rosli, A. and Fanconi, A. (1973) Neonatal hypocalcemia. Early type in low birth weight newborns. *Helv. Paediatr. Acta* 28, 443-457.

Saxena, M.C., Siddiqui, M.K.J. Bhargava, A.K., Seth, T.D., Krishnamurti, C.R., and Kutty, D. (1980) Role of chlorinated hydrocarbon pesticides in abortions and premature labour. *Toxicol.* 17, 323-331.

Saxena, M.C., Siddiqui, M.K.J. Bhargava, A.K., Murti, C.R.K., and Kutty, D. (1981) Placental transfer of pesticides in humans. *Arch. Toxicol.* 48, 127-130.

Saxena, M.C., Siddiqui, M.K.J., Agarwal, V., and Kutty, D. (1983) A comparison of organochlorine insecticides contents in specimens of maternal blood, placenta and umbilical cord blood from stillborn and live born cases. *J. Toxicol. Environ. Hlth.* 11, 71-78.

Shami, Y. and Radde, I.C. (1971) Calcium-stimulated ATPase of guinea pig placenta. *Biochem. Biophys. Acta* 249, 345-352.

Siddiqui, M.K.J. and Saxena, M.C. (1985) Placenta and milk as excretory routes of lipophilic pesticides in women. *Human Toxicol.* 4, 249-254.

Treinen, K.A. and Kulkarni, A.P. (1986) High affinity calcium-stimulated ATPase in brush border membranes of the human term placenta. *Placenta*, in press.

Varner, M.W., Cruikshank, D.P., and Pitkin, R.M. (1983) Calcium metabolism in the hypertensive mothers, fetus and newborn infant. *Am. J. Obstet. Gynecol.* 147, 762-765.

Verman, A.K. and Penniston, J.T. (1981) A high affinity Ca^{2+}-stimulated and Mg^{2+}-dependent ATPase in rat corpus luteum plasma membrane fractions. *J. Biol. Chem.* 256, 1269-1275.

Wassermann, M., Ron, M., Bercovici, B., Wassermann, D., Cucos, C., and Pines, A. (1982) Premature delivery and organochlorine compounds: Polychlorinated biphenyls and organochlorine insecticides. *Environ. Res.* 28, 106-112.

Whitsett, J.A. and Lessard, J.L. (1978) Characteristics of the microvillous brush border of human placenta: insulin receptor localization in brush border membranes. *Endocrinol.* 103, 1458-1467.

Whitsett, J.A. and Tsang, R.C. (1980) Calcium uptake and binding by membrane fractions of human placenta: ATP-dependent calcium accumulations. *Pediatr. Res.* 14, 769-775.

Trophoblast Research 2:343-356, 1987

THE PLACENTA AS A MODEL FOR TOXICITY SCREENING OF NEW MOLECULES

Rebecca Beaconsfield[1], Bojana Cemerikic, Olga Genbacev, and Vojin Sulovic

[1]SCIP Research Unit, University of London
London, England

INEP Department of Endocrinology, Immunology and Nutrition,
11080 Zemun, Banatska 31b, Yugoslavia

INTRODUCTION

Our in vitro model of placental tissue explants is designed to assess the direct acute effect of drugs on trophoblast cell metabolism (Genbacev et al., 1982). The possible clinical implications of the information obtained are indirect, but it could be inferred that alterations in trophoblast cell metabolism affects in turn fetal well being. Our purpose was to evaluate, using the described experimental model, the direct effect of two groups of drugs that are believed to have adverse effects on fetal growth: negative /morphine and morphine-like substances and nicotine/ and positive /beta-mimetic drugs. The parameters selected were glucose uptake, lactate production, ^{14}C-Leucine uptake and incorporation into total soluble proteins, placental hormone release and intracellular cAMP content (Genbacev et al., 1982).

Glucose uptake was studied because the placenta uses glucose for its energy metabolism and actively transports it for fetal needs.

Lactate production is a measure of anaerobic metabolism of glucose by the trophoblast and the amount of lactate produced is important for fetal metabolic requirements. The placenta provides lactate for the fetus in sufficient quantities to account for 25% of fetal oxidative metabolism (Burd et al., 1975).

The incorporation of ^{14}C-Leucine into total soluble proteins was measured because trophoblast cells are active in protein synthesis and produce specific placental protein hormones which are used clinically to assess trophoblast function during gestation. The intracellular concentration of cAMP was also determined to discriminate binding as the first step in the transport mechanisms from receptor-mediated regulation of cell function.

[1]To whom correspondence should be addressed.
Received: 10 October 1985; Accepted: 1 August 1986

The incubation time (3 hr) was short enough to avoid the disadvantages of the "closed system", such as the accumulation of metabolic end products and their possible effects on the tissue, and drug degradation.

MATERIALS AND METHODS

Clinical Material

Fresh human placental tissue was obtained from legal abortions by curettage (at 8-10 wks gestation) and after spontaneous delivery at term. Immediately after removal, the tissue was placed in phosphate buffered saline (PBS) at 4°C, transported to the laboratory, and processed within 2 hrs.

For first trimester placentae, the experimental tissue was taken from a pool of placental tissue fragments from 10 abortions all collected at one operating session. This pool was used for 20 individual incubation flasks: 5 control and 3 x 5 experimental.

Term placentae were obtained within 5 min of vaginal delivery from women with uncomplicated pregnancies (38-41 wks). One placenta was used for each experiment; 5 vials were used as controls and 3 x 5 vials for drug testing. Three individual placentae or placental pools were tested for each drug.

Incubation of Placental Tissue

Tissue fragments (1 x 2 mm) were well washed and then minced in saline. A wet weight averaging 0.5 g of tissue was placed in 10 ml Erlenmeyer flasks with 3 ml of Eagle's minimal essential medium (MEM). After a preincubation period of 1 hr at 37°C, the medium was discarded and replaced with fresh Eagle MEM containing 2.5 µCi of uniformly labeled ^{14}C-Leucine (specific activity 348 mCI/mmol, Radiochemical Centre, Amersham, England). Specific radioactivity in the medium was 0.83 µCi ^{14}C-Leucine per 52 µg Leucine per ml. The drugs under investigation were dissolved in Eagle MEM.

Nicotine (Pfaltz and Bauer, Inc., Research Chemical Division, Stanford, CT.) was dissolved in the incubation medium, the pH adjusted to 7.2 by adding 10-30 µl 7.5% $NaHCO_3$ to 30 ml of the medium. The tested doses ranged from 0.048-2.43 mg per incubation vial, i.e., 0.5 g wwt (0.1-5 mM).

Heroin (Diosynth, Apeljoorn, Holland) was dissolved in Eagle MEM and a dose range of 0.558 ng - 2.22 mg per incubation vial was used in the experiments (0.5 nM - 2 mM).

Ridodrine (Duphar B.V., Amsterdam, Holland) solution was directly pipetted into the incubation medium. The solution, 10-50 µl containing 8.2-41 µg ritodrine, was added to the incubation medium, and the final volume corrected to 3 ml where necessary.

Oxprenolol (Ciba Laboratories, Horsham, Sussex, England) was dissolved in the incubation medium and the concentrations used were 110-580 ng/0.5 g wwt.

Incubation was conducted for 3 hr at 37°C in a Gallenkampf shaking water bath (50 rpm) under an atmosphere of 95% O_2 and 5% CO_2. Checks on pH levels and osmolarity in the incubation medium were made routinely at the end of each experiment.

Processing of Tissue and Incubation Media

The incubation was terminated by immersing the Erlenmeyer flasks into an ice bath. Following incubation, the media and tissue fragments were filtered through 2 layers of cheesecloth. The filtrate was centrifuged at 2000 g for 10 min, the pellet discarded, and the supernatant fraction recentrifuged at 10,000 g and kept at -30°C until analyzed.

The tissue was removed from the cheesecloth and homogenized with a Teflon-coated pestle for 3 min in a 10 ml grinding vessel immersed in ice. PBS, pH 7.2, was used. The crude homogenate was centrifuged at 2000 g for 15 min; the pellet was discarded, and the supernatant fraction was recentrifuged at 10,000 g for 15 min.

The incubation medium was used for glucose and lactate determination, and radioimmunoassay of HPL and beta-hCG, using commercially available INEP-Vinca (Yugoslavia) RIA kits. Lactate and glucose concentrations in the incubation media were routinely measured using Boehringwerke enzyme kits (West Germany).

Tissue extracts were used for the determination of ^{14}C-Leucine uptake and incorporation into total soluble proteins and of cAMP intracellular concentration.

Uptake of ^{14}C-Leucine

^{14}C-Leucine uptake was measured in 0.1 ml tissue extract suspended in Bray's solution using a liquid scintillation counter (Model SL 4000 Intertechnique). The results are expressed as cpm/0.1 g wwt.

Determination of ^{14}C-Leucine

Aliquots of 0.1 ml of the incubation medium and the supernatant tissue fraction (10,000 rpm) were precipitated with 10% cold trichloroacetic acid (TCA), filtered, and washed twice with 2 ml of 5% TCA through Sartorius membrane filters (SM 11406). The radioactivity was analyzed as described above and in detail by Genbacev et al. (1981).

Determination of cAMP

For cAMP determination, 0.5 g wwt were incubated in 3 ml Eagle MEM, with and without the drug under investigation, as described in detail by Cemerikic et al. (1985). After 20 min of incubation, the tissue was boiled in 5 ml of 0.05 M sodium acetate buffer, pH 4, and homogenized according to the double cooking method described by Lahav et al. (1980).

Tissue concentrations of cAMP were determined by an RIA kit (Amersham, England). The results are expressed as pmols of cAMP per 0.1 g of placental wwt tissue. The analyses of significant differences between control and experimental groups were performed by means of Student's-t-test.

The free Leucine intracellular pool was determined using a Beckman Spinco model 120C amino acid analyzer (Lorincz et al., 1968).

Tissue Viability

Tissue viability was tested immediately at the end of the incubation period in one vial of each experimental group. The tissue was washed in the incubation medium and resuspended in an additional 3 ml of Eagle MEM and the incubation performed for another 3 hr. Glucose and lactate concentrations were measured as described, and the results obtained were compared with the values obtained in control vials from the first incubation period. In all the experiments reported, glucose uptake and lactate production after the incubation period were within the range of the control group (Mean ± SD).

RESULTS

Control Group

The control values obtained for each parameter studied are shown in Table 1. For each gestational age, 24 independent experiments with 5 control vials per experiment, that is, N = 24 x 5 = 120 vials in total were performed. The values are expressed as the mean from all experiments ± SD.

Effects of Beta-Mimetics and Beta-Blockers

For the study of beta-mimetics and beta-blockers ampules of ritodrine and oxprenolol for intravenous injection were used. The dose was extrapolated to 1 g of tissue from the maximal single dose taken by an adult weighing 50-70 kg. The dose corresponding to the therapeutic dose for ritodrine was 8.2 μg per incubation vial, and for oxprenolol 115.5 ng per vial. Doses 2, 3, and 5 times higher than the average drug dose were used, as in conventional drug testing.

Glucose uptake, lactate production, ^{14}C-Leucine uptake and incorporation into total proteins and hormone release were not affected by ritodrine (Table 2) in either first or third trimester placentae. A significant dose dependent increase in intracellular cAMP concentration was observed in the first trimester placental tissue slices after 20 min exposure to ritodrine (Figure 1a). The same treatment had no effect on term placenta (Figure 1b).

The effect of the beta-blocker oxprenolol was studied in a dose range of 115.5 - 580 ng per incubation vial. Oxprenolol treatment in vitro did not affect the parameters studies. A slight increase statistically non-significant in ^{14}C-Leucine incorporation into total proteins was observed in both early and term placentae with the dose of 231 ng of oxprenolol per incubation vial - that is twice the dose corresponding to the therapeutic one. This slight increase persisted in the presence of

Table 1

Control values for parameters measured after 3 hr of incubation of first (8-10 wk) and third (38-41 wk) trimester placental tissue explants. All the results are expressed per 0.1 g of tissue and are the mean ± S.E.M. of 120 incubation flasks. One asterisk denotes $p < 0.05$, and two asterisks $p < 0.001$.

Parameters Measured	Tissue explants	
	8-10 Week	38-41 Week
Glucose Uptake (μmol/1h)	0.606 ± 0.07	0.452 ± 0.091
Lactate Production (μmol/1h)	0.348 ± 0.021	0.345 ± 0.08
^{14}C-Leucine Uptake (cpm)	107,435 ± 11,224	108,081 ± 5,928
^{14}C-Leucine Incorporation into Proteins (cpm)	35,800 ± 2,557	19,350 ± 2,202*
HPL (μg/3h)	5.415 ± 0.23	18.795 ± 1.08**
Beta-hCG (IU/3h)	48.4 ± 8.3	0.446 ± 0.062**
cAMP (pmoles/2min.)	30 ± 10.5	22 ± 8.3

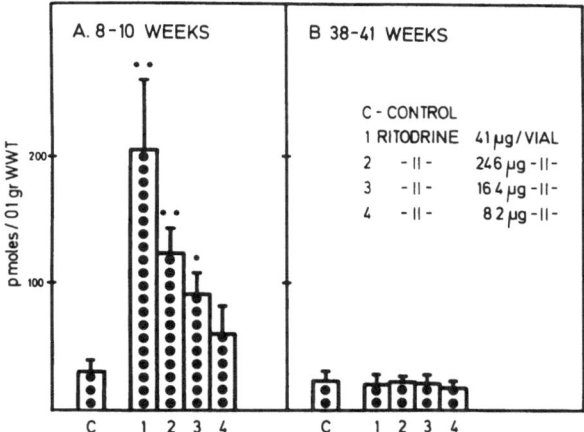

Figure 1. Dose-dependent effect of ritodrine on cAMP concentrations in 8-10 wk (A) and term placentae (B). Each bar is the mean ± S.E.M. of results from 15 independent incubation flasks. Significant differences are indicated with two asterisks for $p < 0.001$ and with one for $p < 0.05$.

Table 2

Ritodrine and oxprenolol effect on ^{14}C-Leucine uptake and incorporation into proteins in first and third trimester placental tissue explants. All the results are expressed as cpm/0.1g wwt and are the mean ± S.E.M. of 15 independent incubation vials for each experimental group and 120 incubation vials for the control group.

	8 - 10 Week		38 - 41 Week	
	^{14}C-Leucine Uptake	^{14}C-Leucine Incorporation	^{14}C-Leucine Uptake	^{14}C-Leucine Incorporation
Control	107,435 ± 11,224	35,800 ± 2,557	108,081 ± 5,928	19,350 ± 2,202
Ritodrine				
8.2 µg	106,917 ± 4,783	37,370 ± 3,087	108,917 ± 4,809	18,906 ± 1,780
16.4 µg	107,512 ± 6,050	36,918 ± 2,860	111,614 ± 2,971	19,507 ± 2,093
24.6 µg	110,492 ± 10,285	40,074 ± 3,713	112,092 ± 2,368	20,389 ± 2,641
41 µg	112,830 ± 7,440	41,463 ± 4,075	112,944 ± 4,653	22,962 ± 2,113
Oxprenolol				
115.5 ng	110,352 ± 8,406	39,711 ± 2,836	107,996 ± 6,350	19,734 ± 1,197
231 ng	119,567 ± 5,888	44,929 ± 3,180	110,085 ± 5,328	23,420 ± 3,412
346.5 ng	118,178 ± 6,602	44,340 ± 5,111	109,661 ± 4,962	22,892 ± 2,256
580 ng	126,658 ± 12,281	46,182 ± 4,505	112,404 ± 7,612	23,955 ± 3,881

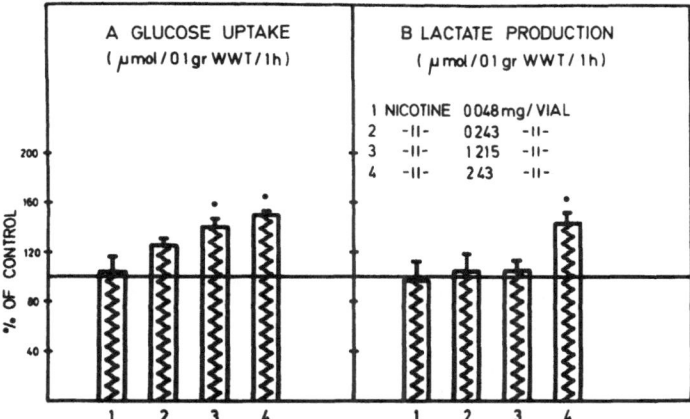

Figure 2. Effect of nicotine on glucose uptake (A) and lactate production (B) by first trimester placentae. The results are expressed as % of control. Absolute concentrations for the control group are shown in Table 1. Each bar is the mean value from 15 independent incubation flasks ± S.E.M. Significant differences are indicated with asterisk for $p < 0.05$.

higher concentrations of oxprenolol in the incubation medium - up to 580 ng per vial (Table 2).

Effects of Nicotine

The doses of nicotine used in this experiment were from 0.048-2.43 mg per incubation vial (0.1-2 mM).

The effect of nicotine on glucose uptake and lactate production in the first trimester trophoblast explants is shown in Figure 2. Nicotine treatment induced a dose-dependent increase of glucose uptake. Lactate production was increased only when the highest dose of nicotine was used. Significant differences ($p < 0.05$) are indicated by asterisks.

[14]C-Leucine uptake and incorporation into total proteins, hormone release (HPL and beta-hCG), and intracellular cAMP content remained within the control range in early placenta.

Term placentae appear more sensitive to nicotine treatment. Glucose uptake and lactate production were significantly increased with higher doses of nicotine ($p < 0.05$, Figure 3).

The incorporation of ^{14}C-Leucine into total proteins (Figure 4b) was significantly depressed by 2.5 and 5 nM nicotine ($p < 0.05$). The uptake of ^{14}C-Leucine however, was not affected by the doses of nicotine studied (Figure 4a).

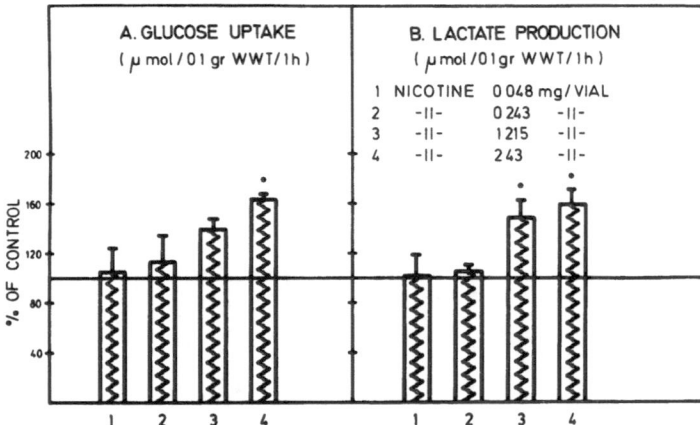

Figure 3. Effect of nicotine on glucose uptake (A) and lactate production (B) by term placental tissue explants. The explanations are as under Figure 2.

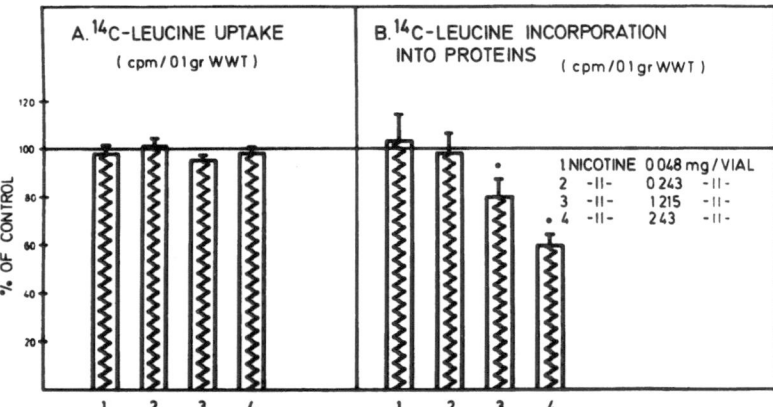

Figure 4. Nicotine effect on ^{14}C-Leucine uptake (A) and incorporation into total soluble proteins (B) by term placental tissue explants. Each bar is the mean from 15 incubation flasks ± S.E.M. One asterisk denotes $p < 0.05$. Absolute values for control group are given in Table 1.

Hormone release (HPL and beta-hCG) and cAMP tissue content remained unchanged, i.e., within control levels.

Effect of Heroin (Diacetylmorphine)

The concentration range of heroin in these experiments was from 0.558 ng - 2.22 mg per incubation vial (0.5 nM - 2mM).

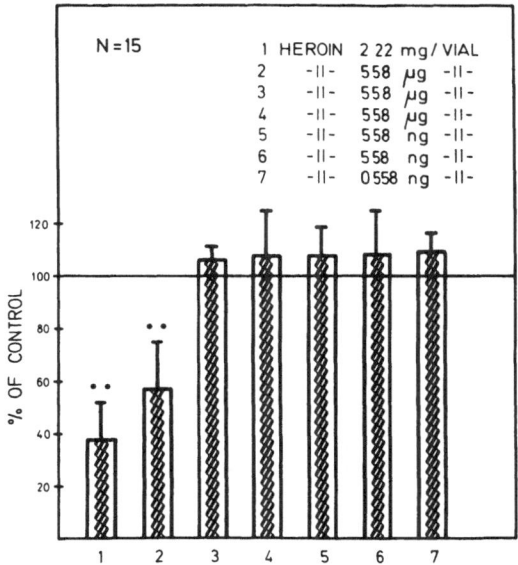

N=15	1 HEROIN	2 22 mg/VIAL	
	2 -II-	558 μg -II-	
	3 -II-	558 μg -II-	
	4 -II-	558 μg -II-	
	5 -II-	558 ng -II-	
	6 -II-	558 ng -II-	
	7 -II-	0558 ng -II-	

Figure 5. Heroin effect on ^{14}C-Leucine incorporation into total proteins by first trimester placental tissue explants. Results are expressed as % of control. Each bar is the mean from 15 incubation flasks ± S.E.M. Asterisks indicate $p < 0.001$.

The highest doses of heroin significantly depressed ^{14}C-Leucine incorporation into total proteins by first trimester placental tissue explants. Doses less than 50 μM were ineffective in all experiments (Figure 5). ^{14}C-Leucine uptake was not affected at any dose.

Heroin treatment in the first trimester placentae was without effect on glucose uptake, lactate production, ^{14}C-Leucine uptake, cAMP tissue content, and hormone release.

In term placentae heroin treatment significantly depressed ^{14}C-Leucine incorporation into total proteins in a dose dependent manner (Figure 6). All other parameters studied remained unchanged, i.e., within the normal range (Table 1).

DISCUSSION

Several drugs are known to cause intrauterine fetal growth retardation in experimental animals and man (Tuchmann-Duplessis, 1975). Morphine and morphine-like drugs and nicotine are known to be associated with retarded intrauterine growth (Stone et al., 1971; Hasselmeyer et al., 1979).

Beta-adrenergic agents during pregnancy are used to prevent uterine contractions and avoid premature labor and to improve placental transport and fetal growth by some direct effect on trophoblast cells (Wansbrough et al., 1968).

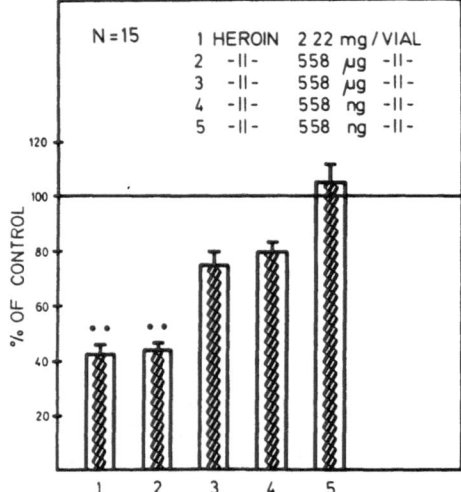

Figure 6. Heroin effect on ¹⁴C-Leucine incorporation into total soluble proteins by third placental tissue explants. Results are expressed as % of control and are the mean ± S.E.M. from 15 independent incubation flasks. Asterisks indicate p < 0.001.

The common link between the drugs studied is that their effects are believed to be receptor-mediated. The presence of specific receptors in term placentae has been demonstrated for beta-adrenergic drugs (Whitsett et al., 1980; Schocken et al., 1980), for opiate-like drugs (Valette et al., 1980; Ahmed et al., 1981), and for nicotine (Olubadewo et al., 1978).

The drug doses used in this study were selected on the basis of two different criteria. For drugs used therapeutically, the dose was calculated per gram of tissue from the average single dose taken by an adult weighing 50-70 kg. To evaluate toxic effects, two to five times the average dose was used, as in conventional drug testing. For drugs of abuse, a wide range of concentrations was tested, since there is no therapeutic baseline to follow.

In our experiments, the beta mimetic, ritodrine, 8.2 µg/incubation vial, which is equivalent to the dose used clinically (150 mg for an adult individual), did not alter the parameters measured in both young and term placentae. In first trimester placenta, the further increase of ritodrine concentration resulted in a dose-dependent stimulation of cAMP accumulation in the tissue. This finding is in agreement with Falkay and Kovacs (1983), who found beta-adrenergic receptors in placental membranes at 8-10 wks gestation.

As the described increase in cAMP concentration was not accompanied by changes in any other parameters studies, it is difficult to interpret these results. The absence of in vitro effects of beta-mimetics on other parameters cannot be attributed to poor drug penetration - a possibility suggested by Shu-Rong et al. (1982), since we have shown a significant increase in cAMP in first trimester placentae.

The fact that ritodrine treatment did not change HPL secretion and glucose metabolism is unexpected, since it has been reported that adrenergic stimulation increases HPL release and glycogenolysis (Belleville et al., 1978; Ginsburg and Jeacock, 1968; Satch and Ryan, 1971). One possible explanation is that the 3 hr incubation period in the presence of high ritodrine concentrations resulted in the desensitization of the beta-andrenoreceptor system (Berg et al., 1985).

The absence of ritodrine effect on cAMP in term placentae could be due to a decrease in the number of the beta-adrenergic receptors in the placenta with advancing gestational age, as has been reported for ovine (Padbury et al., 1981) and human placentae (Falkay and Kovacs, 1983).

It appears, however, that prolonged treatment of pregnant women with beta-mimetics might change to varying extents the beta-adrenergic placental system in the early stages of pregnancy and at term.

The doses of nicotine used in our experiments were in the same range as those used by Barnwell et al. (1983). These workers showed that nicotine at mM concentrations inhibited the uptake of amino acids in placental villi.

Nicotine in doses equivalent to 0.1-5 mM did not affect the uptake of ^{14}C-Leucine in early or term placentae. This finding can be attributed to the insensitivity of the "L" transport system to regulatory influences, as suggested by Barnwell et al. (1983).

However, ^{14}C-Leucine incorporation into total proteins was significantly depressed in term tissue explants by 2.5 and 5 mM, in a dose-dependent manner. This effect was not observed in first trimester trophoblast.

Glucose uptake was stimulated by nicotine in both young and term placentae in a dose-dependent manner. However, lactate production was increased in first trimester placentae only at doses of 5 mM and in term placentae at dose of 2.5 and 5 mM nicotine.

These doses are extremely high (approximately 1000 times the lethal dose), which could indicate that nicotine induces general cellular toxic effects. Against this possibility, nicotine did not affect ^{14}C-Leucine uptake and hormone release in both early and term placentae; ^{14}C-Leucine incorporation into total proteins was not inhibited in first trimester trophoblast; and more than 85% of the nicotine effect on glucose metabolism was reversed after washing in incubation media and then performing an additional 3 hr incubation.

It is known that among heavy smokers, fetal growth retardation is found more frequently than in other women, presumably as the direct result of nicotine on both fetus and trophoblast metabolism. The in vitro effects of nicotine described in this study could explain in part, the mechanism by which chronic exposure to nicotine leads to trophoblast metabolic insufficiency.

The differences in sensitivity to heroin (ED_{50}) between early and term placenta regarding ^{14}C-Leucine incorporation into total soluble proteins was marked, being approximately 0.5-1 mM in early and 50-100 µM in term placentae. Barnwell and Sastry (1983) observed that morphine inhibits amino acid uptake dependent on acetyl choline release, and suggest that such an interaction may be present in human placental villus tissue.

Our results support their conclusion and show that heroin is more potent than morphine, as the ED_{50} for ^{14}C-Leucine incorporation into total proteins in our experiments was about 20 times lower than the ED_{50} for morphine.

The in vitro model using placental tissue explants appears suitable for testing the acute effects of drugs on selected metabolic pathways during gestation. The differences in the sensitivity shown in this study may serve as a partial explanation for the known clinical effects. Selection of relevant parameters that will provide insight into trophoblast metabolism, its regulation and disturbances, remains to be established (Beaconsfield, 1979; 1982). The foregoing study may be some contribution to this task.

SUMMARY

The acute effects on trophoblast metabolism of beta-mimetics, heroin, and nicotine were studied using human placental explants in vitro. These drugs are believed to affect fetal growth.

The parameters of trophoblast metabolism studied were glucose uptake, lactate production, ^{14}C-Leucine uptake and incorporation into total soluble proteins, placental hormone release and intracellular cAMP content. The only effect of the beta-mimetic, ritodrine, (28.2 µg/vial) was on cAMP intracellular concentration in the first trimester placentae; in term placentae there was no effect. Oxprenolol, a beta-blocker, (110-580 ng/0.5 g wwt) had no effect on either first or third trimester placentae. Nicotine affected only glucose uptake and lactate production in the first trimester trophoblast. Term placentae appeared more sensitive to nicotine exposure. In a concentration of 2.43 mg per vial (5 mM), nicotine significantly increase glucose uptake and lactate production and inhibited ^{14}C-Leucine incorporation into total proteins. Heroin treatment significantly affected ^{14}C-Leucine incorporation into total soluble proteins in both first and third trimester placentae. The effective dose of heroin was 558 µg (500 µM) in early and 558 ng (500 nM) in term placentae. This difference in sensitivity between early and term placentae may be due to disparity in the number of opioid receptors at the two stages of gestation. It is proposed that in vitro studies of placental explants can further our knowledge of regulation and disturbances of trophoblast metabolism.

ACKNOWLEDGEMENTS

This work was supported by Serbian Academy of Science and by the Lord Dowding Fund.

REFERENCES

Ahmed, M.S., Byrne, W.L., and Klee, W.A. (1981) Solubilization of opiate receptors from human placenta. *Placenta* (Suppl. 3), 113-122.

Barnwell, S.L. and Sastry, B.V.R. (1983) Depression of amino acid uptake in human placental villus by cocaine, morphine and nicotine. *Trophoblast Res.* 1, 101-120.

Beaconsfield, P. (1979) Drugs and the placenta - A personal view. In: *Placenta - A Neglected Experimental Animal*, (eds.) P. Beaconsfield and C. Villee, Pergamon Press, pp. 105-108.

Beaconsfield, P. (1982) Drugs and placental metabolism - An introduction to the reports of preliminary work. In: *Placenta - The Largest Human Biopsy*, (eds.), R. Beaconsfield and G. Birdwood, Pergamon Press, pp. 79-90.

Belleville, F., Lasbennes, A., Nabet, P., and Paysant, P. (1978) HCS-HCG regulation in cultured placenta. *Acta Endocrinol. (Kbh)* 88, 169-176.

Berg, G., Andersson, R.G.G., and Ryden, G. (1985) Beta-Adrenergic receptors in human myometrium during pregnancy: Changes in the number of receptors after beta- mimetic treatment. *Am. J. Obstet. Gynecol.* 151, 392-396.

Burd, I.L., Jones, M.D., Jr., Simmons, M.A., Makowski, E.L., Meschia, G., and Bataglia, F.C. (1975) Placental production and foetal utilization of lactate and pyruvate. *Nature* 254, 710-711.

Cemerikic, B., Genbacev, O., and Sulovic, V. (1984) In vitro effect of ritodrinum on cAMP concentration in human placentas of different gestational age. *Exp. Clin. Endocrinol.* 87, 72-77.

Falkay, G. and Kovacs, L. (1983) Beta-adrenergic receptors in early human placenta characterization of ($3H$)-dihydroalprenolol binding. *Life Sci.* 32, 1583-1590.

Genbacev, O., Cemerikic, B., Movsesijan, M., and Sulovic, V. (1981) Protein synthesis by human fetal lung. *Am. J. Obstet. Gynecol.* 134, 41-46.

Genbacev, O., Cemerikic, B., Poljakovic, Lj., and Ilic, D. (1982) Incubation studies using placental tissue slices. In: *Placenta - The Largest Human Biopsy*, (eds.), R. Beaconsfield, and G. Birdwood, Pergamon Press, pp. 113-126.

Ginsburg, J. and Jeacock, M.K. (1968) Effect of epinephrine on placental carbohydrate metabolism. *Am. J. Obstet. Gynecol.* 100, 357-368.

Hasselmeyer, E.G., Meyer, M.B., Catz, C., and Longo, L.D. (1979) Pregnancy and infant health. In: *Smoking and Health: A Report of the Surgeon General*, (ed.), J.M. Pinney, United States Department of Health, Education and Welfare Publication No. (PHS) 79-5006.

Lahav, M., Sofer, N., Brandes, J.M., Paldi, E., and Barzilai, D. (1980) Placenta concentrations of progesterone, estradiol-17 beta, and cyclic AMP at delivery. *Obstet. Gynecol.* 56, 616-620.

Lorincz, A.B. and Kuttner, R.E. (1968) Comparative studies on free amino acids in female reproductive tissues. *Am. J. Obstet. Gynecol.* 101, 462-472.

Olubadewo, J.O. and Sastry, B.V.R. (1978) Human placental cholinergic system; Stimulation-secretion coupling for release of acetylcholine form isolated placental villus. *J. Pharm. Exp. Therap.* 204, 433-445.

Padbury, J.F., Hobel, C.J., Diakomanolis, E.S., Lam, R.W., and Fisher, D.A. (1981) Ontogenesis of beta-adrenergic receptors in the ovine placenta. *Am. J. Obstet. Gynecol.* 139, 459-464.

Satch, K. and Ryan, K.J. (1971) Adenyl cyclase in the human placenta. *Biochim. Biophys. Acta* 224, 618-626.

Schocken, D.D., Caron, M.G., and Lefkowitz, R.J. (1980) The human placenta. A rich source of beta-adrenergic receptors: Characterization of the receptors in particulate and solubilized preparations. *J. Clin. Endocrinol. Metab.* 50, 1082-1088.

Shu-Rong, Z., Bremme, K., Enorth, P., and Nordberg, A. (1982) The regulation in vitro of placental release of human chorionic gonadotropin, placental lactogen and prolactin: Effects of an andrenergic beta-receptors agonist and antagonist. *Am. J. Obstet. Gynecol.* 143, 444-450.

Stone, M., Salerno, L., Green, M., and Zelson, C. (1971) Narcotic addition in pregnancy. *Am. J. Obstet. Gynecol.* 109, 716-723.

Tuchmann-Duplessis, H. (1975) Drugs acting on the central nervous system. In: *Drug Effects on the Fetus*, Publishing Sciences Group, Acton, MA, pp. 142-194.

Valette, A., Reme, J.M, Pontonnier, G., and Cros, J. (1980) Specific binding for opiate-like drugs in the placenta. *Biochem. Pharm.* 29, 2657-2661.

Wansbrough, H., Nakanishi, H., and Wood, C. (1968) The effect of adrenergic receptors blocking drugs on the human uterus. *J. Obstet. Gynaecol. Br. Cmwlth.* 75, 189- 197.

Whitsett, J.A., Johnson, C.L., Noguchi, A., Darovec-Beckerman, C., and Costello, M. (1980) Beta-adrenergic receptors and catecholamines sensitive adenyate cyclase of the human placenta. *J. Clin. Endocrinol. Metab.* 50, 27-32.

Trophoblast Research 2:357-366, 1987

THE PHARMACOKINETICS OF CADMIUM IN THE DUALLY PERFUSED HUMAN PLACENTA

Patrick J. Wier[1] and Richard K. Miller[2]

Department of Obstetrics and Gynecology, Pharmacology,
and Radiation Biology and Biophysics,
Division of Toxicology
University of Rochester School of Medicine and Dentistry
601 Elmwood Avenue
Rochester, New York 14642 USA

INTRODUCTION

Cadmium is reported to be a significant occupational hazard to over 100,000 workers employed in the mining and processing of this metal (NIOSH publication number 76-192). Because cadmium is released into the environment by mining and numerous manufacturing processes, the general population is exposed to cadmium in air, food, and water (Friberg et al., 1974; Nriagu, 1982). Cigarette smoking is, in addition, an important route of exposure to cadmium (Friberg et al., 1974). The occurrence of exposure to cadmium in the general population has been documented by analyses of the cadmium content in human tissue (Gross et al., 1976). Cadmium is found in human placentae at levels from 0.04 to 1.6 nmole Cd/g (Miller and Shaikh, 1983)

Low molecular weight, cytosolic, metal-binding proteins (metallothioneins) are important in cadmium metabolism (Webb, 1979). Waalkes and coworkers (1984) have found a metallothionein-like protein in the human placenta, which can be affected by zinc and cadmium exposure.

The rodent placenta also has a high capacity for cadmium binding which may limit the placental transfer of cadmium. One hr after a single subcutaneous injection of 40 μmole Cd/kg to the rat, the placenta accumulated cadmium (60 nmole Cd/g) to levels of 50X the maternal blood concentration of cadmium, while fetal levels were very low (0.2 nmole Cd/g; Levin et al., 1983). Cadmium accumulation by the rodent placenta results in pathological alteration of placental structure (Levin et al., 1981, 1983; di Sant'Agnese et al., 1983) and possibly alteration of placental transport of zinc and vitamin B_{12} but not amino acids (Samarawickrama and Webb, 1979; Webb and Samarawickrama, 1981; Danielsson and Dencker, 1984).

[1]Current Address: Exxon Corporation, PO Box 235, East Millstone, New Jersey 08873 USA
[2]To whom reprints should be requested.
Submitted: 5 July 1986; Accepted: 30 July 1986

The goal of this investigation was to study the pharmacokinetics and tissue binding of cadmium in the dually perfused human placenta lobule in vitro, using the range of cadmium levels reported to be toxic to the rodent placenta. The experiments with the perfused human placenta were designed to determined: (i) the permeability of cadmium to the human placenta, (ii) the degree to which cadmium is accumulated by the placental tissue, and (iii) cadmium binding by human placental proteins.

MATERIALS AND METHODS

The technic of dual recirculating perfusion of isolated human placental lobules has been previously described in detail (Wier et al., 1983). A balanced salt solution (Krebs-Ringer's-phosphate-bicarbonate) supplemented with dextran (4000 daltons) and equilibrated with 95% oxygen/5% carbon dioxide was used as the perfusate. Sodium bicarbonate (0.89 mEq/ml) was added to the maternal or fetal circulation as necessary to maintain the pH at approximately 7.38. Viability of the preparation was documented using hemodynamic endpoints (maternal flow rate $= 14.4 \pm 0.7$ ml/min; fetal flow rate $= 3.5 \pm 0.6$ ml/min, fetal arterial systolic pressure $= 28\text{-}45$ mm Hg, volume loss from the fetal circulation < 5 ml/hr). The average mass of the perfused lobule was 25.0 ± 5.0 g.

To investigate the permeability of cadmium to the human placenta, carrier free $^{109}CdCl_2$ (New England Nuclear) and nonradioactive $CdCl_2$ (reagent grade, Fischer Scientific, Fair Lawn, NJ) were added to the maternal perfusate only to a concentration of 6.5 to 9.5 nmol Cd/ml (specific activity $= 0.2 \times 10^{-4}$ cpm/nmol). In some experiments both maternal and fetal perfusates were supplemented with a crude fraction of human plasma protein. This protein fraction was prepared by exhaustive dialysis (Dialyzer Tubing, 12000 molecular weight exclusion, Fisher Scientific) of outdated human plasma against doubly distilled, deionized water. The protein suspension was lypholized (Virtis Freezemobile II) to obtain the dried product. Protein concentration of the perfusates was determined by the method of Lowry et al. (1951).

Perfusate samples were centrifuged at 1500 x gravity for 10 min and 1.0 ml aliquots of the supernatants were removed for radioactivity counting. Tissue samples were blotted to remove extracellular water and weighed to the nearest 0.1 mg. ^{109}Cd was determined with a Micromedic MS538 gamma counting system using the ^{125}I channel.

The separation of "free" from protein-bound cadmium in perfusates was performed by ultrafiltration through a YMT ultrafiltration membrane (micro-partition system MPS-1; Amicon, Danvers, MA). After ultrafiltration of a solution of ^{109}Cd in Ringer's balanced salt solution, 97% of the ^{109}Cd was recovered in the filtrate.

Tissue samples collected for gel filtration were immediately blotted dry, weighed to the nearest 0.1 mg and homogenized (Tekmar tissue homogenizer) in 4 ml of 5 mM Tris HCl (pH $=8.6$) containing 0.02% sodium azide at 4°C for 1 min. The homogenate was then centrifuged for 90 min at 104000 x gravity (at 4°C). An aliquot (4.0 ml) of the supernatant was applied to a Sephadex G-75 (superfine grade, Pharmacia, Piscataway, NJ) column (1.6 x 70 cpm) equilibrated at 4°C with 5 mM

Tris- HCl containing 0.02% sodium azide, pH = 8.6. The column was eluted at a flow rate of 9.6 ml/hr. The total volume of the packed bed (V_T) was 129 ml; the void volume of the column (V_0) was 41.7 ml. The column was calibrated for molecular weight determination with ovalbumin (45000 daltons), native beta-lactoglobulin (36800), trypsinogen (24000), ribonuclease A (13690), and cytochrome C (12384) (Sigma Chemical, St. Louis, MO). The concentration of radioactive cadmium in the column fractions was determined by the counting method described above.

RESULTS

Within the first hr of perfusion, the concentration of Cd in the maternal compartment rapidly declined due to placental accumulation (Figure 1). After 4-6 hr, the concentration of Cd in the maternal compartment (0.1-1.6 nmole Cd/ml) approached "steady-state". The placental and fetal concentrations of cadmium were 22-37 nmole Cd/g and <0.08 nmole Cd/ml, respectively (Figure 1).

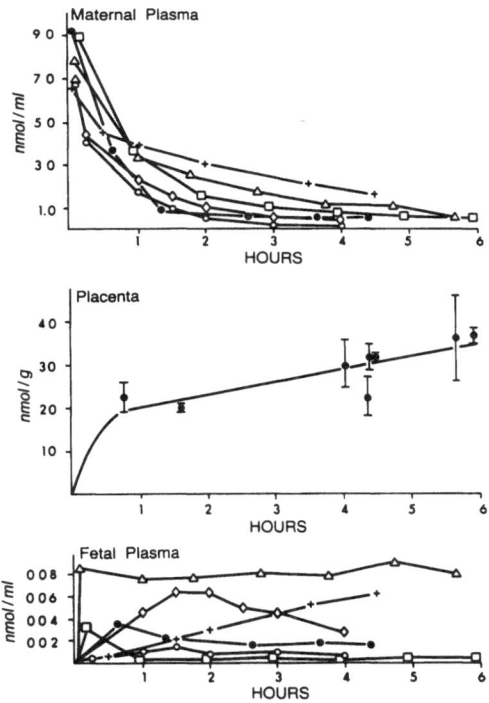

Figure 1. Cadmium concentration in maternal, placental, and fetal compartments during 6 hr of dual placental perfusion. Cadmium was added to the maternal circulation only at 0 hr of perfusion. Placental levels (means ± standard deviation) were measured at the end of perfusions lasting 0.75 to 5.9 hr. The average detection limit for cadmium in the fetal perfusate was 0.03 nmole Cd/ml.

Table 1

Cadmium distribution between the maternal and fetal compartments of the perfused human placenta. Fetal cadmium concentration was initially 0 nmol/ml.

Hours of Perfusion	Perfusate Protein Concentration[1] (mg/ml)		Initial Maternal Cd Concentration (nmol/ml)	Final Maternal Cd Concentration (nmol/ml)	Final Fetal Cd Concentration[2] (nmol/ml)	% Cd Protein Bound	
	Maternal	Fetal				Maternal	Fetal
0.75	nd	nd	7.99	2.28	0.112	nd	nd
1.9	23	24	7.55	2.16	0.068	99%	96%
4.0	5.4	0.02	6.73	0.14	0.006	nd	nd
4.4	16	3.6	9.23	0.55	0.017	nd	nd
4.4	5.1	0.6	6.79	0.36	0.028	nd	nd
4.5	25	22	6.54	1.60	0.066	97%	100%
5.7	6.5	5.3	7.83	0.56	0.081	96%	nd
5.9	4.1	0.2	8.90	0.55	0.009	97%	nd

1   In the experiments of 1.9 and 4.5 hours duration, lypholized human plasma protein was added to the KRP perfusates. In all other experiments protein was not added exogenously, and the low protein concentration was due to the presence of protein washed form the placenta.

2   The average detection limit for fetal cadmium concentration was 0.03 nmol Cd/ml, assuming twice the background counting rate to be the detection.

nd = not determined.

The average detection limit for cadmium in the fetal perfusate was 0.03 nmole Cd/ml (assuming twice background counting rate as the detection limit). Therefore, the rate of cadmium appearance in the fetal circulation could not be accurately determined in these experiments. While apparent, differences among individual experiments in the appearance of cadmium in the fetal circulation should be considered relative to the detection limit of this measurement.

Addition of protein to both maternal and fetal perfusates did not greatly affect the tissue accumulation or transport of cadmium (Table 8). Note that even without addition of exogenous protein, the molar ratio of protein: Cd was over 70 in the fetal perfusate and greater than 150 in the maternal perfusate (assuming an average protein molecular weight of 50000 daltons). Thus, whether the maternal protein concentration was 4 or 245 mg/ml, at least 97% of the cadmium was bound to protein with >10000 daltons molecular weight (Table 1).

Following ultracentrifugation of homogenized placental tissue, it was observed that approximately 45% of the total tissue cadmium was in the cytosol fraction (Table 2). The pattern of cadmium binding to perfused placentae cytosolic proteins is shown in Table 2. Cadmium was observed to bind to high (45000 daltons), intermediate (35000 daltons), and low (12000 daltons) molecular weight proteins as determined by gel filtration chromatography. A representative chromatogram from a single experiment is shown in Figure 2.

The protein(s) with an elution volume suggesting a molecular weight of 12000 daltons may be a metallothionein, although no proof other than chromatographic behavior was obtained. Metallothionein characteristically appears to have a molecular weight of approximately 10000 daltons when determined by gel filtration,

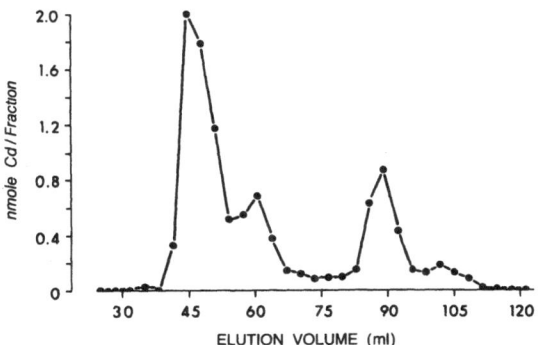

Figure 2. Gel filtration chromatogram of cadmium bound to cytosolic proteins of a placenta perfused for 4 hr. The volume of each fraction was 3.18 ml. For comparison, the elution volume of ovalbumin (45000 daltons) was 52.5 ml; the elution volume of native beta-lactoglobulin (36000) was 63.4 ml; and the elution volume of cytochrome c (124000) was 87.5 ml.

although amino acid analysis of this protein predicts a molecular weight of approximately 6000 daltons (Kagi and Nordberg, 1979). The difference is a result of appreciable dimer formation due to partial oxidation of the protein during isolation procedures (Kagi et al., 1974). Waalkes and coworkers (1984) recently described a metallothionein-like protein in the human placenta at a concentration of approximately 3 µg/g placenta. Assuming this protein (presumed molecular weight of 6050 daltons) can bind 6 nmole Cd/nmole metallothionein (Onosaka and Cherian, 1981), then the capacity for Cd binding to metallothionein in the human placenta would be 3.0 nmole Cd/g placenta. This concentration is within the range found in the current study (Table 2).

A cadmium binding protein of 30000 dalton molecular weight has been isolated from rat testes (Chen et al., 1974). In this study, cadmium in the placenta was observed to bind to intermediate (30000-40000 daltons) molecular weight protein(s) (Table 2).

Cadmium binding to proteins with molecular weight greater than 45000 daltons likely reflects non-specific binding to several cytosolic proteins which co-elute at the exclusion limit under this gel filtration condition.

Table 2

Distribution of cadmium bound to protein in human placental cytosol after in vitro perfusion.

Duration of Experiment (hours)	Placental Cd Concentration (nmol/g)	Cytosol Cd Concentration (nmol/g)	Protein Bound (nmol/g)[1]		
			>45000	35000	12000
0.75	22.5	9.7	5.0	1.4	0.6
1.9	20.2	8.5	3.9	1.7	0.6
4.0	29.4	15	5.1	2.2	4.4
4.4	31.7	14	5.1	2.2	2.2
4.5	31.3	14	4.7	1.8	1.3

[1] Cadmium bound to protein which eluted from the gel filtration column at this molecular weight. The actual elution volumes were: >45000 (41.6-51.2 ml), 35000 (54.4-67.2 ml), and 12000 (83.2-96 ml). For comparison, the elution volume of ovalbumin (45000) was 52.5 ml; the elution volume of native beta-lactoglobulin (36000) was 63.4 ml; and the elution volume of cytochrome c (12400) was 87.5 ml.

The average recovery of Cd in the cytosol applied to the column was 78% (total of all fractions).

DISCUSSION

This investigation of cadmium movement across the human placenta in vitro indicated that, like the rodent placenta, the human placenta is not freely permeable to cadmium. Four to six hr after the addition of cadmium to the maternal ciruclation (initially, 6.5 to 9.5 nmole Cd/ml), the placenta had accumulated cadmium to levels 50X that of the maternal circulation (31 nmol/g vs. 0.6 nmole Cd/ml) while fetal perfusate levels were less than 0.08 nmole Cd/ml. These findings are very similar to observations in the rodent following an acute dose of cadmium to the dam. The rodent placenta accumulates cadmium to levels 60X maternal blood levels (60 nmole Cd/g vs. less than 1 nmole Cd/ml) while the concentration of cadmium in the fetus is less than 1 nmole Cd/g fetus (Levin et al., 1983). Unfortunately, fetal blood levels have not been determined following an acute dose of cadmium to the pregnant rat.

To help understand the mechanisms of extensive accumulation of cadmium but lack of movement across the human placenta, the distribution of cadmium in placental tissue was investigated. A significant fraction (45%) of the cadmium accumulated by the placenta was recovered in the soluble fraction obtained after homogenization and ultrafiltration. This finding agrees with the report by Lucis and coworkers (1972) showing the most of rodent placental cadmium is bound to the cytosolic fraction. Also common to the current study and investigations in the rodent (Lucis et al., 1972; Wolkowski, 1974; LaFont et al., 1976; Arizono et al., 1981; Hanlon et al., 1982), gel filtration chromatography of placental cytosol showed that cadmium binds to high (>45000 daltons) and low (10000 daltons) molecular weight proteins. The role of these proteins in preventing cadmium from moving across the placenta is unknown. Among the cytosolic proteins, modifying metallothionein is of particular interest because this protein appears to function in the regulation of cadmium metabolism (reviewed by Webb, 1979).

Waalkes and coworkers (1984) recently identified a metallothionein in human placenta which was characterized by chromatographic behavior and ultraviolet spectroscopy. In the perfused human placentae, cadmium may be binding to the same protein(s) based on chromatographic behavior. Furthermore, the quantity of cadmium bound to this fraction was consistent with the concentration of metallothionein determined in human placenta by Waalkes and coworkers (1984).

Webb (1983) has reported that maternal (rodent) exposure to either inorganic cadmium or mercury results in an increased placental content of metallothionein, presumably by way of induced protein synthesis. Metallothionein levels in human placental chorion cells cultured in vitro are also reported to increase after 24 hr of exposure to zinc (Waalkes et al., 1984). In the placental perfusion studies, cadmium binding to the low molecular weight fraction appeared to increase as a function of time (1 to 4.5 hr) in vitro. Analyses of cadmium exposed placentae at later time points (8 to 12 hr) using technics for long term perfusion (Miller et al., 1985; Wier et al., 1986) are needed to determine the time dependence of protein binding by cadmium.

The current investigation was undertaken as a preliminary step towards investigating the toxicity of cadmium to the human placenta in vitro. Future work will evaluate the effects of cadmium on hemodynamics, metabolic, and transport functions of the placenta, and placental morphology.

SUMMARY

This investigation has shown that the human placenta can accumulate a large quantity of cadmium from the maternal circulation while transfer of cadmium to the fetal circulation is minor. Analysis of placental tissue exposed to cadmium identified multiple binding proteins for cadmium in the soluble fraction obtained after ultracentrifugation of homogenized tissue. While binding of cadmium to low molecular weight protein, possibly a metallothionein, was identified, the chemical state of most cadmium accumulated by the placenta is unknown. The localization of cadmium not found in the soluble fraction and the identity of intermediate and high molecular weight cadmium binding proteins needs to be established.

ACKNOWLEDGEMENTS

The authors wish to acknowledge the assistance of Jacqulyn White in the preparation of this manuscript and to express appreciation to the Labor and Delivery Staff at Strong Memorial Hospital for assistance in obtaining placentae. This project was funded in part by NIH Grants: ES02774, ES01247 and ES01248.

REFERENCES

Arizono, K., Ota, S., and Ariyoshi, T. (1981) Purification of metallothionein-like protein in rat placenta. *Bull. Environ. Contam. Toxicol.* 27, 671-677.

Chen, R.W., Wagner, P.A., Hoekstra, W.G., and Ganther, H.E. (1974) Affinity labelling studies with 109-cadmium in cadmium-induced testicular injury in rats. *J. Reprod. Fertil.* 38, 293-306.

Danielsson, B.R.G. and Dencker, L. (1984) Effects of cadmium on the placental uptake and transport to the fetus of nutrients. *Biol. Res. PR* 5(3), 93-101.

di Sant'Agnese, P.A., Jensen, K., Levin, A.A., and Miller, R.K. (1983) Placental toxicity of cadmium: an ultrastructural study. *Placenta* 4, 149-163.

Friberg, L., Piscator, M., Nordberg, G., and Kjellstrom, T. (1974) *Cadmium in the Environment II*. Chemical Rubber Company, Cleveland, Ohio.

Gross, S.B., Yeager, D.W., and Middendorf, M.S. (1976) Cadmium in liver, kidney, and hair of humans, fetal through old age. *J. Toxicol. Environ. Hlth.* 2, 153-167.

Hanlon, D.P., Specht, C., and Ferm, V.H. (1982) The chemical status of cadmium ion in the placenta. *Environ. Res.* 27, 89-94.

Kagi, J.H.R., Himmelhock, S.R., Wanger, P.D., Bethone, J.L., and Vallee, B.L. (1974) Equine hepatic and renal metallothioneins. Purification, molecular weight, amino acid composition, and metal content. *J. Biol. Chem.* 249, 3537-3542.

Kagi, J. and Nordberg, M. (eds.), (1979) *Metallothionein and Other Low Molecular Weight Metal Binding Proteins*, Birkhauser, Basel.

Lafont, J., Rouanet, J.M., Besancon, P., and Moretti, J. (1976) Existence d'une metallothioneine das le placenta. *C.R. Acad. Sci. (Paris)* 283, 417-420.

Levin, A.A., Plautz, J.R., di Sant'Agnese, P.A., and Miller, R.K. (1981) Cadmium: placental mechanisms of fetal toxicity. *Placenta* (Suppl. 3), 303-318.

Levin, A.A., Miller, R.K., and di Sant'Agnese, P.A. (1983) Heavy metal alterations of placental function: a mechanism for the induction of fetal toxicity of cadmium. In: *Reproductive and Developmental Toxicity of Metals*, (eds.), T. Clarkson, G. Nordberg, and P.R. Sager, Plenum Press, New York, pp. 633-654.

Lowry, O.H., Roseberg, N.J., Farr, A.L., and Randall, R.J. (1951) Protein measurement with the folin phenol reagent. *J. Biol. Chem.* 193, 265-275.

Lucis, O.J., Lucks, R., and Shaikh, Z.A. (1972) Cadmium and zinc in pregnancy and lactation. *Arch. Environ. Hlth.* 25, 14-22.

Miller, R.K. and Shaikh, Z. (1983) Prenatal metabolism: metals and metallothionein. In: *Reproductive and Developmental Toxicity of Metals*, (eds.), T. Clarkson, G. Nordberg, and P.R. Sager, Plenum Press, New York, pp. 153-204.

Miller, R.K., Wier, P.J., Maulik, D., and di Sant'Agnese, P.A. (1985) Human placenta in vitro: characterization during 12 h of dual perfusion. In: *Contrib. Gynecol. Obstet.*, (eds.), H. Schneider and J. Dancis, Karger, Basel, pp. 77-84.

Nriagu, J.O. (1982) *Cadmium in the Environmental, Parts I and II*, John Wiley and Sons, New York.

Onosaka, S. and Cherian, M.G. (1981) The induced synthesis of metallothionein in various tissues of rat in response to metals. I. Effects of repeated injection of cadmium salts. *Toxicol.* 22, 91-101.

Samarawickrama, G.P. and Webb, M. (1979) Acute effects of cadmium on the pregnant rat and embryo-fetal development. *Environ. Hlth. Persp.* 28, 245-249.

Waalkes, M.P., Poisner, A.M., Wood, G.W., and Klaassen, C.D. (1984) Metallothionein-like proteins in human placenta and fetal membranes. *Toxicol. Appl. Pharmacol.* 74, 179-184.

Webb, M. (1979) The metallothioneins. In: *The Chemistry, Biochemistry and Biology of Cadmium*, (ed.), M. Webb, Elsevier/North Holland, Biomedical Press, Amsterdam.

Webb, M. (1983) Endogenous metal binding proteins in the control of Zn, Cu, Cd, and Hg metabolism during prenatal and post-natal development. In: *Reproductive and Developmental Toxicity of Metals*, (eds.), T. Clarkson and G. Nordberg, Plenum Press, New York, pp. 655-674.

Webb, M. and Samarawickrama, G.P. (1981) Placental transport and embryonic utilization of essential metabolites in the rat at the teratogenic dose of cadmium. *J. Appl. Toxicol.* 1, 270-277.

Wier, P.J., Miller, R.K., Maulik, D., and di Sant'Agnese, P.A. (1983) Bidirectional transfer of alpha-aminoisobutyric acid by the perfused human placental lobule. *Trophoblast Res.* 1, 37-54.

Wier, P.J., Miller, R.K., Maulik, D., and di Sant'Agnese, P.A. (1986) Viability of the human placental lobule during 12 hours of dual perfusion, Submitted.

Wolkowski, R.M. (1974) Differential cadmium-induced embryotoxicity in two inbred mouse strains: analysis of inheritance of the response to cadmium and of the presence of cadmium in fetal and placental tissues. *Teratol.* 10, 243-262.

Trophoblast Research 2:367-393, 1987

CHARACTERIZATION OF MAGNETIC RESONANCE PARAMETERS IN THE PREGNANT UTERUS

Peter J. Thomford[1,5,6], Janet Jordan[1], Teresita Angtuaco[2],
H. Howard Cockrill[3], and Donald R. Mattison[1,4,5]

Departments of Obstetrics and Gynecology[1], Radiology[2],
and Pharmacology[4]
University of Arkansas for Medical Sciences
Radiological Associates[3]
Little Rock, Arkansas
and
The National Center for Toxicological Research[5]
Jefferson, Arkansas

INTRODUCTION

Although not widely appreciated, nuclear magnetic resonance (NMR) spectroscopy has been used as a research tool in obstetrics and gynecology for many years. Odeblad and colleagues have used NMR to characterize the structure of water in human vaginal cells (Odeblad, 1959), human milk (Odeblad and Westin, 1958), myometrium (Odeblad and Ingelman-Sundberg, 1965) and cervical mucus (Odeblad and Bryhn, 1957). Recently, with the growing availability of magnetic resonance imaging (MRI) units there has been an interest in exploring the technique in obstetrics (Smith et al., 1983; Johnson et al., 1984) and gynecology (Thickman et al., 1984). Indeed, a recent report suggests that MRI may be useful in prenatal evaluation of intrauterine growth retardation — by virtue of its ability to quantitate subcutaneous fat (Stark et al., 1985).

Nuclear magnetic resonance imaging and spectroscopy depend on the small magnetic moments which some nuclei possess (Table 1). After the discovery of nuclear magnetic moments in the 1940's, the initial scientific effort by physicists characterized their magnetogyric ratios Magnetogyric ratios of nuclear magnetic moments are important because they define the behavior of the magnetic moments in a magnetic field. Chemical applications followed, with the development of high resolution NMR spectroscopy to determine molecular structure (Pople et al., 1959; Fukushima and Roeder, 1981; Foster, 1984). High resolution NMR spectroscopy is possible because the technique is sensitive to the number of nuclei in a sample, and the detailed chemical environment of those nuclei.

[6]To whom corrrespondence should be addressed: Department of Obstetrics and Gynecology, Slot 518, 4301 West Markham, Little Rock, Arkansas 72205, USA
Submitted: 10 October 1985; Accepted: 2 April 1986

Table 1

Nuclei with magnetic moments which are important in biology

Atom	Magnetogyric Ratio	Natural Abundance (%)	Biological Importance
1_H	42.58	99.98	Water environments; imaging
13_C	10.71	1.11	Macromolecular structure; metabolic pathways; glycogen stores
19_F	40.05	100.00	Labeled substrates; metabolic pathways
23_{Na}	11.26	100.00	State of intra- and extra-cellular Na^+; muscle ischemia; imaging
31_P	17.23	100.00	Energy balance; pH; phospholipids; metabolism; intra-cellular reaction rates
39_K	1.99	93.1	State of intracellular K^+

Over the past decade, high resolution NMR spectroscopy has provided information of considerable importance in biochemistry (Foster, 1984; Cohen et al., 1983). The utility of NMR in providing this information derives from the ability to extract from NMR spectra information concerning the chemical structure and conformation of the macromolecules. In addition, NMR provides biochemical information including substance concentration, chemical state, and kinetics (Bock, 1985; Alger and Shulman, 1984). The ability to measure non-invasively substance concentration in vivo has been used to measure phosphate levels and intracellular pH in living tissues (Foster, 1984; Cohen et al., 1983; Kay and Mattison, 1986; Cady et al., 1983). These spectroscopic techniques for determining substance concentration, have been translated into non-invasive diagnostic tools for muscle disease and neonatal intracerebral pH measurements (Beher et al., 1983; Cady et al., 1983; Prichard et al., 1983; Radda et al., 1984; Maudsley and Hillal, 1984).

Recently, advances in computer software for image processing and reconstruction, along with improvements in magnet design have led to the development of magnetic resonance imaging. The techniques used to form an MR image are similar to those used in NMR spectroscopy. The patient or animal is placed in an intense magnetic field (0.2 to 2.0 Tesla) and exposed to a radiofrequency (RF) pulse (10-100 MHz). One Tesla is 10,000 gauss, and the magnetic field on the earth varies between 0.5 and 1.0 gauss. Information extracted from the region of the organism in the magnet and inside the RF coils is used to construct the MRI (Beall et al., 1984; Steiner and Radda, 1984; Damadian, 1981; Partain et al., 1983; Gadian, 1982). The images formed using MRI are based on entirely different physical principles than images formed by x-ray or ultrasound (Kay and Mattison, 1986; Partain et al., 1983; Damadian, 1981). Because the intensity and contrast which forms the MR image is based on different physical principles, it is instructive to review these factors. With imaging techniques based on ionizing radiation, intensity and contrast are determined by differences in the transmission of radiation by the tissues. With ultrasound, intensity and contrast are determined by the sound wave reflection characteristics of the tissues. In MRI, the intensity and contrast of an image is determined by two separate sets of factors; intrinsic or tissue dependent factors, and extrinsic or machine determined (operator selected) characteristics or parameters.

The intrinsic or tissue dependent characteristics which influence intensity and contrast in MRI include: i) spin-spin relaxation time, ii) spin-lattice relaxation time, iii) concentration of the imaged nucleus, and iv) flow (Kay and Mattison, 1986; Damadian, 1981; Partain et al., 1983; Beall et al., 1984; Mitchell et al., 1984; Moore, 1984). The spin-spin relaxation time (T_2) is a measure of the efficiency of energy transfer between the nuclear spins, or nuclear magnetic moments. The spin-lattice relaxation time (T_1) is a measure of the efficiency of energy transfer between nuclear magnetic moments and their molecular environment (e.g., molecular motion; local magnetic fields, etc.). The concentration of the imaged nucleus can be represented by the number of nuclei per unit volume of the sample. Flow alters image intensity moving the region of tissue studied out of the imaging plane.

The extrinsic parameters, those that are set by the software, or at the console of the MR imager include: i) RF pulse sequence chosen, ii) time interval between RF pulses, and iii) repetition time for multipulse sequences (Partain et al., 1983; Kay and Mattison, 1986; Moore, 1984; Mitchell et al., 1984). Note that the intensity of the magnetic field used for imaging can also influence the intensity and contrast in MRI by effects on T_1 and T_2. However, at the present time it is not possible to alter the magnetic field strength during imaging. There are three RF pulse sequences which have been used in MRI; partial saturation, inversion recovery, and spin-echo.

The partial saturation sequence is a series of 90° RF pulses (Figure 1A). The interval between the 90° pulses is defined as the repetition time (TR). The intensity of the signal from a tissue during a partial saturation sequence is:

$$I = PF \left[1 - \exp(-TR/T_1)\right],$$

where P is the nuclear density, F is a function of flow, TR is the repetition time, and T_1 the overall spin-lattice relaxation time for the sampled region (Kay and Mattison, 1986; Mitchell et al., 1984).

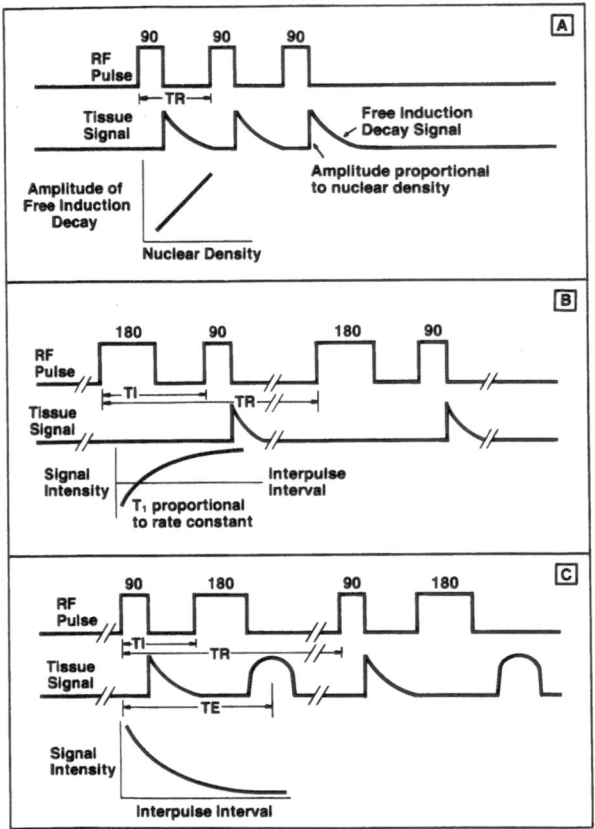

Figure 1A. Partial Saturation RF Pulse Sequence. The pulse sequence used in determining nuclear density and performing a partial saturation image. The upper line illustrates a series of 90° RF pulses separated by the repetition time (TR). The second line illustrates the signal received from the tissue following the 90° pulse. The amplitude of the free induction decay signal is proportional to the nuclear density, as shown on the lower chart (see also Figure 3).

1B. Inversion Recovery RF Pulse Sequence. The pulse sequence used to determine T_1 and perform an inversion recovery image. The upper line illustrates the 180°-90° RF pulse pair used in this technique. The interval between the 180° and 90° pulse is called the interpulse interval (TI), and the interval between successive 180° pulses is called the repetition time (TR). The second line illustrates the tissue signal which is collected after the 90° RF pulse. As the interpulse interval is lengthened, the signal from the tissue goes from a negative value, through zero and then becomes positive. The rate constant of this process is T_1 -- the spin- lattice relaxation time.

1C. Spin-Echo RF Rules Sequence. This pulse sequence is used to measure T_2 and perform spin-echo imaging. The upper line illustrates the RF pulses used -- a 90° pulse followed by a 180° pulse. The time between the 90° and 180° pulse is called the

The inversion recovery pulse sequence consists of two RF pulses, a 180° followed by a 90° pulse (Figure 1B). The interval between successive pairs of pulses is the repetition time (TR) and the time interval between the 180 and 90° pulses is the interpulse interval (TI). The intensity of the signal from a tissue during an inversion recovery pulse sequence is

$$I = PF[1 - 2\,EXP\,(-TI/T_1) + EXP\,(-TR/T_1)],$$

where P is nuclear density, F is a function of flow, TI is the interpulse interval and TR is the repetition time (Kay and Mattison, 1986; Mitchell et al., 1984).

The third imaging pulse sequence is called the spin-echo sequence, consisting of a 90° followed by a 180° RF pulse (Figure 1B). The interpulse interval is the time between the 90 and 180° pulse, and the repetition time is the time between successive pairs of pulses. The intensity of the signal from a tissue during spin-echo imaging is described by

$$I = PF[1 - 2\,EXP(-(TR - TI)/T_1) + EXP(-TR/T_1)] * EXP(-2TI/T_2),$$

where P is nuclear density, F is a function of flow, TR repetition time, TI interpulse interval, and T_2 the spin-spin relaxation time for the sampled region of the tissue (Kay and Mattison, 1986; Mitchell et al., 1984).

It is this ability to alter image intensity and contrast by changing the imaging parameters that makes MRI such a powerful technique. For reviews defining the effect of intrinsic and extrinsic factors on MRI image intensity and contrast consult; Fukushima and Roeder (1981), Steiner and Radda (1984), Kay and Mattison (1986), and Mitchell et al., (1984). An additional strength of MRI is the potential to extract biochemical information from the tissue during the imaging process (Fukishima and Roeder, 1981; Cohen et al., 1983; Bock, 1985). In this respect, MRI is several orders of magnitude more powerful than x-ray or ultrasound based imaging techniques which only reveal anatomical information. This paper reports the results of measurement of T_1 and T_2 in human myometrium, placenta, amniotic fluid, fetal blood, and fetal sheep tissues. The effect of Mn^{++}, a paramagnetic ion, on placental and amniotic fluid relaxation times is described. The effect of meconium of the T_1 of saline is also described.

MATERIALS AND METHODS

All NMR parameters; proton density, free induction decay (FID) amplitude, spin-lattice (T_1) and spin-spin (T_2) relaxation times were measured on a Praxis II

---

interpulse interval (TI). The time between successive 180° RF pulse is called the repetition time (TR). The second line illustrates the signal received from the tissue following both the 90° and the 180° pulses. Note that the time to echo (TE) is twice the interpulse interval. As the interpulse interval is increased, the amplitude of the echo decreases -- as shown on the chart. The rate constant for this process is the spin-spin relaxation time (T_2).

Pulsed Nuclear Magnetic Resonance Analyzer (The Praxis Corporation, San Antonio, Texas). The instrument consists of a 0.25 Tesla permanent magnet with the sample coil and RF pulse generator operating at 10.7 MHz. An Apple IIe microcomputer was used for data acquisition and analysis. MR images of pregnant non-human primates (cynomolgus monkeys) were obtained on a Picker 0.26 Tesla whole body imager using a head coil. MR images of pregnant humans were obtained using a Siemens 0.35 Tesla whole body imager with a body coil. Approval for MRI during pregnancy was given by the Human Investigation Committee, and all pregnancies imaged had ultrasound documented severe fetal anomalies.

Three different pulse sequences were used to determine proton density, T_1, and T_2. Proton density was determined by measuring the amplitude of the FID following a 90° pulse (Figure 1A). Proton density standard curves were derived using D_2O and H_2O (Figure 2).

T_1 was measured with a 90°-t-90° pulse sequence, similar to that used in partial saturation imaging (Figure 1A). In this pulse sequence the magnetization is inverted 90°, and the amplitude of the FID measured. A second 90° pulse is given at a later

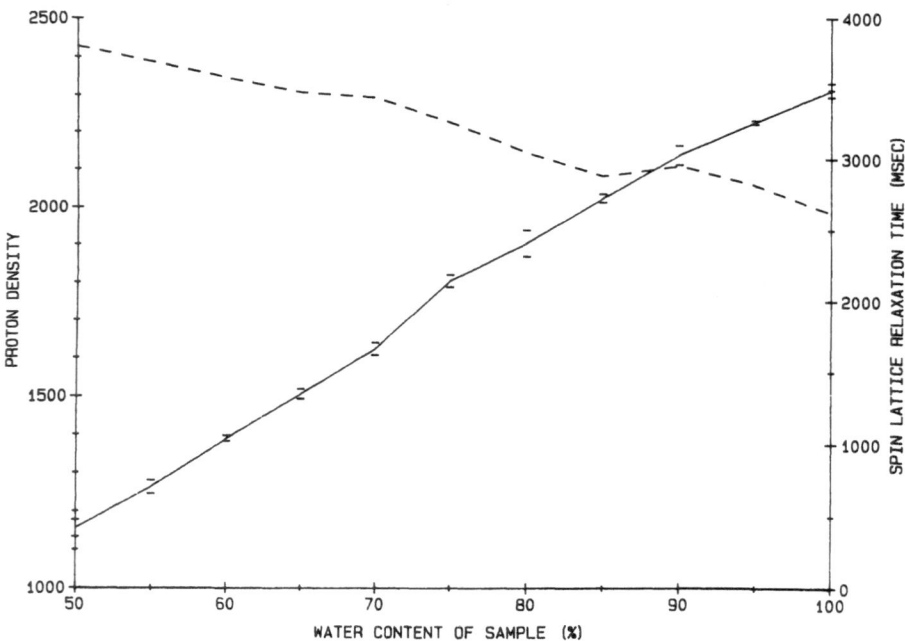

Figure 2. Effect of Proton Concentration on FID Amplitude. H_2O was mixed with D_2O in the indicated concentrations and the FID amplitude and T_1 measured. The FID amplitude is shown on the left axis (solid line) and T_1 is shown on the right axis (dashed line). As the amount of H_2O decreases (i.e., proton density decreases) the FID amplitude also decreases. As the water content decreases, T_1 increases, this occurs because the efficiency of the relaxation processes decreases.

time (t) and the amplitude of the second FID measured. The difference in amplitude following the pair of 90° pulses is then determined over a range of interpulse intervals and fit, using a least squares algorithm, to the equation

$$A = A_0 \exp(-t/T_1),$$

where T_1 is the spin-lattice relaxation time, t is the variable interpulse interval, and A_0 the amplitude of the magnetic moment following the first 90° pulse.

T_2 was measured using a 90°-t-180° pulse sequence (Figure 1C). The amplitude of the magnetization was measured following the 90° pulse and then after an interpulse interval (t) a 180° pulse is given. The 180° pulse inverts the magnetic moments and produces an echo centered around time 2t. The change in amplitude of the echo as a function of interpulse interval is fit using a least square algorithm to the equation

$$A = A_0 \exp(-t/T_2),$$

where T_2 is the spin-spin relaxation time, t is the variable interpulse interval and A_0 the amplitude of the magnetic moment following the 90° pulse.

Placentae were obtained from the labor and delivery suite at the University of Arkansas. Placentae from both vaginal and cesarean section deliveries were placed immediately on ice and transported to the laboratory. Placental villi were dissected from the chorionic plate and cut into 10 mm cubes with a surgical scissors and washed three times with ice-cold normal saline. The villi were then placed in 0.10 M phosphate buffered saline (pH - 7.40) and incubated at 37°C for periods of up to five hrs in buffer containing 0.002, 0.02, 0.2, and 2.2 mM manganese chloride. At times ranging from 30 to 280 min. after the start of incubation 3 g of tissue were removed and placed in a 10 x 75 mm borosilicate glass tube and FID, T_1 and T_2 recorded at room temperature over a period of approximately 15 min. Myometrial tissue was obtained from cesarean section deliveries at the University of Arkansas and kept on ice until FID, T_1, and T_2 were measured.

Amniotic fluid was collected at various stages of gestation and stored frozen at 20°C until the time of analysis. After thawing, and warming to room temperature FID, and T_1 was measured using methods similar to that for placenta.

Meconium was collected shortly after birth and stored frozen until the time of analysis. Meconium was diluted with normal saline to final concentrations of 0.0031, 0.0062, and 0.0125, 0.025, 0.05, 0.1, 0.2, and 0.5 g/ml, and FID, and T_1 measured at room temperature.

Fetal sheep tissue was obtained from twin fetuses (male, female) on approximately 130 days of gestation. The fetuses were removed surgically and placed at 3°C for 1 hr until the tissue could be excised and analyzed. T_1 and T_2 values were measured for leg muscle, kidney, brain, lung, liver, heart, and spleen from one female fetus and leg muscle, kidney, lung, liver, and heart from one male fetus.

Fetal blood samples were collected from human placentae just after the third stage of labor. The blood was heparinized and T_1 measured on whole blood. The blood was then centrifuged and separated into plasma and fetal red blood cells and T_1 measured on those individual components.

Manganese chloride ($MnCl_2$) was used as obtained from Aldrich Chemical Co. (Milwaukee, WI). D_2O was used as obtained from Cambridge Isotope Laboratory (Cambridge, MA). H_2O used in these experiments was purified by a Millipore Milli Q system. Data are reported as mean $\pm$ standard deviation for at least two replicates per sample.

RESULTS

Proton Content

Proton content in the tissues was estimated by comparing the FID amplitude of the tissue with that of the H_2O/D_2O solutions of known composition (Figure 2). These data are shown to demonstrate that altering proton density does alter the amplitude of the signal from the tissue. The amplitude of the signal from the tissue corresponds to the intensity or brightness of the image. Also noted is the increase in T_1 as the H_2O content falls. This occurs because the number of interactions between protons on water molecules decreases as the concentration of D_2O increases, thereby decreasing the efficiency of molecular processes contributing to T_1.

Myometrium

Samples of myometrial tissue were obtained from three patients following cesarean delivery. The spin-lattice relaxation times ranged from 531-882 msec (643 ± 119). The spin-spin relaxation times ranged from 54 to 88 msec (67 ± 10) (Table 2).

Table 2

Human Myometrial T_1 and T_2

Patient	T_1 (msec)	T_2 (msec)
# 1	882	88
# 2	518	59
# 3	531	54

Placenta

The spin-spin and spin-lattice relaxation times for normal placental tissue measured after incubation were 53 ± 4 msec and 670 ± 93 msec, respectively (Figure 3). Following incubation with .002, .02, .2, and 2.0 mM $MnCl_2$ for 60 min at $37°$ the T_1's were 576 ± 26, 499 ± 25, 275 ± 80, and 61 ± 13 msec, respectively (Figure 4). Following incubation with the same concentrations of $MnCl_2$ the T_2 relaxation times were 46 ± 3, 46 ± 2, 43 ± 6, and 26 ± 4 msec, respectively (Figure 5).

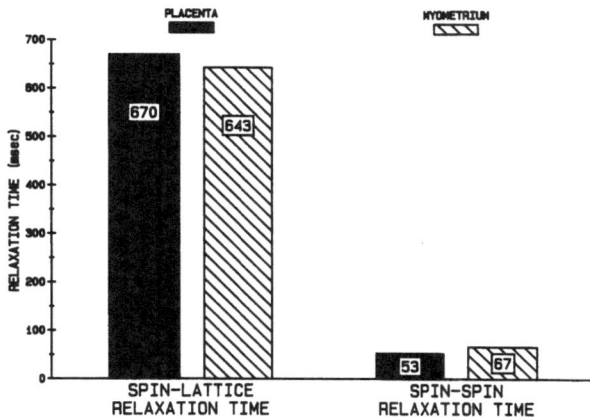

Figure 3. Human Myometrial and Placental T_1 and T_2. Placental and myometrial T_1 and T_2 are very similar. Because image contrast in MRI depends, in part, on these relaxation times, the contrast between these two images will be small and they will be difficult to distinguish on an MR image.

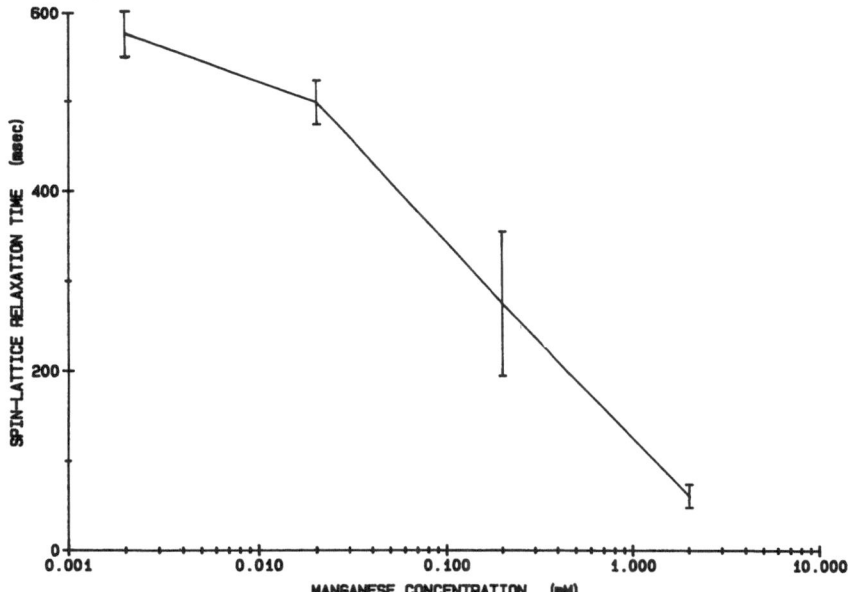

Figure 4. Effect of $MnCl_2$ on Placental T_1. Incubating slices of human placental villi for 60 min. in buffer in $37°$ with increasing concentrations of $MnCl_2$ produced a dose dependent decrease in T_1.

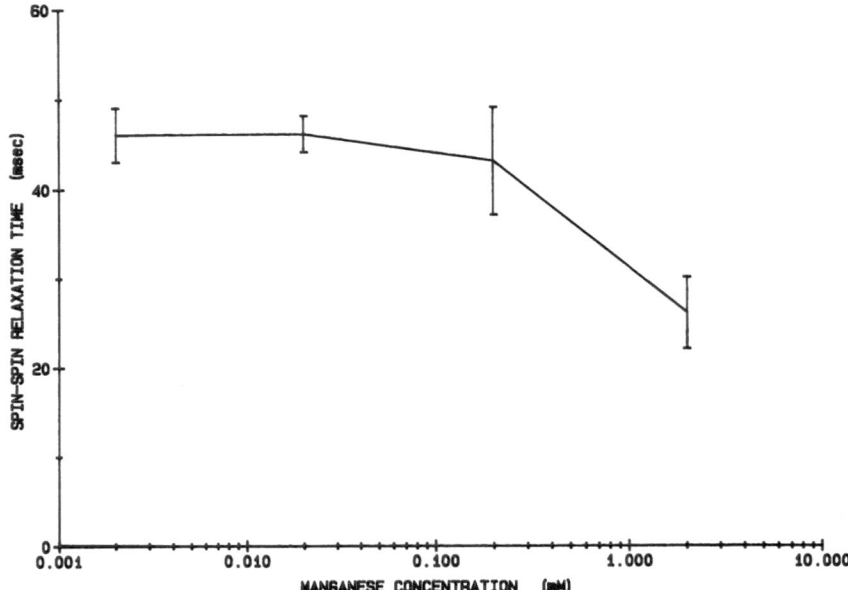

Figure 5. Effect of MnCl$_2$ on Placental T$_2$. Incubating human placental villi with MnCl$_2$ produced no effect on T$_2$ until the MnCl$_2$ concentration reached 1 mM. This blunted effect on T$_2$ is different from the decrease in placental T$_1$ observed at lower MnCl$_2$ concentrations.

Placental villi were incubated with 0, .002, or .02 mM MnCl$_2$ for periods ranging from 30 to 280 min (Figure 6). Control values for placental T$_1$ over this time ranged from 750 to 840 msec. T$_1$ values for the .002 mM MnCl$_2$ placental incubation decreased to 700 msec at 1 hr and had decreased to 660 msec at the end of the incubation period. T$_1$ values for placental villi incubated in .02 mM MnCl$_2$ incubation had decreased to 615 msec within 60 min and reached 410 msec at the end of the incubation period.

Amniotic Fluid

MnCl$_2$ was added to amniotic fluid to bring the final concentration to 0 (control), .0015, 0.16, and 1.66 mM. At these concentrations the T$_1$ relaxation times of amniotic fluid were 2131 $\pm$ 132, 2062 $\pm$ 205, 1262 $\pm$ 313, 294 $\pm$ 92 and 56 $\pm$ 9 msec, respectively (Figure 7). The effect on Mn^{++}, decreasing T$_1$, is similar in both amniotic fluid and saline, however the T$_1$'s for saline are longer at all concentrations except the highest. The inhomogeneity of the magnetic field of the Praxis prohibited the measurement of accurate T$_2$ values in amniotic fluid.

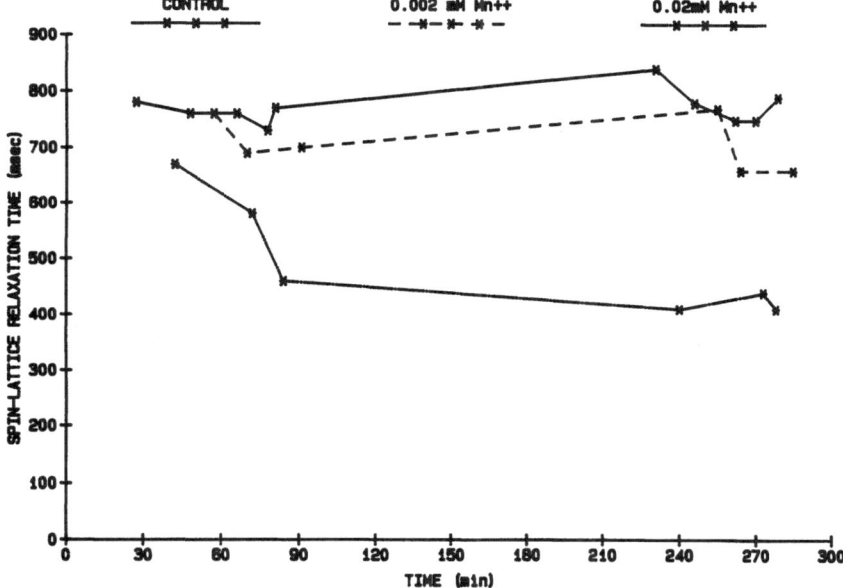

Figure 6. Time Course of $MnCl_2$ Effect on Placental T_1. Placental villi were incubated at 37° in control, 0.002 mM and 0.02 mM $MnCl_2$. No change was observed in control placental T_1 over the course of the experiment. However, in buffer containing $MnCl_2$, placental T_1 was decreased in a time dependent fashion with the optimum effect occurring approximately 90 min after the start of incubation.

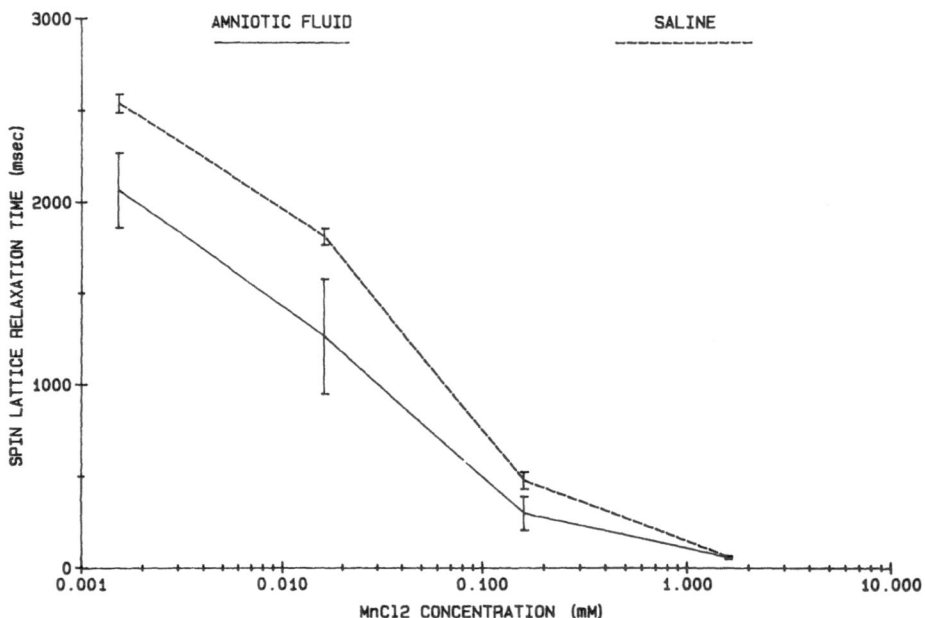

Figure 7. Effect of MnCl₂ on the T₁ of Saline and Amniotic Fluid. Manganese decreased T_1 in both amniotic fluid (solid line) and saline (dashed line) in a dose dependent fashion.

Meconium

Meconium was diluted in normal saline to final concentrations ranging from 7.8×10^{-4} to 5.0×10^{-1} g/ml. Spin-lattice relaxation times ranged from 2747 ± 102 to 503 ± 151 for the lowest and highest concentrations of meconium, respectively (Figure 8).

Fetal Tissues

T_1 values ranged from 267 msec for fetal liver to 837 msec for fetal brain. T_2 values ranged from 53 msec for leg muscle to 136 msec for lung tissue (Table 3).

The T_1 for whole fetal blood (814 ± 92 msec) falls between that for plasma (1673 ± 219 msec) and red blood cells (617 ± 58 msec) (Figure 9). This is an important concept in MRI -- the T_1 or T_2 which influences image intensity is a composite of the relaxation times of the individual biochemical environments in the tissue studies. Fetal blood is composed of two separate proton environments -- protons in plasma, and protons in red blood cells. The equation that expresses the composite T_1 for fetal blood is

$$1/T_{1\text{total}} = P_{\text{plasma}}/T_{1\text{plasma}} + P_{\text{RBC}}/T_{1\text{RBC}},$$

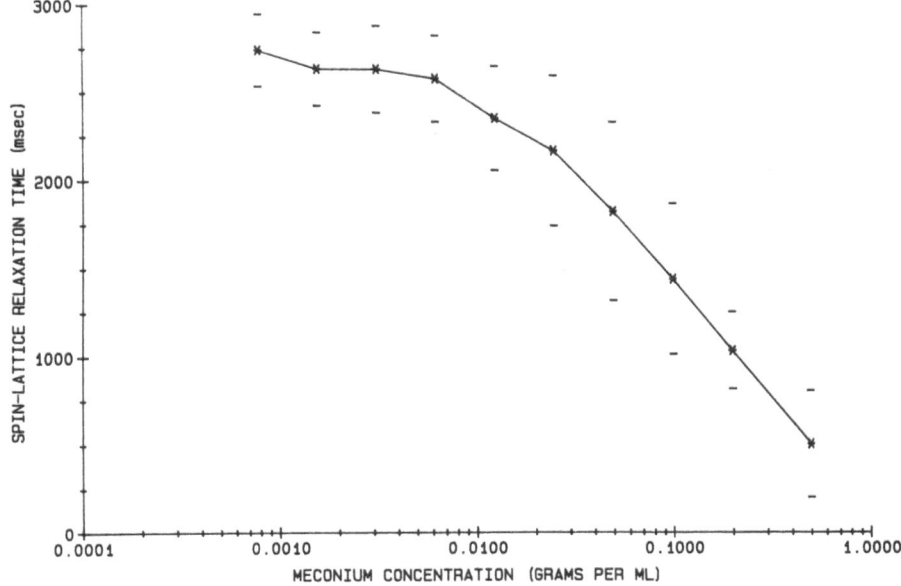

Figure 8. Spin-Lattice Relaxation Time of Meconium in Saline (Mean Values). Meconium decreased the T_1 of saline in a dose-dependent fashion.

where P_{plasma} is the proportion of protons in the plasma, and P_{RBC} the proportion in the fetal red blood cells. If we use the hematocrit (55) to estimate P_{plasma} and P_{RBC} we get

$$1/T_{1total} = .45/1673 + .55/617 = 1/862,$$

The estimated T_{1total} (862 msec) is quite close to the measured T_1 for whole fetal blood (814 $\pm$ 92 msec).

Magnetic Resonance Images

Human MRI were obtained from patients with ultrasound documentation of a fetal anomaly, and were conducted to determine the utility of MRI as an adjunct to ultrasound in prenatal diagnosis (Figures 10-14). Both were thought to have severe renal anomalies (confirmed at autopsy) which were associated with a decrease in the volume of amniotic fluid. Imaging in both patients was conducted to try to define in greater detail the nature of the fetal renal anomaly. Maternal abdominal and pelvic soft tissue structures are well delineated in these images, illustrating the power of MRI. However, as predicted by the similarity in placental and myometrial T_1 and T_2 it is very difficult to define placental implantation site, or internal structure of the placenta.

Table 3

Nuclear magnetic relaxation characteristics of tissues from fetal sheep

Tissue	T_1 (msec)	T_2 (msec)
Female		
Muscle	570	54
Kidney	694	57
Brain	838	77
Lung	824	55
Liver	267	57
Heart	744	54
Spleen	574	83
Male		
Muscle	700	93
Kidney	650	107
Lung	820	136
Liver	320	67
Heart	675	89

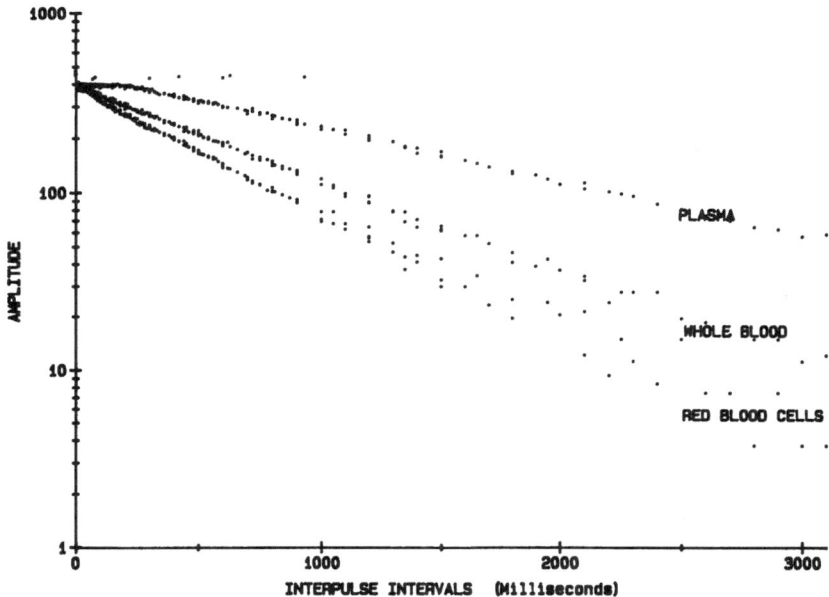

Figure 9. Effect of Interpulse Interval on Signal Amplitude of Fetal Blood, Plasma, and Red Blood Cells. The T_1 of fetal blood, plasma, and red blood cells was measured using the partial saturation technique (Figure 1). The horizontal axis -- interpulse interval -- is the time between successive 90° pulses. The vertical axis -- amplitude -- represents the difference in FID amplitude between the first and second pulse for each pair of 90° pulses. At very short values of the interpulse interval the FID amplitude following the second pulse is very small -- so the difference is large. As the interpulse interval increases so does the amplitude of the FID after the second pulse -- decreasing the difference. The rate of change of amplitude with interpulse interval determines the value of T_1. In this sample, the signal amplitude from fetal red blood cells decreased most rapidly, so this tissue had the shortest T_1.

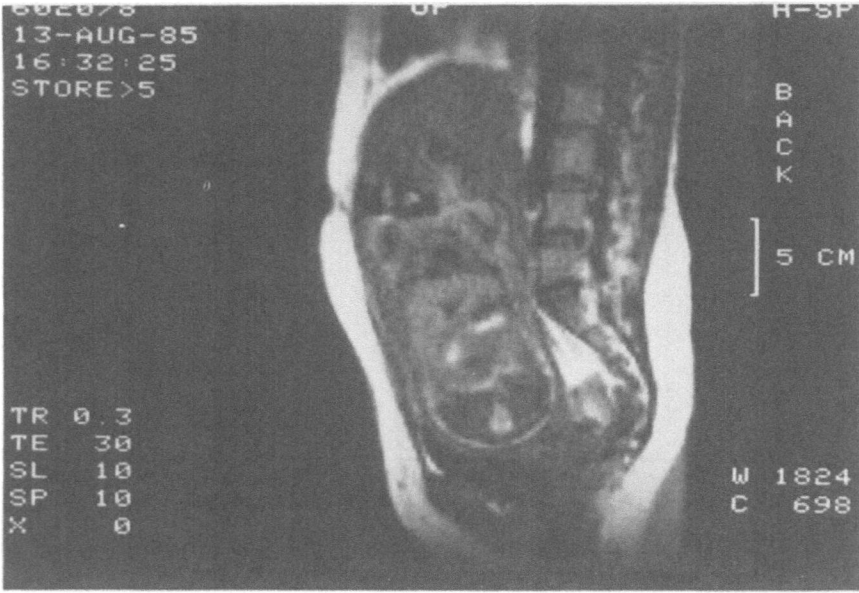

Figure 10. MRI During Human Pregnancy - Sagittal Image. This spin-echo image was obtained in the Siemens MRI; 0.35 Tesla, TR = 300 msec, TE = 30 msec. The section is through the umbilicus which appears as a defect on the anterior abdominal wall. Maternal spine is well delineated. Fetal head is in the pelvis. The placenta appears fundal, but is difficult to separate from the myometrium (13-Aug-85- 16:32:35).

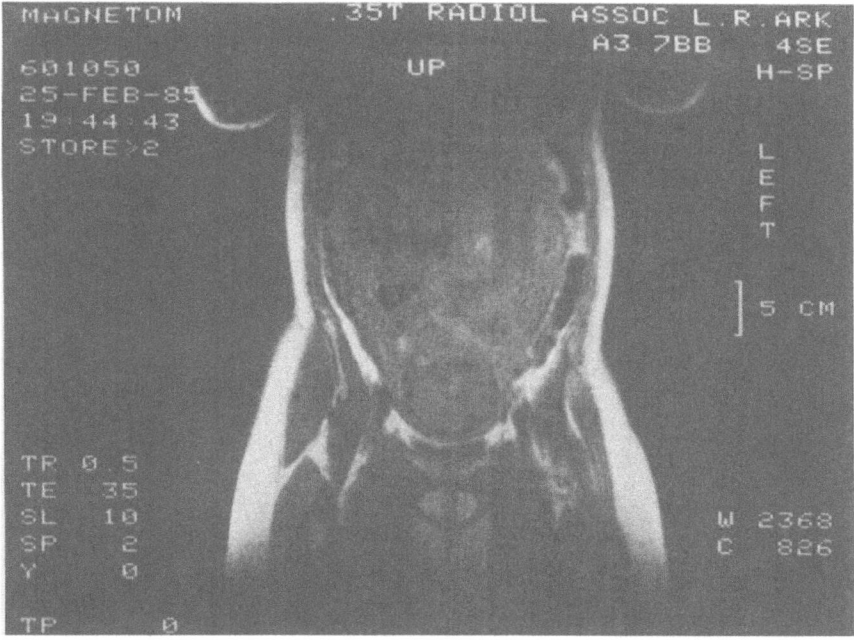

Figure 11. MRI During Human Pregnancy - Coronal Section. This spin-echo image was obtained on a Siemens MRI with the magnetic field strength at 0.35 Tesla, the repetition time (TR) was 500 msec and the time to echo (TE) was 35 msec. Adipose tissue is white, note the maternal subcutaneous adipose tissue. Muscle is darker, and demonstrated by maternal skeletal muscle and myometrium. The fetal head is in the pelvis, pressing against the bladder. Only minimal fetal subcutaneous adipose tissue is apparent on this image -- consistent with the clinical impression of uterine growth retardation. The fundal placenta is not well delineated in this image (25-Feb-85-19:44:43).

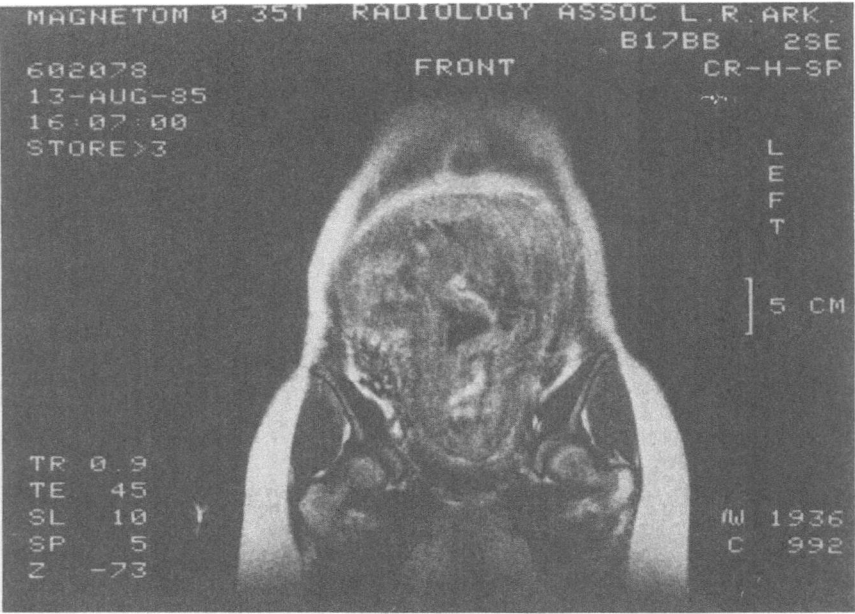

Figure 12. MRI During Human Pregnancy - Control Image. This spin-echo image (TR - 900 msec, TE = 45 msec) demonstrates pelvic architecture very clearly. Bone is black but marrow is lighter. Note especially the articulation of the hip and pelvis sidewalls. The fetal head is in the pelvis pressing against the maternal bladder. The darker area above the fetal head is the fetal heart (13-Aug-85-16:07:00).

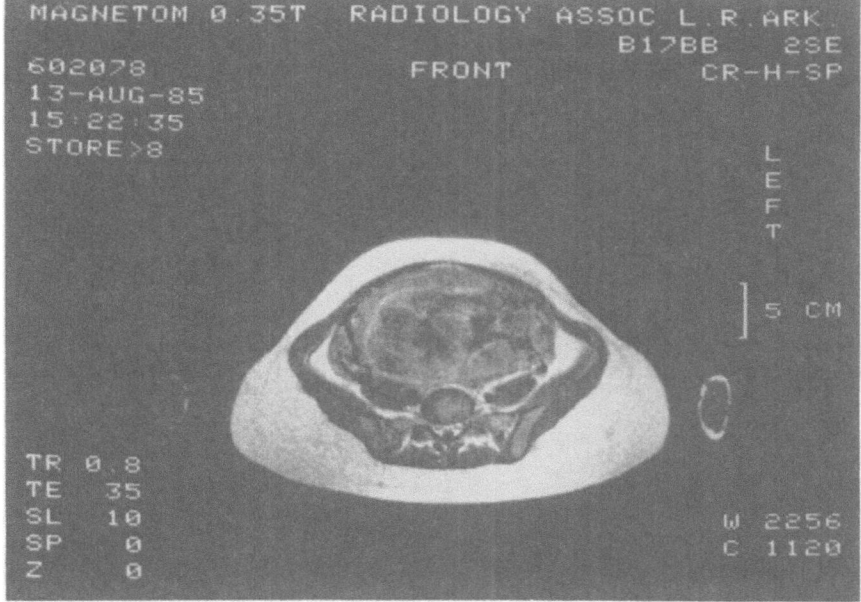

Figure 13. MRI During Human Pregnancy - Transverse Image. This spin-echo image (TE = 800 msec, TR = 35 msec) illustrates the resolution available for maternal structures -- spine and kidneys are well delineated. The uterus, fetus, and placenta are also shown. Contrast between myometrium and placenta is very low (13-Aug-85-15:22:35).

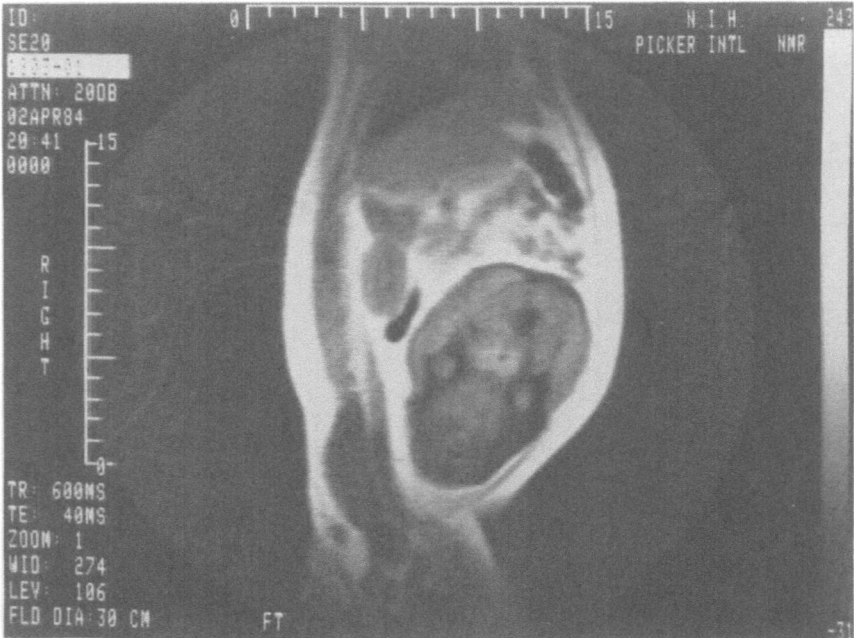

Figure 14. MRI During Cynomologus Pregnancy. This spin-echo image (TR = 600 msec, TE = 40 msec) was obtained late in gestation. The anterior and posterior lobes of the placenta can be seen in the image. Maternal kidney is also visualized. This image was obtained on the Picker MRI at the Clinical Center, NIH in collaboration with Dr. Richard Knop and Dr. Helen Kay (1803-01). The same difficulty in defining placental implantation site and structure is also seen in the MR images from a non-human primate.

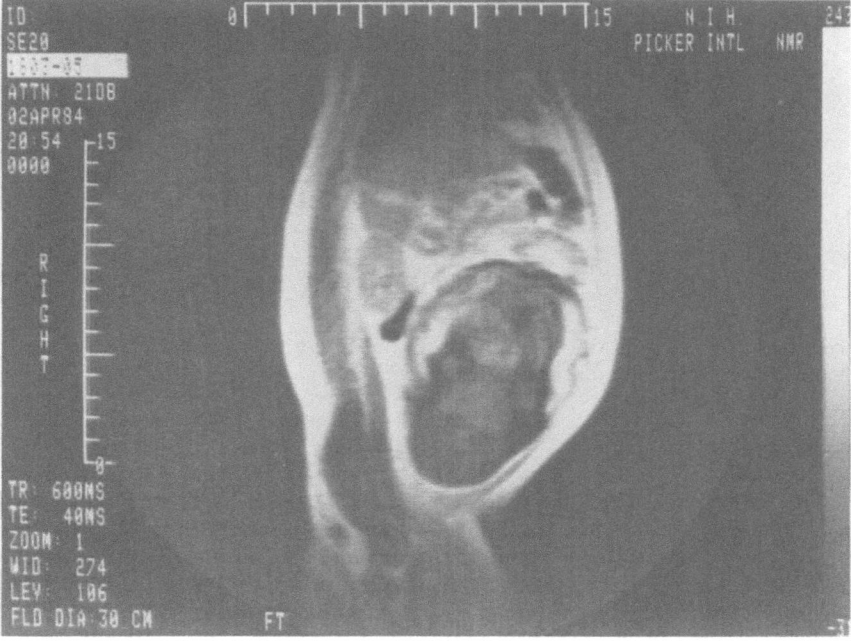

Figure 15. Following infusion of Mn⁺⁺ there is clear definition of the placental disks. In addition, it is possible to define internal structure of the placenta following the infusion and uptake of Mn⁺⁺ by cynomologus placental tissue.

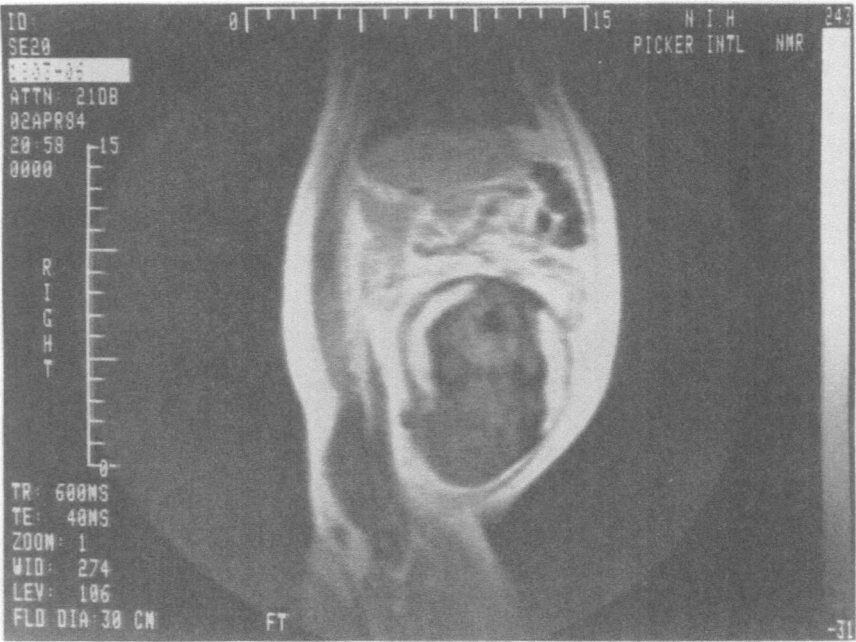

Figure 16. MR image obtained 4 min after the image in Figure 15. Note increased brightness of placenta and maternal kidney.

DISCUSSION

The utility of MRI depends, in part, on differences in T_1 and T_2 in adjacent tissues. The similarity in placental and myometrial T_1 and T_2 (Figure 3) suggested that it would be difficult to define the implantation site and internal structure of the placenta (Kay and Mattison, 1986). This prediction appears to be correct based on spin-echo images obtained from human and non-human primate pregnancies (Figures 10-16). An additional factor which may decrease contrast between placental and myometrial tissue at the implantation site is blood flowing through the intervillous space. Image acquisition times are typically in the range of several minutes so that blood flow through the intervillous space will move the villi, blurring the image. However, because the anchoring villi are fixed between the chorionic plate and the decidua it may be possible to resolve these structures during MRI. In addition, it may be possible to enhance the image of synchronizing the collection of data with the maternal cardiac cycle. Similar collection schemes, using respiratory and cardiac gating are presently being used in imaging the thorax.

Paramagnetic ions like Mn^{++} have been used for many years in high resolution NMR to shorten relaxation times and speed up acquisition of the spectra. More recently, these same paramagnetic ions have been used in MRI to enhance contrast between adjacent tissues (see Wolf, 1984 and Gore and Smith, 1985) for recent journal symposia on contrast enhancement in MRI. Paramagnetic ions, like

Mn^{++}, shorten relaxation times by providing additional dipole moments within the sample to interact with the nuclear magnetic moments (Moore, 1984; Cohen et al., 1983). Contrast enhancement by paramagnetic ions in MRI is dependent on differential tissue uptake of the administered ion. For example, in these experiments we want the placental concentrations of Mn^{++}, the paramagnetic contrast agent, to be higher than myometrial Mn^{++} concentration. Alternatively, the concentrations of the paramagnetic ion may be similar in the two tissues but effect on the relaxation times different. This type of effect is seen with Mn^{++} in saline and amniotic fluid (Figure 7).

Manganese, a paramagnetic ion, is taken up by the placenta in vitro, shortening T_1 and to a lesser extent T_2, in a dose dependent fashion (Figures 4 and 5). The time course for Mn^{++} uptake by human placental villi in vitro in these experiments was rather slow. Optimum effect appeared to require at least 60 min (Figure 6). The in vitro data presented (Figure 3) suggested that little contrast would exist between placenta and myometrium in the absence of a contrast enhancing agent. This theoretical suggestion is supported by the MR images obtained on two separate imaging units using non-human primates and pregnant women (Figures 10-16). The uptake of Mn^{++} and subsequent shortening of placental T_1 suggested that this paramagnetic ion would be useful as a contrast enhancing agent in the pregnant uterus. Monkeys treated with Mn^{++} demonstrated rapid uptake and enhancement of the placental image. There appears to be a discrepancy between the time course of Mn^{++} uptake by placental villi in vitro and the rapidity with which Mn^{++} altered contrast in the MR image of the pregnant cynomologus monkey. The difference may be explained in part by better access of Mn^{++} to placental tissue in vivo. In addition, the procedure used in preparing the placental tissue for the incubation studies may compromise the function of the tissue and impair uptake of the paramagnetic ion.

Manganese also decreased the T_1 of amniotic fluid and normal saline in a dose dependent fashion (Figure 7). No effect was observed in vivo however suggesting that only small amounts of Mn^{++} reached the amniotic fluid during the imaging session (Figures 14-16). Of interest however, is the effect of meconium on the T_1 of normal saline. There was a dose dependent decrease in T_1 of normal saline as the concentration of meconium increased (Figure 8). Bene and co-workers (Bene, 1980; Borcard et al., 1982; Bene et al., 1982) have previously noted that meconium decreases the T_2 of normal saline in a dose dependent fashion. As meconium is thought to be passed into the amniotic fluid in response to fetal distress (Meiss et al., 1978) -- these data suggest that it may be possible to determine, non-invasively, the presence of meconium. In addition, the effect of meconium on amniotic fluid T_1 may obscure MR image contrast between the fetus and amniotic fluid in some cases of fetal distress.

The values for spin-spin and spin-lattice relaxation times measured in fetal sheep tissue demonstrate the range for these parameters (Table 3). These values are consistent with similar data from other fetal animals as reported in a recently published handbook of NMR data (Beall et al., 1984). These values are useful in determining the pulse sequence, interpulse intervals and repetition times which will optimize contrast when imaging the fetus. Several sources present computer programs for the algorithms which can be used to determine contrast and intensity for the different MR imaging sequences (Kay and Mattison, 1986; Mitchell et al., 1984).

The data on T_1 relaxation times of fetal blood, plasma and red blood cells are instructive because they illustrate that the relaxation time measured in a given tissue volume is actually a composite relaxation time -- formed by summing the relaxation times according to proportion of nuclei in a given environment (Weisman et al., 1981). For example:

$$1/T_{1total} = E (P_i/T_{1i}).$$

although it is clear that there are at least two distinct proton environments in fetal blood -- those inside of and those outside of the fetal RBC's, it is clearly more difficult to define the number of proton environments in other tissues. At the present time it is not known how many distinct proton environments are present in the placenta. In addition, it is not known how many of these different proton environments are accessible to Mn^{++}. Further definition of the interaction of Mn^{++} with the placenta may define the number of proton environments as well as provide information on the biochemical status of placental water in normal and abnormal pregnancy.

SUMMARY

The similarity of T_1 and T_2 in normal human placenta and myometrium prevents accurate assessment of implantation site and placental architecture during MR imaging. Mn^{++}, a paramagnetic ion, is taken up by placental slices in vitro shortening T_1, and to a lesser extent T_2, in a dose dependent fashion. Mn^{++} is also taken up by the non-human primate placenta in vivo, with a marked increase in contrast. Mn^{++} or other paramagnetic contrast enhancing agents may increase our ability to visualize the placenta in vivo and enhance the amount of biochemical or functional information we can collect during human pregnancy.

ACKNOWLEDGEMENTS

We would like to thank Ms. Marsha Gordon for her expert assistance in the preparation of this manuscript. Drs. J.F. Young, F. Evans, and D. Miller deserve special thanks for their critical reviews of this manuscript.

REFERENCES

Alger, J.R. and Shulman, R.G. (1984) Metabolic application of high-resolution ^{13}C nuclear magnetic resonance spectroscopy. *Br. Med. Bull.* 40, 160-164.

Beall, P.T., Amtey, S.R., and Kasturi, S.R. (1984) *NMR Data Handbook for Biomedical Applications*, Pergamon Press, New York, pp. 101-153.

Behar, K.L., DenHollander, J.A., Stromshi, M.E., Ogino, T., Shulman, R.G., Petroff, O.A.C., and Prichard, J.W. (1983) High resolution ^{1}H nuclear magnetic resonance study of cerebral hypoxia in vivo. *Proc. Natl. Acad. Sci. USA* 80, 4945-4948.

Bene, B.C. (1980) Diagnosis of meconium in amniotic fluid by nuclear magnetic resonance spectroscopy. *Physiol. Chem. Phys.* 12, 241-245.

Bene, G.J., Borcard, B., Graf, V., Hiltbrand, E., Magnin, P., and Noach, F. (1982) Proton NMR-relaxation dispersion in meconium solutions and healthy amniotic fluid: possible applications to medical diagnosis. *Z. Naturforsch* 37, 394-398.

Bock, J.L. (1985) Recent developments in biochemical nuclear magnetic resonance spectroscopy. *Methods Bioch. Analy.* 31, 259-315.

Borcard, B., Hiltbrand, E., Magnin, P., Rone, G.J., Briguet, A., Duplan, J.C., Delman, J., Guibaud, S., Bonnet, M., Dumont, M., and Fara, J.F. (1982) Estimating meconium (fetal feces) concentration in human amniotic fluid by nuclear magnetic resonance. *Physiol. Chem. Phy.* 14, 189-192.

Cady, E.B., Costello, A.M., Dawson, M.J., Delpy, D.T., Hope, P.L., Reynolds, E.O.R., Tofts, P.S., and Wilkie, D.R. (1983) Non-invasive investigation of cerebral metabolism in newborn infants by phosphorous nuclear magnetic resonance spectroscopy. *Lancet* 1, 1059-1062.

Cohen, J.S., Knop, R.H., Navon, G., and Foxall, D. (1983) Nuclear magnetic resonance in biology and medicine. *Life Chem. Rep.* 1, 281-457.

Damadian, R. (1981) *NMR in Medicine*, Springer-Verlag, New York.

Foster, M.A. (1984) *Magnetic Resonance in Biology and Medicine*, Pergamon Press, New York.

Fukushima, E. and Roeder, S.B.W. (1981) *Experimental Pulse NMR. A Nuts and Bolts Approach*. Addison-Wesley, Reading, MA.

Gadian, D.G. (1982) *Nuclear Magnetic Resonance and its Applications to Living Systems*. Oxford University Press, New York.

Gore, J.C. and Smith, F.W. (1985) Contrast Agents. *Magnetic Resonance Imaging* 3, 1- 97.

Johnson, I.R., Symond, E.M., Kean, D.M., Worthington, B.S., Broughton, Pipkin, F., Hawkes, R.C., and Gyngell, M. (1984) Imaging the pregnant human uterus with nuclear magnetic resonance. *Am. J. Obstet. Gynecol.* 148, 1136-1139.

Kay, H.H. and Mattison, D.R. (1986) Magnetic Resonance Imaging in Nonhuman Primates. In: *Animals in Perinatal Research*, (ed.), P.W. Nathanielsz, in press.

Maudsley, A.A. and Hilal, S.K. (1984) Biological aspects of sodium-23 imaging. *Br. Med. Bull.* 40, 165-166.

Meis, P.J., Hall, M., Marshall, J.R., and Hobel, C.J. (1978) Meconium passage: a new classification for risk assessment during labor. *Am. J. Obstet. Gynecol.* 131, 509-513.

Mitchell, M.R., Conturo, T.E., Gruber, T.J., and Jones, J.P. (1984) Two computer models for selection of optimal magnetic resonance imaging (MRI) pulse sequence timing. *Invest. Radiol.* 19, 350-360.

Moore, W.S. (1984) Basic physics and relaxation mechanisms. *Br. Med. Bull.* 40, 120-124.

Odeblad, E. (1959) Studies on vaginal contents and cells with proton magnetic resonance. *Ann. NY Acad. Sci.* 3, 189-206.

Odeblad, E. and Bryhn, U. (1957) Proton magnetic resonance of human cervical mucus during the menstrual cycle. *Acta Radiol.* 47, 315-320.

Odeblad, E. and Ingelman-Sundberg, A. (1965) Proton magnetic resonance studies on the structure of water in the myometrium. *Acta Obstet. Gynec. Scand.* 44, 117-125.

Odeblad, E. and Westin, B. (1958) Proton magnetic resonance of human milk. *Acta Radiologica* 49, 389-392.

Partain, C.L., James, A.E., Rollo, F.D., and Price, R.R. (eds.) (1983) *Nuclear Magnetic Resonance (NMR) Imaging*, Saunders, Philadelphia.

Pople, J.A., Schneider, W.G., and Bernstein, H.J. (1959) *High Resolution Nuclear Magnetic Resonance*, McGraw-Hill, New York.

Prichard, J.W., Alger, J.R., Behar, K.L., Petroff, O.A.C., and Shulman, R.G. (1983) Cerebral metabolic studies in vivo by ^{31}P NMR. *Proc. Natl. Acad. Sci. USA* 80, 2748-2751.

Radda, G.K., Bore, P.J., and Rajagopalan, B. (1984) Clinical aspects of ^{31}P NMR spectroscopy. *Br. Med. Bull.* 40, 155-159.

Smith, F.W., Adams, A., and Phillips, W.P. (1983) NMR imaging in pregnancy. *Lancet* 1, 61-62.

Stark, D.D., Hricak, H., and Paver, J.T. (1985) Intrauterine growth retardation: evaluation by magnetic resonance. *Radiol.* 155, 425-427.

Steiner, R.E. and Radda, G.K. (1984) Nuclear magnetic resonance and its clinical applications. *Br. Med. Bull.* 40, 113-206.

Thickman, D., Kressel, H., Gussman, D., Axel, L., and Hogan, M. (1984) Nuclear magnetic resonance imaging in gynecology. *Am. J. Obstet. Gynecol.* 149, 835-840.

Weisman, I.D., Bennett, L.H., Maxwell, L.R., and Henson, D.E. (1981) Cancer detection by NMR in the living animal. In: *NMR in Medicine*, (ed.), R. Damadian, Springer-Verlag, New York, pp. 17-37.

Wolf, G.L. (1984) Contrast enhancement in biomedical NMR: A symposium. *Physiol. Chem. Phys. Med.* NMR 16, 89-172.

Trophoblast Research 2:395-404, 1987

THE ACTIONS OF PROSTAGLANDINS AND THE INTERACTIONS OF ANGIOTENSIN II IN THE ISOLATED PERFUSED HUMAN PLACENTAL COTYLEDON

David G. Glance, Murdoch G. Elder, and Leslie Myatt[1]

Institutes of Obstetrics and Gynaecology
Hammersmith Hospital
Du Cane Road
London W12 OHS England

INTRODUCTION

Several lines of evidence suggest that in the fetal-placental circulation, the vasoconstrictive actions of angiotensin II may be modulated by prostaglandins. Firstly, prostaglandins have been shown to be vasoactive within the fetal-placental vaculature (Tulenko 1981; Mak et al., 1984). Secondly, angiotensin II is able to stimulate the release of PGE and prostacyclin (PGI_2) (a potent vasodilator) into the fetal circulation of the isolated perfused human placental cotyledon (Glance et al., 1985). Thirdly, in pathological pregnancies where umbilical blood flow is reduced (such as pre-eclampsia [Makila et al., 1983]), circulating fetal angiotensin II levels are increased (Broughton Pipkin and Symonds, 1977) and PGI_2 levels are decreased (Makila et al., 1983) when compared with levels found in normal pregnancies.

To study the actions of prostaglandins and their interaction with angiotensin II in the fetal-placental circulation, human placental cotyledons were perfused in vitro on both the fetal and maternal sides. Prostaglandins D_2, E_2, E_1, $F_{2\alpha}$, 6_β-PGI_1 (a stable analogue of PGI_2) and U46619 (a thromboxane A_2 mimetic) were injected either on their own or in combination with angiotensin II into the fetal circulation and their effects on fetal perfusion pressure measured. Angiotensin II was also injected after indomethacin treatment to assess the effect of endogenous prostaglandins on the vasoconstrictive actions of angiotensin II.

MATERIALS AND METHODS

Placentae from women with uncomplicated pregnancies were collected within 5 min of vaginal delivery or Cesarean section at term.

Placental cotyledons, free of infarcts or tears, were perfused by a method which has been described in detail elsewhere (Glance et al., 1984). Briefly this perfusion involved an open circuit technique perfusing tissue culture medium 199 (Gibco) containing polyvinylpyrrolidone (PVP-40T, 5% w/v) through a cannulated chorionic artery on the fetal surface of the placenta and allowing the effluent to flow through

[1]To whom correspondence should be addressed.
Submitted: 10 October 1985; Accepted: 31 August 1986

the cannulated, associated chorionic vein. Perfusion to the intervillous space was achieved by the insertion of two butterfly needles (21 G) through the maternal surface of the perfused cotyledon. The whole preparation was maintained at 37°C in a controlled temperature cabinet. The flow rates of the fetal and maternal perfusates were maintained at 4 and 10 ml/min, respectively. Lateral pressure was measured continuously by means of two pressure transducers (Devices, 4-327-L223) attached to the maternal and fetal inflow lines. The pressure contribution of the cannulae and tubing between the pressure transducer and tissue was of the order of 15 and 10 mm Hg on the maternal and fetal sides, respectively. The resulting perfusion pressures for both sides (including the pressure contribution from the tubing and cannulae) were between 30-60 mm Hg.

PDG$_2$, PGE$_2$, PGE$_1$, PGF$_{2\alpha}$ (Sigma), 6$_\beta$-PGI$_1$ and U46619 were stored in ethanol and made up daily in phosphate buffered saline. They were then diluted ten-fold with gassed perfusate to final injection doses of 0.005, 0.05, 0.5, 5.0, and 50.0 µg in 0.5 ml medium. To assess the effects of these drugs on the fetal perfusion pressure, bolus injections of the test drugs (0.5 ml) were made directly and rapidly ($<$ 15 sec) into the fetal inflow line of the perfused cotyledon after an equilibration time of 15-20 min had elapsed. A period of approximately 10 min was allowed between each injection. Doses were administered in ascending order with a maximum of 6 boli per experiment. Each experiment was performed three times in separate placentae. The onset of vasoactive response was approximately 60 sec and pressure had returned to baseline in less than 10 min.

To assess the interactions between the various prostaglandins and angiotensin II on fetal perfusion pressure, the test drugs were injected in increasing doses (0.005-50.0 µg) with a repeated dose of angiotensin II (0.5 µg) after a control injection of angiotensin II (0.5 µg) had been made. In three placentae, angiotensin II (0.5 µg) was injected 6 times to assess the reproducibility of the response.

To examine the interactions between endogenous prostaglandins and angiotensin II on fetal perfusion pressure, indomethacin (10^{-6} M) was infused into the fetal inflow line by means of a peristaltic pump (L.K.B.) after an initial injection of angiotensin II (0.5 µg) had been made. Fifteen min after the infusion of indomethacin had begun, angiotensin II (0.5 µg) was injected another 5 times into the fetal inflow line.

The degree of prostaglandin synthetase inhibition by the infusion of indomethacin (10^{-6} M) into the fetal inflow line was estimated by collecting fetal and maternal perfusates for 10 min periods commencing 15 min after the establishment of perfusion on the maternal side. Further fractions were collected for 10 min periods: i) immediately after injection of angiotensin II (0.5 µg), ii) 15 min after the commencement of indomethacin infusion (10^{-6} M) into the fetal inflow line, and iii) after injection of angiotensin II (0.5 µg) into the fetal side during the indomethacin infusion. Perfusion medium was passed through the tubing without the cotyledon in place and was collected (40 ml) to serve as a blank.

The effluents were centrifuged, extracted on Sep-Pak C$_{18}$ cartridges (Waters) and PGE measured by radioimmunoassay as described previously (Glance et al., 1985). The results were expressed as ng of PGE produced per min. Recovery of primary prostaglandins from the Sep-Paks ranged from 82-96% and no correction was made for recovery. The limit of detection in the radioimmunoassay was 6 pg/tube.

Analysis of Data

For the experiments involving the simultaneous injections of the prostaglandins with angiotensin II, all the results were expressed in terms of a percentage of the initial control injection of angiotensin II (% control response).

Straight lines were fitted to the log(dose) - response graphs of each individual experiment by the principle of least squares and mean $\pm$ SD taken for the gradients and constants for each drug. If the relation was not linear over the entire log(dose) range then the line was fitted to the most linear part where possible. In the case of the repeated injections of angiotensin II with or without indomethacin these were arbitrarily treated as log(dose) - response graphs in order to compare them with the other lines. Tests of significant difference were carried out using paired and unpaired t-tests where appropriate.

RESULTS

Bolus injections of PGD_2 and $PGF_{2\alpha}$ into the fetal side (0.005- 50.0 µg) gave small, dose-dependent increases in fetal perfusion pressure (Figure 1) as compared to the large, dose-dependent increases in fetal perfusion pressure observed with U46619 (0.005-5.0 µg) (Figure 2). However, PGE_1 (0.005-50 µg) gave small, dose-dependent decreases in fetal perfusion pressure when injected on the fetal side and PGE_2 and 6_β-PGI_1 had no discernible effect at all (Figure 1). The slopes of the log(dose)-response lines of PGD_2, $PGF_{2\alpha}$, and PGE_1 were all significantly different from zero (p $<$ 0.05 for $PGF_{2\alpha}$ and PGE_1 and p $<$ 0.01 for PDG_2) (Table 1).

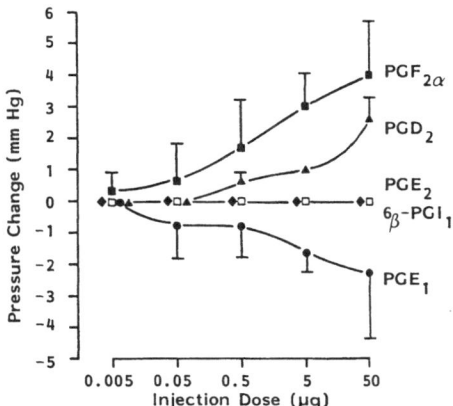

Figure 1. Changes in fetal perfusion pressure after bolus injections of various doses (0.005-50 µg) of PGE_1, PGE_2, PGD_2, PGF_2, and 6_β-PGI_1 into the fetal inflow line (mean $\pm$ SD, n=3 separate placentae). (With permission from the British Journal of Obstetrics and Gynaecology.)

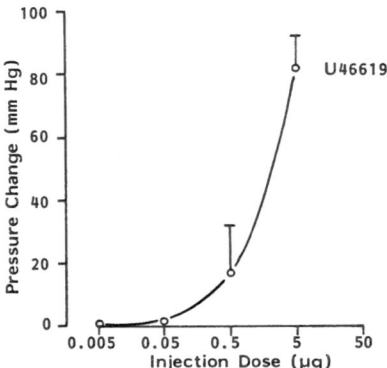

Figure 2. Changes in fetal perfusion pressure after bolus injections of various doses (0.005-5.0 µg) of U46619 into the fetal inflow line (mean ± SD, n = 3 separate placentae). (With permission from the British Journal of Obstetrics and Gynaecology.)

    Tachyphylaxis of the angiotensin II response (either by desensitization or by a time effect) was observed with repeated injections (Figure 3) when the fetal perfusion pressure increases were expressed as a percentage of that caused by the first injection (21.7 ± 5.5 mm Hg (mean ± S.D. n = 27)). Linearization of the angiotensin II response over the 5 injections by rbitrary assignment of increasing log(doses) on the x-axis, gave a line with a slope significantly different from zero (p < 0.01) (Table 1).

    Simultaneous injection of 6β-PGI_1, PGD_2, and PGE_1 (0.005-50.0 µg) with angiotensin II (0.5 µg) gave a dose-dependent attenuation of the angiotensin II response alone (Figure 3). The slopes of the log(dose)-response lines were all significantly less (p < 0.01 for PGD_2 and PGE_1 and p < 0.05 for 6β-PGI_1 (Table 1), than the slope of the line for angiotensin II alone reinforcing the concept of dose dependent attenuation of the angiotensin II response by these prostaglandins. The constants for the lines of PGD_2 and PGE_1 (but not 6β-PGI_1) with angiotensin II were significantly less than that for angiotensin II alone (p < 0.01 for PGD_2 and PGE_1) implying that the attenuation is significant at the lowest doses of PGE_1 and PGD_2 but not for 6β-PGI_1 (Table 1).

    Prostaglandin E_2 in increasing doses (0.005-50 µg) with angiotensin II (0.5 µg) gave a greater response than for angiotensin II alone but was more marked at the lower doses of PGE_2 (Figure 4). This is reflected by the similar slopes but greater constant of the lines for PGE_2 with angiotensin II compared to angiotensin II alone (Table 1). PGF_{2a} (0.005-50.0 µg) with angiotensin (0.5 µg) only gave a greater response than that for angiotensin II (0.5 µg) alone at the higher doses (Figure 4) where the slope of the log(dose)-response line was significantly greater than that of angiotensin II line alone (p < 0.01) but the constants were similar (Table 1).

Table 1

Substance	Parameters of Log dose-response lines	
	Slope	Constant
A II	-5.7 ± 1.2++	81.2 ± 5.1 (15)
A II + PGE_2	-4.3 ± 3.9	95.7 ± 4.6* (15)
A II + $PGF_{2\alpha}$	5.6 ± 1.0**	87.3 ± 5.5 (12)
A II + 6_β-PGI_1	-11.9 ± 2.5*	66.9 ± 8.6 (15)
A II + PGE_1	-23.0 ± 3.4**	41.9 ± 11.0** (12)
A II + PGD_2	-20.3 ± 0.8**	31.2 ± 4.3** (9)
A II + Indomet	-3.7 ± 8.7	105.7 ± 10.7* (15)
$PGF_{2\alpha}$	0.9 ± 0.2+	2.2 ± 1.2 (15)
PGD_2	0.6 ± 0.1++	0.9 ± 0.2 (15)
PGE_1	-0.7 ± 0.4+	-1.4 ± 0.9 (15)

Slopes and constants of log (dose)-response lines shown in Figures 1-6. The lines were calculated by principle of least squares. The lines for the repeated injections of angiotensin II (0.5 µg) either with or without indomethacin infused into the fetal inflow line (indomet, 10^{-6} M) were derived by arbitrarily taking the same x-axis as for the other lines. Numbers in brackets indicate the number of points used in calculating the lines by principle of least squares. [Statistical significance as compared with angiotensin II on its own by unpaired t-test (* = $p < 0.05$, ** = $p < 0.01$) or tested for difference with zero by t-test (+ = $p < 0.05$, ++ $p < 0.01$). Results are expressed as mean ± SD for three experiments.]

U46619 (0.005-5.0 µg) with angiotensin II gave a significantly greater response than for angiotensin II alone at the 5 µg dose ($p < 0.01$, Figure 5).

Infusion of indomethacin (10^{-6} M) into the fetal inflow line did not cause any change in baseline fetal perfusion pressure but led to greater increases in fetal perfusion pressure with repeated injections of angiotensin II (0.5 µg) (Figure 6). The constancy of the increased angiotensin II response following indomethacin is reflected by the similar slopes but significantly different constants ($p < 0.05$) (Table 1). Indomethacin infusion (10^{-6} M) into the fetal inflow line caused a significant decrease in the amount of PGE released into the fetal effluent ($p < 0.01$) but did not significantly affect maternal release (Table 2).

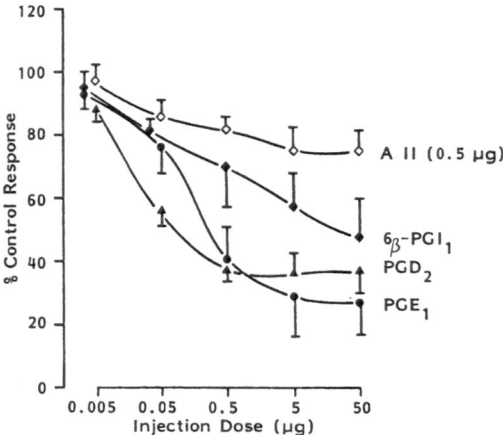

Figure 3. Percentage change in fetal perfusion pressure increase of initial bolus injection of angiotensin II (0.5 µg) when injected repeatedly either on its own or in combination with various doses (0.005-50.0 µg) of 6_β-PGI$_1$, PGE$_1$, and PGD$_2$ (mean $\pm$ SD, n = 3 separate placentae). (With permission from the British Journal of Obstetrics and Gynaecology.)

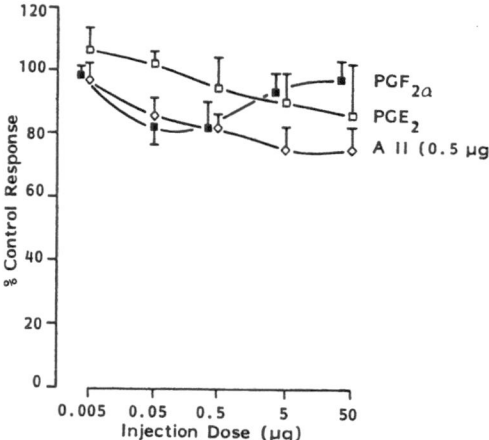

Figure 4. Percentage change in fetal perfusion pressure increase of initial bolus injection of angiotensin II (0.5 µg) when injected repeatedly either on its own or in combination with various doses (0.005-50.0 µg) of PGF$_{2\alpha}$ and PGE$_2$ (mean $\pm$ SD, n = 3 separate placentae). (With permission from the British Journal of Obstetrics and Gynaecology.)

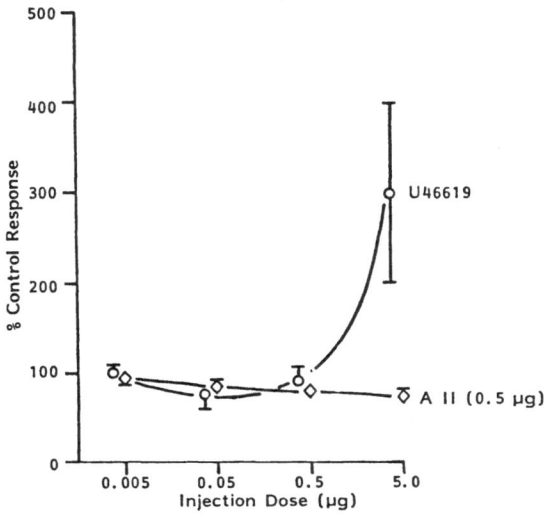

Figure 5. Percentage change in fetal perfusion pressure increase of initial bolus injection of angiotensin II (0.5 µg) when injected repeatedly either on its own or in combination with various doses (0.005-5.0 µg) of U46619 (mean ± SD, n = 3 separate placentae). (With permission from the British Journal of Obstetrics and Gynaecology.)

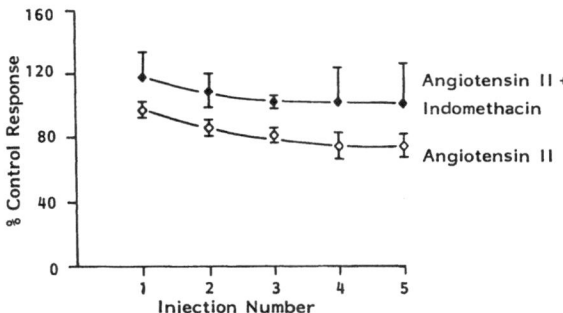

Figure 6. Percentage change in fetal perfusion pressure increase of initial bolus injection of angiotensin II (0.5 µg) when injected repeatedly either on its own or after infusion of indomethacin (10^{-6} M) into the fetal inflow line (mean ± SD, n = 3 separate placentae). (With permission from the British Journal of Obstetrics and Gynaecology.)

Table 2

	Control	Indomethacin $(10^{-6}$ M)	Angiotensin II $(0.5$ μg)	Indomethacin $(10^{-6}$ M) + Angiotensin II $(0.5$ μg)
Fetal PGE Release	0.37 ± 0.06	0.1 ± 0.05	4.7 ± 0.6	0.7 ± 0.2
(ng·min^{-1})		$(p < 0.01)^*$	$(p < 0.01)^*$	$(p < 0.01)^+$
Maternal PGE Release	1.3 ± 0.13	1.07 ± 0.2	2.3 ± 0.7	1.5 ± 0.45
(ng·min^{-1})		$(p > 0.05)^*$	$(p > 0.05)^*$	$(p > 0.05)^+$

Table 2. PGE release from both sides of perfused cotyledons. Samples were taken after a control period, after bolus injection of angiotensin II (0.5 μg), after indomethacin infusion (10^{-6} M) into the fetal inflow line and after injection of angiotensin II (0.5 μg) during infusion of indomethacin (10^{-6} M). [Statistical significance as compared with respective control release (*) and release by angiotensin II (+) by paired t-test. Results are mean $\pm$ SD from 3 placentae.]

Bolus injection of angiotensin II (0.5 µg) into the fetal side which significantly increased the amount of PGE released into the fetal effluent when injected prior to the infusion of indomethacin ($p < 0.01$) (Table 2) failed to increase the release of PGE to the same level following indomethacin (Table 2). Injection of angiotensin II (0.5 µg) into the fetal side did not significantly alter the release of PGE into the maternal effluent either before or after indomethacin infusion on the fetal side (Table 2).

DISCUSSION

Bolus infusion of agents known to be local hormones released on demand for paracrine or autocrine action may be more analogous to the in vivo situation than constant infusion. Prostaglandins E_1, D_2, and 6_β-I_1 were able to significantly attenuate the vasoconstrictive response of the fetal-placental vasculature to angiotension II at doses which did not cause very large changes in fetal perfusion pressure if injected alone. Similarly, PGE_2 and $PGF_{2\alpha}$ increased the response to angiotensin II at doses which did not cause very large changes (when injected independently). The actions of $PGF_{2\alpha}$, with angiotensin II were additive and not synergistic. The effect of U46619 on the angiotensin II response was not surprising in view of its own potent vasoconstrictive actions and again appeared to be additive.

There was no change in baseline fetal perfusion pressure when endogenous production of prostaglandins in the fetal-placental vasculature was decreased by indomethacin treatment suggesting that prostaglandins do not maintain fetal vascular tone but modulate the vasoconstrictive actions of angiotensin II on fetal placental vessels. This view is reinforced by the observation that inhibition of endogenous prostaglandin production on the fetal side by indomethacin was able to significantly increase the vasoconstrictive actions of angiotensin II when injected into the fetal circulation. Attenuation of the vasoconstrictive responses to angiotensin II by prostaglandins would preserve the low vascular resistance of the fetal-placental circulation in the face of vasoconstriction by circulating angiotensin II.

Previously, we have shown that angiotensin II can stimulate the release of PGE and PGI_2 into the fetal circulation of the isolated perfused human placental cotyledon (Glance et al., 1985). Although the radioimmunoassay employed in these studies could not differentiate PGE_1 and PGE_2 it is presumed that PGE_2 is the major product formed by the placenta since arachidonic acid (the precursor of PGE_2) is present in much greater quantities than dihomo-γ-linolenic acid (the precursor of PGE_1) (Robertson et al., 1968). Angiotensin II may then cause increased release of both PGI_2 and PGE_2 which have opposing actions on the angiotensin II response in the fetal-placental vessels. The property of indomethacin to potentiate the actions of angiotensin II suggests that the balance of the prostaglandins stimulated by angiotensin II lies in favor of those which are attenuative in action.

Umbilical blood flow correlates with levels of 6-oxo-$PGF_{1\alpha}$ found in cord blood (Makila et al., 1983) and umbilical arteries show a lack of sensitivity to the actions of angiotensin II when compared with the fetal-placental vasculature (Tulenko, 1979) suggesting that the same modulatory actions of PGI_2 operate within the umbilical vessels. This mechanism would support the argument that reduction in umbilical blood flow in pre-eclampsia is due to a defect in prostaglandin production by the fetal vessels.

SUMMARY

PGE_1, PGE_2, PGD_2, $PGF_{2\alpha}$, U46610, and 6_β-PGI_1 were administered as bolus injections both separately and in combination with angiotensin II into the fetal circulation of isolated perfused human placental cotyledons. $PGF_{2\alpha}$ and PGD_2 caused small dose-dependent increases in fetal perfusion pressure when compared to U46619 which acted as an extremely potent vasoconstrictor of the fetal-placental vasculature. PGE_1·caused very small dose-dependent decreases in fetal-perfusion pressure when injected on its own. PGE_1, PGD_2, and 6_β-PGI_1 caused significant, dose-related attentuations of the angiotensin II vasoconstrictive response whereas PGE_2, $PGF_{2\alpha}$, U46619, and indomethacin potentiated the response. The results indicate that prostaglandins may exert their effects on the fetal-placental circulation by modulating the actions of angiotensin II.

ACKNOWLEDGEMENTS

We are grateful to Action Research for the Crippled Child for their financial support.

REFERENCES

Broughton Pipkin, F. and Symonds, E.M. (1977) Factors affecting angiotensin II concentrations in the human infant at birth. *Clin. Sci. Mol. Med.* 52, 449-456.

Glance, D.G., Bloxam, D.L., Elder, M.G., and Myatt, L. (1984) The effects of the components of the renin-angiotensin system on the isolated perfused human placental cotyledon. *Am. J. Obstet. Gynecol.* 149, 450-454.

Glance, D.G., Elder, M.G., and Myatt, L. (1985) Prostaglandin production and stimulation by angiotensin II in the isolated perfused human placental cotyledon. *Am. J. Obstet. Gynecol.* 151, 387-391.

Mak, K.K.W., Gude, N.M., Walters, W.A.W., and Boura, A.L.A. (1984) Effects of vasoactive autacoids on the human umbilical-fetal placental vasculature. *Br. J. Obstet. Gynaecol.* 91, 99-106.

Makila, V.M., Jouppila, P., Kirkinen, P., Viinikka, L., and Ylikorkala, O. (1983) Relation between umbilical prostacyclin production and blood-flow in the foetus. *Lancet* 1, 728-729.

Robertson, A., Sprecher, H., and Wilcox, J. (1968) Free fatty acid patterns of human maternal plasma, perfused placenta and umbilical cord plasma. *Nature* 217, 378-379.

Tulenko, T.N. (1979) Regional sensitivity to vasoactive polypeptides in the human umbilicoplacental vasculature. *Am. J. Obstet. Gynecol.* 135, 629-636.

Tulenko, T.N. (1981) The actions of prostaglandins and cyclo-oxygenase inhibition on the resistance vessels supplying the human fetal placenta. *Prostaglandins* 21, 1033-1043.

TROPHOBLAST CULTURE

Trophoblast Research 2:407-421, 1987

DIFFERENTIATON OF HUMAN CYTOTROPHOBLAST INTO SYNCYTIOTROPHOBLAST IN CULTURE

Harvey J. Kliman[1], Michael A. Feinman, and Jerome F. Strauss, III

Departments of Pathology and
Laboratory Medicine, and
Obstetrics and Gynecology
School of Medicine
University of Pennsylvania
3400 Spruce Street
Philadelphia, Pennsylvania 19104 USA

INTRODUCTION

The trophoblast performs many important functions during pregnancy: it mediates the transport of nutrients between maternal and fetal circulations, and secretes steroid and protein hormones (Loke and Whyte, 1983). It is also believed to play an active role in immunoglobulin transport from mother to fetus (Wood et al., 1982), and in the prevention of rejection of the fetus by the mother (Lala et al., 1983). The regulation of these activities is poorly understood. The availability of a system to study purified trophoblasts in vitro would facilitate research on the cell biology of this tissue. In an earlier report (Kliman et al., 1986), we described such a system for 1) the isolation of cytotrophoblasts from human term placentae by enzymatic digestion and Percoll-gradient-centrifugation, and 2) the transformation of these mononuclear cells into syncytial structures in culture. In this report, with the aid of scanning electron microscopy, we document the morphological changes occurring as cytotrophoblasts form syncytiotrophoblasts in vitro. We also use immunocytochemistry to study the appearance of human chorionic gonadotropin (hCG) and pregnancy-specific β_1-glycoprotein (SP$_1$), as the trophoblasts differentiate from single cells to syncytiotrophoblasts. We also describe double antibody immunocytochemical localization of hCG and human placental lactogen (hPL). Finally, we examine the pattern of hCG and progesterone secretion by trophoblasts in culture.

MATERIAL AND METHODS

Preparation of Cytotrophoblasts

Cytotrophoblasts were isolated as described by Kliman et al. (1986). Briefly, a Percoll gradient centrifugation step was added to the enzymatic dispersion method

[1]To whom correspondence should be addressed: 2 Gibson, Department of Pathology and Laboratory Medicine, Hospital of the University of Pennsylvania, 3400 Spruce Street, Philadelphia, Pennsylvania 19104 USA
Received: 10 October 1985; Accepted: 6 June 1986

described by Hall et al. (1977). Normal term (36-42 wks of gestation) placentae were obtained immediately following spontaneous vaginal delivery or uncomplicated Cesarean section. After three trypsin-DNase digestions, the cell suspensions were pooled and placed on a pre-formed Percoll gradient. The gradient was made from 5 to 70% Percoll (vol/vol) in 5% steps of 3 ml each by dilutions of 90% Percoll (nine parts Percoll to one part 10X Hanks') with calcium and magnesium-free Hanks' and layered in a 50 ml conical polystyrene centrifuge tube. The gradient was centrifuged at 1200 x g at room temperature for 20 min. Following centrifugation, three distinct bands were noted (Figure 1): a bottom band (density = 1.075 g/ml) made up of erythrocytes and occasional polymorphonuclear leukocytes and lymphocytes, a top band (density = 1.017 to 1.033 g/ml) made up primarily of villous cores missing the trophoblast layer, and a middle band (density = 1.048 to 1.062 g/ml) made up of a homogeneous population of mononuclear cells (40-70 million cells/30 grams starting wet placental tissue) with an average diameter of 10 microns. These cells were shown to possess the structural and functional characteristics of cytotrophoblasts (Kliman et al., 1986). In our previous report, we utilizied immunocytochemistry to demonstrate that this cytotrophoblast preparation was not contaminated with 1) endothelial cells or fibroblasts, due to the absence of immunostaining for the intermediate filament protein vimentin; 2) macrophages, due to the absence of immunostaining with an antibody against the macrophage marker $alpha_1$ anti-chymotrypsin; or 3) fragments of syncytiotrophoblasts, due to the absence of staining for various hormones including hCG, hPL, and SP_1. The viability of the purified cytotrophoblasts was greater than 95% as assessed by trypan blue exclusion. We also showed by DNA-content analysis using 4',6-diamidino-2-phenylindole-stained cells that 15% of these Percoll-gradient-purified cytotrophoblasts were in G_2 or mitosis, consistent with evidence that the cytotrophoblasts are the mitotically active trophoblast (Galton, 1962). The middle band containing the cytotrophoblasts (d = 1.048-1.062 g/ml) was removed, washed once with Dulbecco's modified Eagle's medium containing 25 mM Hepes and 25 mM glucose (DMEM-H-G) and then resuspended in medium for tissue culture.

Culture of Cytotrophoblast

Purified cytotrophoblasts were diluted to a concentration of 4 or 8 x 10^5 cells/ml (for 16 or 35 mm dishes, respectively) with DMEM-H-G containing 4 mM glutamine, 50 µg/ml gentamicin and 20% (vol/vol) calf serum, heat inactivated (56°C for 30 min) calf serum, or heat inactivated fetal calf serum. The cells were plated into 16 mm or 35 mm Nunclon (Nunc, Roskilde, Denmark) culture dishes and incubated in humidified 5% CO_2-95% air at 37°C. Some 35 mm dishes contained a 22 mm square glass coverslip (#1). At selected time points the culture medium was removed, frozen at -20°C, and replaced with fresh medium. Coverslips were removed at selected time points for histological evaluation, immunocytochemical staining or scanning electron microscopy. Cells were cultured up to five days.

Transmission Electron Microscopy

Thin (1-5 mm) slices of fresh placenta or Percoll-gradient-purified cytotrophoblasts were suspended in half-strength Karnovsky's fixative (Karnovsky, 1965) at room temperature. Cells were resuspended in the same fixative, pelleted at 1000 x g for 10 min and both preparations were allowed to fix overnight at 4°C. The slices of placenta were minced to 1 mm cubes and then both preparations were transferred to

0.1 M cacodylate buffer, pH 7.4, post-fixed with 1% osmium-tetroxide for 1 hr, dehydrated in graded ethanols and propylene oxide and embedded in Epon. Silver sections were stained with uranyl acetate and lead citrate. For transmission electron microscopy of cytotrophoblasts in culture, cells were plated on aclar dishes and processed according to Paavola et al. (1985). Pale gold to silver thin sections were cut perpendicular to the bottom of the culture dish, and then stained with uranyl acetate and lead citrate. All sections were viewed at 50 kV in a Hitachi 600 electron microscope.

Scaning Electron Microscopy

Cells cultured on glass coverslips were processed at the times indicated by first rinsing with warm DMEM-H-G followed by fixation with half-strength Karnovsky's fixative (Karnovsky, 1965) for 1 hr at room temperature. The cultures were then dehydrated in graded ethanols, critical point dried in a Tousimis Autosamdri unit (Tousimis, Rockville, MD), using absolute ethanol and liquid carbon dixoide, coated with a 30 nm layer of gold in a SPI sputter coater (Structure Probe Inc., West Chester, PA) and viewed in an ETEC scanning electron microscope (Perkin-Elmer ETEC, Heywood, CA).

Histological and Immunohistochemical Staining of Cultured Trophoblast

At the indicated times coverslips were washed twice with 150 mM NaCl-10 mM phosphate buffer, pH 7.4 (PBS), fixed for 15 min with Bouin's solution and washed twice with PBS at room temperature. Coverslips were immersed in PBS at 4°C for up to 7 days prior to staining. All subsequent steps were performed at 24°C. While still in the culture dishes, the coverslips were washed twice with PBS and once with PBS-1% bovine serum albumin (PSB-BSA). Endogenous peroxidase activity was quenched by a 15 min incubation with 0.6% H_2O_2 in PBS. Non-specific IgG binding sites were blocked by a 30 min incubation with 5% goat serum (Dako, Santa Barbara, CA) in PBS. Prior to the next and all subsequent immune reactions, the coverslips were washed twice with PBS and once with PBS-BSA. Primary antisera raised in rabbits were diluted in PBS and applied for 45 min. These included antibodies against the beta chain of human chorionic gonadotropin (beta-hCG; Dako) at 1:1000 dilution, human placental lactogen (hPL; Calbiochem-Behring, LaJolla, CA) at 1:800, pregnancy-specific β_1-glycoprotein (SP$_1$; Boehringer-Manheim, Indianapolis, IN) at 1:800, and the alpha chain of hCG (alpha-hCG; a gift from Drs. Canfield and Birken, Columbia Medical Center, NY) at 1:800. The primary antibodies were visualized using an avidin-biotin peroxidase detection method (Hsu et al., 1981) with a kit purchased from Vector Laboratories (Burlingame, CA). In most experiments, 3,3'-diaminobenzidine (DAB) was used as the color reagent (producing a dark brown color), while in the double staining method, 3-amino-9-ethyl carbazole (AEC) was used as the second color reagent (producing a red color). For single primary antibody immunocytochemistry, following the reaction with DAB, the coverslips were rinsed twice with tap water, counterstained with hematoxylin, dehydrated in absolute ethanol, dipped in xylene until cleared and mounted on glass slides with Preservaslide (E.M. Science, Cherry Hill, NJ).

For dual immunocytochemistry, following the reaction with DAB, coverslips were washed twice with PBS and stored overnight at 4°C. Subsequently, the coverslips were again washed twice with PBS, followed by a 30 min quenching step with 0.6% H_2O_2 in PBS. The coverslips were then processed as described above for application of the second primary antibody starting with the PBS and PBS-BSA washes for 5% goat serum incubation, except that at the final step, AEC was substituted for DAB. The AEC solution was prepared as follows: 4 mg AEC was dissolved in 1 ml dimethyl formamide, diluted with 19 ml 30 mM acetate buffer, pH 5.0, and filtered first with a 0.45 μm syringe filter and then with a 0.22 μm syringe filter to remove undissolved material. Eighteen μl of 30% H_2O_2 was added and the solution applied to the coverslips for 20 min. The coverslips were then washed three times with PBS, counterstained with hematoxylin, washed with tap water three times, and mounted on glass slides with Gelvatol (Monsanto, Indian Orchard, MA), a water soluble mounting media. Photomicrographs were taken with a Nikon Optiphot microscope.

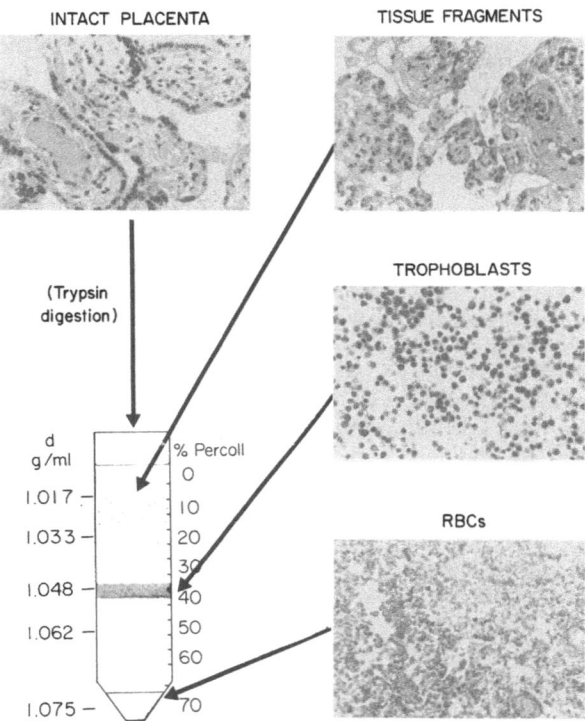

Figure 1. Purification of cytotrophoblasts from human term placentae. 30 g of soft villous tissue are minced and digested with three consecutive exposures to trypsin-DNase. The dispersed tissue is pooled and applied to a 5-70% Percoll gradient and centrifuged at 1200 x g for 20 min. Three bands can be identified: 1) a top band made up of debris, and villous cores minus the surrounding trophoblast layer; 2) a middle band of cytotrophoblasts; and 3) a bottom band made up primarily of erythrocytes. Between the middle and lower bands a haze is present that consists of a mixture of lymphocytes and polymorphonuclear leukocytes.

Analytical Methods

Progesterone was quantitated by radioimmunoassay as described previously (DeVilla et al., 1972). hCG was assayed by radioimmunoassay using reagents purchased from Corning Medical (Medfield, MA). Protein content of cells solubilized in 0.2% Triton X-100 in 1 N NaOH was quantitated by the method of Bradford (Bradford, 1976) using bovine serum albumin as a standard.

RESULTS

Fine Structural Appearance of Cytotrophoblasts

Examination of cytotrophoblasts 1) in situ, 2) after Percoll-gradient-purification and 3) 3.5 hr after plating revealed cells with similar fine structural features (Figure 2). The cytotrophoblasts of the term placenta exhibited sparse stacks of rough endoplasmic reticulum (RER), perinuclear Golgi, few mitochondria, desmosomal connections to the syncytium, occasional coated pits, and interdigitating microvilli (Figure 2a). The freshly isolated cytotrophoblasts had many of these features (Figure 2b), with some cells showing occasional lipid droplets. Trophoblasts

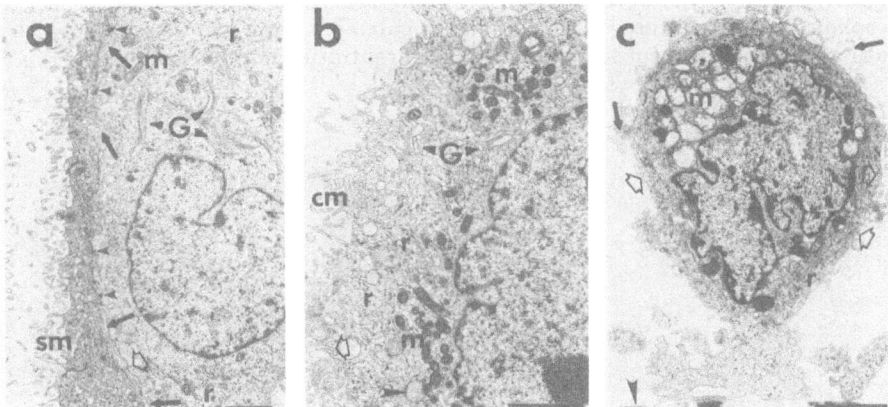

Figure 2. Transmission electron microscopy. Ultrastructural features of cytotrophoblasts in situ (a), Percoll-gradient-purified (b), and 3.5 hr following plating (c). (a), portion of chorionic villous of a term placenta showing syncytium with microvilli (sm) and underlying cytotrophoblast. Desmosomes (arrow heads) join the cytotrophoblast to the syncytial trophoblast. Folded cytotrophoblast microvilli can be seen filling in small spaces between cells (arrows). (b), portion of isolated Percoll-gradient-purified cytotrophoblast with surface microvilli (cm) and one cytoplasmic lipid droplet (arrow head). (c), cytotrophoblast beginning to adhere to collagen layer of aclar dish (arrow head). Cytoplasmic surface exhibits microvilli (arrows). Mitochondria (m), Golgi (G), segments of rough endoplasmic retriculum (r), coated pits (open arrows). Bars represent one micron.

examined 3.5 hr after plating again showed scattered segments of RER, perinuclear Golgi, few mitochonrdia, coated pits, and surface microvilli. Desmosomes were occasionally seen between clusters of cytotrophoblasts. These results further verify that the cells we purified on the Percoll-gradient and subsequently cultured are in fact cytotrophoblasts.

In Vitro Morphological Differentiation

When the Percoll-gradient-purified cytotrophoblasts were placed into culture in the presence of 20% serum, they underwent a series of morphological changes corresponding to differentiaton from mononuclear cytotrophoblasts to multinucleated syncytiotrophoblasts (Figure 3). The cytotrophoblasts adhered to the glass coverslips as early as 2.5 hr after plating. At these early time points the cells were still rounded and compact (Figures 3a and 2c). After 24 hr, the cells flattened out. As we have shown previously by time lapse cinematography (Kliman et al., 1986), these trophoblasts migrated towards each other to form cellular aggregates (Figure 3b), which subsequently fuse to form multinucleated syncytiotrophoblasts. Increasing numbers of syncytial structures were evident in our cultures between 24 and 72 hr (Figure 3c).

Serum supplementation appears necessary for the initial adherence of the cytotrophoblasts to the culture dishes, but the cell structures remained intact by morphological criteria when serum was removed from the medium after 48 hr of culture (data not shown). When heat-inactivated fetal calf serum, calf serum, and heat-inactivated calf serum were compared, few differences were noted with respect to the morphology of the cultures. Occasionally, a batch of serum was associated with poor plating efficiency, and subsequently, poor differentiation of the trophoblasts.

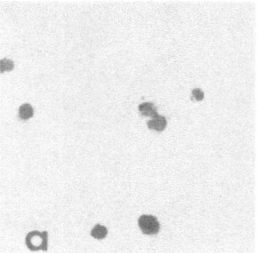 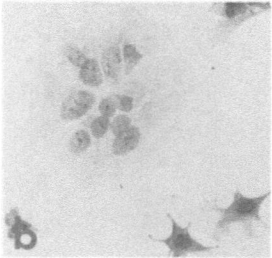 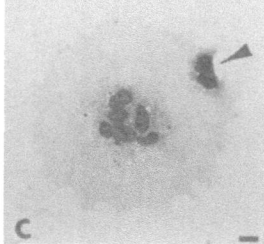

Figure 3. In vitro differentiation of human cytotrophoblasts. Purified cytotrophoblasts were cultured as described in Material and Methods. (a) 3.5 hr after plating, single cells and occasional groups of trophoblasts can be seen. The cells are still round (see Figure 2c for comparison) and thus appear dense. (b), 24 hr after plating, aggregates are the dominant trophoblast form, but single cells and occasional syncytia are present. (c), multinucleated syncytia are observed beginning at 24 hr after plating and steadily increase in numbers up to 72 hr after plating (Figure 5A). Two nuclei, representing either two cytotrophoblasts or one binucleated trophoblast, have recently been incorporated into the syncytium (arrow head). Bar represents 10 microns.

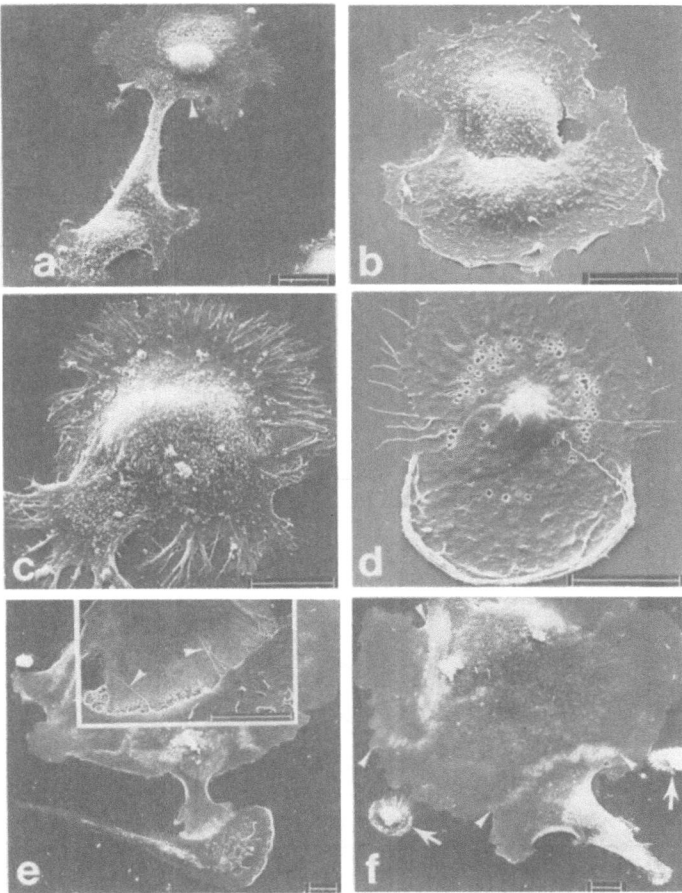

Figure 4. Scanning electron microscopy of cultured trophoblast. Cytotrophoblasts were prepared, cultured, and processed as described in the Materials and Methods. Cultures were examined 24 (a,b), 48 (c,d), and 96 hr (e,f) after plating. (a), two mononuclear trophoblasts approaching each other and making cell contact across a broad area of cell membrane (arrow heads). Note microvilli covering the surface of the lower cell. (b), two mononuclear trophoblasts in close apposition, similar to light microscopic images of the aggregate stage (Figure 3b). (c), trophoblast with large nuclear mound consistent with a recent fusion event. (d), flattened trophoblast with fine microvilli and ruffled membrane indicative of cell movement. Breaks in surface are an artifact of the freeze-drying process and appear to occur over areas of lipid droplets or other peri-nuclear vesicles. (e), large syncytial mass extending cytoplasm to make contact with mobile cell. Lower cell (inset) extends microvilli (arrowheads) over surface of syncytia near site of cell-to-cell contact. (f), forming syncytium showing central confluent area with at least two fusing cell masses (between arrow heads). Two apparent mononuclear trophoblasts at periphery of syncytia (arrows), extending microvilli to make initial contact. See Figure 6f for comparison. Bars represent 10 microns.

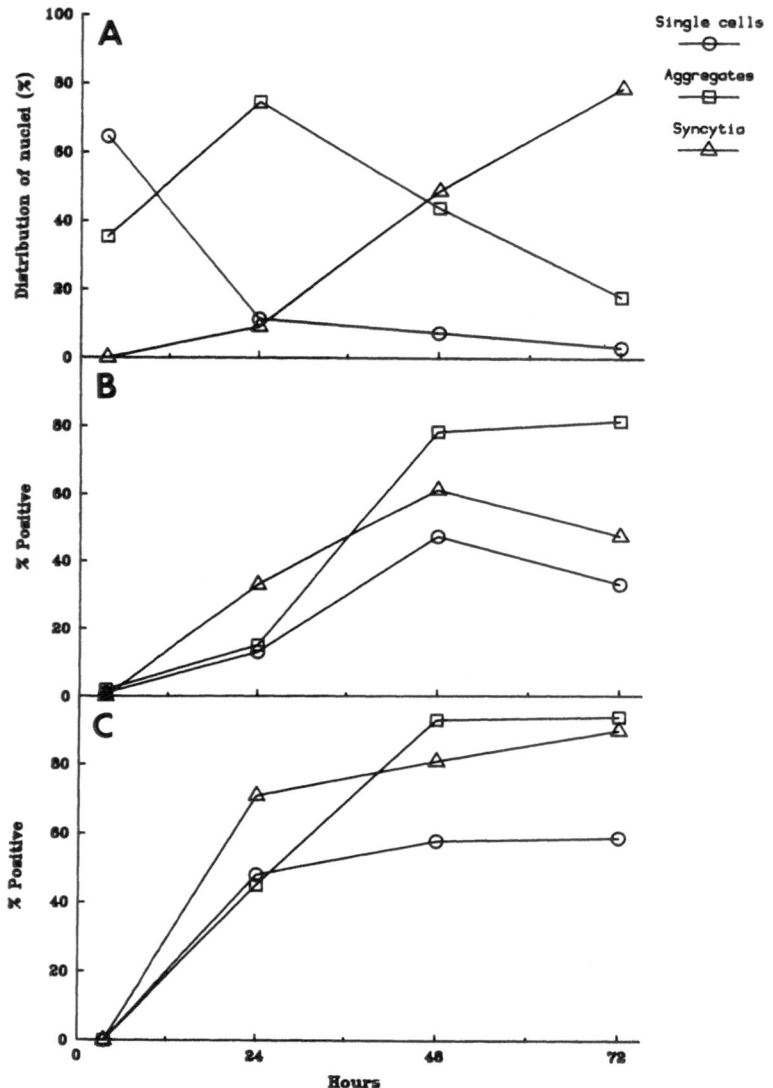

Scanning electron microscopy performed between 24 and 96 hr following plating revealed cellular interactions consistent with our previously reported concept of in vitro syncytiotrophoblast formation, that is, cellular aggregation followed by fusion to form syncytia (Figure 4). At 24 hr the beginning stages of cell aggregation could be seen (Figures 4a and 4b). The trophoblasts extend portions of cytoplasm to allow cell-to-cell contact (Figures 4a and 4e). Pairs and groups of closely applied cells could also be seen at 24 hr, consistent with the aggregate stage (compare Figure 4b to Figure 3b). At 48 hr, cellular masses with large central mounds were seen, consistent with multinucleated cells (Figure 4c). Also at this time, flattened cells consistent with the syncytiotrophoblasts seen at the light level (Figures 4d-f) could be identified, some with ruffled membranes (Figure 4d). At 96 hr, the cellular masses appeared even larger (Figures 4e and 4f). In Figure 4e, a cytoplasmic extension was seen in close contact with a cell which appeared to have been moving from left to right. At the point of contact, fine microvillous-like extensions were seen eminating from the lower cell. Figure 4f may represent a slightly later stage of cellular incorporation, with a cell mass joining a large flat syncytium in its lower right corner. Two individual cytotrophoblasts could also be seen probing the border of the large syncytium.

In Vitro Functional Differentiation of Cytotrophoblast

In addition to a morphological progression from single cells to multinucleated syncytiotrophoblasts, the cytotrophoblasts functionally differentiate in culture. Immunocytochemistry performed on cell cultures between 3.5 and 72 hr showed a progressive increase in the number of cellular elements producing two known markers of the syncytiotrophoblast: hCG and SP_1 (Figure 5). At 3.5 hr, when single cells were the dominant form, no staining for these markers was seen. At 24 hr, when aggregates became the dominant cell type, about 20% of the forms stained for hCG, while about 60% stained for SP_1. These numbers increased to about 55% and 80% at 72 hr for hCG and SP_1, respectively. These data demonstrate that the syncytial state is not necessary for the synthesis of these products since they appear at similar rates in all three cellular forms (single cells, aggregates, or syncytia).

Do these cultured trophoblasts make only one hormone product at one time, or can they simultaneously produce more than one? To answer this question, we performed a double antibody immunocytochemical study to examine structures for the presence of the alpha-subunit of hCG and hPL (Figure 6). Alpha-subunit of hCG

Figure 5. Acquisition of hCG and SP_1 by cultured trophoblasts. (A) At the times indicated following plating, coverslips were fixed and stained as described in the Materials and Methods. Beginning at a random point near the middle of each coverslip, 500 nuclei were counted in sequential high power fields and assessed as to whether they were in single isolated cells, single cells of an aggregate (two or more cells attached to each other or single cells attached to presumed syncytia), or apparent syncytia. (B) Percent cellular staining for beta-hCG. At the times indicated, coverslips were stained for beta-hCG by immunoperoxidase methods. Cells were evaluated as described above but in addition, cells were scored for immunoperoxidase staining. (C) Percent cellular staining for SP_1. Cells were evaluated as described above for beta-hCG.

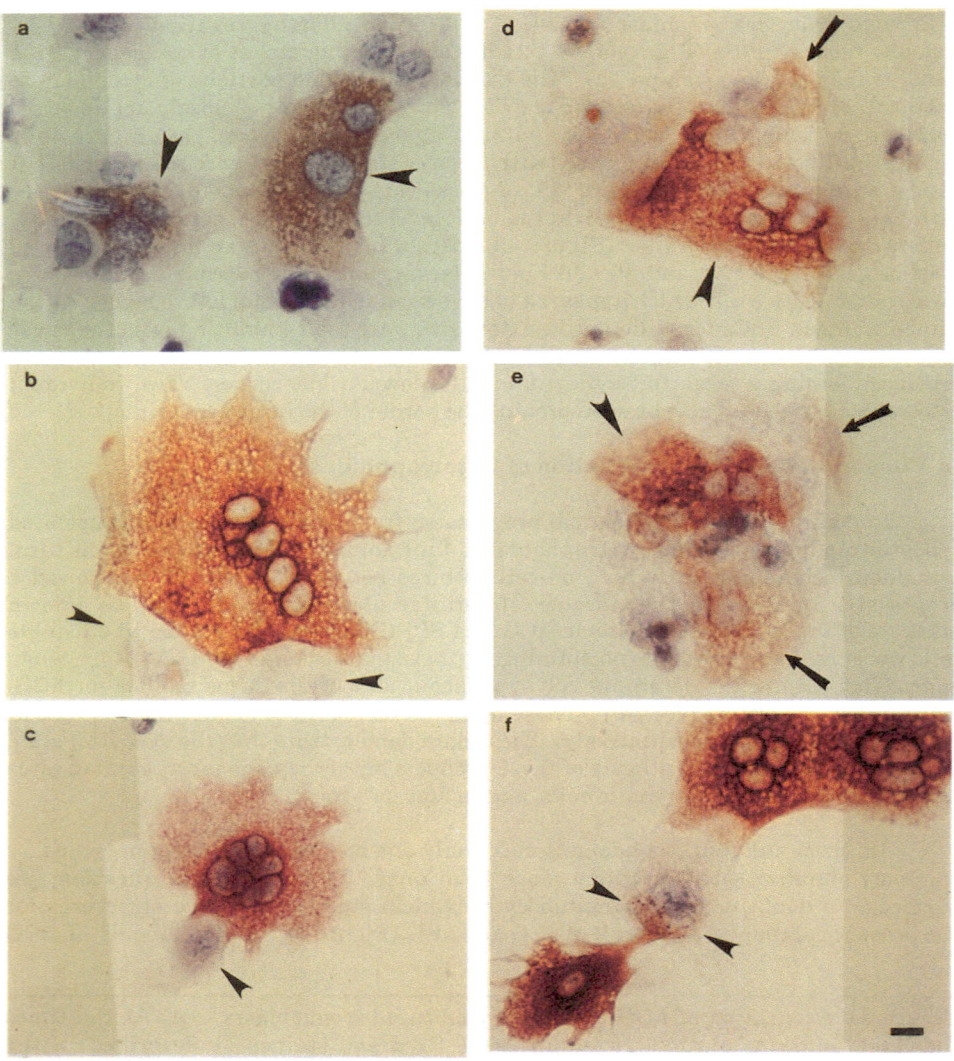

was visualized using DAB as the chromogen, producing a rich, dark, brown deposit (Figure 6a), while AEC, which produces a bright red deposit, was used to visualize hPL (Figure 6c). When the cultured trophoblasts were first reacted with anti-alpha-subunit antisera and DAB followed by anti-hPL and AEC, the stained cells appeared red-brown (Figures 6b, 6d-f). The majority of the reactive cells in this double immunostaining experiment appeared to be red-brown; i.e., they contained both the alpha-subunit and hPL. Occasionally, cells were seen which were only brown (Figures 6d and 6e), suggesting that cells can elaborate the alpha subunit without making hPL.

hCG and Progesterone Secretion

Do the cultured trophoblasts synthesize protein and steroid hormones at different rates as they differentiate from mononuclear cytotrophoblasts to multinucleated syncytiotrophoblasts? To begin to answer this question, we examined the daily secretion of hCG and progesterone between days 1 and 4 of culture. The results of a representative experiment are presented in Figure 7. During the first two days hCG was secreted at relatively low rates, corroborating our immunocyto-chemistry findings (Figure 5b). After two days, hCG secretion increased appreciably, reaching levels which were 14-fold above those on day 1 levels. In marked contrast, progesterone synthesis remained essentially constant throughout this same period.

DISCUSSION

The investigation of the biochemistry and cell biology of the human trophoblast can be greatly facilitated by the availability of a method which generates a highly purified population of cytotrophoblasts from the human placenta. We recently demonstrated that the addition of a Percoll-gradient-centrifugation step to standard enzymatic dispersion procedures yields such a preparation (Kliman et al., 1986).

Figure 6. Double immunostaining for alpha-hCG and hPL. Cultured trophoblasts were processed as described in the Materials and Methods. Times indicate hours in culture at time of fixation. (a), 24 hr, two syncytiotrophoblasts (arrow heads) stained for alpha-hCG alone with DAB as chromogen. Note brown staining of positive cells and absence of stain in surrounding trophoblasts. (b), 72 hr, syncytiotrophoblast stained for both alpha-hCG and hPL with DAB and AEC as chromogens, respectively. Staining is red-brown and appears distinct from either DAB (a) or AEC (c) stainnig alone. Negative trophoblasts (arrow heads). (c) 72 hr, syncytiotrophoblast stained for hPL alone with AEC as chromogen. Positive cell is red. A negative single cell can be seen closely opposed to the syncytium (arrow head). (d), 48 hr, an aggregate stained for both alpha-hCG and hPL showing that a portion of this cellular mass is a red-brown syncytium (arrowhead). A single brown cell (arrow) can also be identified. (e) 72 hr, another cellular aggregate composed of a syncytium which is red-brown (arrow head) and two single cells which appear brown (arrows). (f), 72 hr, four trophoblast groups, three of which are red-brown, and one which is unstained. The lower stained cell appears to be extending fine cytoplasmic processes towards the unstained cell (arrow heads). Note that these cytoplasmic extensions appear to end in small circular structures and extend over a wide portion of the cell surface. All photomicrographs are at the same magnification. Bar represents 10 microns.

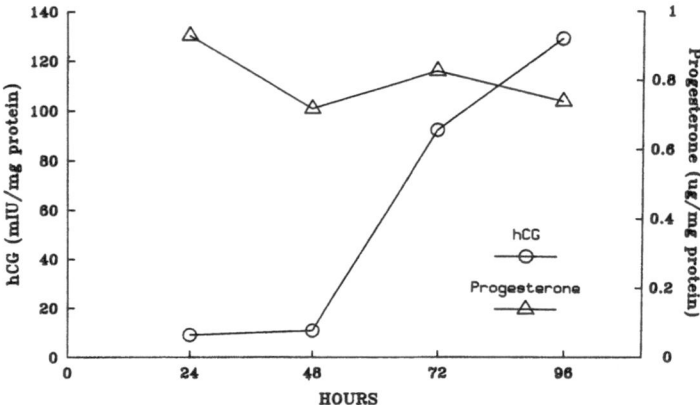

Figure 7. hCG and progesterone secretion by cultured trophoblasts. Cytotrophoblasts were plated in 16 mm diameter culture dishes as described in the Materials and Methods. At the times indicated, the culture media were removed, stored at -20°C, and replaced with fresh media. The media were then assayed for hCG and progesterone. Values are normalized to protein content in each well at the conclusion of the experiment. Values presented are means of triplicate determinations. Replicates varied less than 10%. This experiment was repeated on two other occasions with qualitatively similar results.

Using this method (Figure 1), we have further characterized the in vitro differentiation of these mononuclear trophoblasts through an aggregate stage and finally to mature multinucleated syncytiotrophoblasts which result from cell fusion (Figure 3). Transmission electron microscopy (Figure 2) of the purified cells and newly adherent cytotrophoblasts demonstrates that these cells have many of the characteristics seen in the cytotrophoblasts of the intact term placenta. Interestingly, the cytotrophoblasts generated by our method show a spectrum of cytoplasmic differentiation, with some having no lipid, little rough endoplasmic reticulum (RER) and few coated pits, indicative of an immature state, while other cells are more like syncytiotrophoblasts, containing many lipid droplets, extensive RER, and many coated pits. As others have noted (Dearden et al., 1983), this spectrum of cytoplasmic differentiation can be observed in cytotrophoblasts of the intact placenta.

In order to better characterize the formation of syncytiotrophoblasts in vitro, we performed scanning electron microscopy on the cultured trophoblasts between 24 and 96 hr (Figure 4). The images illustrate the progression of in vitro trophoblast transformation: cell contact (Figures 4a and 4e), approximation (Figure 4b), and finally fusion (Figure 4f). These stages can also be seen clearly at the light microscopic level (Figures 3 and 6). For example, Figure 6c illustrates a syncytium in close contact with a single cell which presumably would have become incorporated into the syncytium. Figure 6f illustrates cell-to-cell contact at the light microscopic level very similar to that observed in Figures 4a, 4e, and 4f by scanning electron microscopy. These data support our hypothesis that as the trophoblasts come in contact, fine cytoplasmic processes palpate neighboring cells, possibly to establish identity.

Electron microscopic (Boyd and Hamilton, 1966) and immunocytochemical studies (Kurman et al., 1984) have suggested the existence of trophoblasts which are intermediate between the cytotrophoblast and syncytiotrophoblast. The distinction between these two types of trophoblasts is morphologic: cytotrophoblasts are mononuclear whereas syncytiotrophoblasts are multinucleated. This morphologic distinction, however, does not help to discern functional differentiation. Our immunocytochemical data (Figure 5) support the concept that isolated mononuclear trophoblasts differentiate in culture at the same rate as mononuclear trophoblasts in aggregates, and as syncytiotrophoblasts. In other words, the syncytial state is not necessary for the capacity to synthesize hCG and SP_1. Thus, syncytium formation is only part of the process of trophoblast differentiation, not the initiator of differentiation.

To answer the question - do trophoblasts aquire the ability to form various secretory products at different times - we performed two experiments: double immunochemical staining for the alpha subunit of hCG and hPL, and measurement of secretion of progesterone and hCG between 1 and 4 days in culture.

The double immunostaining experiments (Figure 6) support the concept that mature trophoblasts can elaborate more than one protein hormone at a given time. Greater than 98% of all cellular structures stained in these experiments contained both the alpha-subunit of hCG and hPL. Very rarely, individual cells were found which stained only for the alpha-subunit (Figures 6d and 6e). Brown staining of syncytial groups was not noted. This observation suggests that there might be a differential acquisition of the ability to synthesize the alpha-subunit of hCG and hPL. This finding is consistent with the work of Boime and his colleagues (Hoshina et al., 1983), who have elegantly demonstrated, using choriocarcinoma cell lines, that syncytiotrophoblasts and some cytotrophoblasts contain mRNA coding for the alpha and beta-subunit of hCG, but hPL mRNA is found only in the syncytium. They have proposed that the alpha-subunit is expressed first, followed by the beta-subunit gene and then the hPL gene. This suggests that the sequential activation of genes is correlated with trophoblast differentiation.

Unlike hCG, SP_1 and hPL, progesterone secretion did not undergo major alterations during the culture period (Figure 7). Since cytotrophoblasts can synthesize progesterone from the time they are isolated from the placenta (Figure 7 and Kliman et al., 1986), the steroidogenic machinery is clearly expressed in cytotrophoblasts before peptide hormones are elaborated in quantity. In contrast to the pattern of progesterone secretion we observed, the capacity to form estrogens may increase with trophoblast differentiation, since Lobo et al. (1985) have recently reported aromatase activity increases during culture of dispersed placental tissue. However, it is notable that these authors found that aromatase activity increased prior to an increase in hCG secretion.

SUMMARY

We have developed a system whereby purified cytotrophoblasts can be isolated from human placenta. By examining these cells in culture we have elucidated the process by which syncytial structures are formed through cell aggregation and subsequent fusion, and have described the sequence in which various differentiated functions of the trophoblast are expressed. We have also observed that progesterone secretion precedes increased secretion of peptide hormones, and that elaboration of hCG subunits may preceed synthesis of hPL.

ACKNOWLEDGEMENTS

We gratefully acknowledge Dr. L. Paavola and members of her laboratory for assistance in preparing samples for transmission and scanning electron microscopy and for the use of the ETEC scanning electron microscope, Drs. Canfield and Birken for the anti-alpha-hCG, and Ms. Janet Brennar for the skillful preparation of the manuscript. This work was supported by a research grant from the Diabetes Research and Education Foundation to Dr. Kliman, USPHS Grant HD-06274 to Dr. Strauss and a grant from the Mellon Foundation to Dr. Feinman.

REFERENCES

Boyd, J.D. and Hamilton, W.J. (1966) Electron microscopic observations on the cyto-trophoblast contribution to the syncytium in the human placenta. *J. Anat.* 100, 535-548.

Bradford, M.M. (1976) A rapid and sensitive method for the quantitation of micro-gram quantities of protein utilizing the principle of protein-dye binding. *Anal. Biochem.* 72, 248-254.

Dearden, L. and Ockleford, C.D. (1983) Structure of human trophoblast: correlation with function. In: *Biology of Trophoblast*, (eds.), Y.W. Loke and A. Whyte, Amsterdam, Elsevier Science Publishers B.V., pp. 69-110.

DeVilla, G.O., Roberts, K., Wiest, W.G., Mikhail, G., and Flickinger, G. (1972) A specific radioimmunoassay of plasma progesterone. *J. Clin. Endocrinol. Metab.* 35, 458-460.

Galton, M. (1962) DNA content of placental nuclei. *J. Cell Biol.* 13, 183-191.

Hall, C. St. G., James, T.E., Goodyer, C., Branchand, C., Guyda, H., and Giroud, C.J.P. (1977) Short term tissue culture of human midterm and term placenta: parameters of hormonogenesis. *Steroids* 30, 569-580.

Hoshina, M., Hussa, R., Pattillo, R., and Boime, I. (1983) Cytological distribution of chorionic gonadotropin subunit and placental lactogen messenger RNA in neoplasms derived from human placenta. *J. Cell Biol.* 97, 1200-1206.

Hsu, S.M., Raine, L., and Fanger, H. (1981) Use of avidin-biotin peroxidase complex (ABC) in immunoperoxidase techniques: A comparison between ABC and unlabeled antibody (PAP) procedures. *J. Histochem. Cytochem.* 29, 557-580.

Karnovsky, M. (1965) A formaldehyde-glutaraldehyde fixative of high osmolarity for use in electron microscopy. *J. Cell Biol.* 27, 137a.

Kliman, H.J., Nestler, J.E., Sermasi, E., Sanger, J.M., and Strauss, J.F. III (1986) Purification, characterization and in vitro differentiation of cytotrophoblasts from human term placentae. *Endocrinol.* 118, 1567-1582.

Kurman, R.J., Main, C.S., and Chen, H.-C. (1984) Intermediate trophoblast: a distinctive form of trophoblast with specific morphological, biochemical, and functional features. *Placenta* 5, 349-370.

Lala, P.K., Chatterjee-Hasrouni, S., Kearns, M., Montgomery, B., and Colavincenzo, V. (1983) Immunobiology of the feto-maternal interface. *Immunol. Rev.* 75, 87-116.

Lobo, J.O., Bellino, F.L., and Benkert, L. (1985) Estrogen synthesis activity in human term placental cells in monolayer culture. *Endocrinol.* 116, 884-895.

Loke, Y.W. and Whyte, A. (eds.), (1983) *Biology of Trophoblast*, Amsterdam, Elsevier Science Publishers, B.V.

Paavola, L.G., Strauss, J.F. III, Boyd, C.O., and Nestler, J.E. (1985) Uptake of gold- and [^{3}H]cholesteryl linoleate-labeled human low density lipoprotein by cultured rat granulosa cells: cellular mechanisms involved in lipoprotein metabolism and their importance to steroidogenesis. *J. Cell Biol.* 100, 1235-1247.

Wood, G.W., Bjerrum, K., and Johnson, B. (1982) Detection of IgG bound within human trophoblast. *J. Immunol.* 129, 1479-1484.

Trophoblast Research 2:423-445, 1987

ISOLATION AND CHARACTERIZATION OF HUMAN CYTOTROPHOBLAST CELLS

Susan Daniels-McQueen[1], Alexander Krichevsky[2],
and Irving Boime[1,3]

[1]Departments of Pharmacology and Obstetrics/Gynecology
Washington University School of Medicine
St. Louis, Missouri 63110

[2]Department of Medicine
Columbia University
College of Physicians and Surgeons
New York, New York 10021

INTRODUCTION

One of the unique features of the placenta is its continued differentiation during gestation. We have previously shown by in situ hybridization, that human choriogonadotropin (hCG) α and β subunit mRNAs are expressed in both the cytotrophoblast and syncytiotrophoblast regions of the placenta whereas human placental lactogen (hPL) mRNA was expressed only in the syncytium. HCGα mRNA is reduced 6-fold between first trimester and term, and hCGβ mRNA falls to undetectable levels (Boothby et al., 1983). HPL, on the other hand, attains maximal levels at term although the amount of mRNA per gram of tissue is not greatly altered (Hoshina et al., 1982). Based on these data we proposed that the hormonal genes are activated at different stages of placental differentiation (Hoshina et al., 1982, 1983, 1985). We also suggested that hCG α and β subunit expression is dependent on the presence of proliferating cytotrophoblasts whereas hPL expression is sustained in differentiated syncytium.

Our findings in normal placentae were supported by probing trophoblastic neoplasms with hPL and HCGα and β subunit cDNAs. Choriocarcinoma consists of clusters of cytotrophoblast-like cells and large multinucleated cells. Signals corresponding to hCGα and β mRNAs were seen predominantly in the latter. By contrast, no hPL signal was observed. Hydatidiform mole tissue, which maintains villous morphology, gave positive signals for all three mRNAs (Hoshina et al., 1983). Based on these findings, we proposed the following model: hCG and hPL genes remain unexpressed in proliferating cytotrophoblast. The next stage, commitment of cytotrophoblast to syncytiotrophoblast, is associated with activation of the hCGα and β subunit genes. Such intermediates, especially those bearing the hCGβ subunit, are transient; they differentiate further at which point hCG α and β mRNAs decline while maximal hPL expression is coincident with syncytial function.

[3]To whom correspondence should be addressed.
Submitted: 10 December 1985; Accepted: 1 July 1986

        To test the predictions derived from this model we needed a cell culture system
which would allow us to examine the differentiation of cytotrophoblasts in vitro. Here
we describe a cell culture system which we have used to show that initial aggregated
cytotrophoblasts produce large amounts of hCG. With increasing time in culture hCG
production ceases, largely due to a rapid fall in β subunit production; decreased
amounts of α subunit continue to be secreted in the uncombined form.

MATERIALS AND METHODS

Culture of Cytotrophoblasts

        Placentae (8-12 wk) which were washed with Hank's balanced salt solution
(HBSS) (Hanks and Wallace, 1949) and phosphate buffered saline (PBS), were covered
with a solution containing 0.05% trypsin and 0.02% EDTA in 50 ml Falcon tubes and
kept under ice for 2 hr. These tubes were then incubated at 37°C for 30 min. The
initial trypsin solution was discarded, and tissue was again covered with
trypsin/EDTA and incubated another 30 min at 37°C.

        Trypsin-treatment was repeated 1 to 3 times until the syncytium had been
removed as assessed by phase contrast microscopy (Figure 1). During the above
procedures the tissue was periodically scraped gently with a rubber policeman.
Finally, the tissue was treated with type XIV protease (isolated from Streptomyces
griseus) from Sigma (0.4 mg/5 ml of HBSS) for 20 min at 37°C. This enzyme

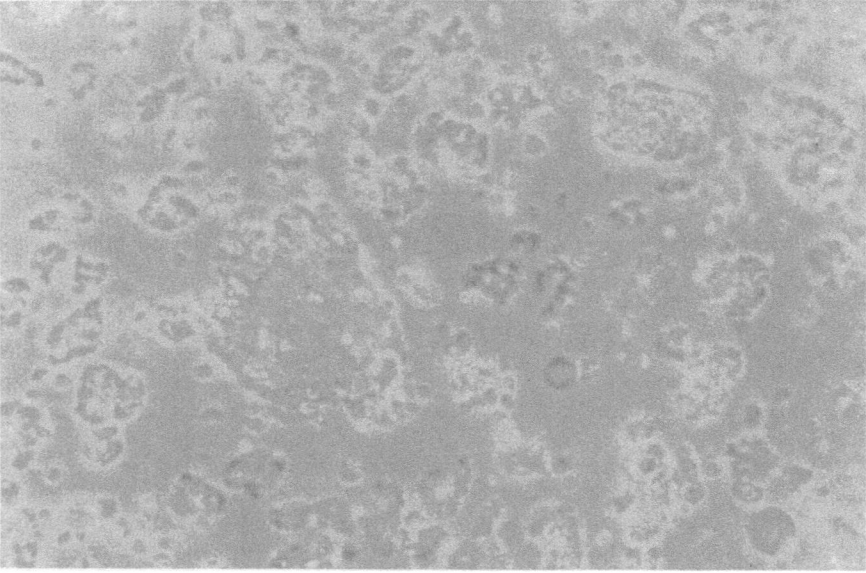

Figure 1. Appearance of syncytium by phase contrast microscopy following removal
from first trimester placenta with trypsin. (X150)

efficiently removes cytotrophoblast sheets (when present) which are composed of mononucleated cytotrophoblasts joined by desmosomes (Figure 2). Since released DNA prevents efficient recovery of cells, type III DNAase 1 from Sigma (0.04 mg/ml) was added when necessary and the tissue was incubated for 20 min at 37°C. The remaining tissue, which consisted of villi with exposed villous cytotrophoblasts, was covered with HBSS and incubated overnight at 0°C. Villous cytotrophoblasts are relatively resistant to removal by trypsinization and, this step facilitates trypsin treatment. The HBSS was discarded, and the tissue trypsinized for several 30 min cycles at 37°C. Most of the villous attached cells release during 2 to 3 cycles of trypsinization. The villous core was still intact, when we ceased collecting cells.

Cells were pelleted and washed twice with L15 (Leibovitz, 1963) medium containing 10% fetal calf serum (fcs), 100 μ/ml of penicillin and 100 μ/ml of streptomycin. Combined cell pellets were separated from tissue fragments/debris by short-burst centrifugation, washed with F12 (Ham, 1965) minus valine medium and cultured in this medium supplemented with 92 mg/L D-valine, 10% fcs, penicillin, and streptomycin. Cells were plated at a density of 4×10^6 cells/ml. The recovery of cells from a placenta at 10 wk of gestation was about 24×10^6 cells. After 3 to 4 hr, this media was removed and replaced with fresh D-valine supplemented F12 containing 10% dialyzed fcs and antibiotics or D-valine supplemented serum-free defined media (SFDM) (Murakami et al., 1982). All media were supplemented with EGF (0.1 μg/ml). Plating in the absence of EGF resulted in decreased survival of most cytotrophoblasts within the first 2 days of culture.

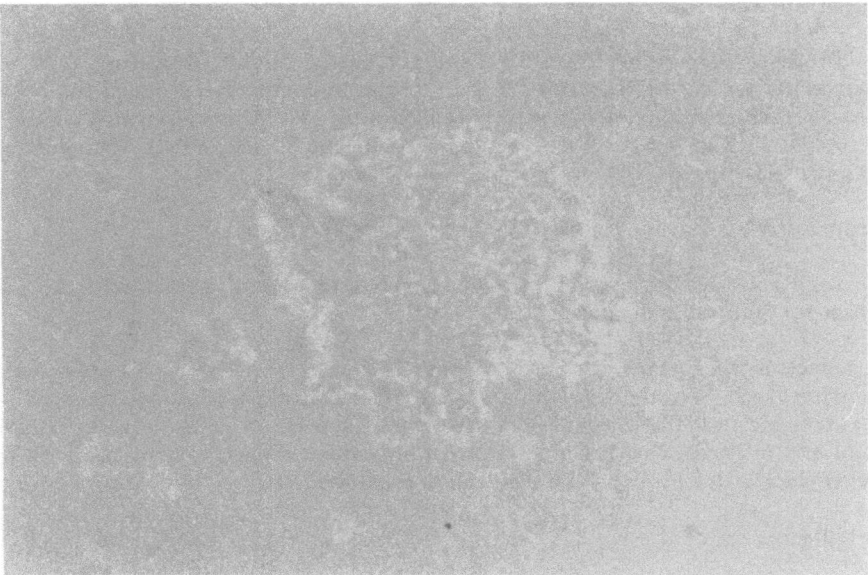

Figure 2. Sheets of cytotrophoblast cells released from a 12 wk placenta by protease. (X150)

Preparation of Tissue Culture Plates

Fibronectin coated wells were prepared by wetting wells with PBS containing 1 mg/ml of fibronectin and allowing the dishes to set at room temperature for 90 min with occasional swirling. The fibronectin solution was aspirated, and the wells were covered with PBS and refrigerated until needed.

Labeling of Trophoblasts

Cells were labeled for 24 hr in media consisting of equal volumes of Dulbeccos modified Eagles medium (Smith et al., 1960) containing high bicarbonate/high glucose (DMEHGHB) and F12 medium, supplemented with nonessential amino acids, L-glutamine (2 mM), unlabeled cysteine (20 µM), D-valine (92 mg/L), and [35S] cysteine (25 µCi/ml).

The medium was adjusted to 5 mM EDTA, 0.5% Triton, 20 mM Tris (pH 8), and 0.1% SDS. Equivalent volume of media were immunoprecipitated with antisera against the native hCGα or hCGβ subunits, and normal rabbit serum; reaction mixtures were incubated at 4°C for 20 hr. Pansorbin® was added (final concentration 50 µl/ml), and the samples were agitated for 3 hr at room temperature. Pellets were collected, washed once with PBS and dissolved in 25 µl of sample buffer [125 mM Tris HCl (pH 6.8), 4% SDS, 20% glycerol, 10% BME, and bromphenol blue] and boiled for 5 min. Equivalent volumes of the immunoprecipitates were resolved by gel electrophoresis.

Electrophoresis

Proteins were resolved on 20% polyacrylamide gels containing 0.1% SDS (Weber and Osburn, 1969) at 100-125 V until the indicator dye reached the bottom. Gels were removed and fixed in a solution containing 25% ethanol and 10% acetic acid then soaked with PPO by the APEX method (Jen and Thach, 1982). The gels were dried and exposed at -70°C to preflashed Kodak XAR-5 x-ray film.

Photography

Light microscopy: cells were observed and photographed with a Nikon microscope equipped with phase optics.

Electron microscopy was performed at the Electron Microscopy Center for Basic Cancer Research, Washington University Medical School. Cells were fixed in 2-5% glutaraldehyde in 0.1 M cacodylate buffer, post-fixed in 1% osmium, dehydrated in ethanol and infiltrated with Nadic methyl anhydride (NMA). Tissue sections were photographed with a Philips #201 electron microscope.

Materials

Cell culture media was obtained from the Cancer Center at the Washington University Medical School. Serum-free defined medium (SFDM) contained DME and F12 in a ratio of 1:1, 25 mM Hepes (pH 7.4), 1.2 g/L NaHCO$_3$, 2 mM L-glutamine, nonessential amino acids, 100 µ/ml penicillin, 100 µg/ml streptomycin, 35 mg/L human transferrin, 5 mg/L Bovine insulin, 20 μM ethanolamine and 25 μM NaSelenite. Mouse epidermal growth factor was prepared according to Savage and Cohn (Savage et al., 1972). Fetal bovine sera was purchased from KC Biological, fibronectin from the New York Blood Center, tissue culture flasks from COSTAR and Pansorbin® from Calbiochem Companies. [35S]Cysteine was obtained from Amersham Corporation. Dialyzed fetal bovine serum was prepared by dialyzing 20 ml of sera against 1 liter of PBS for 24 hr at 4°C with one change of dialysate. Antisera to the α and β subunits of hCG were gifts from Robert Canfield and Steven Birken (Columbia University). Anti-cytokeratin monoclonal antibody (PKK1) and antibodies directed against cytotrophoblast antigens were gifts from Charles Loke (University of Cambridge, Cambridge, England). The Vectastain mouse ABC kit was used for immunohistochemical staining.

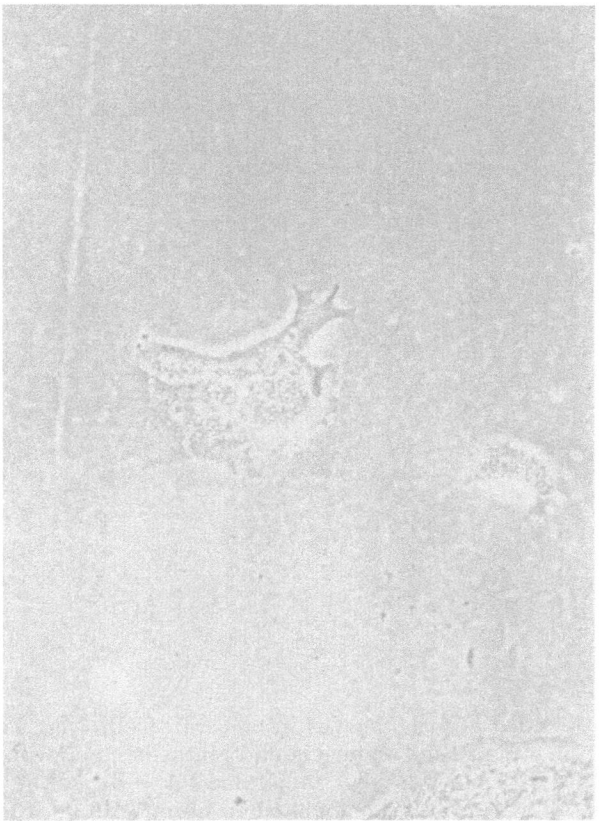

Figure 3. First trimester cytotrophoblast cell colony after 17 days in culture. Cells were plated on fibronectin and maintained in F12ssm/D-Val with EGF. Note the heavy "granulation" which seems characteristic of these cells. (X220)

RESULTS

The Cytotrophoblast in Culture

Cytotrophoblasts isolated from a first trimester placenta were transferred to fibronectin-coated tissue culture flasks (Figure 3). Plating on untreated flasks resulted in very poor cytotrophoblast recovery although other cell types will adhere. If cells are plated on untreated tissue culture grade plastic, the media can be poured off in 24 hr and again in 48 hr and floating cells replated on matrix. This replating will recover some of the cytotrophoblasts lost from the untreated surface. Replating was performed up to 6 days after the initial plating with resultant recovery of viable cytotrophoblasts.

Cytotrophoblasts were continuously maintained in D-valine media since cells cultured in L-valine media were eventually overgrown with fibroblasts (Gilbert and

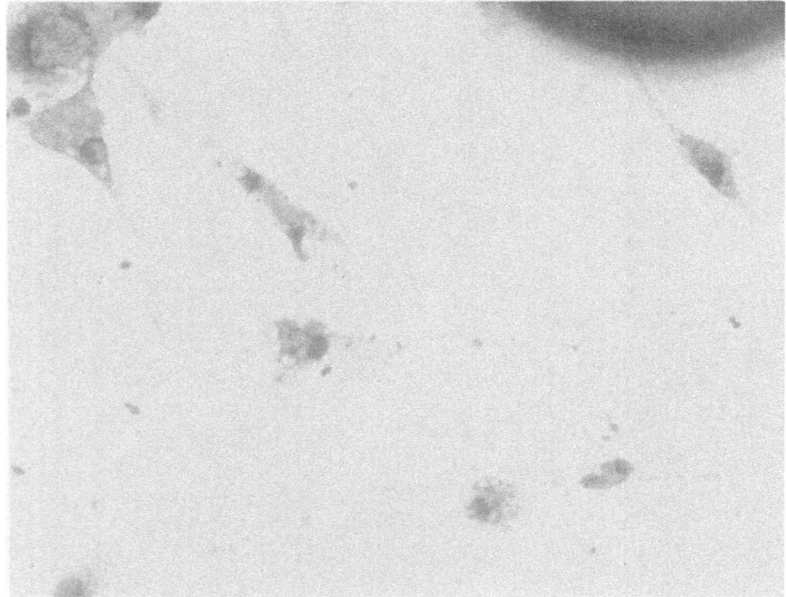

Figure 4. Identification of macrophages in culture. Cells were isolated from first trimester tissue as outlined in "Materials and Methods". Following removal of the cytotrophoblasts the tissue was kept under ice for 3 days, the remaining cells adhering to the villi were removed and plated on fibronectin in SFDM supplemented with D-Valine. The majority of the cells recovered appeared to be macrophages. After 8 days in culture the cells were fed zymosan. Forty-eight hr later they were fixed and stained with PAS and counter stained with hematoxylin Gill #3 (Sigma). All cells in the field were macrophages. They display both a stellate and fibroblast form. (X320)

Migeon, 1975). Cultures of established cytotrophoblast contain macrophages, some of which can be visualized by zymosan uptake. Figure 4 shows a typical field in a culture where the method of cell recovery enriched for a macrophage population (see Legend). The cells were allowed to engulf zymosan before fixing and staining. With phase contrast optics it may be difficult to differentiate macrophages from cytotrophoblasts since they are similar in size and appearance. These macrophages, however, do not divide; moreover in our normal culture procedure only a few are seen and thus are not a problem.

Differentation of Cytotrophoblasts in Culture

Freshly isolated cytotrophoblast cells transferred to fibronectin coated plates quickly undergo major transitions - from single cells to small aggregates and eventually to larger multinucleated ("syncytial") aggregates - with increasing time in culture (Figure 5, Panels A-E).

A

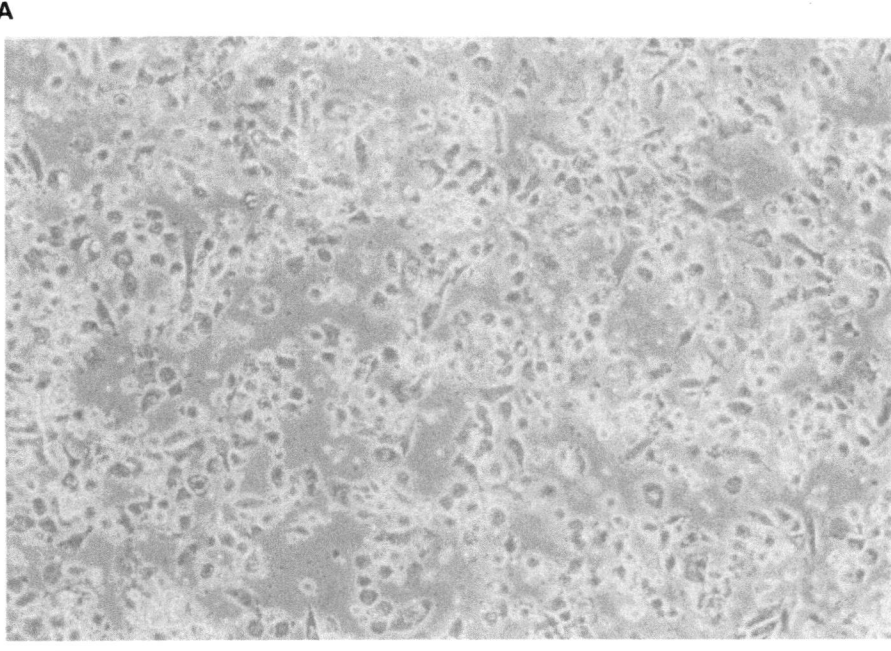

Figure 5. Morphological changes of cytotrophoblasts plated in serum-free defined medium containing D-Val (SFDM/Dval) or F12 serum supplemented medium containing D-Val (F12ssm/D-Val). A, Appearance of cytotrophoblasts after 24 hr in culture. At this time there is no obvious morphological difference between cells plated in SFDM/D-Val or F12ssm/D-Val (X80); B, 3 days in culture (X150); C, 13 days in culture (X75); D, 33 days (X150); and E, 43 days (X150). ———————————▶

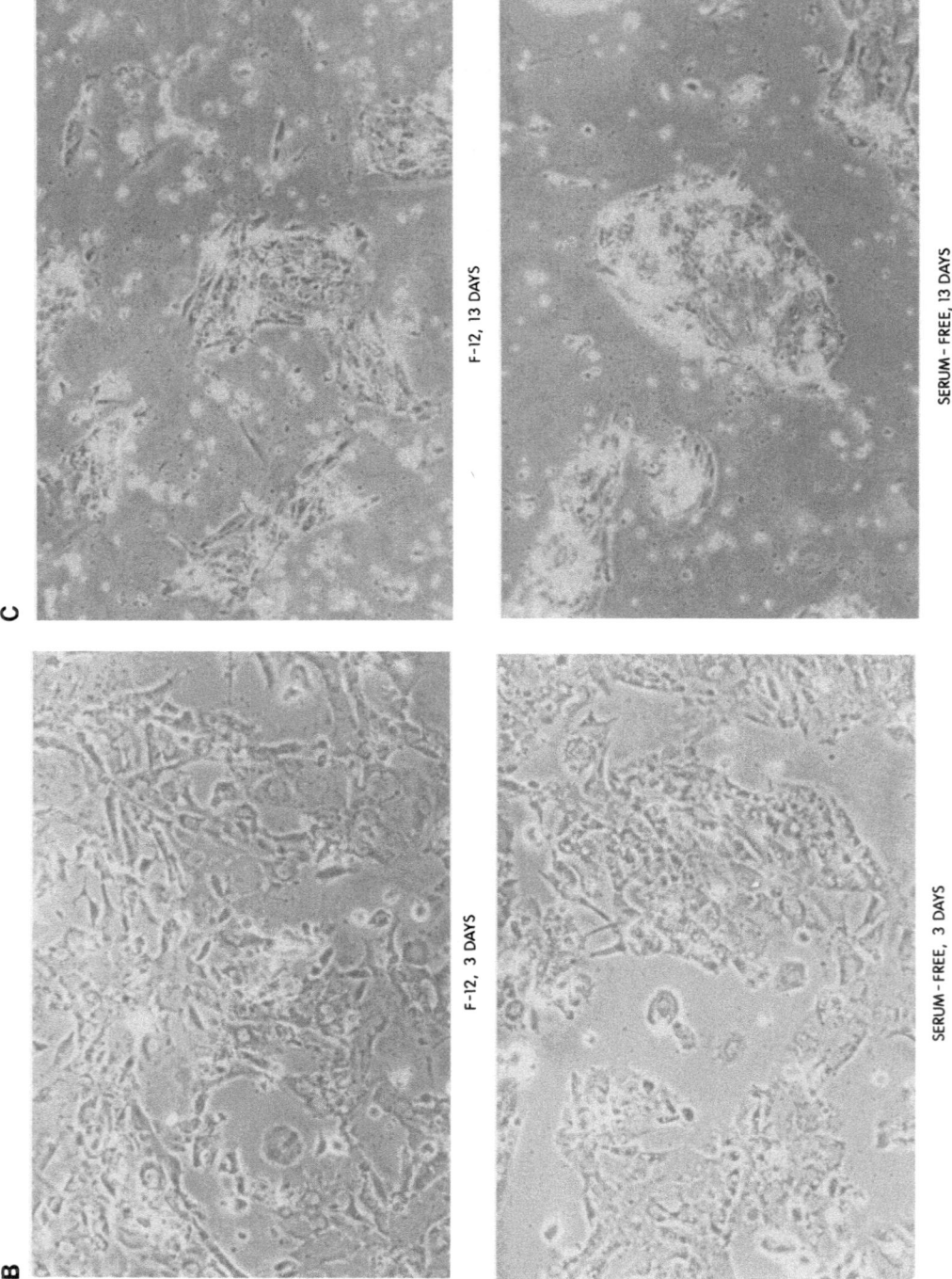

C

F-12, 13 DAYS

SERUM - FREE, 13 DAYS

B

F-12, 3 DAYS

SERUM - FREE, 3 DAYS

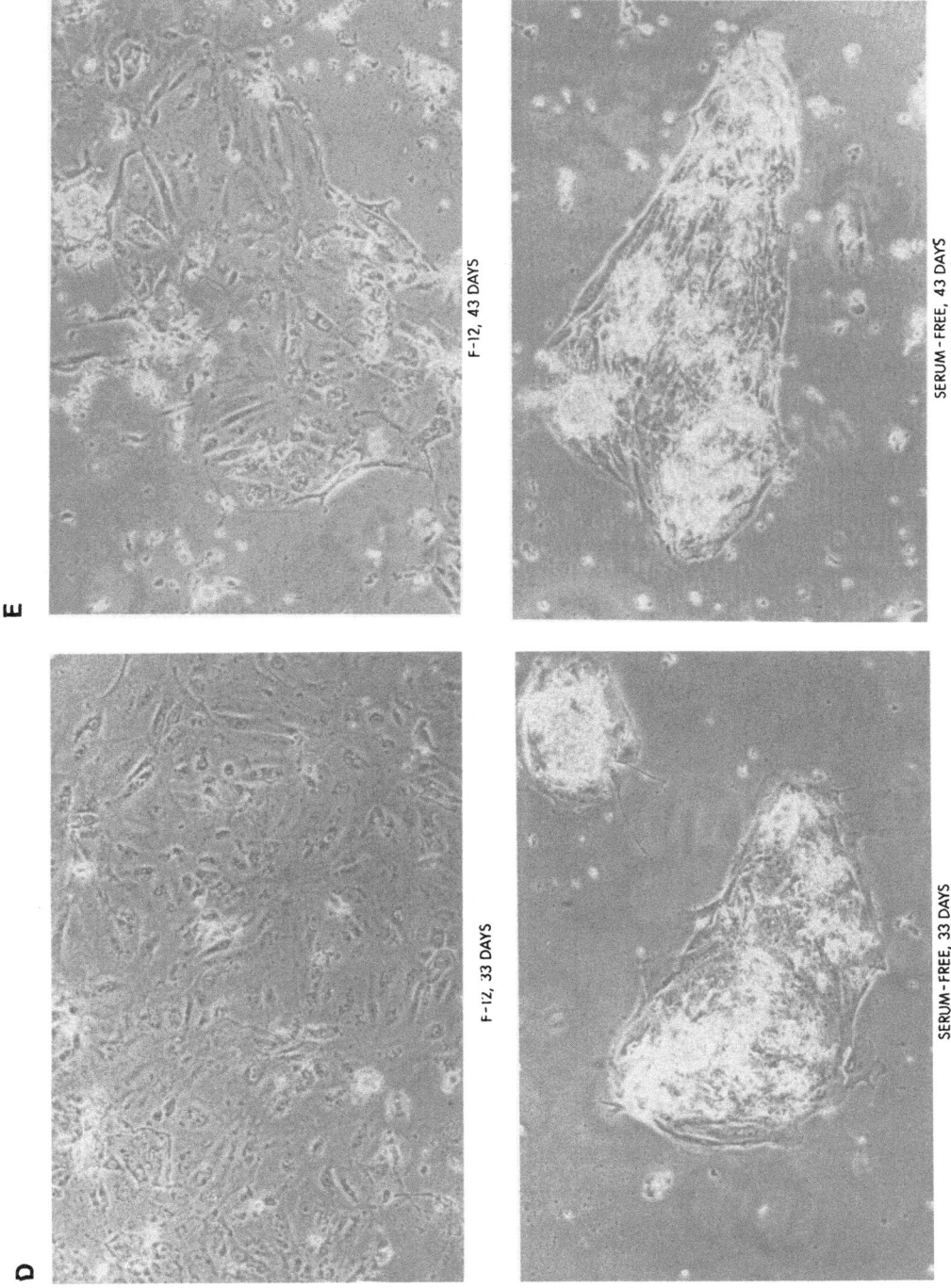

E

F-12, 43 DAYS

SERUM - FREE, 43 DAYS

D

F-12, 33 DAYS

SERUM - FREE, 33 DAYS

Cells 24 hr in Culture (Panel A)

Cells appears to aggregate rapidly after plating. This photomicrograph shows cells following 24 hr in culture. Each individual cell is discernible but the initiation of aggregates or island formation is apparent. Cells in either SFDM or F12 serum supplemented media (F12ssm) appear similar at this point. Cells illustrated here are in SFDM.

3 Days (Panel B)

The cells have now come together to form discrete islands. The entire culture well contains numerous cell aggregates some of which are linked through cell chains. A few individual cytotrophoblasts can still be seen migrating between the islands.

13-33-43 Days in Culture (Panels C-E)

With increasing time in culture, the difference between F12ssm and SFDM is clear. The SFDM cells form islands with more of a tertiary structure and are rounded up (compare to cells in F12ssm). Consequently, these islands are tighter and have a domed appearance. The cells in F12ssm are spread out and form aggregates which are larger and more two dimensional. In either case, the cells are viable since the rates of protein synthesis remain unchanged during this period (Table 1). From EM data the islands formed in SFDM appear to consist of 1 to 3 layers of cells.

Table 1

Trichloroacetic acid precipitable counts in 10 μl aliquots of the media used to generate the data for Figure 10

	Time in Culture (days)	CPM
SDFM	3	10,128
	7	16,270
	13	13,114
	21	13,102
F12ssm	3	15,632
	7	20,600
	13	16,458
	21	30,884

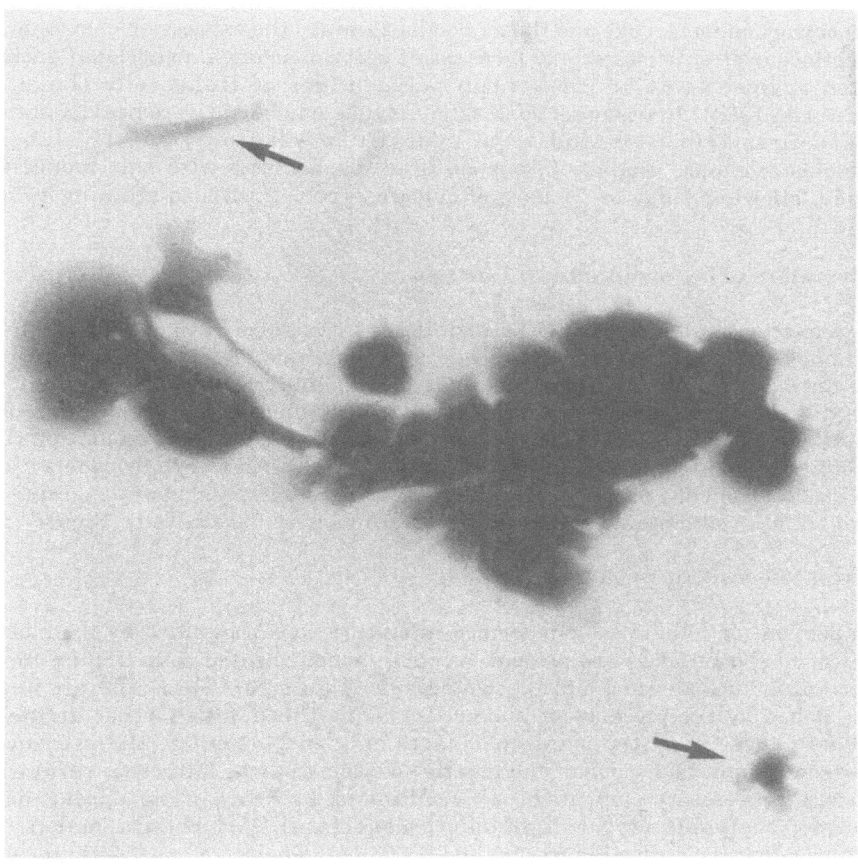

Figure 6. Cytokeratin staining of cytotrophoblast cultures. Light micrograph of immunoperoxidase stained cytotrophoblast colony. Cells were plated as outlined under "Materials and Methods" and maintained in SFDM/D-Val. After 21 days they were fixed and stained with anticytokeratin (pKKI). Note unstained macrophages in the field (arrows). (X400)

Cytokeratins

Cytokeratins are a family of intermediate filament proteins expressed by epithelial cells with molecular weight of 40-68 kd. The type of cytokeratin synthesized is dependent on the epithelial cell and its stage of differentiation. By immunocytochemistry, Loke and Butterworth examined the expression of cytokeratin polypeptides in first trimester and term tissue sections using a monoclonal antibody directed against 44-54 kd cytokeratin polypeptides of HeLa cells (Loke and Butterworth, 1985). In tissue sections, villous cytotrophoblasts were heavily stained, whereas first trimester and term syncytium were only faintly labeled. Immunohistochemical analysis of our cultured trophoblasts with this monoclonal antibody following 1 day or 21 days of culture revealed intense staining of cells (Figure 6).

Confirmation of Cytotrophoblast Lineage

Monoclonal antibodies 18B/A5 and 18A/C4 recognize antigens unique to the cytotrophoblast or on cytotrophoblast and syncytiotrophoblast cells, respectively. These antibodies were generated against cell membrane preparations, but the identity of the antigens recognized has not been established (Loke and Day, 1984). The staining pattern indicates cytoplasmic and membrane antigen recognition (Loke and Day, 1984). Figure 7B depicts the reaction of 18B/A5 to cytotrophoblast cells following 1 day of culture. Either antibody recognized our cells, but no staining was seen when cells were incubated with an irrelevant monoclonal antibody (Figure 7A).

Electron Microscopy of Cell Aggregates

Sections of cellular islands formed in culture were examined by the electron microscope. Desmosomes are present between mononucleated cells (Figure 9), and between mononucleated and multinucleated cells (Figure 10). These cells are further distinguished by the presence of microvilli. Thus these data further define the epithelial character of the cytotrophoblasts and, in particular, the presence of desmosomes implies adhesion of single cells (Weissman and Clairborne, 1975) which represents a necessary step in the differentiation pathway. The appearance of vacuoles, resolvable at the light microscope level, has been a marker for multinucleated structures (Figures 8A and B and 10). These are apparently the same vacuoles observed by Boyd and others in their histologic studies of the syncytium (Boyd and Hamilton, 1970).

Figure 7. Cytotrophoblasts following 16 hr of culture were fixed and stained with monoclonal antibody 18B/A5 using the avidin-biotin-peroxidase technique. This antibody recognizes undifferentiated mononucleated cytotrophoblasts in tissue sections of first trimester placenta. (X75)

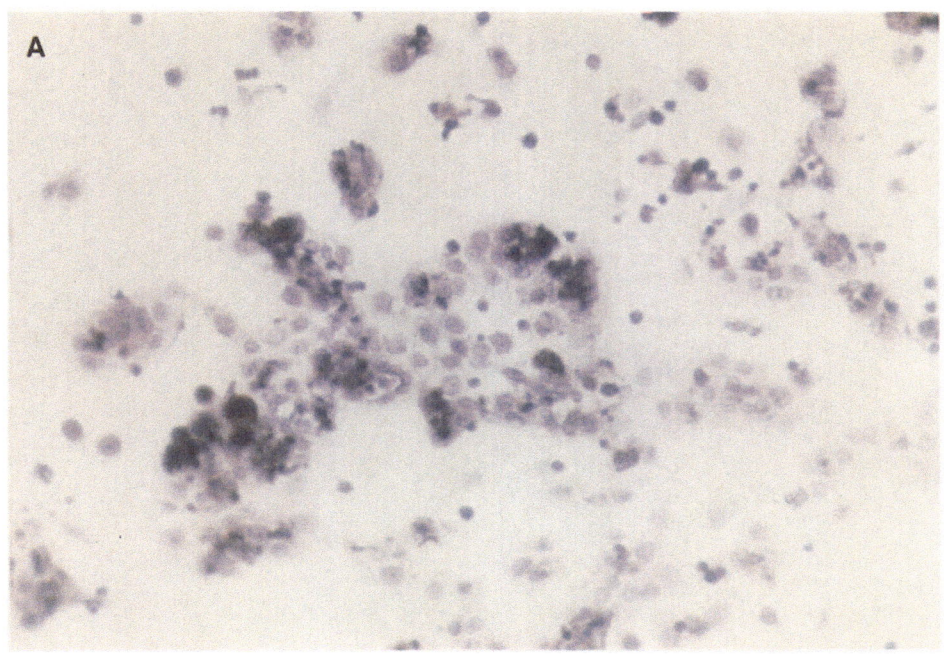

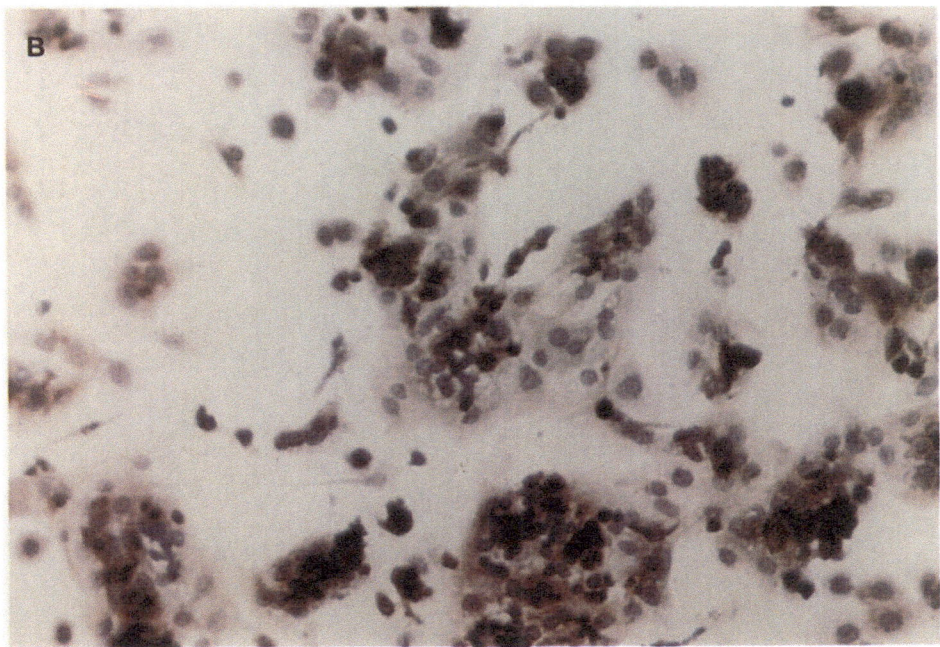

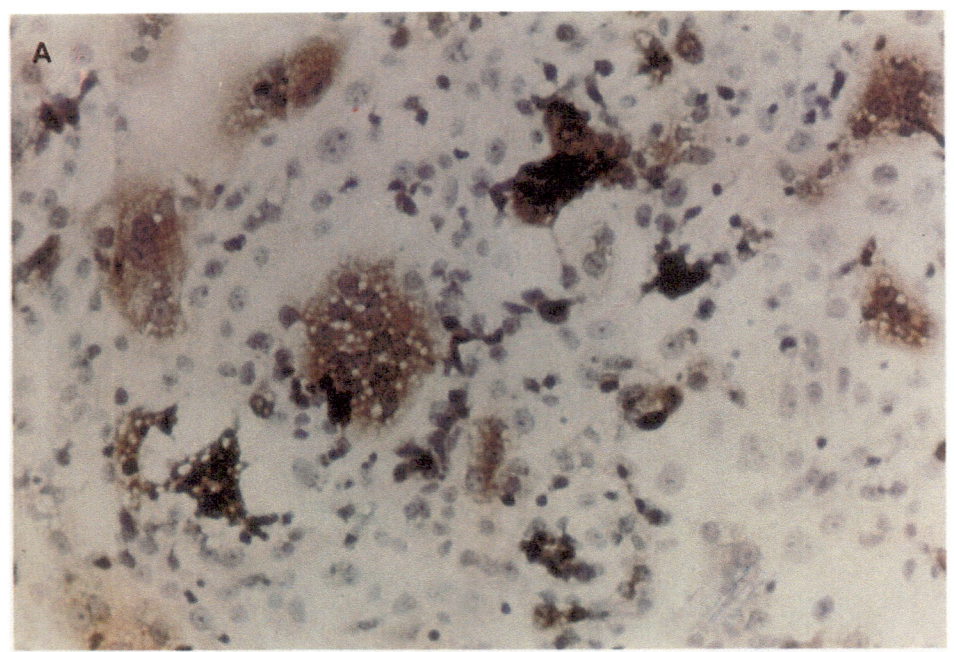

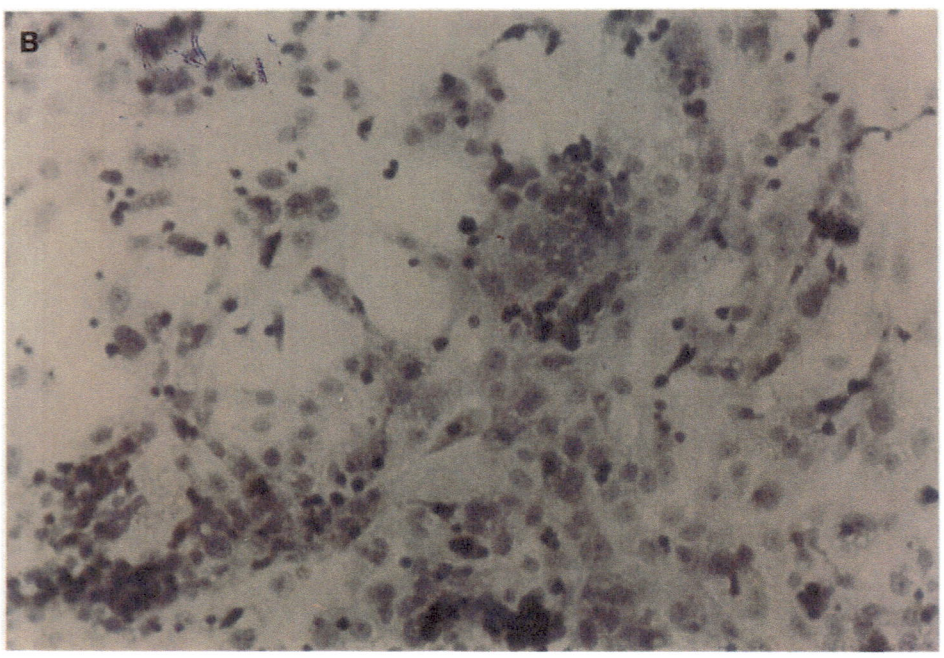

HCG Production by Cytotrophoblasts in Culture

Synthesis of the α and β subunits of hCG was examined by labeling the trophoblast cultures 4 hr, 2-4 days in culture and at 7-10 day intervals thereafter. Cells growing in SFDM were compared to cells in F12ssm (Figure 11). Aliquots of the labeled media (from cells 3 days or longer in culture) were precipitated with either antiserum against native α or β subunits. Precipitation with β subunit antiserum resulted in the appearance of the β and α subunits, which precipitate via recognition of the β in the dimer (Figure 11B). Antiserum against native α subunit precipitates three proteins, dimer α or β subunits, and the uncombined form of the α subunit (α*). The larger size of α* reflects the presence of heterogeneous asparagine-linked sugars (Corless et al., in preparation). As expected, no bands were precipitated with non-immune serum (Figures 11, 12; lane N). Media from 4 hr and two day cultures contain primarily α*. Synthesis of β subunit (and hCG dimer) was detected after 1 to 2 days of culture (Figure 12). With increasing time in culture, cytotrophoblasts lost their ability to synthesize β subunit but continued to synthesize free α subunit. The data suggests that our cells have gone through a functional differentiation in vitro. Immediately after plating α subunit synthesis precedes synthesis of β subunit. With increasing time in culture, β subunit synthesis is markedly attenuated or turned off but α synthesis continues, albeit at a lower rate. The switch from large amounts of dimer to excess free α mimics the situation seen during gestation.

If cells on fibronectin are maintained in medium containing serum they lose their ability to synthesize both α and β subunit much faster than cells plated in SFDM (Figure 11B). Thus cells maintained in SFDM produce more hCG and sustain that production for a longer time.

To compare protein secretion by the cultures at various time points 10 μl aliquots of labeled media were precipitated with trichloroacetic acid and counted. There was very little change throughout 21 days of culture (Table 1). It is possible that continued synthesis is from the mononucleated cytotrophoblasts which persist throughout the culture period and that synthesis of hCG ceases because the multinucleated cells lose their viability. Alternatively protein synthesis from the

---

Figures 8A and B. Cells were stained by the avidin-biotin-peroxidase technique after exposure to antibody B105 (Panel A) (recognizes hCG and monomer β). Panel B, trophoblast cells exposed to media containing 20% horse serum. Similar negative results were seen when trophoblast cells were exposed to media from mouse cells expressing a monoclonal antibody to sheep rbc antigen. Note the large vacuoles surrounding the nuclei in the multinucleated areas.

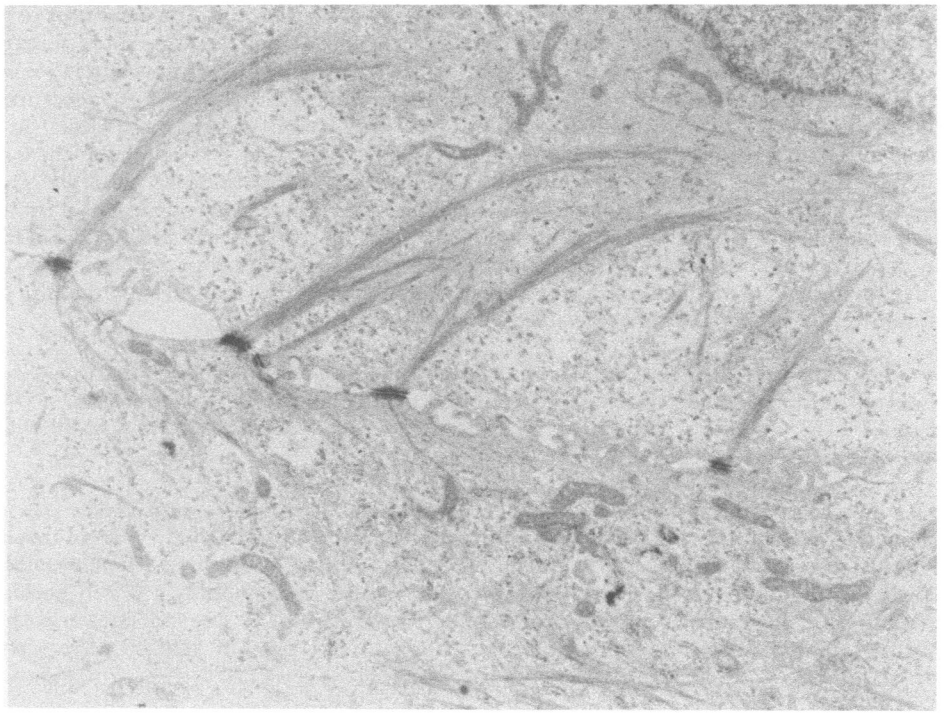

Figure 9. Electron micrograph of a section of the cellular island, after 21 days in culture. These cells were maintained in SFDM supplemented with 90 mg/L D-Valine and 5 mg/L L-Valine. The presence of numerous desmosomes and microvilli can be seen. (X18000)

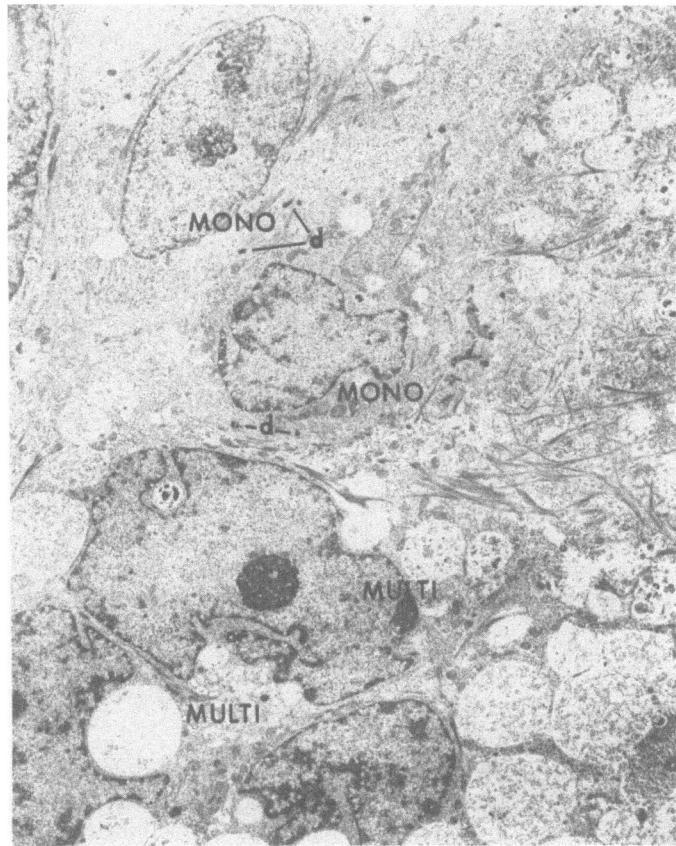

Figure 10. Electron micrograph of a mononucleated and neighboring multinucleated cytotrophoblast. Cells were maintained in SFDM with L-Valine and fixed after 5 days of culture. (X11500)

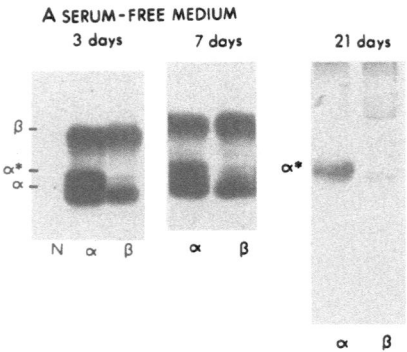
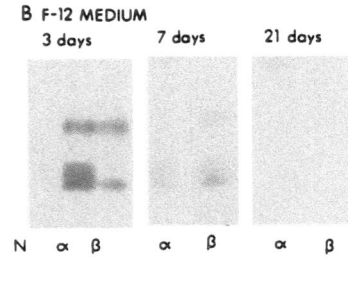

Figure 11. Synthesis of hCGα and β subunits in first trimester cytotrophoblasts with increasing time in culture. Cells were labeled with [35S]cysteine (see Methods) and the media from 3, 7, and 21 day cultures were immunoprecipitated with the following: normal rabbit serum (N); hCGα (α); and hCGβ antisera (β). Panels A and B denote cultures in serum free defined and F12, serum supplemented media, respectively. The 3 and 21 day signals were obtained after 3 days of exposure, while the 7th day signal was obtained following 4 days of exposure.

multinucleated cell may revert to the primitive pattern seen in its progenitor cell. The increase in certain proteins which are associated with syncytial formation appear, in our culture system, to coincide with maximal hCG secretion (i.e., hPL, data not shown).

Immunohistochemistry

It was proposed in 1962 by Midgley and Pierce that hCG dimer is synthesized in the syncytial region of the placenta. This hypothesis has since been examined by in situ hybridization and immunohistochemistry on tissue sections and isolated cells in culture. Based on earlier work from our laboratory and others, we proposed that the time lag seen in culture, which is necessary for dimer production, is due to the time necessary for formation of multinucleated cytotrophoblasts. To confirm this we labeled cells with monoclonal antibodies generated against hCG and subunits (Ehrlich, 1985) using the avidin-biotin-peroxidase technique. We found that only the multinucleated cell areas stained with antibody which recognized hCG dimer or β subunit (Figure 8). Control cultures containing an irrelevant antibody were negative (Figure 8B). These data further support the hypothesis that hCG is expressed subsequent to the initial expression of the α subunit gene and in a multinucleated structure.

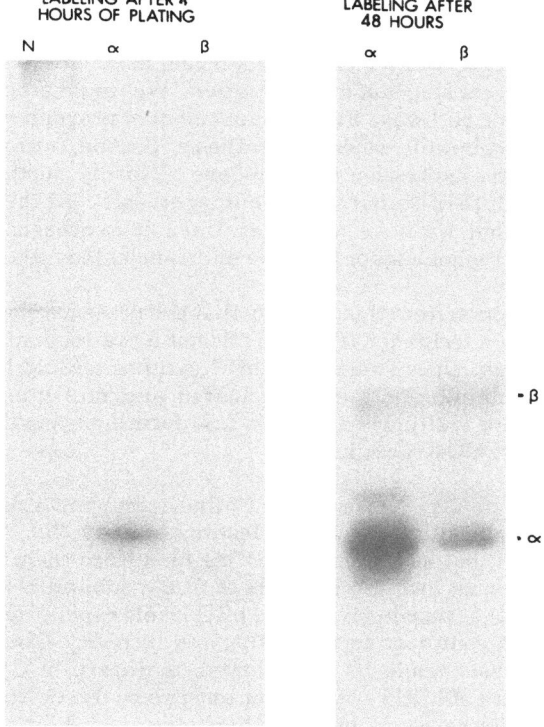

Figure 12.   Cells were labeled 4 hr and 2 days post plating and media immunoprecipitated (Figure 11). The gels containing labeled media after 4 and 48 hr in culture were autoradiographed for 5 and 6 days, respectively.

DISCUSSION

Here we describe a trophoblast culture system using first trimester tissue as a source of cytotrophoblasts. There are several components in this system which we feel are critical for efficient plating of differentiating cells. First, use of first trimester tissues rather than term is preferable since the former are replete with cytotrophoblasts whereas term villi are sparsely populated with such cells. Second, fibronectin coated plates are essential for efficient plating of cells. Thirdly, serum-free defined medium containing EGF and D-Valine is critical for sustaining the morphological and hCG biosynthetic parameters which appear to be associated with at least one step in the differentiation pathway. Cells in SFDM may represent a more differentiated phenotype since there is evidence that epithelial cells lose differentiated functions and appearance in serum supplemented media (Hammond et al., 1984; Orly et al., 1980; Barnes and Sato, 1980).

Using this culture system we wished to test predictions derived from previous models to explain the role of trophoblast differentiation in regulating synthesis of hCG and hPL. The essential feature of this model associates activation of hCG and hPL genes with different stages of differentiation. We would propose that there are at least three steps in the pathway. First, commitment of progenitor cytotrophoblasts to syncytium; these cells initiate a subunit synthesis. Second, fusion of cytotrophoblasts occurs which activates expression of the β-gene. Thirdly, further differentiation of these fused cells leads to cessation of β subunit synthesis. (At this time, the a subunit is still synthesized but we have not determined if synthesis is from remaining mononucleated cytotrophoblasts or from the multinucleated cells.)

One key prediction from this model is that the presence of syncytium as defined in our cell preparation technique (Figure 1), is not a pre-requisite for hCG synthesis. This was verified here since substantial hCG synthesis could be achieved in early cultures (24 hr) containing only mononucleated and multinucleated cells. With longer time in culture multinucleated cells fuse forming syncytial-like areas, which produce low levels of hCG β subunit.

Our culture data also support the hypothesis that the β subunit is synthesized in a transient trophoblast intermediate (Hoshina et al., 1985). This is evident from the data which show that as a function of time in culture there is a diminution of β subunit synthesis but an increase in levels of free a subunit relative to hCG dimer. This is analogous to that seen in vivo where hCG levels rapidly decrease beyond 12 wk while free a subunit continues to be synthesized to term. Thus a pool of differentiating cytotrophoblasts would be necessary to maintain hCGβ production. This might explain the lack of hCGβ synthesis at term since few cytotrophoblasts are seen at this time. Our data however, do not exclude other possibilities. For example, two villous cytotrophoblast populations may exist: one that is present only during first trimester which is capable of forming multinucleated cells with maximal expression of hCG dimer and a second population, maintained throughout pregnancy, that is capable of synthesizing hCGa and hPL but not hCGβ. Another possibility is a dosage effect operable on the β subunit. In first trimester tissue (with cytotrophoblasts thickly coating the villous core) neighboring mononucleated cells contact each other and form multinucleated cells which eventually fuse with the terminal syncytial layer. If that step is by-passed, as in term, and cytotrophoblast cells fuse to the overlying terminal syncytial layer, then the β subunit gene will not be expressed but other genes such as hPL, will be activated.

It will be necessary to isolate the early progenitor cells and to collect homogeneous populations of cytotrophoblasts to further analyze the differentiation pathway. This might be achieved using fluorescence activated cell sorting. By such methods, precise intermediates could be identified and related functionally by the proteins produced.

SUMMARY

Previously, we demonstrated that hPL and hCG subunit genes were activated at different stages of trophoblast differentiation. To examine the factors, including cell-types that are associated with this process, we employed an in vitro culture system that would support one or more of the steps in the pathway. First trimester placenta was treated with trypsin/DNAse to release cytotrophoblasts. Washed cells were cultured on fibronectin-coated plates using serum-free media containing D-Valine to reduce fibroblast contamination. Cytokeratin staining demonstrated that the cultures contained primarily epithelial cells. Monoclonal antibodies directed ·against cytotrophoblast and syncytiotrophoblast antigens confirmed that these epithelial cells were trophoblasts. Initially, the cells were discrete and then subsequently formed closely knit colonies containing multinucleated syncytial-like cells. As a functional test, hCG subunit synthesis was examined in cultures labeled with [^{35}S]cysteine. Over 3 wk in culture, synthesis of the β subunit was reduced to background and only a free α subunit was seen. This is similar to the in vivo observations; hCG declines during pregnancy and free α subunit persists.

We have identified at least one step in the differentiation pathway. The data support the hypothesis that mature syncytium is not required for hCG production and β subunit synthesis occurs in a transitory intermediate.

ACKNOWLEDGMENTS

This work was supported by a grant from the Monsanto Corporation.

REFERENCES

Barnes, D. and Sato, G. (1980) Serum-free cell culture: A unifying approach. *Cell* 22, 649-655.

Boothby, M., Kukowska, J., and Boime, I. (1983) Imbalanced synthesis of human choriogonadotropin α and β subunits reflects the steady-state levels of corresponding mRNAs. *J. Biol. Chem.* 258, 9250-9253.

Boyd, J.D.N. and Hamilton, W.J. (1970) (eds.), In: *The Human Placenta*, W. Heffer and Sons Ltd. Publishing Co., Cambridge, England, pp. 165-167.

Corless, C., Bielinska, M., and Boime, I., in preparation.

Ehrlich, P.H., Moustafa, Z.B., Kirchevsky, A., Birken, S., Armstrong, E.G., and Canfield, R.E. (1985) Characterization and relative orientation of epitopes for monoclonal antibodies and antisera to human chorionic gonadotropin. *Am. J. Reprod. Immunol.* 8, 48-54.

Gilbert, S.F. and Migeon, B.K. (1975) D-Valine as a selective agent for normal human and rodent epithelial cells in culture. *Cells* 5, 11-17.

Ham, R.G. (1965) Clonal growth of mammalian cells in a chemically defined, synthetic medium. *Proc. Nat. Acad. Sci.* 53, 288-293.

Hammond, S.L., Han, R.G., and Stampfer, M.R. (1984) Serum-free growth of human mammary epithelial cells: Rapid clonal growth in defined medium and extended serial passage with pituitary extract. *Proc. Natl. Acad. Sci.* 81, 5435-5439.

Hanks, J.H. and Wallace, R.E. (1949) Relation of oxygen and temperature in the preservation of tissues by refrigeration. *Proc. Soc. Exp. Biol. Med.* 71, 195-205.

Hoshina, M., Boothby, M., and Boime, I. (1982) Cytological localization of chorionic gonadotropin α and placental lactogen mRNAs during development of human placenta. *J. Cell Biol.* 93, 193-198.

Hoshina, M., Hussa, R., Pattillo, R., and Boime, I. (1983) Cytological distribution of chorionic gonadotropin subunit and placental lactogen mRNA in neoplasms derived from human placenta. *J. Cell Biol.* 97, 1200-1206.

Hoshina, M., Boothby, M., Hussa, R., Pattillo, R., Camel, M., and Boime, I. (1985) Linkage of human chorionic gonadotropin and placental lactogen biosynthesis to trophoblast differentiation and tumorigenesis. *Placenta* 6, 163-172.

Jen, G. and Thach, R.E. (1982) Inhibition of host translation in encephalomyocarditis virus infected L cells: a novel mechanism. *J. Virol.* 43, 250-261.

Laemmli, U.K. (1970) Cleavage of structural proteins during the assembly of the head of bacteriophage T4. *Nature (London)* 227, 680-685.

Leibovitz, A. (1963) The growth and maintenance of tissue-cell cultures in free gas exchange with the atmosphere. *Am. J. Hyg.* 78, 173-180.

Loke, Y.W. and Day, S. (1984) Monoclonal antibody to human cytotrophoblast. *Am. J. Reprod. Immunol.* 5, 106-108.

Loke, Y.W. and Butterworth, B.H. (1986) Heterogeneity of human trophoblast populations, in press.

Murakami, H., Masui, H., Sato, G.H., Sueoka, N., Chow, T.P., and Kano-Sueoka, T. (1982) Growth of hybridoma cells in serum-free medium: ethanolamine is an essential component. *Proc. Nat. Acad. Sci.* 79, 1158-1162.

Midgley, A.R., Jr., and Pierce, G.B., Jr. (1962) Immunohistochemical localization of human chorionic gonadotropin. *J. Exp. Med.* 115, 289-294.

Orly, J., Sato, G., and Erickson, G.F. (1980) Serum suppresses the expression of hormonally induced functions in cultured granulosa cells. *Cell* 20, 817-827.

Pappas, G. (1975) In: *Cell Membrane, Biochemistry and Pathology*, (eds.), G. Weissmar and R. Clairborre, H.P. Publishing Co., New York, NY, pp. 87-94.

Savage, C.R., Jr. and Cohen, S. (1972) Epidermal growth factor and a new derivative. *J. Biol. Chem.* 247, 7609-7611.

Smith, J.D., Freeman, G., Vogt, M., and Dulbecco, R. (1960) The nucleic acid of polyoma virus. *Virol.* 12, 185-196.

Weber, K. and Osborne, M. (1969) The reliability of molecules weight determinations by dodecyl sulfate-polyacrylamide gel electrophoresis. *J. Biol. Chem.* 244, 4406-4412.

Trophoblast Research 2:447-460, 1987

CHARACTERIZATION OF ISOLATED CELLS IN PRIMARY CULTURE FROM HUMAN TERM PLACENTA BY ELECTRON MICROSCOPY AND IMMUNOHISTOCHEMISTRY

Juliet O. Lobo, Francis L. Bellino and Alan Siegel

Dept. of Biological Sciences
State University of New York at Buffalo
Buffalo, New York 14260 USA

INTRODUCTION

During pregnancy the syncytiotrophoblast layer of the human placenta elaborates both protein and steroid hormones (Gaspard et al., 1980; Simpson and MacDonald, 1981, Winkel et al., 1980). The mitotically inactive but biochemically functional syncytiotrophoblasts are formed in vivo by the fusion of the underlying mitotically active, mature, mononucleated cytotrophoblast cells (Boyd and Hamilton, 1967; Enders, 1965; Pierce and Midgley, 1963; Wislocki and Dempsey, 1955; Wynn, 1972).

Studies of the regulation of hormone secretion in human placenta have utilized tissue slices, perfused placenta, organ and cell cultures. The advantage of using isolated cells is that hormone synthetic and secretory processes can be identified with specific cell types. Cells released by trypsin treatment of villi fragments from human term placenta are predominantly small (10-25 µm diam.) and mononucleated (Kawata et al., 1984; Lobo et al., 1985). Compared to the tissue from which these cells are derived, their ability to synthesize estrogen is low (Lobo et al., 1985).

Under the appropriate culture conditions (Huot et al., 1979; Martinez et al., 1983; Stromberg et al., 1978), placental cells adhere to tissue culture dishes and grow, exhibiting at least three well defined cell types: small polygonal cells, spindle-shaped fibroblasts and granular, highly multinucleated giant cells (Thiede, 1960), while hormone synthesizing/secretion rates increase drastically (Lobo et al., 1985). All three cell types have been characterized according to their morphology (Foldes and Schwartz, 1972; Jogee et al., 1983; Loke, 1983; Thiede, 1960), histochemistry (Fox and Kharkongor, 1970a; Jogee et al., 1983; Thiede, 1960), presence of Fc and C_3 receptors as well as vimentin and keratin filaments (Contractor, 1979; Contractor et al., 1984), and ability to produce hCG (Fox and Kharkongor, 1970b; Loke et al., 1972).

Although many investigators describe similar changes in morphological and functional properties of fresh and cultured cells, the postulated mechanism of these changes differ. Some propose that trypsin releases cytotrophoblast cells which differentiate in culture into syncytiotrophoblast (Cotte et al., 1980; Fox and

Submitted: 6 October 1985; Accepted: 7 July 1986

Kharkongor, 1970a; Lobo et al., 1985). Alternatively trypsin may release mononucleated "cells" derived from the syncytiotrophoblast layer; these cells then recover from the trypsin treatment to reform multinucleated cells (Deal and Guyda, 1983; Foldes and Schwartz, 1972; Loke, 1983; Winkel et al, 1980). The ability of trypsin to remove the syncytial layer from the placental villi fragments has been directly demonstrated (Carr, 1972; Cotte et al., 1980; Kaspi and Nebel, 1974).

To gain insight into the mechanism of multinuclear giant cell formation in culture as a possible model of syncytiotrophoblast formation in vivo, we examined fresh and cultured cells by electron microscopy for evidence of cell-cell fusion and by immunohistochemistry for trophoblast marker antigens. We also report here the effect of various culture media on the morphology of placental cells.

MATERIALS AND METHODS

Materials

Hanks Balanced Salt Solution (HBSS), amphotericin B (fungizone), penicillin-streptomycin, medium 199 (M199) and fetal bovine serum (FBS) were purchased from Grand Island Biological Co. (Grand Island, NY). Trypsin (porcine, 1:250) came from Difco Laboratories (Detroit, MI). Deoxyribonuclease 1 (bovine pancreas), and goat anti-rabbit IgG-TRITC conjugate were obtained from Sigma Chemical Co. (St. Louis, MO). NuSerum was obtained from Collaborative Research Laboratories (Lexington, MA).

Antisera to human chorionic gonadotropin (hCG) and beta subunit of hCG (β-hCG) were obtained from Dr. O.P. Bahl. Human placental lactogen (hPL) antiserum was a gift from the National Hormone and Pituitary Program, NIADDK.

Preparation of cells

The procedure for obtaining the cells was described previously (Lobo et al., 1985). About 2×10^6 cells were plated in 30 x 15 mm petri dishes (Falcon) containing sterile coverslips and incubated in a humidified atmosphere of 5% CO_2 in air at 37°C. Unattached cells and debris were poured off at 24 hr. The medium was changed at 24 hr intervals. Plates removed for immunostaining and electron microscopy were washed in medium without FBS.

Cell culture conditions

Studies to determine optimal culture conditions were carried out using two different media and with several different lots of FBS or NuSerum. Cells were plated in medium M199 or Dulbecco's modified Eagle medium (DMEM) containing high (4.5 g/l;DMEMh) or low (1 g/l;DMEMl) glucose. All three media had different sera added at 20% (v/v). At 24 hr, unattached cells and debris were removed. The attached cells were washed and fixed in 4% glutaraldehyde. Cells plated in the different media were scored for their ability to attach only or to attach and spread, and photographed using a Leitz microscope with a fitted camera (E. Leitz, Inc., Rockleigh, NJ). Cells were classified as mononucleated (cells appearing to contain 1 nucleus) and apparent multinucleated (cells appearing to contain 2 or more nuclei and, in some cases, aggregates of mono-or multinucleated cells).

Electron microscopy

Cells cultured on carbon coated coverslips were washed in buffer, fixed in 4% glutaraldehyde in 0.1M Na cacodylate buffer (pH 7.2), post-fixed in 1% OsO_4-0.1M Na cacodylate, dehydrated, embedded in epon resin araldite, sectioned and examined on a Hitachi H500.

Immunostaining

Coverslips with smears of fresh cells and attached cultured cells were fixed in 3% paraformaldehyde containing 0.5% Triton-X100. The coverslips were washed several times in 67 mM phosphate buffer (pH 7.2) containing 0.1% Triton-X100 and then preincubated with normal rabbit serum for 10-30 min. After decanting, the coverslips were exposed to the respective antisera (diluted from 1:500 to 1:1000) for 16 hr at 4°C with rocking. Control sections were exposed to normal rabbit serum. On the following day, the coverslips were washed with repeated changes of phosphate buffer and incubated with goat anti-rabbit gamma globulin conjugated to rhodamine (1:50) for 1 hr at room temperature. The sections were washed, mounted in glycerol-phosphate (1:9, v/v) buffer, observed, and counted in a Zeiss Photomicroscope III (Carl Zeiss, Oberkochen, West Germany) with an exposure time of 1.25 mins, and printed under identical conditions.

Smears of the fresh cell population were evaluated for fluorescence. At 24 hr and beyond, only attached and spread mono- and apparent multinucleated cells were scored. Over 400 fresh cells, and at least 100 mono- and 50 apparent multinucleated cells in culture were scored for fluorescence from a single coverslip at each time point. All fluorescent cells were counted, regardless of the staining intensity.

RESULTS

Cell culture conditions

Mononucleated or apparent multinucleated cells obtained from a single placenta were assessed for attachment and spread as described. Examples of cell morphology after 24 hr in culture are shown in Figure 1. With two different lots of serum, DMEMh or DMEMl gave similar results in terms of cell attachment and spread (Figure 2). With M199 and the same sera, either a greater proportion (serum 28) or a smaller proportion (serum 30) of cells attached and spread relative to DMEM. But in either case, the extent of multinucleated cell formation was considerably less than with DMEM. Heat inactivated serum (serum 28h) reduced attachment and spread by almost 50% in DMEM-grown cells. Cells grown in M199 containing NuSerum attached but remained mostly mononucleated. While the general pattern of cell attachment and spread described here remained similar, interplacental differences in the extent of multinuclear cell formation were observed (data not shown).

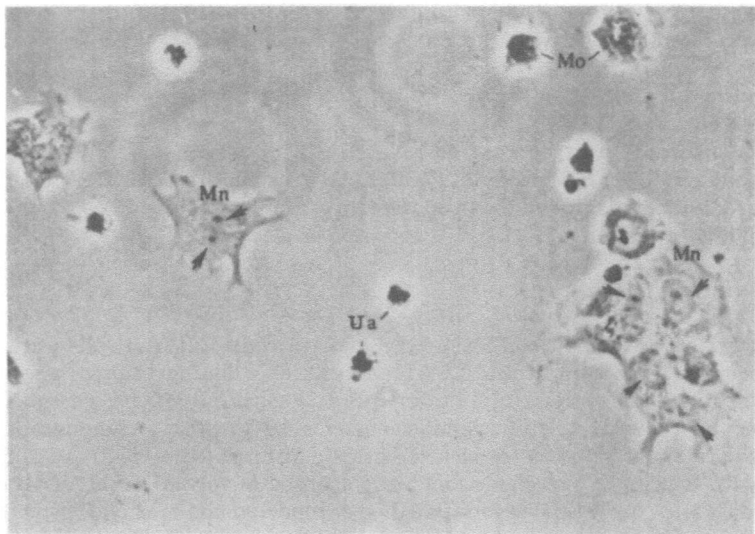

Figure 1. Phase contrast micrograph of placental cells at 24 hr. Cells were grown on coverslips with DMEMh containing FBS. At 24 hr, the coverslips were scored for attached (a) and attached and spread mono- (Mo) and apparent multinucleated (Mn) cells. Arrows indicate nuclei (X360).

Electron microscopy

Our previous study (Lobo et at., 1985) using light microscopy suggests that placental monolayer cells in culture over a period of 72 hr contain aggregates of mononucleated cells and true syncytia. Examples of changes in cell morphology from fresh cells to cultured cells at 24 and 72 hr are shown in Figure 3. Figure 4 shows more detailed changes occurring at 48 hr in culture.

Fresh cells were characterized by a well developed endoplasmic reticulum, numerous mitochondria and a nucleus that occupied most of the cytoplasm (Figure 3A). The surface of these cells contained numerous microvilli that varied in size, shape and number.

Placental cells in tissue culture for 24 hr (Figure 3B) showed mononuclear cells with irregular nuclei. Aggregates were sometimes composed of cells of varying electron densities. The outer surface of the clump was elaborated into cytoplasmic folds and well developed microvilli. The surface of cells at the sites of apposition lacked microvilli but was characterized by smooth and gently contoured intercellular membranes, punctuated by coated pits (Figure 3B).

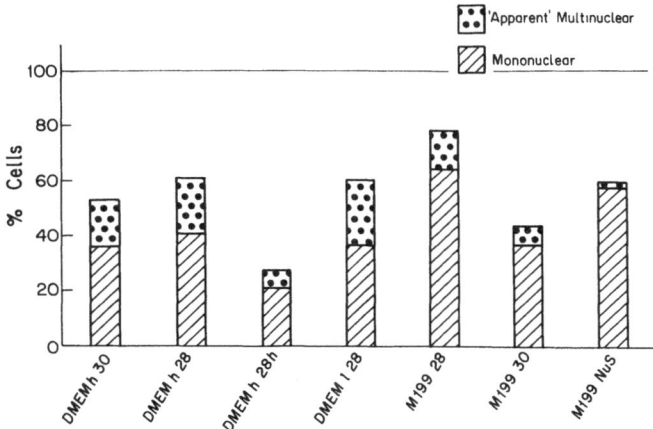

Figure 2. Relative proportion of cells that attach and spread in various media and serum combinations. Cells isolated from a single placenta were grown on coverslips for 24 hr with the indicated medium-serum combinations. The number of cells attached only or attached and spread were determined. For the latter category, the number of mononucleated and apparent multinucleated cells were counted for each culture condition. The bar graphs show the distribution of mono- and multinucleated cells for those cells attached and spread. The proportion of cells attached only is given by the difference between the top of the individual bar graphs and the line indicating 100%. The FBS lot number is indicated by the two-digit number following the medium designation. The designation "h" following the two-digit FBS lot number indicates heat inactivation.

At 48 hr (Figure 4A) the exposed surface of the multicellular clumps retained their extensive microvillus features. By this time, the internal surface of these nascent syncytia developed prominent desmosomal connections (Figure 4C). Adjacent to the surface of these cells were numerous bundles of intermediate filaments (Figure 4B). Internalized desmosomal plaques still associated with the filaments were visible around this area of the cytoplasm.

At 72 hr, multinucleated cells with no sign of plasma membrane between nuclei were readily located (Figure 3C). Numerous mitochondria were distributed throughout the cytoplasm. Large, electron dense granules were more apparent at this time and the nuclei contained prominent nucleoli.

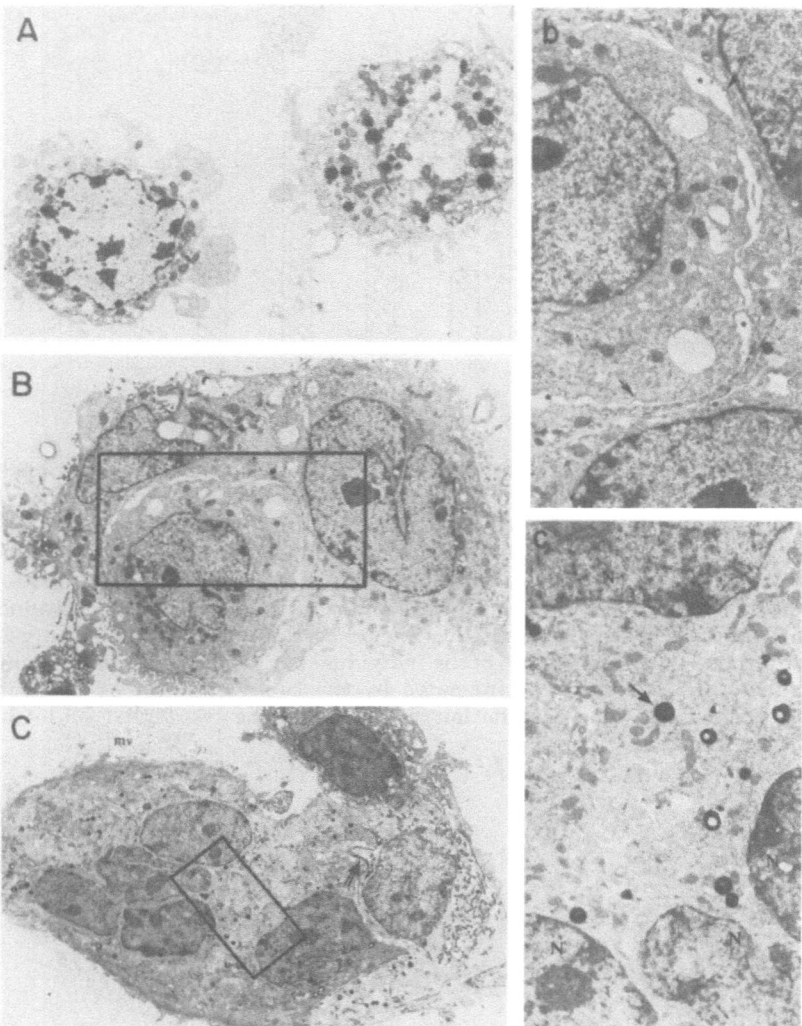

Figure 3. Electron micrographs of placental cells in primary culture. Freshly prepared cells (A, X4900), attached cells after 24 hr (B, X4900) and 72 hr (C, X2100) grown in M199 and 20% FBS were fixed and processed as described in Materials and Methods. Photographs are from representative areas of the entire population. Enlargements (X8700) are shown of sections of B (b) and C (c), as indicated by the rectangles.

       In (b), arrows indicate coated pits. The intercellular space between adjacent cells is indicated by a star. In (C), open shafted arrow points to a bundle of intermediate filaments in an area where a mononucleated cell is beginning to fuse with a multinucleated cell. In (c), N denotes individual nuclei of a multinucleated cell. The arrow indicates a dense granule. Note the absence of clearly defined cell membranes between nuclei.

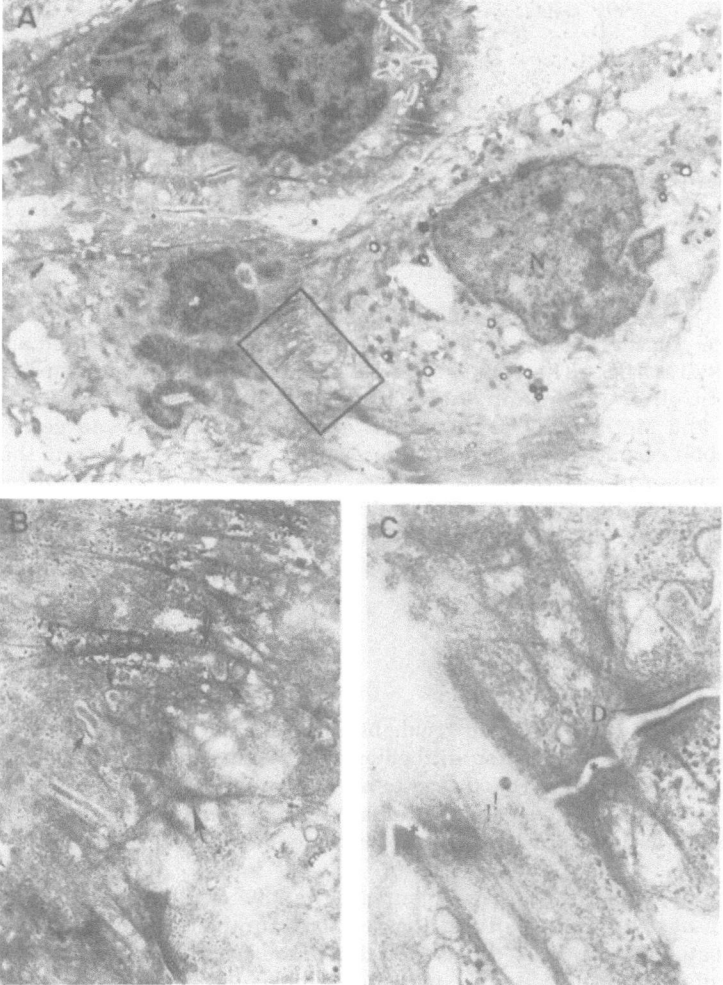

Figure 4. An electron micrograph of cells in culture at 48 hr.
Cells were grown and treated as described.

(A) Section through a nascent syncytium showing different degrees of fusion. The two bottom cells show regions where filamentous bundles extend from one cell to the other; definite cell boundaries can be seen between the upper and lower cells. The star symbol highlights the intercellular spaces (X4900).

(B) A higher magnification (X21000) of a section of A showing intermediate filaments extending through the cytoplasm (shafted arrow), internalized coated pits (small arrowhead) and remnants of desmosomes (large arrowhead) contiguous with intermediate filaments.

(C) A section outside the field of view in (A) which shows desmosomal connections (D). The intercellular space is shown by the star symbol. The exclamation marks depict a region of apparent cytoplasmic fusion.

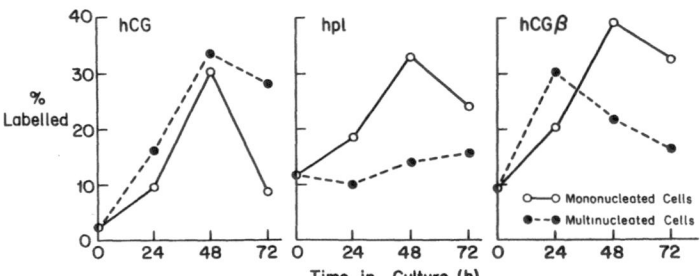

Figure 5. Distribution patterns of various trophoblast marker antigens in primary cell culture. Placental cells were grown on coverslips for the indicated period of time in DMEM-20% FBS. The cells were then fixed and incubated with normal rabbit serum (control) or rabbit antisera against hCG, β-hCG, or hPL, as described in Materials and Methods. The proportion of attached and spread cells exposed to normal rabbit serum and containing fluorescence was subtracted from the proportion of those exposed to the trophoblast marker antisera, and the difference is given for each antigen as a function of time in culture. The mononucleated cells were scored separately from the apparent multinucleated cells.

Immunostaining

We localized the different trophoblast antigens (hCG, β-hCG, hPL) in these cells using antigen specific first antibodies and rhodamine-labeled second antibody. Cells fixed to coverslips were exposed to normal rabbit serum (control) or antiserum against the various antigens, and the relative proportion of fluorescent cells in excess of the control was determined as a funtion of time in culture (Figure 5).

In analyzing for the hCG antigen, less than 5% of the fresh cells were positive. During the next 48 hr in culture, the relative number of fluorescent mono- and multinucleated cells increased and then decreased from 48 to 72 hr. This decrease was particularly pronounced in the mononucleated cells. Specific uptake of the β-hCG antiserum in the fresh cells was about 10%. This increased in the mononucleated cells over the next 48 hr, and then decreased by 72 hr. The same general pattern was observed for the multinucleated cells, except that the maximum value occurred at 24 hr and the subsequent decrease was more drastic. With the hPL antiserum, the percentage of fluorescent mononucleated cells increased up to 48 hr, and then declined by 72 hr. However, fluorescent positive multinucleated cells (at about 10-15% of the population) remained constant over the period of study.

Figure 6 shows cells at 48 hr by phase contrast microscopy and corresponding fluorescent microscopy. Normal rabbit serum showed non-specific staining in 15% of all cells examined (Figures 6A, a). Fresh cells that stained with any of the antisera showed a uniform fluorescence throughout the cytoplasm. With time in culture this fluorescence became more localized to small and large granules, irregardless of which antisera was used (Figures 6B-D, b-d).

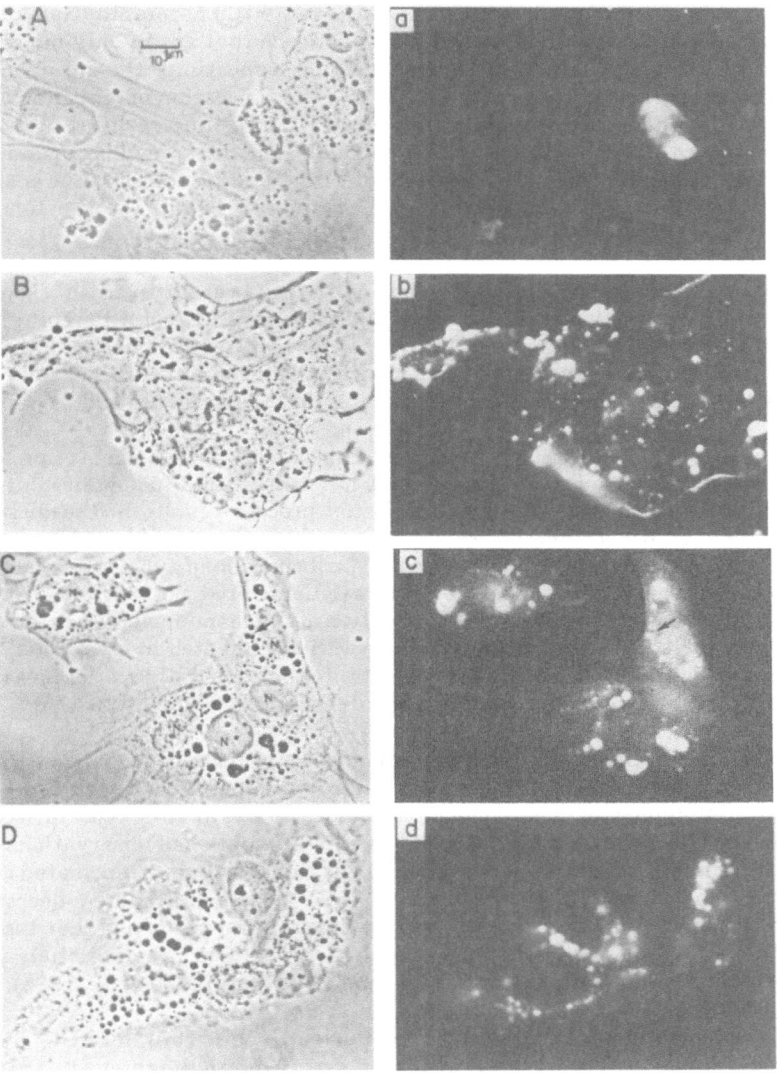

Figure 6. Phase contrast and corresponding fluorescence micrograph of cultured cells at 48 hr. Cells were prepared as described, incubated with the specific antiserum, reacted with fluorescent-labeled second antibody, counted and photographed. The same field of view was photographed under phase contrast (A - D) or fluorescence (a - d) conditions. The specific antisera used were nonimmune (A,a), hCG (B,b), β-hCG (C,c), or hPL (D,d).

DISCUSSION

Treatment of human term placental villi with a combination of trypsin-DNAase releases a heterogeneous population of cells that are mainly mononucleated (Lobo et al., 1985). Under appropriate culture conditions the number of these mononucleated cells decreases with time, while the population of multinucleated cells increases in both size and number (Lobo et al., 1985). Successful establishment of primary culture of these cells in terms of cell adhesion and spread is dependent on the type of culture medium used and the batch of FBS. We find that DMEM is superior to M199, the medium we used previously (Lobo et al., 1985), for the formation of multinuclear cells.

No attempts to identify the various cell types present in freshly isolated cell population have been made in this study. In our hands, use of a Percoll gradient to separate these cells (Kliman et al., 1986; Zeitler et al., 1983) has met with limited success. In culture we have been able to identify the three categories of cells described previously (Thiede, 1960; Foldes and Schwartz, 1972; Jogee et al., 1983).

By electron microscopy, evidence was obtained that formation of multi-nucleated cells in vitro occurs through cell-cell contact and fusion/dissolution of the intervening membranes. The dynamic fusion process results in the dissolution of intervening membranes and shows remnants of desmosomes (desmosomal plaques) and internalization of surface membranes. The formation of syncytiotrophoblast cells in vitro has also been demonstrated by others (Kliman et al., 1986; Nelson et al., 1986) as a cell-cell fusion process. The transformation of the mononucleated cytotrophoblast to the multinucleated syncytiotrophoblast is well documented in vivo, where division, maturation and fusion of the cytotrophoblast into the syncytial layer occurs (Boyd and Hamilton, 1967; Carter, 1964; Enders, 1965; Pierce and Midgley, 1963).

The distribution of the various trophoblast marker antigens in cultured mono- and multinucleated placental cells shows different temporal patterns for the different antigens. Immunoperoxidase-coupled second antibody in this type of experiment using fresh cells gave virtually identical results (unpublished observations). Except for the multinucleated cells stained for hPL, all mono- and multinucleated cells show an increase in the fraction of cells staining at 24-48 hr, followed by a decrease by 72 hr. One possible explanation for this decreasing staining pattern is that the cells are losing their hormone storage due to hormone secretion. Our previous study suggested that hormone secretion increases sharply from 48 to 72 hr (Lobo et al., 1985).

In this system, both mono- and multinucleated cells stain for hCG, β-hCG, and hPL. Other investigators (Fox and Kharkongor, 1970b; Kliman et al., 1986; Thiede and Choate, 1963) also show that both cell types stain for hCG. However, using immature placenta, Loke (1983) and Loke and Borland (1970) found hCG to be localized only in the small round or polygonal cells in culture. Cotte et al.(1980) found β-hCG only in multinucleated cells that formed in vitro, and Kurman et al. (1984) have evidence for β-hCG in the intermediate cytotrophoblast in vivo. hPL staining has been localized in the intact placenta to the syncytiotrophoblast layer (de Ikonikoff and Cedard, 1973; Gaspard et al., 1980), and intermediate cytotrophoblasts (Kurman et al., 1984). Our observation that a fraction of the mononucleated cells contain these marker antigens is consistent with the concept of cytotrophoblast maturation in the process of syncytiotrophoblast formation.

The morphological development of the multinucleated cell in culture seen in this study, coupled with the increase in the fraction of the population staining for hCG and β-hCG, is consistent with a maturation leading towards differentiation in vitro. Electron microscopy strongly suggests that the mechanism of multinucleated cell formation is by fusion, rather than nuclear division as proposed by Cotte et al.(1980) and Loke (1983). However our results do not rule out a recovery process in multinucleated cell formation. A study using antibodies shown to be completely specific to cytotrophoblast or syncytiotrophoblast placental layers are necessary to reach a definitive conclusion.

SUMMARY

Trypsin dispersion of human term placental tissue releases primarily mononucleated cells (Lobo et al., 1985). At 24 hr after plating, some attached cells remain small and mononucleated while others become large and multinucleated (Lobo et al., 1985). We found that the extent of these changes depends on the culture medium and the batch of fetal bovine serum used. Electron microscopic studies reveal a change from tight clumps of mononucleated cells at 24 hr in culture to true multinucleated cells by 72 hr. The observation of desmosomal and other membrane fragments as well as extensive filamentation in regions of apparent fusion between mono- and multinucleated cells at 48 hr in culture suggests that formation of at least some of the multinucleated cells is by cell-cell fusion. Freshly isolated and cultured cells were examined for content of trophoblast markers human chorionic gonadotropin (hCG), beta subunit of hCG (β-hCG), and human placental lactogen (hPL) by immunocytochemistry. Except for multinucleated cells stained for hPL, the proportion of cells containing these antigens rose from about 5% in fresh cells to about 30 to 40% over 24-48 hr in culture, then decreased by 72 hr. In the former instance, the fraction of stained cells remained constant at about 10% of the population over 72 hr.

ACKNOWLEDGEMENTS

We gratefully acknowledge the cooperation of the Delivery Room Staff at the Millard Fillmore Suburban Hospital, Dr. O.P. Bahl for use of facilities, Dr. Barry Eckert for advice in the immunohistochemical studies, and Mr. Jim Stamos for photographic assistance. This work was supported by NIH grant HD-16593.

REFERENCES

Boyd, J.D., and Hamilton, W.J. (1967) Development and structure of the human placenta from the end of the third month of gestation. *J. Obstet. Gynecol. of Brit. Commonwealth* 74, 161-226.

Carr, M.C. (1972) Study into the mechanisms of damage of human placental syncytiotrophoblast. ii. Histologic comparison of plasma-induced damage and specific enzyme-induced damage. *Gynecol. Invest.* 3, 203-214.

Carter, J.E. (1964) Morphologic evidence of syncytial formation from the cytotrophoblastic cells. *Obstet. Gynecol.* 23, 647-656.

Contractor, S.F. (1979) Receptors in cultured human trophoblast cells. In *Protein Transmission Through Living Membranes*, Hemmings, W.A. (ed.), Elsevier/North Holland Biomedical Press, Amsterdam, pp 3-75.

Contractor, S.F., Routledge, A. and Sooranna, S.R. (1984) Identification and estimation of cell types in mixed primary cell cultures of early and term human placenta. *Placenta* 5, 41-54.

Cotte, C., Easty, G.C., Neville, A.M., and Monaghan, P. (1980) Preparation of highly purified cytotrophoblast from human placenta with subsequent modulation to form syncytiotrophoblast in monolayer culture. *In Vitro* 16, 639-646.

Deal, C.L., and Guyda, H.J. (1983) Insulin receptors of human term placental cells and choriocarcinoma (JEG-3) cells: characteristics and regulation. *Endocrinol.* 112, 1512-1523.

de Ikonicoff, L.K. and Cedard, L. (1973) Localization of human chorionic gonadotropic and somatomammotropic hormones by the peroxidase immunohistoenzymologic method in villi and amniotic epithelium of human placentas (from six weeks to term). *Am. J. Obstet. Gynecol.* 116, 1124-1132.

Enders, A.C. (1965) Formation of syncytium from cytotrophoblast in the human placenta. *Obstet. Gynecol.* 25, 378-386.

Foldes, J.J., and Schwartz, J. (1972) Some new aspects in human trophoblast cultures. *Experientia* 28, 1468-1470.

Fox, H., and Kharkongor, F.N. (1970a) Morphology and enzyme histochemistry of cells derived from placental villi in tissue culture. *J. Path.* 101, 267-276.

Fox, H., and Kharkongor, F.N. (1970b) Immunofluorescent localization of chorionic gonadotrophin in the placenta and in tissue cultures of human trophoblast. *J. Path.* 101, 277-282.

Gaspard, U.J., Hustin, J., Reuter, A.M., Lambotte, R., and Franchimont, P. (1980) Immunofluorescent localization of placental lactogen, chorionic gonadotropin and its alpha and beta subunits in organ cultures of human placenta. *Placenta* 1, 135-144.

Huot, R.I., Foidart, J.M. and Stromberg, K. (1979) Effects of culture conditions on the synthesis of human chorionic gonadotropin by placental organ cultures. *In Vitro* 15, 497-502.

Jogee, M., Myatt, L., Moore, P. and Elder, M.G. (1983) Prostacyclin production by human placental cells in short term culture. *Placenta* 4, 219-229.

Kaspi, T. and Nebel, L. (1974) Isolation of syncytiotrophoblast from human term placentas. *Obstet. and Gynaecol.* 43, 549-557.

Kawata, M., Parnes, J.R., and Herzenberg, L.A. (1984) Transcriptional control of HLA- A,B,C antigen in human placental cytotrophoblast isolated using trophoblast- and HLA-specific monoclonal antibodies and the fluorescence-activated cell sorter. *J. Exp. Med.* 160, 633-651.

Kliman, H.J., Nestler, J.E., Sermasi, E., Sanger, J.M. and Strauss, J.F. III (1986) Purification, characterization and in vitro differentiation of cytotrophoblasts from human term placentae. *Endocrinol.* 118, 1567-1582.

Kurman, R.J., Main, C.S., and Chen, H.C. (1984) Intermediate trophoblast: a distinctive form of trophoblast with specific morphological, biochemical and functional features. *Placenta* 5, 349-370.

Lobo, J.O., Bellino, F.L., and Bankert, L. (1985) Estrogen synthetase activity in human term placental cells in monolayer culture. *Endocrinol.* 116, 889-895.

Loke, Y.W. (1983) Human trophoblast in culture. In *Biology of Trophoblast*, Loke, Y.W. and Whyte, A. (eds.), Elsevier Science Publishers, Amsterdam, pp 663-701.

Loke, Y.W., and Borland, R. (1970) Immunofluorescent localization of chorionic gonadotropin in monolayer cultures of human trophoblast cells. *Nature* 228, 561- 562.

Loke, Y.W., Wilson, D.V. and Borland, R. (1972) Localization of human chorionic gonadotropin in monolayer cultures of trophoblast cells by mixed agglutination. *Am. J. Obstet. Gynecol.* 113, 875-879.

Martinez, F., Cheung, S.W., Crane, J.P. and Arias, F. (1983) Use of trophoblast cells in tissue culture for fetal chromosomal studies. *Am. J. Obstet. Gynecal.* 147, 542-547.

Nelson, D.M., Meister, R.K., Ortman-Nabi, J., Sparks, S. and Stevens, V.C. (1986) Differentiation and secretory activities of cultured human placental cytotrophoblasts. *Placenta* 7, 1-6.

Pierce, G.B. and Midgley, A.R. (1963) The origin and function of human syncytiotrophoblastic giant cells. *Am. J. Pathol.* 43, 153-172.

Simpson, E.R. and MacDonald, P.C. (1981) Endocrine physiology of the placenta. *Ann. Rev. Physiol.* 43, 163-188.

Stromberg, K., Azizkhan, J.C. and Speeg, K.V., Jr. (1978) Isolation of functional human trophoblast cells and their partial characterization in primary culture. *In Vitro* 14, 631-638.

Thiede, H.A. (1960) Studies of the human trophoblast in tissue culture. I. Cultural methods and histochemical staining. *Am. J. Obstet. Gynecol.* 79, 636-647.

Thiede, H.A., and Choate, J.W. (1963) Chorionic gonadotropin localization in the human placenta by immunofluorescent staining. II. Demonstration of hCG in the trophoblast and the amnion epithelium of immature and mature placentas. *Obstet. Gynecol.* 22, 433-443.

Winkel, C.A., Snyder, J.M., MacDonald, P.C. and Simpson, E.R. (1980) Regulation of cholesterol and progesterone synthesis in human placental cells in culture by serum lipoproteins. *Endocrinol.* 106, 1054-1060.

Wislocki, G.B. and Dempsey, E.W. (1955) Electron microscopy of the human placenta. *Anat. Rec.* 123, 133-149.

Wynn, R.M. (1972) Cytotrophoblastic specializations: An ultrastructual study of the human placenta. *Am. J. Obstet. Gynecol.* 114, 339-353.

Zeitler, P., Markoff, E. and Handwerger, S. (1983) Characterization of the synthesis and release of human placental lactogen and human chorionic gonadotropin by an enriched population of dispersed placental cells. *J. Clin. Endocrinol. Metab.* 57, 812-818.

Trophoblast Research 2:461-464, 1987

TROPHOBLAST CELL CULTURE

- A Workshop Report -

Y.W. (Charles) Loke[1,3] and Robert O. Hussa[2]

[1]Department of Pathology
Cambridge University
Tennis Court Road
Cambridge CB2 1QP, England

[2]Department of Obstetrics/Gynecology
Medical College of Wisconsin
8700 West Wisconsin Avenue
Milwaukee, Wisconsin 53226

Current State of the Art

Using procedures currently available it is possible to obtain viable trophoblasts from both early and term placenta. In accomplishing this it is important to use fresh placental tissue. Dispersion of cells from the primary tissue can be achieved with some preparations of crude collagenase or with some preparation of trypsin. There was wide agreement that one's ultimate success in obtaining healthy trophoblast cells is dependent on choosing the proper lot of either protease. There was general agreement in the workshop that syncytium is lost during sequential trypsinization of placenta, probably by lysis, though one participant suggested that the syncytium may end up in the red blood cell pellet in a Percoll gradient. Calf serum should also be screened, since some lots are superior to others.

Several investigators have successfully prepared trophoblast cell cultures using the procedure described by Hall et al. (Steroids 30:569-580, 1977). For example, Donald Morrish (University of Alberta, Edmonton) favors the Hall method with term placenta to prepare monolayer cell cultures that contain viable trophoblast cells as measured by the continuous secretion of both hPL and hCG for 6 wk. Morrish indicated that the optimum secretion of these hormones occurred with media containing epidermal growth factor (EGF), high density lipoprotein (HDL), insulin and transferrin, using plates coated with extracellular matrix (ECM) (the plates can be purchased from Accurate Chemicals Co.). Hormone secretion is promoted longer in medium containing 10% fetal calf serum than in serum-free medium (Dulbecco's Minimal Essential Medium, DMEM).

[3] To whom correspondence should be addressed.
Received: 1 December 1985; Accepted: 15 February 1986

Harvey Kliman (University of Pennsylvania, Philadelphia) has modified the Hall procedure by introducing a discontinuous Percoll density gradient to purify trophoblasts. In his method, term placenta (30 g soft villous tissue) is minced and digested in an Erlenmeyer flask on a rotator for 30 min at 37°C in 150 ml Hank's Medium (Mg^{2+}- and Ca^{2+}-free) containing 0.125% trypsin and 0.2 mg/ml DNAse I (DNAse I is added again after 20 min). The solids are allowed to settle and the supernatant fluid is withdrawn. The entire process is repeated five more times. The supernatant fluids are added to 1.5 ml calf serum (to inactivate proteolytic enzymes) and centrifuged at low speed. The pooled pellet is resuspended in serum-free DMEM containing 25 mM glucose, and layered onto a discontinuous Percoll gradient (0-70% in 20 mM HEPES, pH 7.4, 5% intervals, 5 ml each interval). Following centrifugation the trophoblast-enriched layer is found between 34 and 42% Percoll. Subsequent culture of these cells is in DMEM containing 25 mM glucose and 20% fetal calf serum, with gentamycin added as an antibiotic. The cells aggregate within 18 hr, and by 24 hr a "syncytium" is formed which secretes hCG and hPL.

Susan Daniels-McQueen (Washington University, St. Louis) has developed her own method for isolation of trophoblast from first trimester placenta. The tissue is first digested on ice for several hours with 0.05% trypsin, 0.02% EDTA. The supernatant fluid is discarded and the tissue is covered with fresh trypsin-EDTA and incubated for 30 min at 37°C. This process is repeated, with DNAse addition as needed. Protease is added, followed by further incubation for 30 min at 37°C. The cells are washed with medium containing antibiotics and 10% fetal calf serum, covered with medium, and kept on ice for 17 hr. At this point the tissue (now free of syncytium) is incubated repeatedly (20-30 min, 37°C) with trypsin-EDTA and centrifuged. The washed pellet is resuspended (1-2 x 10^6 cells/ml) in serum-free medium containing D-Valine (to selectively reduce the fibroblast population), and cells are plated (1 ml per well) on a 12-well plate coated with fibronectin or with biomatrix from endothelial cells. Dr. Daniels-McQueen incubates the cells in either F12/D-Val medium (Ham's Medium F12 containing antibiotics, 10% dialyzed fetal calf serum, and 92 mg/l D-Valine) or serum-free defined medium (Ham's F12:Dulbecco's Modified Eagles HG/HB [1:1] minus valine; antibiotics; 92 mg/l D-Valine; supplemented with EGF, transferrin, insulin, ethanoline, and sodium selenite). By 33-45 days of culture, cells grown in F12/D-Val attach and spread out, whereas the cells grown in serum-free medium form large multinucleated islands. That these islands may represent differentiated forms of trophoblast is supported by the observation that these cultures produce more hCG than the cells grown in Medium F12/D-Val, even though the cells grown in Medium F12/D-Val make more protein than do the cultures maintained on serum-free medium.

Like Daniels-McQueen, Charles Loke (Cambridge University, Cambridge) favors using first trimester placenta rather than term placenta as a source of trophoblast cells. Based on immunocytochemical staining with monoclonal antibody A5 (specific for cytotrophoblast), Loke obtains only 11% cytotrophoblast cells from term placenta, compared to 40% cytotrophoblast from first trimester placenta. Loke dissociates the trophoblast cells from the placenta with trypsin-EDTA (with or without DNAse) for 30 min at 37°C using magnetic stirrer, followed by Ficoll gradient centrifugation.

Laurence Cole (University of Michigan, Ann Arbor) obtains organ cultures from either early or term placenta that remain viable for 2 to 3 wk. Villous tissue is dissected from the placenta, placed in phosphate-buffered saline on ice, and rocked mechanically. The rocking causes fine villous threads to be extruded from the tissue, and these are embedded in gel foam glued onto 60 mm plastic culture dishes. Cole's culture medium consists of equal parts of Ham's F12 Medium and Dulbecco's Modified Eagle's Medium supplemented with hydrocortisone, insulin, and transferrin.

In the case of malignant trophoblast cell cultures, well-established cell lines derived from choriocarcinoma (e.g., BeWo, JAr) are routinely subcultured by incubation for 10 min at 37°C with 0.05% trypsin, 0.02% EDTA in Earle's Ca^{2+}- and Mg^{2+}-free Medium. A Pasteur pipet is used to disperse minor cell clumps, and cells are centrifuged for 3 min at 800 rpm. The supernatant fluid is removed and the cells are resuspended in culture medium (Medium 4510 or RPMI 1640, containing 10% fetal or newborn calf serum). Split ratios vary between 1:2 and 1:8 for the BeWo and JAr cell lines, which are routinely subcultured weekly, with daily medium replenishment. The conditions of trypsinization should be watched carefully as the cells may be destroyed by even a slight excess of proteolytic treatment. The malignant trophoblast cells in culture appear to be predominantly cytotrophoblast-like in that they contain a single nucleus. However, occasional giant cells and multinucleated cells can be observed in most culture flasks.

The greatest difference of opinion in the workshop related to the estimate of the proportion of cytotrophoblast cells in the final preparation. Estimates ranged from 11% to 97% cytotrophoblast cells when term placenta was the starting material. There was considerable opinion that macrophages and maternal cells may constitute a significant proportion of cells in the preparation, and that the use of monoclonal antibodies specific for such cells may be the best way to estimate their contribution.

Toward Cell Isolation Procedures That Select for Trophoblast

The standard procedures currently available for culturing human trophoblast cells need to be improved because they lead to a significant degree of contamination by other cell types beside trophoblast. With the help of functional and cell surface markers, there is evidence that the major non-trophoblast population consists of macrophages, but fibroblasts and endothelial cells released from the villous mesenchyme are also involved. From chromosomal studies, it would appear that even maternal cells must be considered as a potential source of some of the cells found in culture.

Therefore, for investigators who need pure trophoblast cultures or even cultures which are predominantly trophoblast, technical refinements are necessary. The recent availability of appropriate monoclonal antibodies which can be used to identify trophoblast and other cell types isolated from the placenta has been very helpful in this context in that the success or otherwise of any new techniques can now be evaluated objectively.

New Developments in Cell Isolation Procedures That Select For Trophoblast

A method was described employing differential centrifugation over discontinuous Percoll gradients which claimed to result in a cell suspension of which over 90% were trophoblast. Other possible methods like panning and the use of a cell sorter in conjunction with monoclonal antibodies directed at trophoblast cells were also discussed but no definitive data were as yet available.

Refinement of Culture Conditions

Inclusion of an appropriate collagenous matrix like fibronectin or extracellular matrix (ECM) was found to improve plating efficiency and to increase the proliferative capacity of placental cells although this modification alone did not select for trophoblast.

Substances could be added to the culture medium which would selectively enhance trophoblast growth. These included transferrin and epidermal growth factor.

Alternatively, alterations could be made to the medium in order to inhibit the proliferation of non-trophoblast populations. The replacement of L-valine with D-Valine has been tried with some success, the rationale for this being that most cells of epithelial origin have the appropriate enzyme to utilize D-Valine while mesenchymal cells cannot do so.

In order to define culture conditions with a greater stringency, serum-free medium has also been tried with the claim that trophoblast differentiation could be induced.

Conclusion

In conclusion, what has emerged from the workshop is that the most likely procedures in future for the selective propagation of human trophoblast cells in vitro will have the following protocol. Trophoblast cells are selectively enriched by isolation methods such as density gradient centrifugation or by using monoclonal antibodies against trophoblast in conjunction with panning or the cell sorter. These trophoblast cells are then plated out on culture dishes coated with an appropriate collagenous matrix-like fibronectin or ECM. Finally, the cultures are then maintained in well defined media (serum-free is needed) which contains both trophoblast enhancing as well as mesenchymal cell inhibitory substances.

ANIMAL PLACENTAL PERFUSIONS

Trophoblast Research 2:467-479, 1987

PLACENTAL STEROID DEHYDROGENASES: ASSESSMENT WITH A NONHUMAN PRIMATE IN SITU PLACENTAL PERFUSION MODEL[1]

William Slikker, Jr.[2], John R. Bailey, George W. Lipe, Zelda Althaus, and Julian E.A. Leakey

Pharmacodynamics Branch
Reproductive and Developmental Toxicology
National Center for Toxicological Research
Jefferson, Arkansas 72079
and
Department of Pharmacology and Interdisciplinary Toxicology
University of Arkansas for Medical Sciences
Little Rock, Arkansas 72205

INTRODUCTION

The oxidoreductase activity of 17β-hydroxysteroid dehydrogenase (17β-HSD) has been demonstrated in human placenta and catalyzes the interconversion of estradiol-17β (E_2) and estrone (E_1) (Langer and Engel, 1958; Jarabak and Sack, 1969; Strickler and Tobias, 1982). Preimplantation rat and mouse embryos have also been reported to possess 17β-HSD activity (Wu and Lin, 1982; Wu and Matsumoto, 1985). It has been postulated that 17β-HSD is responsible for the conversion of the potent estrogen, E_2, to the less potent E_1 in later-term non-human primate pregnancy (Slikker et al., 1982a) and regulates the ratio of E_2 and E_1 throughout the preimplantation period in the rodent (Wu and Matsumoto, 1985).

Another placental oxidoreductase enzyme, 11β-hydroxysteroid dehydrogenase (11β-HSD) is reported to catalyze the interconversion of cortisol and cortisone (Osinski, 1960; Bernal et al., 1980; Murphy, 1979; Murphy et al., 1980). The conversion of the potent endogenous primate glucocorticoid, cortisol, to its physiologically inactive counterpart, cortisone, has been described in the human and nonhuman primate placenta (Berliner and Ruhmann, 1969; Murphy, 1979; Slikker et al., 1982b).

Since it is postulated that delivery of the biologically active hormone to the conceptus is proportional to subsequent developmental toxicity (Murphy, 1979; Slikker et al., 1982a, 1984a), alteration of placental enzyme activities responsible for hormone metabolism (17β-HSD and 11β-HSD) would influence the extent of developmental insult.

[1]This work was presented in part at the 1984 Primatology Society Meetings.
[2]To whom correspondence should be addressed.
Received: 10 October 1985; Accepted: 1 May 1986

467

The goal of the present study was to determine if the enzymatic conversion of E_2 or cortisol to their less active metabolites in the placenta could be enhanced or diminished. To achieve this goal, the in situ, late term rhesus monkey placental model was selected. This nonhuman primate model has the advantage of sharing many common features with the human placenta in that both are hemochorial, usually support the development of a single conceptus and function in a coordinated manner with the fetus to synthesize pregnancy hormones (Stolte, 1975). In addition, information describing the metabolism and distribution of cortisol and E_2 in the intact pregnant monkey is available for comparison (Slikker et al., 1982a, b; 1984a).

Because our earlier studies have shown a nearly complete conversion of E_2 to E_1 upon a single pass through the in situ monkey placenta, we attempted to diminish this conversion by substrate saturation of the enzyme with pharmacological doses of E_2 and the competitive inhibitor, diethylstilbestrol (DES) (Slikker et al., 1982c; Jarabak and Sack, 1969). In a second set of experiments, we attempted to diminish the availability of required cofactor, oxidized NAD or NADP, by pretreating the monkeys with ethanol (EtOH), a substrate which utilizes these same oxidized cofactors for its metabolism (Zahltan et al., 1982; Salaspuro et al., 1981; Jarabak and Sack, 1969). In another group of studies, we attempted to enhance the minimal placental conversion of cortisol to cortisone by the pretreatment of the late term pregnant monkey with dexamethasone, a known inducer of several fetal liver enzymes (Leakey et al., 1985, 1986).

MATERIALS AND METHODS

Chemicals

Radiolabeled $[6,7-^3H]E_2$ (42.0 Ci/mmol) was purchased from Amersham/Searle Corporation (Arlington Heights, IL) and purified by a normal phase HPLC system consisting of a Waters Associates (Milford, MA) μPorasil column (3.9 x 250 mm) and a mobile phase of hexane-chloroform-methanol (70:27.5:2.5). The source and preparation of E_2 and derivatives have been reported (Slikker et al., 1981). The standards included E_1, E_2, estrone-glucuronide (E_1G), and estrone-sulfate (E_1S).

The radioisotope $[1,2-^3H(N)]$-hydrocortisone (50.7 Ci/mmol) was purchased from New England Nuclear (Boston, MA). It was purified to >98% purity by a previously reported HPLC method (Althaus et al., 1982). Dexamethasone (DEX) was obtained from Sigma Chemical Co. (St. Louis, MO) and used as received. No ultraviolet absorbing (254 nM) contaminants were found with HPLC analysis. The sesame seed oil used as a vehicle was laboratory grade from Fisher Scientific (Fairlawn, NJ). The source and preparation of cortisol (hydrocortisone), cortisone, 6βOH-cortisol and cortisol-21-sulfate have been reported (Althaus et al., 1982; Slikker et al., 1984a).

Surgical Preparation and Sampling

Twenty time-mated pregnant rhesus monkeys in late gestation (145-157 days, 88-95% term, Table 1) were used in this work. The monkeys were fed a standard Purina Monkey chow diet supplemented with fruit and water ad libitum. The dose of steroid and associated radioactivity is also indicated in Table 1. A trace dose of

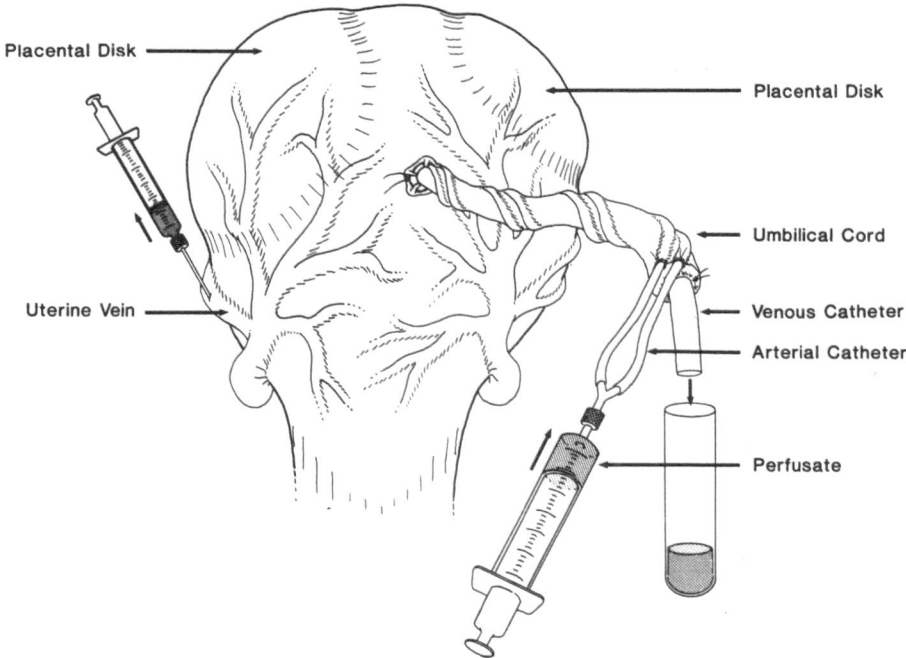

Figure 1. Schematic view of a rhesus monkey uterus with cannulas in place for in situ placental perfusion. Dosing was accomplished via the umbilical arteries (UA) at 15 ml/min x 8 min. Samples were obtained from the umbilical vein (UV) at 1 min aliquots x 9 min and from the uterine vein (UTV) at 1 min intervals x 10 min.

[14]C-dexamethasone was administered along with the ^{3}H-cortisol. The comparison of data from a preliminary study with ^{3}H-cortisol alone did not differ from the present results. Food was withheld for 16 hr prior to surgery. Atropine sulfate (0.02 mg/kg, im) was administered to retard respiratory tract secretions. Anesthesia was induced with Ketamine hydrochloride (10 mg/kg im). The monkeys were fitted with a cuffed tracheal airway and anesthesia was maintained with halothane/N_2O/O_2 throughout the experiment. Gaseous anesthesia was used to enhance uterine relaxation. The rhesus monkeys were fetectomized and the viable neonate placed in an isolette. Care was taken to minimize amniotic fluid loss by utilizing a purse-string suture around the uterine surgical os prior to fetal head delivery. As the fetus was delivered, the suture was sufficiently tightened around the umbilical cord to prevent amniotic fluid loss but not to obstruct flow through the cord. As shown in Figure 1, the placental umbilical arteries (UA) were cannulated with No. 3.5 French umbilical artery catheters, the umbilical vein (UV) with a No. 5 French umbilical artery catheter and a maternal uterine vein (UTV) with a 20 gauge intercath. The placenta was perfused with 6 ml of heparinized saline (3.3 units/ml) and 20 ml of 37°C sterile Hank's Balanced Salt Solution (HBSS; adjusted to pH 7.4 with $NaHCO_3$; Cat. No. 310-4065, Gibco Lab., Grand Island, NY) followed by the perfusion with the test compound in

120 ml of HBSS via the UA. HBSS was selected as the perfusate in lieu of blood or plasma because it does not contain plasma binding proteins and yet is iso-osmotic. The dose was perfused with a syringe pump (Sage Instrument Co., White Plains, NY) at 15 ml/min x 8 min. Samples were obtained from the UV (1 min each x 9 min) and from the UTV at 1 min intervals for 10 min. All perfusions were begun 30 min after fetectomy and all mothers recovered fully from the surgery. All blood samples were immediately centrifuged, the plasma removed and frozen at -70°C until analysis.

Three animals received EtOH via catheters implanted in a maternal radial vein. A 20% (v/v) saline solution containing 1.5 gm EtOH per kg body weight was administered over a period of 30 min with a Sage Md/351 syringe infusion pump (Sage Instruments, Inc., White Plains, NY). Maternal blood samples were collected and analyzed for EtOH concentration as previously described (Hill et al., 1983). The surgical preparation was begun 1 hr after the end of the infusion and the placental perfusion experiment was begun 30 min later. Earlier studies from our laboratory demonstrated that 80-240 min after the start of the 30 min EtOH infusion, the fetal plasma concentrations were equal to the maternal concentrations (Hill et al., 1983).

Sample Analysis

Aliquots of the perfusate and maternal plasma were put into 5 ml centrifuge tubes with the appropriate estrogen or cortisol standards and then treated in one of two ways. (1) To identify and quantitate E_2, E_1, cortisol, cortisone or their conjugates, equal volumes of MeOH/EtOH (1:1) was added to precipitate proteins and salts. The precipitate was pelleted by centrifugation and the supernatants transferred to fresh tubes. This procedure was repeated three times and the supernatants pooled. The pools were reduced in volume under a stream of N_2 to 1.0 ml. A 25 μl aliquot was taken from each for estimation of total radioactivity. The supernatant was reduced to near dryness and the residue dissolved in 0.1 ml MeOH/H_2O (65/35) for HPLC analysis on a LiChrosorb column. (2) To quantify E_2, E_1, cortisol or cortisone, equal volumes of ethyl acetate (Burdick and Jackson) were used to extract the samples three times. Aliquots were taken for estimation of total radioactivity and the samples were reduced in volume to near dryness. Residues were dissolved in 0.1 ml ethyl acetate and injected onto the Chromegabond Diol column.

The HPLC systems used in this work consisted of a Rheodyne 7125 sample injector from Rheodyne, Inc. (Cotati, CA), model 6000-A pumps, model 660 solvent programmers and model 440 dual channel ultraviolet (U.V.) detector equipped with 254 and 280 nm U.V. filters from Waters Associates (Milford, MA). Two commercial pre-packed columns were used, a LiChrosorb RP-18 (5 μm), 10 x 250 mm from MCB Manufacturing Chemists, Inc. (Cincinnati, OH) and a Chromegabond Diol (10 μm), 4.6 x 300 mm ES Industries (Marlton, NJ). An Altex model 210 Injection Valve from Altex Scientifics, Inc. (Berkeley, CA) was used with the LiChrosorb column.

A 50 min convex gradient program (no. 5 on the model 660 solvent programmer) was used with the LiChrosorb column to quantify cortisol and cortisone and to isolate E_2 and E_1. It began with 10% MeOH in 0.01 M ammonium acetate buffer (pH 6.9) and concluded with 100% MeOH. The flow rate was 2.0 ml per min and 1.0 ml fractions were collected every 0.5 min. A 100 min linear gradient program (no. 6 on the model 660 solvent programmer) was used with the Chromegabond Diol

Table 1

Summary of Experimental Animals and Dosage of Radiolabeled Steroid

Monkey No.	Gestation Age (days)	wt (kg)	Dose (µCi)	Dose (µg)
^{3}H-Estradiol				
1925	157	7.6	85	0.7
4015	153	6.8	78	0.7
4410	156	8.6	86	0.7
306	156	5.6	75	50
1925[a]	156	9.1	93	50
4048	155	6.5	95	50
4144	153	9.5	85	50
1268[b]	156	7.1	105	100
4022[b]	154	7.9	66	100
4150[b]	154	7.9	92	100
4012[c]	154	7.4	94	0.7
4013[c]	155	7.4	92	0.7
4158[c]	153	6.2	17	0.7
^{3}H-Cortisol				
304	145	7.1	104	0.7
4010	145	8.6	106	0.7
4150[a]	145	5.8	100	0.7
305[d]	146	6.7	111	0.8
4060[d]	144	11.7	97	0.7
4144a[d]	144	9.1	114	1.0
4046[d]	147	6.8	118	0.8

[a]   These monkeys were used twice during the 2 year period.

[b]   A combination of 50 µg E_2 and 50 µg DES.

[c]   Maternal animal administered 1.5 gm/kg EtOH over a 30 min period ending 1.5 hr before placental perfusion.

[d]   Predosed with DEX 10 mg/kg sc on each of 3 days preceeding the experiment.

column to quantify E_2 and E_1. It began with 100% hexane and ended with hexane/isopropanol (80/20). The flow rate was 1.5 ml per min and fractions were collected at 1 min intervals (Slikker et al., 1981; Althaus et al., 1982). The relative abundance of the steroids was determined by comparing the percent radioactivity which co-migrated with the reference standards on each chromatogram.

Chromatograms were recorded on Fisher Recordall Series 5000 Dual Pen recorders (Fisher Scientific, Pittsburg, PA). Radioactive fractions from the LiChrosorb column were collected on a microfractionator model FC-80K (Gilson Medical Electronics, Inc., Middletown, WI). An Isco model 328 Retriever III (Instrumentation Specialties Company, Lincoln, NE) was used to collect fractions from the Chromegabond Diol column. The scintillation fluid was Scintisol from Isolab, Inc. (Akron, OH). Scintillation spectrometry was accomplished with a Mark III 6881 Liquid Scintillation System (Tracor Analytical, Inc., Atlanta, GA).

The hexane and methanol used were distilled-in-glass quality from Burdick and Jackson (Muskegon, MI) and the isopropanol was from Fisher Scientific (Pittsburg, PA). The ammonium acetate buffer (pH 6.9) was made with an ultra-pure product from Mallinckrodt (St. Louis, MO). The water used in the aqueous solution was first deionized then filtered through a Millipore Milli-Q System (Bedford, MA). Aqueous solvents were filtered through a Millipore BDWP 0.6 µM filter and degassed before use.

RESULTS

The HPLC analysis of perfusate samples collected from the UV and UA perfusion with a trace dose of ^{3}H-E_2 revealed a greater than 95% conversion of E_2 to E_1 (Table 2a). Over the 1-9 min perfusion period, the percent conversion did not change significantly nor were any other metabolites observed. Samples collected from the maternal side of the placenta (UTV) showed a lesser but consistent percent of E_2 to E_1 conversion (75%). In an attempt to saturate placental 17β-HSD, the enzyme responsible for the E_2 conversion, 50 µg of E_2 was added to the 120 ml of perfusate. As shown in Table 2b, no significant change in the conversion of E_2 and E_1 resulted from the higher concentration of substrate. The addition of 50 µg of DES, a competitive inhibitor or 17β-HSD (Jarabak and Sack, 1969; Blomquist et al., 1984), also failed to change the nearly complete oxidation of E_2 (Table 2c).

Because the 17β-HSD conversion of E_2 to E_1 requires NAD or NADP cofactor, the same cofactor required for EtOH metabolism, three monkeys were dosed via the maternal radial vein with 1.5 gm/kg EtOH 2 hr before placental perfusion with ^{3}H-E_2. As shown in Figure 2, the maternal blood EtOH concentrations achieved peak values (718 mg%) at the end of the 30 min EtOH infusion and declined to 420 mg% by 2 hr when the ^{3}H-E_2 was perfused through the placenta. Even though EtOH metabolism by alcohol dehydrogenase is known to reduce the NAD:NADH and NADP:NADPH cofactor ratios as demonstrated in the baboon and rat (Salaspuro et al., 1981; Badawy and Evans, 1981), no significant change in E_2 metabolism was observed in samples obtained from the fetal side of the placenta (UV) after ^{3}H-E_2 perfusion of EtOH pretreated monkeys. Samples collected from the UTV, however, indicated a 15% reduction of the conversion of E_2 to E_1 (Table 2d).

Table 2

The Percentage of E_2 to E_1 Conversion by the In Situ Rhesus Monkey
Perfused-Placenta (Mean ± SD)

		1-2 min	3-4 min	5-6 min	8-9 min	Mean 1-9 min
2a		Control (N = 3): 100 μCi^{3}H-E$_2$[a]				
	UV	95.8	95.9	97.4	98.0	96.8
		3.6	4.5	1.5	1.7	1.1
	UTV	79.3	75.3	76.5	70.7	75.5
		3.1	7.1	3.0	6.5	3.6
2b		E_2 (N = 4): 100 μCi^{3}H-E$_2$ + 50 μg E$_2$				
	UV	98.5	98.8	98.7	98.5	98.6
		0.7	0.7	0.4	0.8	0.2
	UTV	76.8	78.8	77.6	85.3	79.6
		5.2	7.9	12.5	5.4	3.9
2c		E2 + DES (N = 3): 100 μCi^{3}H-E$_2$ + 50 μg E$_2$ + 50 μg DES				
	UV	96.6	97.7	97.6	98.1	97.5
		2.9	1.6	2.0	1.6	0.6
	UTV	74.2	79.0	81.4	85.9	80.1
		3.2	1.6	3.8	4.6	4.9
2d		E_2 + EtOH (N = 3): 100 μCi^{3}H-E$_2$ + 50 μg E$_2$ + 1.5 gm/kg EtOH				
	UV	97.7	98.6	98.6	98.7	98.4
		0.8	0.3	0.5	0.5	0.5
	UTV	63.9	57.7	58.1	59.6	59.8[c]
		a	9.8	12.2	11.3	2.8

a   Taken from Slikker et al., 1982c

b   N = 2

c   Significantly less than respective control conditions ($p < 0.05$)

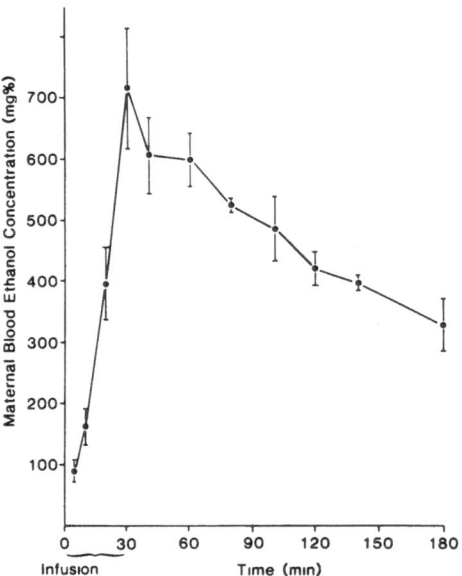

Figure 2. Maternal plasma ethanol concentrations after 1.5 gm/kg dose of ethanol infused over a 30 min period iv (mg%, mean ± SE of three monkeys).

Conversion of cortisol to cortisone by the in situ perfused placenta, as indicated by HPLC resolved samples collected from the UV, was only 8-24% (Table 3a). Samples collected from the maternal side of the placenta (UTV), however, demonstrated a larger percentage of conversion (38-64%). In contrast to the estrogen studies, the percentage of cortisol oxidation increased as the perfusion proceeded from 2 to 8 min. In an attempt to enhance the cortisol to cortisone conversion, four monkeys were dosed on three consecutive days prior to placental perfusion with dexamethasone (10 mg/kg/day, sc). As shown in Table 3b, dexamethasone pretreatment significantly increased UTV cortisol to cortisone conversion. A similar, but not statistically significant, increase was also seen in the UV samples. Over this 8 min perfusion and sampling period, only 2% of the total radioactivity was not accounted for as cortisol or cortisone.

DISCUSSION

Although placental dehydrogenases have been characterized with in vitro methodologies by several investigators, verification of their metabolic capacity in situ is less well described (Blomquist et al., 1984; Bernal et al., 1980; Jarabak and Sack, 1969). In the present studies, both metabolic conversion of E_2 to E_1 and cortisol to cortisone has been demonstrated in the in situ nonhuman primate placenta. In the

Table 3

The Percentage of Cortisol to Cortisone Conversion by the In Situ Rhesus Monkey
Perfused-Placenta (Mean ± SD)[a]

		1-2 min	3-4 min	7-8 min
3a		Control (N = 3)		
	UV	7.8	17.6	24.0
		2.0	8.8	14.1
	UTV	38.4	47.3	64.4
		11.3	18.7	1.6
3b		DEX Pretreatment (10 mg/kg 1 day sc, 3 doses, N = 4)		
	UV	29.0	34.5	49.4
		31.3	20.8	27.0
	UTV	67.2[b]	76.1[b]	73.9
		8.9	5.5	8.2

[a] Taken in part from Althaus et al., 1986
[b] Significantly different from the respective control conditions (p < 0.05)

late term placenta, greater than 95% of the E_2 was oxidized to E_1, presumably by 17β-HSD. This nearly complete metabolism of E_2 was not sensitive to saturation by perfusate substrate concentrations approximately 400-fold higher than endogenous E_2 plasma concentrations (Slikker et al., 1984b). The attempt to inhibit E_2 17β-HSD activity by DES, a reported in vitro competitive inhibitor of 17β-HSD (Jarabak and Sack, 1969) was likewise unsuccessful. Ethanol, infused to provide blood concentrations of over 400 mg%, also failed to alter E_2 metabolism as measured in UV samples but did produce a 15% reduction of conversion as measured in UTV samples. Studies in the rodent (Badawy and Evans, 1981) had previously demonstrated that blood ethanol concentrations in this range resulted in a significant reduction in the ratio of NADP/NADPH cofactors. It was reasoned that a reduction in the cofactors necessary for E_2 oxidation (i.e., NADP or NAD) would result in a reduction in placental E_2 conversion. Reasons for only a "maternal side" and not a "fetal side" response may include: 1) the fetal side of the placenta contains a greater store of oxidized cofactors; 2) the fetal side of the placenta contains a different 17β-HSD enzyme with a lesser sensitivity to cofactor alteration or; 3) tissue on the fetal side of the placenta does not contain, or contains little, alcohol dehydrogenase so that ethanol pretreatment cannot affect cofactors ratios. Although a recent report provided evidence that the latter explanation may be correct, additional studies will be necessary to resolve this question (Pares et al., 1984). Together, these results indicate that the nearly complete conversion of E_2 to E_1 by the in situ nonhuman primate placenta is resistant to perturbation by either substrate load or EtOH pretreatment.

In contrast to E_2, less than 15% of the perfused cortisol was oxidized to cortisone as measured on the fetal side of the placenta under control conditions. Pretreatment of the maternal monkey with the synthetic glucocorticoid, DEX, resulted in a significant increase of cortisol to cortisone conversion in UTV samples and a similar but nonsignificant trend for UV samples. This apparent enhancement of placental 11β-HSD is consistent with the observed induction of fetal liver glucuronyltransferase and cytochromes P-450 activities after similar DEX pretreatment (Leakey et al., 1985, 1986) and suggests that the placenta is able to respond to environmental influences.

Speculation as to the role of placental conversion of the endogenous hormones has centered on the regulation of the exposure of the embryo/fetus to these potent endogenous steroids. We previously postulated from studies in the intact, fetal catheterized pregnant rhesus monkey that the conversion of the potent estrogen, E_2, to the less potent E_1 results in reducing the toxicity of prenatal, naturally occurring estrogen exposure, a phenomenon which is circumvented by the synthetic estrogen, DES (Slikker et al., 1982a). Based on clinical studies, Murphy (1979) postulated that placental interconversion of cortisol and nonbiologically active cortisone controlled glucocorticoid exposure to the developing conceptus. This report is consistent with the results from our laboratories in the intact pregnant monkey which demonstrated the greater fetal exposure to active glucocorticoid after maternal administration of synthetic hormone (Slikker et al., 1982b, 1984a) as compared to the naturally occurring hormone, cortisol. Recent studies by Wu and Matsumoto (1985) have demonstrated the day to day regulation of E_2 and E_1 interconversion in the preimplantation rat and mouse embryo. Taken together, these studies suggest that the regulation of potent, parent hormone concentration in the conceptus is possible throughout pregnancy. Direct proof that placental regulation of estrogen and glucocorticoid exposure to the conceptus is necessary for normal development will have to await further study.

SUMMARY

An in situ placental perfusion system was developed in the late gestational age rhesus monkey. After delivery of the fetus, the umbilical arteries and vein were fitted with catheters. Radiolabeled steroid hormones (estrogen or glucocorticoid) were then perfused at 15 ml/min in Hank's Balanced Salt Solution (pH 7.4) for 8 min. Time course perfusate samples were collected from the umbilical vein catheter and maternal blood was collected via a catheter in an uterine vein. Studies with ^{3}H-estradiol-17β (E_2) indicated a 97% conversion to estrone (E_1) with a single pass through the placenta. The percentage of E_2 to E_1 conversion was not modified by pharmacological doses of E_2 or diethylstilbestrol. Pretreating the pregnant monkey with 1.5 gm/kg ethanol 2 hr before placental perfusion with E_2 did not alter E_2 and E_1 conversion as measured by umbilical vein samples but did significantly decrease E_1 percentages in the maternal uterine vein samples. In similar placental perfusion studies with ^{3}H-cortisol, a 38-64% placental conversion to cortisone was observed in the maternal uterine vein samples but only a 8-24% conversion to cortisone in umbilical vein samples. After three days of pretreatment with 10 mg/kg dexamethasone sc, the percentage of cortisol to cortisone conversion increased 10-29%. These data indicate the 17β- and 11β-hydroxysteroid dehydrogenase oxidative capacity is present in the in situ nonhuman primate placenta near term. They also suggest that 17β-dehydrogenase activity is resistant to saturation by pharmacological doses of estrogen and only partially reduced by EtOH pretreatment. Placental 11β-dehydrogenase activity towards cortisol is less than 17β-dehydrogenase activity towards E_2. Pretreatment with glucocorticoid, however, did result in a significant enhancement of cortisol to cortisone conversion.

ACKNOWLEDGEMENTS

The authors thank Roberta Mittelstaedt for kindly preparing the HBSS, Glenn Newport for his technical assistance, and Tina Sykes for her clerical assistance.

REFERENCES

Althaus, Z.R., Bailey, J.R., Leakey, J.E.A., and Slikker, W., Jr. (1986) Transplacental metabolism of dexamethasone and cortisol in the late gestational age monkey (Macaca mulatta). *Develop. Pharmacol. Ther.*, in press.

Althaus, Z.R., Rowland, J.R., Freeman, P., and Slikker, W., Jr. (1982) Separation of some natural and synthetic corticosteroids by high performance liquid chromatography. *J. Chromatog.* 227(1), 11-23.

Badawy, A.A.-B. and Evans, M. (1981) The mechanism of the antagonism by naloxone of acute alcohol intoxication. *Br. J. Pharmacol.* 74, 514-516.

Berliner, D.L. and Ruhmann, A.G. (1969) Comparison of the growth of fibroblasts under the influence of 11β-hydroxy and 11-keto corticosteroids. *Endocrinol.* 78, 373-382.

Bernal, A.L., Flint, A.P.F., Anderson, A.B.M., and Turnbull, A.C. (1980) 11β-hydroxysteroid dehydrogenase activity (E.C. 1.1.1.146) in human placenta and decidua. *J. Steroid Biochem.* 13, 1081-1087.

Blomquist, C.H., Lindemann, N.J., and Hakanson, E.Y. (1984) Inhibition of 17β-hydroxysteroid dehydrogenase (17β-HSD) activities of human placenta by steroids and non-steroidal hormone agonists and antagonists. *Steroids* 43(5), 571-586.

Hill, D.E., Slikker, W., Jr., Goad, P.T., Bailey, J.R., Szisazk, T.J., and Hendrickx, A.G. (1983) Maternal, fetal, and neonatal elimination of ethanol in nonhuman primates. *Develop. Pharmacol. Ther.* 6, 259-268.

Jarabak, J. and Sack, G.H., Jr. (1969) A soluble 17β-hydroxysteroid dehydrogenase from human placenta. The binding of pyridine nucleotides and steroids. *Biochemistry* 8, 2203-2211.

Langer, L.J. and Engel, L.L. (1958) Human placental estradiol-17β dehydrogenase. I. Concentration, characterization and assay. *J. Biol. Chem.* 233, 583-588.

Leakey, J.E.A., Althaus, Z.R., Bailey, J.R., and Slikker, W., Jr. (1985) Dexamethasone increases UDP-glucuronyltransferase activity towards bilirubin, oestradiol and testosterone in foetal liver from rhesus monkey during late gestation. *Biochem. J.* 225, 183-188.

Leakey, J.E.A., Althaus, Z.R., Bailey, J.R., and Slikker, W., Jr. (1986) Dexamethasone induces hepatic cytochrome P-450 content and increases certain monooxygenase activities in rhesus monkey foetuses. *Biochem. Pharmacol.* 35, 1389-1391.

Murphy, B.E.P. (1979) Cortisol and cortisone in human fetal development. *J. Steroid Biochem.* 11, 509-513.

Murphy, B.E.P., Sebenick, M., and Patchell, M.E. (1980) Cortisol production and metabolism in the human fetus and its reflection in the maternal urine. *J. Steroid Biochem.* 12, 37-45.

Osinski, P.A. (1960) Steroid 11β-ol dehydrogenase in human placenta. *Nature* 187, 777.

Pares, X., Farres, J., and Vallee, B.L. (1984) Organ specific alcohol metabolism: Placental X-ADH. *Biochem. Biophys. Res. Comm.* 199(3), 1047-1055.

Salaspuro, M.P., Shaw, S., Jayatilleke, E., Ross, W.A., and Lieber, C.S. (1981) Attenuation of the ethanol-induced hepatic redox change after chronic alcohol consumption in baboons: metabolic consequences in vivo and in vitro. *Hepatol.* 1(1), 33-38.

Slikker, W., Jr., Lipe, G.W., and Newport, G.D. (1981) High-performance liquid chromatographic analysis of estradiol-17β and metabolites in biological media. *J. Chromatog.* 224(2), 205-219.

Slikker, W., Jr., Hill, D.E., and Young, J.F. (1982a) Comparison of the transplacental pharmacokinetics of 17β-estradiol and diethylstilbestrol in the subhuman primate. *J. Pharmacol. Exp. Ther.* 221(1), 173-182.

Slikker, W., Jr., Althaus, Z.R., Rowland, J.M., Hill, D.E., and Hendrickx, A.G. (1982b) Comparison of the transplacental pharmacokinetics of cortisol and triamcinolone acetonide in the rhesus monkey. *J. Pharmacol. Exp. Ther.* 223(2), 368-374.

Slikker, W., Jr., Bailey, J.R., Newport, G.D., Lipe, G.W., and Hill, D.E. (1982c) Placental transfer and metabolism of 17α-ethynylestradiol-17β and estradiol-17β in the rhesus monkey. *J. Pharmacol. Exp. Ther.* 223(2), 483-489.

Slikker, W., Jr., Althaus, Z.R., Rowland, J.M., Hendrickx, A.G., and Hill, D.E. (1984a) Comparison of the metabolism of cortisol and triamcinolone acetonide in the early, mid and late gestation age rhesus monkey (<u>Macaca</u> <u>mulatta</u>). *Develop. Pharmacol. Ther.* 7, 319-333.

Slikker, W., Jr., Lipe, G.W., Sziszak, J., and Bailey, J.R. (1984b) Changes in estrogen metabolism after chronic oral contraceptive administration in the rhesus monkey. *Drug Metab. Dispos.* 12(2), 148-153.

Stolte, L.A.M. (1975) Pregnancy in the rhesus monkey. In: *The Rhesus Monkey*, vol. II, (ed.), G.W. Bourne, pp. 171-230.

Strickler, R.C. and Tobias, B. (1982) 20α-hydroxysteroid dehydrogenase and 17β-estradiol dehydrogenase localize in cytosol of human term placenta. *Am. J. Physiol.* 242(3), 178-183.

Wu, J.-T. and Lin, G.-M. (1982) The presence of 17β-hydroxysteroid dehydrogenase activity in preimplantation rat and mouse blastocysts. *J. Exp. Zool.* 200, 121-124.

Wu, J.-T. and Matsumoto, P.S. (1985) Changing 17β-hydroxysteroid dehydrogenase activity in preimplantation rat and mouse embryos. *Biol. Reprod.* 32, 561-566.

Zahlten, R.N., Nejtek, M.E., and Jacobsen, C. (1982) Ethanol metabolism in guinea pig: ethanol oxidation and its effect on NAD/NADH ratios, oxygen consumption, and ketogenesis in isolated hepatocytes of fed and fasted animals. *Arch. Biochem. Biophys.* 213(1), 200-231.

Trophoblast Research 2:481-499, 1987

NUCLEOTIDE INTERCONVERSION AND BREAKDOWN IN THE DUALLY PERFUSED GUINEA PIG PLACENTA

Bernard van Kreel, J.P. van Dijk, and A.M.C.M. Pijnenburg

Department of Chemical Pathology
Erasmus University Rotterdam
PO Box 1738
3000 DR Rotterdam, The Netherlands

INTRODUCTION

Alterations in placental purine metabolism can occur due to diminished oxygen supply during labor (van Kreel and Wallenburg, 1980; Wallenburg and van Kreel, 1980; van Kreel and van Dijk, 1982). To study purine metabolism in the placenta without interference from other organs perfusion of the isolated placenta, or organ culture of trophoblastic cells can be used (Vettenranta and Raivo, 1984). The first objective of this study was to investigate nucleotide breakdown in placental tissue, and to quantitate the extraction of nucleosides and purines during perfusion under various experimental conditions of oxygen supplementation.

Quantitative assessment of placental purine metabolism during perfusion is also of importance for other studies that are performed by means of the isolated perfused placenta. A complete picture of nucleotide breakdown and extraction can only be obtained when all known nucleotides, nucleosides, and purines are analyzed simultaneously in the placental tissue and the perfusion medium. To achieve this purpose, a chromatographic analytical method was devised which also allows measurement of nucleotide synthesis using labeled precursors.

When a deproteinized tissue sample is analyzed in a chromatographic system without further treatment a large number of bases, nucleosides and nucleotides and perhaps deoxynucleotides can be expected to be present in the sample in varying concentrations. Also a number of other compounds, which absorb at the same wave length, are likely to interfere with the measurements. For that reason, we separated the mixture of bases, nucleosides, and nucleotides into three groups. Interfering substances were then removed by absorption of the nucleosides on a boronate affinity gel and absorption of the bases and nucleotides on mercury salts. Following this, the three groups of substances were quantitatively assessed separately using HPLC (for a review of HPLC chromatographic of purines and pyrimidines, see Gehrke and Kuo, 1978; Zakara and Brown, 1981).

When differentiation among mono-, di-, and tri-nucleotides was not necessary, the group consisting of the nucleotides was treated with alkaline phosphatase and then processed the same way as the nucleosides group.

Submitted: 10 October 1985; Accepted: 31 July 1986

In this fashion a considerable reduction in complexity of the chromatogram and improvement of the sensitivity is obtained, since mono-, di-, and tri-nucleotides all appear in one nucleoside fraction. Because more than one step is involved, the recovery of each of the nucleotides, nucleosides, and bases up to the final chromatography must be established. This recovery can be achieved by addition of radioactive markers at the beginning of the procedure.

MATERIALS AND METHODS

Chemicals

All chemicals used were of analytical reagent grade and purchased from either Sigma (St. Louis, MO) or Merck (Darmstadt, Germany). The enzymes used were obtained from Boehringer Mannheim, Germany. Ion exchange resins A6 and A25 used for HPLC were obtained from BioRad, Richmond, Virginia.

Aminex A25 was regenerated in the following way: First it was left for 20 min in a mixture of one part concentrated HCl and two parts ethanol. Following centrifugation it was added to 1 M/l HCl for about 4 hr. Then 1 M NaOH solution was added to the dry resin. After 1 hr, it was washed several times with 1 M/l citric acid until acid. The resin was then equilibrated with the elution buffer by repeated washings (modified after Khym, 1975). Reversed phase material Lichrosorp R.P. 18 was obtained from Merck. The affinity gel Affigel 601 (boronate) was also purchased from BioRad. Ultrafiltration was performed with collodion membrane filters obtained from Sartorius, Gottinger, Germany. The radioactive purine base, nucleosides, and nucleotides were obtained from Amersham (Buckinghamshire, United Kingdom).

HPLC

HPLC was performed on two separate systems. One system (A) was a modified Siemens chromatograph model S100, consisting of the following elements: Orlita high pressure pump; Valco injection value with 3.0 ml sample loop; stainless steel columns, internal diameter 0.3 cm; length 30 cm; and stainless steel frits.

The optical density at 260 nm was monitored with a Zeiss variable wavelength spectrophotometer equipped with a 20 µl flow cuvet. The eluents from the cuvet were collected in a fraction collector (LKB redirac 2112). The spectrophotometer was connected to a recorder. The event marker of the fraction collector was connected in parallel with a recorder input to mark the changes of the fractions on the chromatogram. The calculator output of the spectrophotometer was connected with an A/D interface card (ADALAB interactive microware), which was part of an Apple IIe microcomputer. The chromatographic data were saved by the computer and calculations were performed using a software program (chromatochart) supplied by Interactive Microware.

The other HPLC system (B) was established in our laboratory from the same elements, except that a ten way Valco value was used. The interconnection of this valve with the other elements of the system is shown in Figure 1. The ion exchange columns were slurry packed to a pressure of 100 Bar in a packing column with a 1 cm diameter. The reverse phase material was suspended in one part buffer/one part

LOAD   INJECT

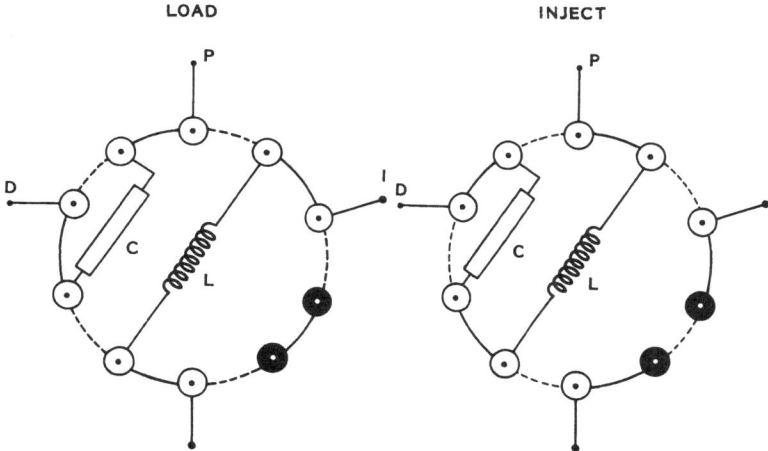

Figure 1. Diagram showing the operation of a 10-way Valco valve during group separation on Aminex A6. Small open circles indicate inlets into the valve. Dark circles are inlets not in operation. Segments in the big circle are solid when conducting P, pump; I, injection ports; L, injection loop 0.5 µl; C, column; D, detector.

alcohol, and packed in this medium packing. When the packing was complete, the column was put into place. The composition of the eluting phase was now altered stepwise to the desired composition.

Perfusions

The placenta and part of the uterus of pregnant guinea pigs, with an estimated duration of gestation between 55 and 65 days, were isolated and perfused on both sides. Gestational age was estimated by weighing the fetuses separately as described by MacNeil and Koong (1983). The surgical procedures used were those described by Leichtweiss and Schröder (1971). The equipment used was described by van Kreel and van Dijk (1982). The fetal and maternal outflows were connected to fraction collectors and sampled at one fraction per min. At the end of the experiment, the fractions were either pooled and freeze-dried or separately freeze-dried. Placentae were separated from the uterine tissue and subplacenta as quickly as possible and stored in liquid nitrogen.

In Vivo Experiments (Uterine Arterial and Venous Concentrations)

In addition to the perfusion experiments performed with the isolated guinea pig placenta, in vivo experiments were also performed in the following manner (Peeters and Wallenburg, 1983; Peeters, 1984). A polyvinyl catheter was inserted into a carotid artery and advanced into the descending aorta. A polyethylene "guide catheter" was advanced retrograde under fluoroscopy from the right jugular vein into the inferior vena cava to approximately 1 cm above the renal veins. A polyvinyl catheter was passed through this "guide catheter" and advanced about 3 cm beyond its top. Both catheters were tunneled subcutaneously and exteriorized between the shoulder blades. The third day after the operation, the inner venous catheter was removed from the I.V.C., cleared, and inserted back through the "guide catheter". Under fluoroscopic examination, the tip was advanced approximately 2 cm into one of the ovarian veins. Blood samples (about 1 ml) were taken from the maternal artery and ovarian vein and added to ice cold 5% TCA solution and the precipitate was processed as described to determine purine concentrations.

Concentrations in the Placenta, and Maternal and Fetal Arteries

Several pregnant guinea pigs were anesthetized with 1.0 ml Vetalar supplemented with 0.1 ml Vetranquil (Bristol Laboratories, New York). The abdomens were opened and the fetuses were exposed. Blood was taken by syringe from the mother and one of the fetuses by cardiac puncture and processed as indicated. Thereafter the placenta was separated from the uterus and immediately placed in liquid nitrogen.

Extraction

Placentae obtained after perfusion was frozen in liquid nitrogen, weighed, and ground in a precooled mortar. The ground tissue was transferred to a blender containing a volume twenty times of the tissue weight of ice cold 5% TCA, and the tissue was homogenized. Known amounts of radioactive bases, nucleosides, and nucleotides were then added. The contents of the blender were transferred to a centrifuge tube and centrifuged. The TCA supernatant was extracted four times with ether and gassed with nitrogen. The sample was then freeze-dried and taken up in a small volume of water.

The fractions of perfusion medium that were collected at the maternal and fetal outflows were combined and freeze-dried after radioactive labeled purines and nucleosides were administered. When non-perfused placentae were analyzed, the placenta was frozen, and part of the frozen tissues was used to determine the hemoglobin concentration after homogenizing the tissue. A sample of fetal and maternal blood was taken with a syringe, weighed, and approximately 1 ml was added to 10 ml ice cold 5% TCA. Two other blood samples were taken to determine hemoglobin concentration (Thunell, 1965).

Group Separations

Ion Exclusion Chromatography on Aminex A6

In Figure 1, it is shown how the elements comprising the chromatographic system are connected. A 0.5 ml sample obtained by dissolving freeze-dried extracted material in water is injected into the sample loop of chromatograph B with the injection valve in position "load". The sample was then injected onto the column by turning the valve into position inject. The column was filled with the cation exchange resin Aminex A6, equilibrated with a 0.1 M sodium citrate buffer, pH 4.2, to which 0.2 M sodium chloride per liter was added. Operating pressure was 50 bar; temperature was ambient. Negative charged nucleotides immerged in one peak from the column and were collected as such. This collection constituted the nucleotides group.

The nucleosides and bases eluted subsequently, but it appeared to take considerable time to elute the adenine and guanine cations. To reduce the time of elution for the bases and nucleosides and to keep the elution volume as small as possible, the value was turned in position "load" again. This modification reverses the flow direction and the bases and nucleosides, including adenine and guanine, were eluted together in one peak shortly afterwards. This fraction constitutes the bases/nucleosides group which is purified further by adsorption on Affi-gel.

Adsorption on Affi-gel

The fraction containing the purine and pyrimidine bases/nucleosides was adjusted to pH 8.7, with 1 N sodium carbonate and applied to a small column containing 1.5 ml boronate gel (Affi-gel 601) at a flow of 0.5 ml/min by means of a peristaltic pump (Autoanalyzer). The eluate containing the bases and thymidine was collected, the other nucleosides remained on the column. After washing the column with 5 ml 0.1 M sodium carbonate buffer, pH 8.7, the nucleosides were eluted with 0.1 M formic acid. This fraction (the nucleosides group) was also collected and freeze-dried subsequently.

At this point the mixture was separated into three groups - nucleotides, nucleosides, and bases. The freeze-dried nucleosides were added to 0.5 ml of the buffer used for the final chromatography. The nucleotides, however, were present in about 3 ml of buffer (0.1 M citrate; 0.2 M NaCl; pH 4.2). When this sample was freeze-dried, taken up into the appropriate buffer and injected onto the column, the high salt concentration resulted in the appearance of an irregular change in optical density during chromatography. This finding also applied to the fraction containing the bases. However, the salt can be removed before the sample is concentrated by adsorption onto mercury carbonate.

Concentration and Buffer Change

Ten ml of a 1 M sodium carbonate solution was added to a tube containing 1.5 ml of 5% mercury acetate. The yellow precipitate of mercury carbonate was collected by centrifugation and washed with water. The fraction containing the bases was added, and after shaking for 1 min, the supernatant was aspirated off. All purine and pyrimidine bases are adsorbed. The mercury carbonate containing the adsorbed bases was washed several times with water until the washings were free of chloride. The mercury carbonate was then solubilized by the addition of 5 ml of 10% acetic acid. Hydrogen sulfide gas was bubbled through the solution to remove the mercury. After centrifugation, the supernatant was collected, and excess hydrogen sulfide was removed by a stream of nitrogen. The acetic acid was extracted with ether and the sample was then freeze-dried.

The fraction containing nucleotides was adjusted to pH 10 with 1 M sodium carbonate and 10 IU of alkaline phosphatase. After 30 min, hydrolysis was complete, and the fraction was applied to the Affi-gel column. After washing with 5 ml of 0.1 M sodium carbonate, the nucleosides were again eluted with 0.1 M formic acid. This fraction was collected and freeze-dried subsequently. When it was necessary to obtain a distinction between mono-, di-, and tri-nucleotides, part of the nucleotides fractions were adjusted to 7.8 with sodium carbonate and adsorbed onto mercury carbonate as explained.

Chromatography

Several chromatographic systems were used depending on the kind of sample under investigation. Actual chromatogram of the purine bases and nucleosides are shown in Figure 2. Additional information about the chromatographic systems used can be obtained from the legends. The freeze-dried samples were taken up in 0.5 ml of the appropriate buffer used in chromatography to which a known amount of internal standard was added. The identity of the peaks appearing on the chromatogram was determined by comparing the retention times with the retention times of a mixture of reference compounds and confirmed by rechromatography on a different system, or by the enzymatic peak shift method. In all instances, the enzyme solution was administered to an ultrafiltration unit and washed with buffer by repeated ultrafiltration. Washings were repeated until no U.V. absorbing material was present in the enzyme solution. When this was the case, the sample to be treated was

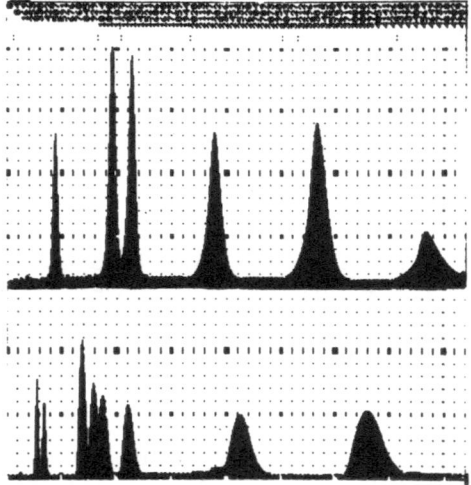

Figure 2. Computer display of chromatograms of purine bases and internal standard (upper drawing) and nucleosides (lower drawing). Retention times in seconds can be obtained from the upper scale. Computer calculated baseline is shown as the dashed horizontal lines. Separation between peaks is indicated by unbroken vertical lines. Elution sequence for purines. Uric acid, xanthine, hypoxanthine, guanine, adenine, aniline, (internal standard). Separated on Aminex A6: elution buffer 0.1 M, citric acid pH 5.6, temp. 40°C, pressure 50 Bar. Elution sequence for nucleosides: cytidine, uric acid, inosine, guanosine, xanthosine, thymidine, adenosine, indol acetic acid (internal standard). Separated on lichrosorp RP 18: elution buffer 0.01 M, citric acid, 2.5% methanol, pH 5.1, room temperature, pressure 50 Bar.

added to the ultrafiltration bag and after reaction had occurred, the ultrafiltrate was applied to the column.

Calculations

The unknown concentrations in the sample injected on the column were determined by comparing the peak areas of the compounds and added standard with the peak areas of a mixture of reference compounds using the following formula:

$$\text{Conc}_x = \left[\text{area X} \times \frac{\text{conc. st.}}{\text{area st.}}\right] \begin{array}{l} \text{sample} \\ \text{mixture} \end{array} \times \left[\frac{\text{area st.}}{\text{conc. st.}} \times \frac{\text{conc. X}}{\text{area X}}\right] \begin{array}{l} \text{reference} \\ \text{mixture} \end{array}$$

Conc_x: concentration purine X in sample before chromatography
Conc.st.:   concentration of added standard

To obtain the concentration of X in the original sample, this value must be corrected according to the recovery of the method as follows: A known amount of radioactive compound X has been added at the beginning into a known volume, so the amount of dpm/ml of X can be calculated. Before chromatography, the total amount of dpm/ml is measured. During chromatography the fractions are collected and the distribution of radioactivity is determined as dpm/ml of X. The following formula is then used:

$$\text{Conc. X in original sample} = \frac{\text{dpm/ml of X at the start}}{\text{dpm/ml of X before chromatography}} \times \text{Conc. X [M.l}^{-}]$$

When tissues are analyzed the concentration has to be multiplied by TCA volume in ml divided by placental weight in g to obtain the concentration in units of $M.Kg^{-1}$.

RESULTS

The recovery obtained when a mixture of labeled compounds was added to a blood sample and analyzed subsequently was approximately 50% of all fractions relative to the initial amount of radioactivity (Figure 2). These results differed slightly for different experiments. It is obvious that the distribution of radioactivity in the different fractions did not change appreciably during the analysis as were seen when the first and last rows of the figure were compared. This analysis meant that the recovery of each kind of purine was the same. This finding allowed one to add one labeled base, nucleoside, and nucleotide to the original mixture instead of the whole mixture as was done in the case depicted in Figure 3. The perfusion experiments were separated in two groups. The results of the first group are shown in Table 1. The placenta was perfused under anaerobic conditions (I) or under aerobic conditions (II and III). At the end of perfusion, the placental tissues were analyzed and were in experiment nr. I and II, the fractions of perfusion medium obtained at the maternal and fetal outflows were combined and analyzed. To assess the variation in the extraction of purines in one experiment (III), three fractions were pooled and analyzed after 2, 10, and 20 min.

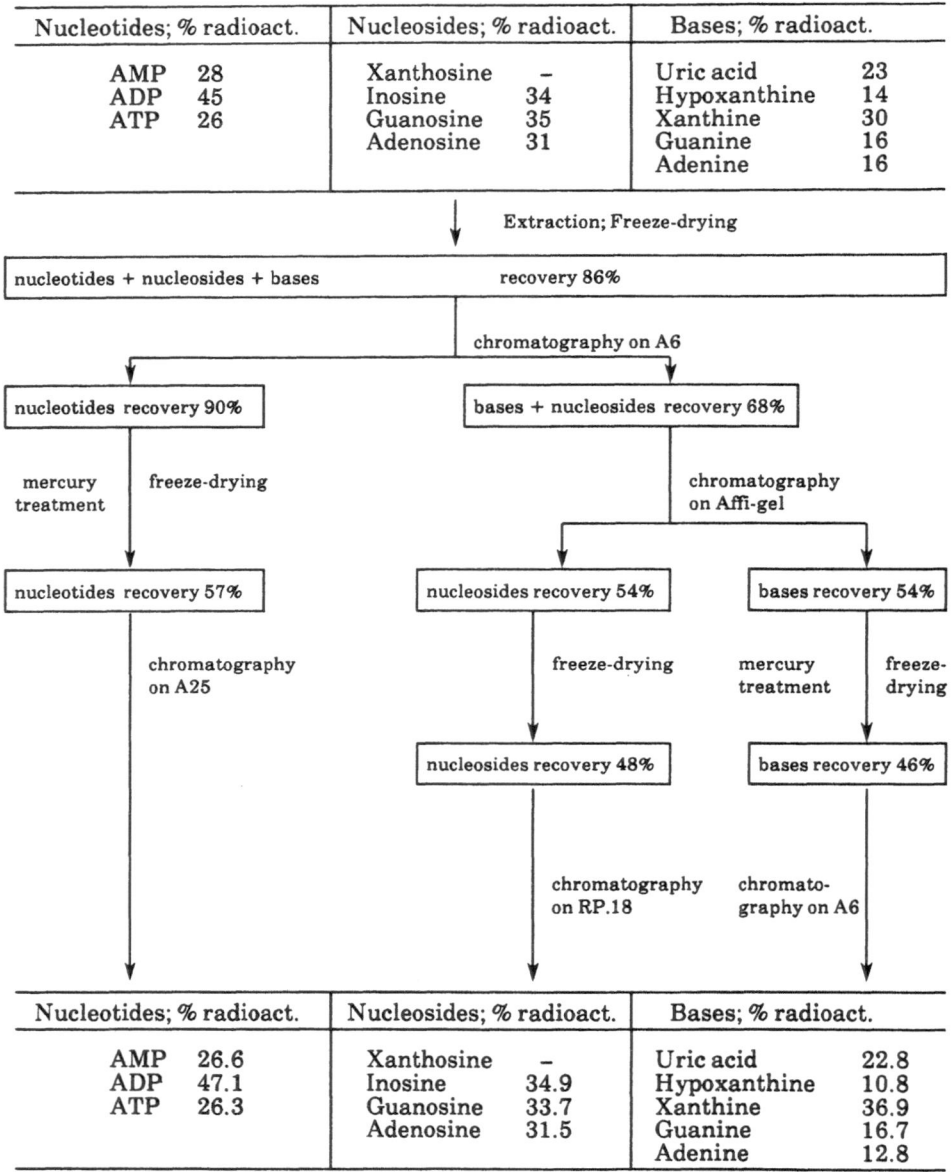

Figure 3. Schematic representation of the analytical procedure. Upper row composition of the mixture of tracers added to the TCA extract. Subsequent rows recovery obtained: 86% recovery of the total activity after extraction with ether and freeze-drying; recovery (90%) of nucleotide group after A6 chromatography and recovery of bases and nucleosides together (68%), etc. Last row composition of tracers in the three groups obtained after the final chromatography.

Table 1

Results of the analysis of perfusion fluids and placental tissues after perfusion under different conditions as indicated in the first row. Second row M, maternal venous outflow; F, fetal outflow. Concentrations are given in µM/l. In the first column the following symbols were used: INO, inosine; GNO, guanosine; XTO, xanthosine; ANO, adenosine; UA, uric acid; X, xanthine; HX, hypoxathine; GU, guanine; AD, adenine; ALL, Allantoine. In the fifth row concentration of nucleotides are given after reduction to nucleosides. The values in parenthesis given the ratios of mono-, di-, and tri-nucleotides as analyzed directly without hydrolysis to nucleosides.

	Perfusion (I) with N₂			Perfusion (II) with O₂			Placenta	Perfusion with O₂ (III)					
	Placenta	M.	F.	Placenta	M.	F.		M 2 min	M 10 min	M 20 min	F 2 min	F 10 min	F 20 min
Nucleosides													
INO	24	1.5	4.2		1.0	1.5	20	1.0	0.7	0.7	0.6	0.7	0.8
GNO		0.5	0.5										
XTO													
ANO													
Purines													
UA	35	13.6	4.6	18	10.2	6.9	10	4.5	5.7	5.3	1.0	3.2	4.5
X	100	8.0	10.3	47	0.2	0.9	7	1.7	0.9	0.9	0.5	0.7	0.5
HX	120	1.0	17.6	65	0.2	1.1	8	–	0.3	–	0.2	0.1	0.3
GU	52	–	0.5	135	0.7	0.5	7	–	–	–	–	–	–
AD	72	–	0.7	–	–	0.2	10	0.7	0.4	0.4	0.3	0.4	0.2
ALL	ND	9.0	4.5	ND	7.3	6.8	ND	7.0	8.4	7.5	4.4	7.2	8.0
Nucleotides													
INO	150			357			258						
GNO	85 (40/60/0)			328			310 (23/52/25)						
XTO	–			–			–						
ANO	303 (50/30/20)			1042			1345 (36/34/24)						
Placental weight	5.7 g			4.5 g				4.5					
Perfusion time	60 min			50 min				20 min fractions taken as indicated					
Maternal flow	2.4 ml.min⁻¹			3.2 ml.min⁻¹				1.6 ml.min⁻¹					
Fetal flow	2.6 ml.min⁻¹			2.8 ml.min⁻¹				2.6 ml.min⁻¹					

Table 2

Results of analysis of these experiments where the conditions of perfusion were altered during the experiments. The symbols used are those explained in the legends to Table 1.

	First N₂ then O₂ (IV)					First N₂ then O₂ (V)					First O₂ then N₂ (VI)					First O₂ then N₂ (VII)		
	Placenta	M O₂	M N₂	F O₂	F N₂	Placenta	M O₂	M N₂	F O₂	F N₂	Placenta	M O₂	M N₂	F O₂	F N₂	Placenta	M O₂	M N₂
Nucleosides																		
INO	8	0.8	–	0.2	–	8	–	0.2	–	11.1	5.0	–	–	–	0.2	4	–	–
GNO	–	–	–	–	–	–	–	–	–	1.5	–	–	0.6	–	–	2	–	–
XTO	–	0.2	–	0.2	–	–	0.3	0.4	0.4	0.5	–	–	–	–	–	–	–	–
ANO	11	0.4	0.7	0.6	1.4	4	0.6	0.4	0.3	0.8	7	0.8	0.6	1.0	1.3	8	2.5	1.6
Purines																		
UA	27	14.3	4.0	2.3	–	–	2.2	3.8	–	0.9	–	0.8	2.1	–	–	17	.3	4.8
X	18	2.6	0.4	3.6	0.1	7	3.1	7.9	–	3.2	–	–	–	–	–	11	.2	1.4
HX	28	1.0	–	2.8	1.2	14	3.7	6.8	0.5	21.3	7	–	–	–	–	29	0.2	8.6
GU	6	–	–	–	–	9	–	–	–	1.2	6	–	–	–	–	6	–	–
AD	5	1.0	0.8	0.7	1.0	17	0.2	0.6	0.5	0.6	19	0.8	0.9	0.8	1.0	10	0.4	4.5
ALL	ND	ND	ND	ND	ND	ND	ND	ND	ND	ND	ND	ND	ND	ND	ND	ND	ND	ND
Nucleotides	Total																	
INO	16					4					163	–	–			53		
GNO	182 (24/0/76)					46					219	–	–			201		
XTO	–					–					–					–		
ANO	520 (7/33/59)					200 (19/30/50)					992 (24/34/41)					1147 (22/34/38)		
Placental weight	6.2 g					6.0 g					3.9 g					2.9 g		
Perfusion time	20 min on N₂ 20 min on O₂					30 min on N₂ 30 min on O₂					30 min on O₂ 30 min on N₂					40 min on O₂ 40 min on N₂		
Maternal flow	3.0 ml.min⁻¹					2.4 ml.min⁻¹					2.5 ml.min⁻¹					2.4 ml.min⁻¹		
Fetal flow	2.5 ml.min⁻¹					2.4 ml.min⁻¹					2.5 ml.min⁻¹					2.3 ml.min⁻¹		

Table 3

Concentration of nucleosides, purines, and nucleotides in the blood of pregnant guinea pigs (mother) and those of the fetus (μm/l). Together with the concentrations (μm/kg wet weight) in the placenta. Corrected for the presence of blood by means of measurement of Hb in fetal and maternal blood and in the placenta (concentrations in mM Hb/l (kg)). The symbols used are those given in the legends of Table 1. Values in parenthesis are the ratios of mono-, di-, and tri-nucleotides as analyzed directly without hydrolysis to nucleoside.

	Placenta μm/Kg	Fetus μm/l	Mother μm/l	Placenta	Fetus	Mother	Placenta μm/Kg	Fetus μm/l	Mother μm/l
Nucleosides									
INO	27.9	9.5	12.3	24.4	4.28	5.04	6.0	6.3	6.5
GNO	–	–	–	–	–	–	3.1	0.6	–
XTO	–	–	–	–	–	–	–	–	–
ANO	–	–	◂	–	–	–	–	–	–
Purines									
UA	23.0	61.8	17.4	12.5	*	9.4	–	–	–
X	7.4	5.9	–	16.6	–	4.4	4.2	1.0	0.4
HX	20.2	4.3	–	33.6	–	3.5	1.0	0.3	0.2
GU	17.4	–	–	9.6	–	–	–	–	–
AD	20.5	25.7	12.0	21.2	–	4.9	1.0	0.6	–
ALL	+ ND	ND	ND	ND	–	ND		–	–
Nucleotides									
INO	78	135	40	44	2.7	1.2	12	12	13.7
GNO	403 (26/32/41)	–	–	326 (3/97/0)	74.9	13.6	120	92	28.4
XTO	–	–	–	–	–	–	–	–	–
ANO	2245 (71/13/16)	411	308	2258 (54/42/4)	359	235	1060 (12/37/51)	414 (6/25/69)	235 (0/24/76)
Hb	2.26	9.10	6.77	1.93	9.8	7.63	2.32	9.7	8.2
Placental weight	5.7 g			3.9 g			4.8 g		
Total concentration	2841 μm/Kg			2745 μm/Kg			1207 μm/Kg		

* The purines were not determined in this sample. + ND; not determined.

Table 4

Concentrations of purines and nucleotides in the uterine vene (vene) and artheri (art.) catheterized in three pregnant guinea pigs with the calculated difference between arterial and venous concentrations (ΔAV). The symbols used are those given in the legends to Table 1.

	animal nr. 1			animal nr. 2			animal nr. 1		
	Art.	Vene	ΔAV	Art3	Vene	ΔAV	Art.	Vene	ΔAV
Purines									
UA	33.5	47.2	−13.7	22.5	28.6	−6	19.1	22.3	−3.2
X	−	−	−	−	−	−	−	−	−
HX	−	−	−	0.9	−	−	−	−	−
GU	−	−	−	−	15.6	−	−	3.7	−
AD	−	−	−	−	−	−	−	−	−
Nucleotides									
INO	10.3	16.4	−6.3	3.8	5.3	−1.5	8.8	14.4	−5.6
GNO	25.7	20.5	+5.2	27.4	33.0	−5.6	25.0	23.7	+1.4
XTO	−	−	−	−	−	−	−	−	−
ANO	228	252	−24	285	297	−12	249	235	+14

In the second group of experiments, the placenta was first perfused under anaerobic conditions and then under aerobic conditions (Table 2, experiments IV and V), while the medium was collected in both cases. In experiments VI and VII, these conditions were reversed. The results of the in vivo experiments are shown in Tables 3 and 4. The following qualitative conclusions were based on all tables regarding the presence or absence of nucleotides, nucleosides, and bases in the tissues and the perfusion media together with fetal and maternal blood.

Nucleotides

No xanthine nucleotides were detected in any of the placentae, while the other nucleotides were present in varying amounts. In the nonperfused placentae, only a very small percentage of the nucleotides was in the form of IMP (Table 3), in contrast to the perfused placentae where often a higher percentage was found. The experiments with the chronically catheterized pregnant guinea pigs (Table 4) showed that in all three animals, the IMP concentration was higher in venous than in arterial blood. No such constant difference was found for the other nucleotides.

Table 5

Number of Exp.	Percentages of nucleotides			Total extracted	Total remaining	Total nucleotides concentration		Adenine nucleotides concentration
	INO	GNO	ANO			after	before	
I	27	15	56	10.3	5.3	538	2736	303
II	20	18	60	4.5	8.9	1727	2977	1042
III	13	16	70	1.1	8.9	1913	2222	1345
IV	3	20	77	4.5	4.8	652	1500	487
V	2	18	80	10.3	1.8	250	2016	200
VI	12	16	72	1.6	4.1	1374	1461	992
VII	3	13	84	-	7.7	1227	-	1033

First column: number of the experiment according to Tables 1 and 2.
Second column: percentages of nucleotides.
Third column: total extracted nucleosides and purines in µMols.
Fourth column: total remaining amounts of nucleotides, nucleosides and purines in µMols.
Fifth column: concentration µM/l of total nucleotides remaining after perfusion (left); and calculated total concentration before the perfusion (right).
Sixth column: Concentration of adenine nucleotides remaining after perfusion in µM.

Nucleosides

Inosine was always present in the placenta; it was the only nucleoside detectable in the nonperfused placenta. In the perfused placenta, adenosine was detected in all but three placentae. Xanthosine was absent in all placentae and in the media obtained from perfusion with oxygen. Guanosine was present in only one perfused placenta and was once detected in the medium.

Purines

All purines were present in the placenta. Guanine was absent in most media and in most blood samples. From Table 4, it can be concluded that uric acid is delivered from the fetus to the mother. When the quantitative aspects are considered, it was clear that the variation in the concentration of extracted purines and nucleosides made it difficult to consider each compound separately. However, Tables 1 and 2 show that the concentration of adenine nucleotides remaining in the tissues after perfusion varied between 200 µM and 480 µM in cases on which the perfusion was anaerobic during the time the experiment lasted or was first anaerobic and later aerobic. In contrast, this concentration varied between 1000 µM and 1300 µM when the experiment was conducted completely aerobically or during the first period.

From Tables 1 and 2 it appeared that the ratios among AMP, ADP, and ATP were variable with a higher ATP percentage in those experiments performed under aerobic conditions compared with the anaerobic experiments. It is also obvious that the percentage of ADP was fairly constant at a value of about 35%.

Another striking point was observed in Table 5 (second column). The concentration of guanine nucleotides was fairly constant in all cases. In columns 3 and 4, the total extraction was shown together with the total amount remaining. These values together with the values from the remaining two columns are shown again in Figure 4. From this figure, there was observed a linear relationship between the total amount extracted and the total concentration of nucleotides remaining. A linear relationship also existed between the concentration of adenine nucleotides and the concentration of total nucleotides. It also appeared that the data points could be separated into two groups. The data points from the group of experiments that were performed anaerobically were separated from another cluster of points belonging to data points obtained when the experiments were done under aerobic conditions. When the last row of Table 3 (total concentration of nucleotides, nucleosides, and purines in the nonperfused placenta) is compared with column 5 of Table 4 (concentration calculated), the total concentrations calculated to be present before the perfusion experiments were in the range of concentrations for nonperfused placentae.

DISCUSSION

The analytical method described here should be adequate to measure the concentrations of purines and pyrimidines, bases, nucleosides, and nucleotides accurately. Because radioactive tracers are added at the beginning of the extraction step, the recoveries for the next steps can be calculated. However, there is one uncertainty, the recovery of the extraction of the intracellular compounds. This can,

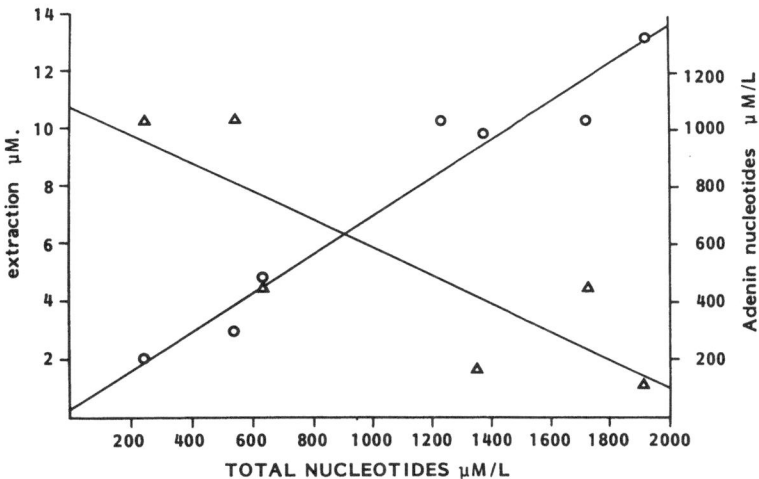

Figure 4. Dependence of total extracted purine bases and nucleosides (y-axis) occuring during the time the perfusion lasted in μM on the sum of the concentration of all nucleotides present in the placenta after perfusion (x-axis). y = 10.67 - 0.005 x P = 0.825. Dependence of the concentration of adenine nucleotides remaining in the placenta after perfusion (y-axis) on the concentration of total nucleotides (x-axis). y = 33.07 + 0.673 x P = 0.971.

in principle, only be measured when the tracers are also intracellular and therefore should be administered in vivo before the experiment starts. Because this is associated with several practical difficulties we did this tracer analysis only in a few pilot studies and found that the extraction was dependent on the extraction volume. When this volume was about 10 times the amount of tissue or blood, the recovery did not increase following a further increase in volume. For practical reasons, the extraction volume was made 20 times the tissue weight for placental tissue samples.

The group separation introduced after the extraction reduced the number of compounds to be chromatographed to 1/3, while the adsorption to the vic-glycol specific Affi-gel and the less specific mercury carbonate purified the sample considerably before the actual chromatography resulted in a very flat baseline. An analysis of five equal blood samples produced an accuracy of about 10%. This may seem high at first but a closer look at the formulas given to calculate the concentrations showed that a considerable summation of (relative) error occurred:

$$\left| \frac{\Delta C_x}{C_x} \right| = \sum_{i=1}^{11} \left| \frac{\Delta X_i}{X_i} \right|$$

Where X_i are the measured parameters such as area, concentration, dpm/ml, and the ΔX_i are the standard deviations of those parameters, $\Delta C_x/C_x$ is the ultimate error in the concentration to be determined. When all these parameters were measured to 1% accuracy, a total error of about 11% resulted.

The total nucleotide concentration calculated to be present before perfusion ranged from 1468 to 2977 µg/kg. This rather wide variation might well be the result of the different gestational ages of the animals used. The concentrations measured in the nonperfused placentae fell between these extreme values. This made it unlikely that during the perfusion, considerable net nucleic acid breakdown or de novo synthesis took place to replenish the nucleotides extracted during perfusion. When the concentrations found in the guinea pig placenta were compared to those found in the human placenta, the total adenine nucleotide concentration varied between 790 µM/kg and 1290 µM/kg in the human placenta obtained following elective operation (Simmons, 1982; Bloxam, 1984), whereas values between 1100 and 2258 µM/kg were found for the guinea pig placenta. Although these values were obtained from only three placentae it seemed reasonable to assume that concentrations in the guinea pig are higher than in the human. Values for the nucleosides, adenosine, and inosine found by Simmons (1982) were 12.7 $\pm$ 5 and 15.2 $\pm$ 13 µM/kg, respectively. These values appeared to be in the same order of magnitude as found in the guinea pig placenta.

The most important purine nucleotide interconversions and their standard free energy changes are given in Figure 5 (middle drawing). From the thermodynamic data (van Kreel, 1985), it becomes clear that under normal conditions of energy charge (Atkinson, 1968) and NAD+/NADH ratio, IMP has a tendency to be oxidized to XMP and deaminated to AMP. This will keep the IMP concentration low as was actually observed in vivo in the placenta. The AMP formed will be partly phosphorylated to ADP and ATP. The XMP formed will be transformed almost entirely to GMP because the standard free energy change made this reaction completely irreversible. This may be expected to also be true when the conditions do not allow normal ATP concentration. These explain the fact that no measurable concentration of XMP was ever found in the tissues. When the placenta is perfused aerobically, an increase in the IMP concentration and a decrease in total adenine nucleotide concentration as indicated in the figure (the concentrations called AMP and GMP are actually the total nucleotide concentrations). This was obviously the result of an increased deamination reaction relative to the amination, probably as a result of an increased AMP concentration and a decrease in oxidative phosphorylation occuring in a perfusion even under aerobic conditions.

As can be observed from Table 4, there was an increase in IMP concentration in the uterine vein relative to the artery in all three animals. This difference was possibly the result of uptake of oxypurines from the placenta and conversion to IMP as observed for other tissues (Henderson and Lepage, 1959; Pritchard et al., 1970). 5'Nucleotidase was inhibited by ATP, and it was likely that this inhibition was partly relieved during perfusion. This will result in the formation of nucleosides at rates proportional to the nucleotide concentration. The total effect will be a fall in the concentration of nucleotides with the percentage of GMP remaining constant and with variable IMP and AMP concentration as was actually observed.

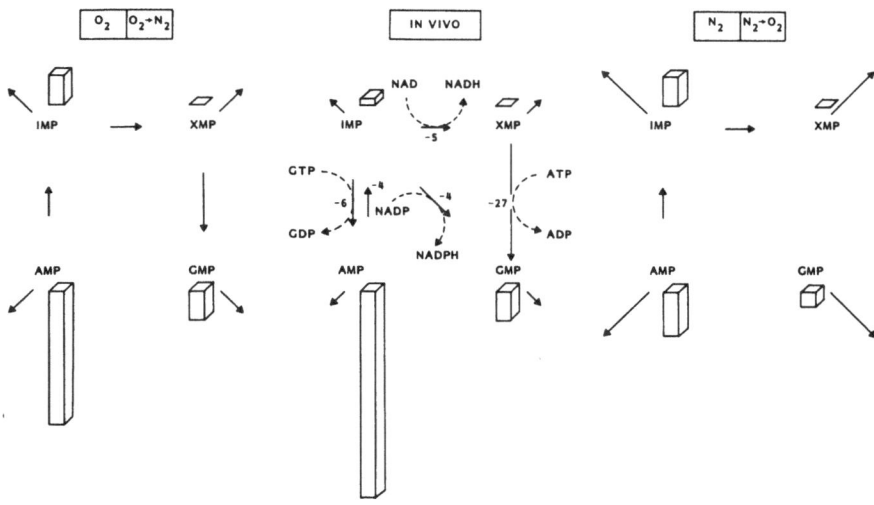

Figure 5. Pathways of nucleotide interconversions and breakdown (breakdown products not indicated) in vivo and under different experimental conditions during perfusion. Solid arrows indicated the direction of the reactions. The values of the standard free energy changes are given along side the arrows. The length of the arrows correlated to the proposed reaction velocity. The values of the concentrations of the different nucleotides (sum of mono-, di-, and tri-phosphates) found at the end of the experiments are proportional to the lengths of the vertical bars.

When the perfusion is performed under anaerobic conditions (Figure 5, right side) ATP concentration will fall steeply and formation of nucleosides and their reaction products will increase. The transformation of XMP to GMP will be hampered. This will force the XMP concentration to rise slightly resulting in an increased breakdown into xanthosine. As in Tables 1 and 2, xanthosine can be detected in the perfusion medium when the experiment is performed anaerobically. The figure shows the difference in nucleotide concentrations between the experiments, where perfusion was done in two phases, first aerobic, then anaerobic $(O_2 \rightarrow N_2)$; and when the sequence was reversed $(N_2 \rightarrow O_2)$. The $O_2 \rightarrow N_2$ sequence resembles the experiment done aerobically (O_2) while the reverse experiment $(N_2 \rightarrow O_2)$ resembles the perfusion under nitrogen. This effect has also been observed in other tissues (Katz, 1981; Neely, 1984).

During anaerobic perfusion some damage is done to the tissues that cannot be repaired by changing to aerobic conditions. For that reason when the first phase occurs anaerobically the tissue will not respond to an increase in oxygen tension in the second phase and will behave like it was perfused anaerobically during both phases.

The distribution between mono-, di-, and tri-nucleotides as given between parenthesis in Tables 1, 2, and 3 is thought not to be representative of the actual values during perfusion or in the nonperfused tissue because very unphysiological ratios are obtained, even with the nonperfused placenta. The reason for this is that degradation of ATP occurs very rapidly and in our experiment no freeze-clamp techniques were used. As the breakdown of mononucleotides to nucleosides occurs relatively slowly we expect to find the actual total nucleotides values to be unaltered when using our protocol of first disconnecting the organ from its surrounding and after that rapid freezing in liquid nitrogen.

SUMMARY

Purine nucleotide interconversions and breakdown in the perfused guinea pig placenta were studied. To this end, guinea pig placentae were perfused under normal and anaerobic conditions. The placentae and perfusates were analyzed by means of a newly developed analytical technique consisting of group separation and subsequent chromatography on HPLC columns. Purine nucleotide concentrations were also determined in nonperfused placentae as reference organs. It was found that a considerable breakdown of purine nucleotides occurs during perfusion. It appears that the changes in nucleotide concentrations resulting from perfusion under anaerobic conditions cannot be reversed by changing to aerobic perfusion. The observed effects are discussed taking into account the known reaction pathways and the changes in thermodynamic potentials.

REFERENCES

Bishop, C., Rankine, D.M., and Talbott, J.H. (1958) The nucleotides in normal human blood. *J. Biol. Chem.* 234, 1233-1237.

Bloxam, D.L. and Bobinski, P.I.N. (1984) Energy metabolism and glycolysis in the human placenta during ischaemis and in normal labor. *Placenta* 5, 381-394.

Gehrke, C.W., Kuo, K.C., Dares, G.E., and Suits, R.D. (1978) Quantitative high performance liquid chromatography of nucleosides in biological materials. *J. Chromat.* 150, 455-476.

Henderson, J.F. and Lepage, G.A. (1959) Transport of adenine 8-14C among mouse tissues by blood cells. *J. Biol. Chem.* 234, 3219-3223.

Katz, A.M. and Messino, F.C. (1981) Lipid-membrane interactions and the pathogenesis of ischemic damage in the myocardum. *Circ. Res.* 48(1), 1-16.

Khym, J.X. (1975) An analytical system for rapid separation of tissue nucleotides at low pressures on conventional anion exchanger. *Clin. Chem.* 21(9), 1245-1252.

Leitchweiss, H.G. and Schröder, H. (1971) Untersuchungen uber den glucosetransaport durch die isolierte beiderseits kunstlich perfundierte meerschweinchen placenta. *Pflügers Arch fur die Gesamet Physiol.* 325, 139.

MacNeil, M D. and Koong, L.J. (1983) Improvements to the mathematical description of prenatal growth. *Growth* 47, 371-380.

Neeley, J.R. and Grotyohann, L.W. (1984) Role of glycolytic products in damage to ischemic myocardum. *Circ. Res.* 55(2), 816-824.

Peeters, L.L. and Wallenburg, H.C.S. (1983) Technique for chronic blood sampling from the ovarian vein in the pregnant guinea pig. *Bio. Res. Preg. Prenatal.* 5, 188-120.

Peeters, L.L., Martensson, L., van Kreel, B.K., and Wallenburg, H.C.S. (1984) Uterine arterial and venous concentrations of glucose, lactate, ketons, free fatty acids and oxygen in the awake pregnant guinea pig. *Ped. Res.* 18(11), 1172-1175.

Pritchard, J.B., Chaves-Leon, F., and Berlu, R.D. (1970) Purines supply by liver to tissues. *Am. J. Physiol.* 219, 1263-1267.

Simmons, R.J., Coade, S.B., Harkness, R.A., Drury, L., and Hytten, T. (1982) Nucleotide, nucleoside and purine base concentration in human placentae. *Placenta* 3, 29-38.

Thunell, S. (1965) Determination of incorporation of ^{59}Fe in hemin of peripheral red blood cells in bone marrow cultures. *Clin. Chim. Acta* 11, 321-333.

van Kreel, B.K. and Wallenburg, H.C.S. (1980) Hypoxanthine metabolism and transfer in the pregnant Rhesus monkey. *J. Develop. Physiol.* 2, 365-372.

van Kreel, B.K., van Dijk, J.P., and Pijnenburg, A.M.C.M. (1982) Placental transfer and metabolism of purines and nucleosides in the pregnant guinea pig. *Placenta* 3, 127-136.

van Kreel, B.K. (1985) The estimation of the apparent standard free energy change of a biochemical reaction from the standard free energy of formation and apparent free energy of ionization of the participating molecules and its application to the reactions of the purine metabolism. *Biochem. Education* 13(3), 125-130.

Vettenranta, K. and Raivo, K.O. (1984) Purine reutilization in normal and malignant cells of human placental origin. *Placenta* 5, 315-322.

Wallenburg, H.C.S. and van Kreel, B.K. (1980) Maternal and umbilical plasma concentration of uric acid and oxypurines at delivery in normal and hypertensive pregnancy. *Arch. Gynecol.* 229, 7-11.

Zakaria, M. and Brown, P.R. (1981) High-performance liquid column chromatography of nucleotides, nucleosides and bases. *J. Chromat.* 226, 267-290.

Trophoblast Research 2:501-514, 1987

UPTAKE OF ASCORBIC ACID AND ITS OXIDIZED PRODUCTS IN THE ISOLATED GUINEA PIG PLACENTA

Heinz-Peter Leichtweiss, Belisario Lisbôa,
and Christiane Steinborn

Universitätsfrauenklinik
Abtlg. exper. Med.
Martinistr. 52
D-2000 Hamburg 20, FR Germany

INTRODUCTION

Placental transfer of ascorbic acid has been studied in man (Hensleigh and Krantz, 1966; Streeter and Rosso, 1981) and guinea pig (Raiha, 1958; Norkus et al., 1979, 1982). In both, the transfer is assumed to the carrier-mediated. Because the concentrations of total ascorbic acid in plasma and tissue are higher in the fetus than in the mother (Wahren and Rundqvist, 1937; Moller-Christensen and Thorup, 1940; Lund and Kimble, 1943; Raiha, 1958) two different modes of transport are assumed: 1) The transport of ascorbic acid is energy dependent and moves uphill (Hensleigh and Krantz, 1966) and 2) Only the oxidized form of ascorbic acid, that is dehydroascorbic acid, is transferred into the fetus by passive carrier mediated diffusion, where it is reduced to ascorbic acid (Raiha, 1958; Norkus et al., 1979). These hypotheses have not been sufficiently verified until now. It is not known whether ascorbic acid, its oxidized products, or both are transported in the placenta.

Two methodological peculiarities must be considered in this context. Vitamin C exists in two readily convertible forms of ascorbic (AA) and dehydroascorbic (DAA) acids, and the latter can irreversibly be converted to diketogulonic acid (DKG), both in vitro and in tissues (Penny and Zilva, 1943). This oxidative conversion occurs within minutes under physiological conditions. On the other hand, in most papers dealing with this problem, ascorbic acid was analyzed as total ascorbic acid after oxidation using the method reported by Bessey et al. (1947). Therefore, the discriminiation between AA, DAA, and DKG could not be performed (Roe, 1961). For the study of the transport mechanism of vitamin C it is important to know which of these substances is offered to the carriers.

This paper demonstrates the trophoblastic uptake of defined and purified substances in the isolated guinea pig placenta using the paired tracer dilution technique (Yudilevich et al., 1979). Boli containing ^{14}C-labeled ascorbic acid, dehydroascorbic acid, and diketogulonic acid were injected into the perfusion fluid immediately before the tissue.

Received: 10 October 1985; Accepted: 15 May 1986

Recently it has been reported that during maternal hyperglycemia in the guinea pig the placental transport of ascorbic acid is reduced (Norkus et al., 1982). Since the transport of ascorbic acid or dehydroascorbic acid can be inhibited by glucose in white and red blood cells (Manns and Newton, 1975; Moser and Weber, 1984), this paper also shows the effect of high glucose concentration on the placental uptake of vitamin C.

MATERIALS AND METHODS

Guinea pig placentae (55 - 68 day of gestation) were completely isolated and artificially perfused on both sides as described elsewhere (Leichtweiss and Schröder, 1971; 1981a,b). Ringer solution was used as perfusion fluid. It was saturated with 95% O_2/5% CO_2 at 37°C and contained in addition 20 g/l Dextran T40, 3 g/l bovine serum albumin (Behring) and 20 mg/l ascorbic acid resp. dehydroascorbic acid (Fluka, Dreieich, FRG). The perfusion flow rates were kept constant at 3.2 ml/min on both sides with syringe pumps. The fluid used to prepare the injection boli was Ringer solution gassed with nitrogen. The presence of dextran and albumin impairs thin layer and column chromatography (see below). The placental uptakes of L-(^{14}C) ascorbic acid (AA) and of its labeled oxidation products were measured using the paired tracer single injection dilution technique as described previously (Yudilevich et al., 1979; Leichtweiss and Schröder, 1981). As reference to the ^{14}C-labeled test substances, we used L-(^{3}H)glucose. Boli of 100 μl each were injected within one second into either the maternal or the fetal arterial cannulae of the placenta, in such a way it took about 3 seconds until the bolus reached the tissue. Twenty samples of 4 drops (maternal) or 3 drops (fetal) each were collected consecutively from the venous outflow at the injection side. Final samples were accumulated until the total collection time was 5 min on the maternal and 4 min on the fetal side. Radioactivity of the samples and the replicas of the boli were measured using a Packard Spectrometer (460C). The SPLINE (this splitted line procedure is a frequently used method to fit a curve out of several points) method was used to determine the respective activities (dpm) of ^{3}H and ^{14}C.

The uptake value U (%) calculated from each venous sample is:

$$U(\%) = (1 - {}^{14}C\text{-test substance (\% of dose)}/L\text{-}({}^3H)\text{glucose (\% of dose)}) \times 100.$$

The maximal uptake value is the mean from the 5-10 highest consecutive U values. From the contralateral (acceptor) side, aliquots of venous pooled samples were counted in order to determine the placental transfer of both the test and reference substance. To investigate other transport functions of the placentae, L-(^{3}H)alanine and D-(^{3}H)glucose together with L-(^{14}C)glucose were used. In some experiments, phloretin (5 x 10^{-4} mol/l) was added to the perfusion fluid to inhibit placental uptake.

Preparation of L-(^{14}C)ascorbic acid and of their oxidation products: In order to obtain ^{14}C-labeled DAA and DGK, L-(^{14}C)ascorbic acid (NEN, Dreieich, FRG) (specific activity: 9.9 mCi/mml) was oxidized with 2.6 dichlorophenolindophenol (Merck, Darmstadt, FRG) at 17°C for 15 seconds. Ethylacetate then was added for extraction of the oxidant. The volume of the separated aqueous phase contains the oxidation products DAA and DKG and is reduced to 40-50 μl by incubation in a water bath for 45 min at 40°C in a nitrogen atmosphere.

Column chromatography was used to gain purified DAA and DKG: 6 g silica gel (particle size 63-200 m ASTM, Merck) were filled into a glass column (length ca. 15 cm, diameter 1.3 cm) using a mixture of methanol and toluene (4:6). The oxidation product was dissolved in water/methanol (1:9), and the 400 µl were applied to the column. After addition of 600 µl toluene, separation was achieved using methanol/toulene/acetic acid/acetone (40:50:5:5) as a solvent. Thirty fractions of 1.6 ml each were collected at a flow rate of 120 ml/h. Fractions 6-15 containing DAA were dried at 40°C under nitrogen atmosphere and dissolved in 1-1.5 ml Ringer's solution. Aliquots were taken for measurement of radioactivity. After a transient perfusion of the column with 8 ml of a solution containing methanol, water, acetic acid, and acetone (18:72:5:5) the diketogluonic acid was then eluted from the column using methanol/water/acetic acid/acetone (9:81:5:5). Twenty samples of 1.6 ml each were collected.

In experiments where DKG was injected, the injection solutions (boli) were prepared as described for DAA.

Analysis of substances was done by high performance thin layer chromatography (HPTLC): precoated 10 x 10 cm silica gel 60F 254 layers (Merck, Darmstadt, FRG) were used, the solvent system was methanol/toulene/acetic acid/acetone (30:60:5:5). The fronts were developed in a glass chamber within 20 min at 20°C. The mobility values amounted to 1-2 mm for DKG, 10 mm for AA and 17 mm for DAA for a migration front of 7.5 cm. The radioactivity was detected by a TLC scanner (LB 2723, Berthold, FRG). The areas of the peaks were scraped out of the layer and the silica gel was dissolved in scintillation liquid (Biofluor, NEN) for counting. The migration of the non-labeled product could be made visible by heating to 110°C for 5 min.

RESULTS

L-(^{14}C)ascorbic acid and its oxidation products DAA when dissolved in perfusion fluid with oxygen are unstable. As shown in Table 1, the rate of spontaneous conversion amounts to 50% or more after 2 min for both and increases further with time. After 15 min, only a residuum of AA exists which is mostly converted to DAA and DKG.

When AA was kept in oxygen-free Ringer solution (gassed with nitrogen) more than 90% of the activity remained as AA for at least 2 hrs. Immediately after initial oxidation (but without separation by column chromatography), the ratio of DAA to DKG was about 9:1 (Tables 1 and 2). Even in nearly oxygen-free Ringer solution, this ratio was changed to about 6:4 after 1 hr (Table 2a). When the oxidation products were separated by column chromatography, the portion of DAA amounted to 85-88% and remained nearly constant for the perfusion time of 2 hrs (Table 2b).

As shown in Table 3, no significant placental uptake of pure ^{14}C-AA could be detected in relation to L-(^{3}H)glucose. The U_{max} values were 5% or less on both sides in 12 placentae investigated. The transferred activities of AA and L-glucose were 6.1% $\pm$ 6.4 and 7.0% $\pm$ 7.9 of dose in materno-fetal direction resp. 7.7% $\pm$ 6.8 and 8.7% $\pm$ 7.3 in feto-maternal direction.

Table 1

Spontaneous conversion of AA and DAA after transfer
into perfusion fluid containing oxygen
(% of total ^{14}C-label)

	Stock solution	After 2 min	After 15 min	After 60 min
(a)				
AA	98.5	33.1	8.1	0.0
	± 1.1	± 8.9	± 7.1	
DAA	0.0	59.2	45.4	11.5
		± 10.6	± 16.7	± 19.9
DKG	1.5	7.7	46.5	88.5
	± 1.2	± 2.6	± 23.0	± 19.9
(b)				
DAA	86.6	49.6	33.1	13.1
	± 0.6	± 3.0	± 6.7	± 9.8
DKG	13.4	50.4	66.9	86.9
	± 0.6	± 3.0	± 6.7	± 9.8

Means ± SD. Data from 3 different stock solutions (a)
or 3 different purified DAA fractions (b)

In 8 placentae, the simultaneously measured maximal uptakes of D-(^{3}H)glucose were about 70% on both sides (Table 3). When ^{14}C-AA had been oxidized (cf. Table 2a) but the oxidation products had not yet been separated then the U_{max} values in 11 placentae were relatively low with 17.4% ± 10.7 on the maternal and 25.7% ± 12.7 on the fetal side.

The ^{14}C-AA oxidation product DKG which was separated by column chromatography was injected into 5 placentae. The uptake of DKG in relation to L-(^{3}H)glucose is negligible (Table 4). The transferred activities of the bolus of DKG and L-glucose amounts to 11.6% ± 5.2 and 16.4% ± 6.7 (materno-fetal) resp. 17.9% ± 6.9 and 2.14% ± 8.9 (feto-maternal).

The maximal uptake values of purified DAA (without AA and DKG) were about 70% and in the same range of simultaneously measured D-glucose uptakes on both sides in 11 placentae (Table 5). On the injection side, the recovered activity of DAA was about 20% of the dose whereas the recovered L-(^{3}H)glucose amounted to

Table 2

Stability of DAA in Ringer solution under nitrogen atmosphere
(in % of total ^{14}C-label)

(a)

without purification by column chromatography

	Control	After 5 min	After 1 h	After 2 h	After 3 h	After 20 h
DAA	90.1	73.5	59.3	54.8	53.6	
	± 3.9	± 5.1	± 4.8	± 7.0	± 7.1	
DKG	9.9	26.5	40.7	45.2	46.4	
	± 3.9	± 5.1	± 4.8	± 7.0	± 7.1	

(b)

after purification by column chromatography

	Control	After 5 min	After 1 h	After 2 h	After 3 h	After 20 h
DAA	92.1	90.9	87.8	86.4	84.9	76.2
	± 3.3	± 2.7	± 1.6	± 0.4	± 2.1	± 3.0
DKG	7.9	9.1	12.2	13.6	15.1	23.8
	± 3.3	± 2.7	± 1.6	± 0.4	± 2.1	± 3.0

Mean values and SD from 3 oxidation procedures each.
Control: immediately after oxidation of ascorbic acid.

Table 3

Maximal uptake values (%) of ^{14}C-AA and D-(^{3}H)glucose

Placenta No.	1	2	3	14	15	16	17	18	20	21	22	23	x	SD
^{14}C-AA														
maternal	3	1	3	3	2	4	3	4	0	5	0	0	2.3 ± 1.7
fetal	5	1	4	2	0	4	5	3	0	2	0	0	2.2 ± 2.0
^{3}H-D-gluc														
maternal	n.d.	60	n.d.	68	72	n.d.	n.d.	80	70	n.d.	75	69	70.6 ± 6.2
fetal	n.d.	55	n.d.	n.d.	75	n.d.	n.d.	76	77	76	74	70	71.9 ± 7.8

n.d.: not determined

Table 4

Maximal uptakes (%) of DKG (the polar product
of the ^{14}C-AA oxidation)

Placenta No.	14	15	16	17	18	x	SD
Maternal	0.0	4.0	0.0	0.0	2.0	1.2	± 1.8
Fetal	0.0	3.0	2.0	0.0	1.0	1.2	± 1.3

more than 80%. On the acceptor side 7.7% $\pm$ 5.8 of DAA and 12.5% $\pm$ 12.5 of L-glucose were recovered when the bolus was injected into the maternal flow. In the opposite direction 8.2% $\pm$ 6.6 of DAA and 10.9% $\pm$ 7.0 of L-glucose could be recovered. The missing ca. 70% of ^{14}C-DAA total remained in the tissue.

In Figure 1, venous dilution curves of the activities for ^{14}C-DAA and L-(^{3}H)glucose after bolus injection into one placenta are shown. The uptake values (Figure 1b) were nearly constant during the collection period. A decline of the curves at the end typical for tracer backflux was missing on the maternal as well as on the fetal side of the trophoblast.

Phloretin, when added in 4 experiments, inhibits the uptake of ^{14}C-DAA incompletely as shown in Figures 1c, 1f, and Table 5. In these experiments only about 60% of DAA and 70% of L-glucose activities were recovered on the injection side. On the acceptor sides 26, 1% [14]C and 29.6% [3]H label indicated that during the phloretin runs the unspecific leak was increased (Figures 1c, f).

The influence of elevated glucose concentration of the perfusion medium on the maximal uptake of ^{14}C-DAA was investigated. As shown in Figure 2 an inhibition of this uptake becomes significant only when the glucose concentration was increased to 100 mmol/l and higher (See also Figures 1b and 1e). Only at 250 mmol/l D-glucose the uptake values were reduced to 50% of control (2.5-5.0 mmol/l glucose) although the uptake values of L-alanine remained nearly unaffected (maternal uptake: 80%, fetal uptake: 76%, cf. Carstensen et al., 1982).

We investigated the DAA uptakes at increasing concentrations of glucose in 13 placentae. From the U_{max} values the DAA fluxes J are calculated (Cunningham and Sarna, 1979) as:

$$J \, (mol/min) = c \times -F \times \ln (1 - U_{max}/100)$$

c = concentration of DAA (mol/l); F = flow rate (ml/min). From the Hill plot (Figure 3) the K_i values (mmol/l) could be estimated. They were approximately 190 on the maternal and 180 on the fetal side.

Table 5

Maximal uptake values (%) of DAA (the apolar product
of ^{14}C-AA oxidation) and of D-(^{3}H)glucose

Placenta No.	Maternal			Fetal		
	DAA	D-gluc.	DAA (phlor.)	DAA	D-gluc.	DAA (phlor.)
25	58	63	n.d.	63	77	n.d.
26	73	68	n.d.	75	77	n.d.
27	71	70	n.d.	68	72	n.d.
28	72	79	n.d.	75	73	n.d.
30	75	77	n.d.	80	74	n.d.
31	75	70	n.d.	80	70	n.d.
32	70	70	n.d.	61	75	n.d.
33	70	80	15[xx]	83	77	10[xx]
34	80	78	46[x]	80	76	42[x]
35	75	55	30[xx]	62	75	18[xx]
36	62	80	15[xx]	63	79	15[xx]
x:	71.0	71.80	24.0	71.80	75.0	21.30
SD:	± 6.2	± 8.0	± 17.9	± 6.0	± 2.6	± 14.2

[x] phloretin: 1 x 10^{-4} M/l
[xx] phloretin: 5 x 10^{-4} M/l
     only added with ^{14}C-DAA
     n.d.: not determined

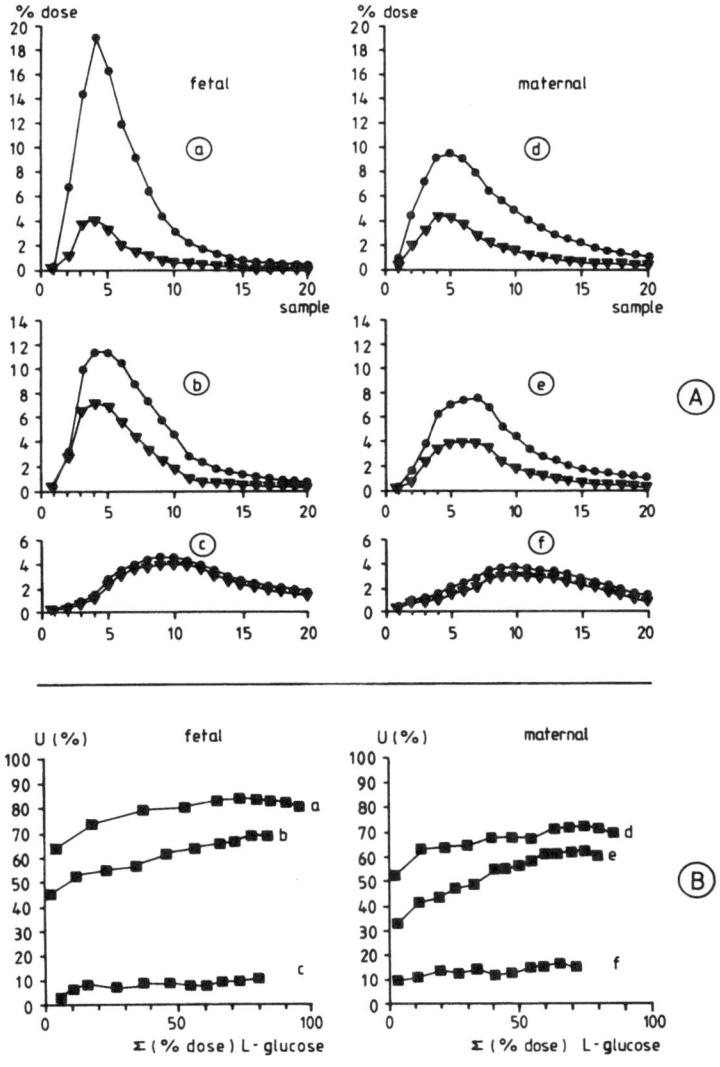

Figure 1A. Venous outflow tracer dilution curves on the injection side. Values for ^{14}C-DAA (∇) and L-(^{3}H)glucose (●) of each of the sequential samples in relation to the bolus dose are plotted.

    a, d at glucose concentration of 2.5 mmol/l
    b, e at glucose concentration of 100 mmol/l
    c, f at glucose concentration of 2.5 mmol/l and 0.5 mmol/l phloretin.

Figure 1B. Uptake curves of injections a-f shown in A. The uptake values of the samples were plotted versus the accumulated amount of L-(^{3}H)glucose.

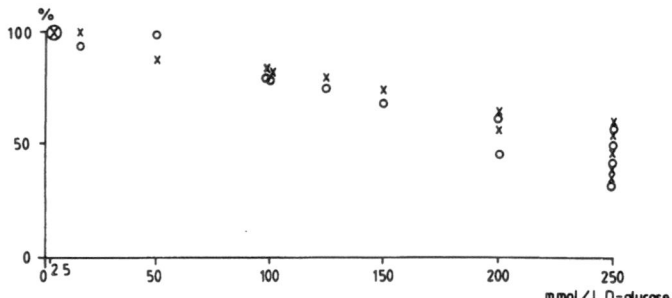

Figure 2. Decrease of maximal uptake values U_{max} of ^{14}C-DAA with increasing concentration of D-glucose (abscissa). Ordinate: ratio of U_{max}: U_{max} at 2.5 mmol/l D-glucose. Maternal (x) and fetal (o) data.

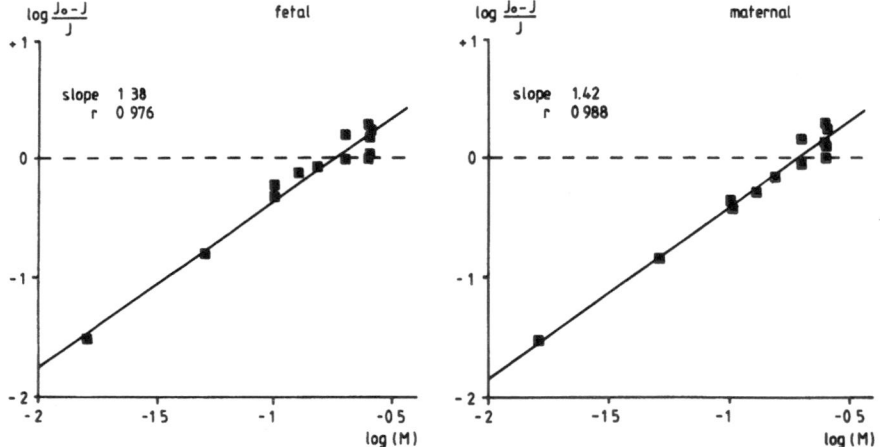

Figure 3. Inhibition of ^{14}C-DAA uptake into the trophoblast by D-glucose. DAA fluxes (J) were obtained at glucose concentrations of 15, 50, 100, 125, 150, 200, and 250 mmol/l. J_O = DAA flux at 2.5 mmol/l glucose.

DISCUSSION

Because AA will spontaneously be oxidized to DAA and to DKG in solutions in vitro (Penny and Zilva, 1943), it is necessary to control this process in experiments which deal with the transport of vitamin C. Under our experimental conditions AA was converted within minutes. DAA itself also is unstable in perfusion fluid containing oxygen and is changed to DKG rapidly. Assuming that the transport properties of these substances in the placenta may be very different, methods to supply AA and its oxidation products separately to the tissue were developed. In our hands, AA and DAA purified by column chromatography and then preserved in oxygen-free Ringer solution at $16°C$, remained stable for at least 2 hrs which is the time of a perfusion experiment. From this stock solution portions were injected into the cannulae connected with the placenta. This technique almost prevents the exposure of the substances to oxygen in the perfusion fluid, since the bolus reached the tissue within 3-5 sec and is not mixed with the fluid. To investigate the uptake of ^{14}C-DAA which is generated by oxidation of ^{14}C-AA it was necessary to purify ^{14}C-DAA by column chromatography. Without this procedure, the ratio of DAA to DKG is not stable and decreases with time (cf. Table 2a) even when the oxygen tension is low. This probably is due to a residuum of the oxidant (2,6-dicholorophenolindophenol) which could not be extracted completely by ethylacetate, but only by column chromatography.

The results presented in this paper demonstrate that in the guinea pig placenta the uptake of pure AA in relation to L-glucose is negligible. The very small uptake of 5% or less which occurs in some placentae (Table 3) can be explained by the possibility that a small amount of AA was oxidized before it reached the membranes. The transfer rates of AA and L-glucose are nearly equal which means that AA behaves like the extracellular marker L-glucose. A paracellular diffusive transfer through "unspecific leaks" seems to be typical for the isolated guinea pig placenta (Leichtweiss and Schröder, 1981b). However, this pathway is without significance under in situ conditions (Bissonette et al., 1979).

The form of vitamin C which is transported in the guinea pig placenta is mainly DAA. Thereby we can prove that the assumption put forward by Raiha (1958) and supported by others (Norkus et al., 1979) holds true. Thus, the placenta is similar to red blood cells, but different than white blood cells and intestine where AA is transported (Moser and Weber, 1984; Stevenson and Brush, 1969).

The uptake of ^{14}C-DKG also is negligible. (This seems reasonable because KG cannot be reconverted to vitamin C in tissue, which is in contrast to DAA). A contamination of DAA with DKG must lower the uptake values of oxidized ^{14}C-AA in experiments like these. The uptake of purified DAA in the guinea pig placenta looks like that of D-glucose: uptake values are similar and the uptake properties of each are the same on both sides of the trophoblast. Moreover in both cases, the uptakes are inhibited by phloretin. However, the uptake curves as shown in Figures 1a and 1d represent the characteristics of cellular retention since the uptake values remain at the same level over the entire collection time. A backflux of tracer (like that of D-glucose; Yudilevich et al., 1979) at the end of the passage is missing.

The retention of DAA usually was complete because the chemical load is low. In order to fill tissue stores, 20 mg/l DAA was added to the perfusion fluid. This addition, however, proved insufficient because DAA was converted to DKG within the perfusion period (Table 1b). Thus, the real DAA load to the placenta was contained only in the bolus, the quantity was close to 0.005-0.01 umol. When a higher chemical load of DAA was applied with the bolus a significant release of labeled substance, e.g., effluxes to both sides, was observed (preliminary results). Considering the nature of the retained substance the form of vitamin C inside the cells is not known. Further analysis of the tissue will hopefully give an answer.

Following the proposal of Raiha (1958), DAA might also be reduced to AA inside the trophoblastic cells. Because AA cannot pass the membranes it would then be locked inside the cell. Thus accumulation would be performed without energy dependent transport.

The transplacental transfer rates for DAA, AA, and DKG are not different from the transfer of L-glucose. It could be concluded therefore that the transfer pathways for these substances are the same and paracellular. This is in agreement with the negligible uptake values of AA and DKG. The uptake of DAA, however, indicates a specific transport into the cell which does not lead to a corresponding transplacental transfer. This inconsistency can be explained by the low DAA load which does not prevent the intracellular accumulation of labeled DAA. The transfer rate of ^{14}C-DAA label does indeed increase with the chemical concentration of DAA in the bolus, and exceeds then the transfer rates of L-glucose (preliminary observation).

We find an incomplete inhibition of ^{14}C-DAA uptake by increasing concentrations of D-glucose in the perfusion fluid. The inhibition, however, can be observed only at relatively high glucose concentration in the isolated guinea pig placenta. The inhibitory effect seems to be specific because the uptake of L-alanine is not reduced under this condition. The K_i values of 190 mmol/l (maternal) and 180 mmol/l (fetal) substantially lie above those reported for red and white blood cells (Mann and Newton, 1975; Moser and Weber, 1984). In these cells already physiological concentrations of D-glucose impair the transport of DAA. This is in contrast to the intestine of guinea pig where a three-fold increase of normal glucose load is without effect (Stevenson and Brush, 1969), since the transported form of vitamin C is supposed to be AA. Our results suggest that the decrease of fetal AA concentration at hyperglycemia (350 mg%) in dams as reported by Norkus et al. (1982) cannot be caused only be competitive inhibition of DAA transport. These authors gave no information about the DAA levels of their in situ preparation. Our results suggest that transient evaluations of blood glucose concentration in the patho-physiological range should not influence the transplacental DAA transport essentially.

SUMMARY

The uptakes of L-(^{14}C) ascorbic acid (AA) and of its oxidation products were measured on both the maternal and the fetal side of the isolated guinea pig placenta. The single passage double tracer dilution technique was used. Because AA is readily oxidized to dehydroascorbic acid (DAA) and diketogulonic acid (DKG) thin layer chromatography was necessary to identify the substance which is offered to the placenta   The maximal uptakes (U_{max}) of pure AA was 5% or less on both sides whereas the uptake of D-glucose was 70%. Experimental oxidation of AA leads to two different products. One more polar than AA is DKG and one less polar is DAA. The U_{max} of DAA purified by column chromatography amounts to 70% on both sides and is equal to that of D-glucose. The uptake of the purified DKG is negligible.

At high D-glucose concentration the uptake of DAA is inhibited incompletely. The K_i values gained from the Hill's plot are 190 mmol/l on the maternal and 180 mmol/l on the fetal side. Elevation of D-glucose in the patho-physiological range does not inhibit the uptake of DAA significantly.

REFERENCES

Bissonette, J.M., Hohimer, A.R., Cronan, J.Z., and Black, J.A. (1979) Glucose transfer across the intact guinea pig placenta. *J. Dev. Physiol.* 1, 415-426.

Bessey, O.A., Oliver, H.L., and Brock, M.J. (1947) The quantitative determination of ascorbic acid in small amounts of white blood cells and platelets. *J. Biol. Chem.* 168, 197-205.

Carstensen, M.H., Leichtweiss, H.-P., and Schröder, H. (1982) Uptake and transfer of labelled L-alanine and insulin in the perfused guinea pig placenta. *Pflügers Arch.* 392, R17.

Cunningham, V.J. and Sarna, G.S. (1979) Estimation of kinetic parameters of unidirectional transport across the blood-brain barrier. *J. Neurochemics* 33, 433-437.

Hensleigh, P.A. and Krantz, K.E. (1966) Extracorporeal perfusion of the human placenta. I. Placental transfer of ascorbic acid. *Am. J. Obstet. Gynecol.* 96, 5-13.

Leichtweiss, H.-P., and Schröder, H. (1971) Untersuchungen uber den Glucosetransport durch die isolierte, beiderseits künstlich perfundierte Meerschweinchenplacenta. *Pflügers Arch.* 325, 139-148.

Leichtweiss, H.-P. and Schröder, H. (1981a) L-lactate and D-lactate carriers on the fetal and the maternal side of the trophoblast in the isolated guinea pig placenta. *Pflügers Arch.* 390, 80-85.

Leichtweiss, H.-P. and Schröder, H. (1981b) Dual perfusion of the isolated guinea-pig placenta. *Placenta Suppl.* 2, 119-128.

Lund, C. and Kimble, M. (1943) Some determination of maternal and fetal vitamin C levels. *Am. J. Obstet. Gynecol.* 46, 635-647.

Mann, G.V. and Newton, P. (1975) The membrane transport of ascorbic acid. *Ann. NY Acad. Sci.* 258, 243-252.

Moller-Christensen, E. and Thorup, C. (1940) Über das Vorkommen von Vitamin C in Plazenta, Nabelstrangblut, Venenblut und Colostrum. *Zentralbl. Gynäkol.* 64, 1858-1861.

Moser, U. and Weber, F. (1984) Uptake of ascorbic acid by human granulocytes. *Int. J. Vit. Nutr. Res.* 54, 47-53.

Norkus, E.P., Bassi, J., and Rosso, P. (1979) Maternal-fetal transfer of ascorbic acid in the guinea-pig. *J. Nutr.* 109, 2205-2212.

Norkus, E.P., Bassi, J.A., and Rosso, P. (1982) Maternal hyperglycemia and its effect on the placental transport of ascorbic acid. *Pediat. Res.* 16, 746-750.

Penny, J.R. and Zilva, S.S. (1943) The chemical behaviour of dehydro-L-ascorbic acid in vitro and in vivo. *Biochem. J.* 37, 403-417.

Raiha, N. (1958) On the placental transfer of vitamin C. An experimental study on guinea pigs and human subjects. *Acta Physiol. Scand.* 45, *Suppl.* 155, 27-42.

Roe, J.H. (1961) Appraisal of methods of the determination of L-ascorbic acid. *Ann. NY Acad. Sci.* 92, 277-283.

Stevenson, N.R. and Brush, M.K. (1969) Existence and characteristics of Na+-dependent active transport of ascorbic acid in guinea pig. *Am.J. Clin. Nutr.* 22, 318-326.

Streeter, M.L. and Rosso, P. (1981) Transport mechanisms for ascorbic acid in the human placenta. *Am. J. Clin. Nutr.* 34, 1706-1711.

Wahren, H. and Rundqvist, O. (1937) Uber den Ascorbinsäuregehalt des Blutes von Mutter und Frucht. *Klin. Wchnschr.* 16, 1498-1499.

Yudilevich, D.L., Eaton, B.M., Short, A.H., and Leichtweiss, H.-P. (1979) Glucose carriers at maternal and fetal sides of the trophoblast in guinea pig placenta. *Am. J. Physiol.* 237, C205-C212.

Trophoblast Research 2:515-522, 1987

CRITERIA FOR STANDARDIZING AND JUDGING VIABILITY PLACENTAL PERFUSIONS IN ANIMALS

- A Workshop Report -

Bruce J. Kelman[1] and Bernard K. van Kreel[2]

[1]Biology and Chemistry Department
Battelle Pacific Northwest Laboratories
Richland, Washington 99352

[2]Department of Chemical Pathology
Erasmus University
3000 Dr. Rotterdam, The Netherlands

Participants: Susan Augello-Vaisey, Raymond B. Baggs, Mary M.L. Lee, Zulf Mughal, Jannice E. Polifka, Richard Ross, and William Slikker.

The recommendations in this manuscript are derived from a workshop which was held on October 7, 1985. In addition to input supplied by the participants, a number of articles published between 1967 and 1985 were examined for trends in criteria reported. These articles did not comprise an exhaustive literature search; rather they represented a very selective survey. The purpose of this manuscript is to describe criteria which the participants felt should used to determine the viability of placentae during perfusions.

It must be recognized that criteria are somewhat dictated by the materials under study. For example, criteria measurements are likely to be different in studies designed to examine movements of lipophilic materials across the placenta as opposed to the effects of drugs on blood gases. Nevertheless, it is possible to define common characteristics which can be applied to most perfusion techniques.

Throughout this document, reference is made to findings from the publications listed following the discussion. These publications are included in the reference list but not called out individually since the participants felt that no one publication includes all of the criteria listed and, therefore, all publications lack some of the criteria discussed during the workshop. The participants were particularly willing to release this report since their own publications were representative of the degree to which deficiencies were found in the literature.

Mechanical Aspects

While authors always specify whether maternal or fetal circulations are perfused, reports do exist in which the perfusion circuitry (open or closed) is not listed.

Occasionally this is a frank omission, but more often, the circuitry was referenced without any hint of configuration listed in the manuscript. Workshop participants felt that salient facts about the mechanical arrangements in the perfusion should always be listed even when the technique is referenced.

Measurement of Maternal Blood Flow to the Placenta

Perfusions carried out in vitro have the advantage that both maternal and fetal circulations of the placenta can be cannulated. It is possible to measure inflow and outflow volumes as well as flow pressure, so that direct measurements of blood flow and subsequent pressure relationships can be obtained (Carstensen et al., 1982a,b; Schröder and Leichtweiss, 1981; Schröder et al., 1972; Schröder et al., 1977; van Dijk and van Kreel, 1978, 1982; van Dijk et al., 1985; van Kreel and van Dijk, 1977a,b). In these types of studies, data should be normalized to reference materials so that artifacts caused by changes in permeability during perfusions can be minimized.

For perfusions carried out in vivo, it is possible to measure uterine blood flow directly with flow meters in larger species. However, in smaller species this is not possible. In all species, changes in maternal blood flow to the placenta can have considerable effects on measurements of movements of materials across the placenta when the maternal circulation is left intact. In those manuscripts which deal with perfusion studies in the guinea pig, 50% of the articles examined contained no mention of efforts to measure changes in maternal blood flow to the placenta. In other cases, authors used antipyrine clearance and water clearance as indicators of changes in maternal blood flow to the placenta (McKerchen et al., 1982; Bailey et al., 1979; Kelman, 1977; Kelman and Sikov, 1981; Kelman and Springer, 1982; Kelman and Walter, 1977, 1980a,b). The manner in which changes in maternal blood flow to the placenta can lead to erroneous assumptions about the actions of physical and pharmacological agents on movements of particular materials across the placenta have been previously presented in the literature by one of the participants (Kelman, 1979; Kelman and Sikov, 1986). We must therefore conclude that some measure of maternal blood flow to the placenta is necessary in all experiments.

Perfusion Rates and Pressures

Perfusion rates reported in the literature vary from extremely small fractions (Powers and Jenkins, 1975) to the lower range of blood flows measured in the intact animal (e.g. Carstensen et al., 1982a,b; Faber et al., 1968; Schröder et al., 1972; Kelman and Sasser, 1977). There is no consistent manner in which to extrapolate from low flow rates to higher flow rates if one is seeking to determine physiologically relevant transfer characteristics and amounts. Higher flow rates often generate greatly increased pressures. In nearly all preparations, it appears that vascular resistance in the artificially perfused placenta is considerably greater than in the intact placenta. Therefore, arterial pressures much higher than those found in the intact animal are necessary in order to maintain physiological flow rates. Increased arterial pressure can lead to placental edema or physical damage to the maternal-fetal barrier. Such physical damage can be monitored by using a large molecule to monitor changes in the size of diffusion channels in the placenta.

Some measure of net water flows should be included in all experiments. In an open system, one can measure the amount of perfusate entering and leaving to determine that there is no net flow across the placenta. In a closed system, the reservoir volume can be monitored for changes during experiments. Venous pressure is rarely reported, but it is an important contributing factor towards maintaining physiological water fluxes. Pressures and flow rates should always be reported in all experiments in a number which allows interpretation of data in terms of some standardized measurement (e.g., milliliters per minute per gram of placenta).

Perfusion Media

Perfusion media reported in the literature ranges from physiological saline to autologous fetal blood. Because the composition of perfusion medium is dictated by such diverse considerations as economic feasibility, availability, and the nature of the materials being measured, it is not likely that greater consistency between laboratories will occur in the future. However, thorough descriptions of the composition of the medium used in perfusions, including pH (both entering and leaving the placenta), should be included in each publication. Description by reference to previous publication does not appear to be adequate to meet this criterion.

Pharmacological Agents Administered During Experiments

In a few cases, anesthetics used in experiments are not reported; in others they are reported only by reference and not by name. Because pharmacological agents can have potential interactions with the measurements obtained during a perfusion experiment, it is important to report in detail all pharmacological agents which are used during the course of experiments.

Criteria Used to Assess the Condition of Placental Preparations

In addition to the measurements of the material under investigation, the following measurements have been reported in the literature as indications of maternal viability: maternal carotid blood pressure, blood gases, electrocardiograms, heart rates, respiratory rates, clearance of specific materials, movements of radioiodine and serum albumin (either within the maternal circulation into the fetal, or the fetal to the maternal), changes in perfusion rate and pressure, measures of perfusion volume, oxygen consumption, electropotential difference between mother and fetus, inulin spaces, production of lactate and pyruvate, and sodium-potassium ratios in venous effluent. All of these characteristics are of value in determining maternal viability.

Participants felt that it is important to report enough data about the preparation to allow the reader to make an independent assessment of the viability of the preparation. One suggestion was that important data could be obtained on the condition of preparations by histologically examining placentae after the completion of perfusion. This is rarely done despite the ease and the minimal cost involved.

Species and Strain

Species used appeared to be uniformly reported in perfusion experiments, but strain and breeds often were not. Since, at least in rodents, it has been shown that strain differences can exist, the strain of the animal used should also be reported. In experiments where strain or breed is not known, that fact should be reported.

Gestational Ages

Gestational ages should be reported as precisely as possible. In a number of publications, gestational age has been reported for nonhuman species in terms of trimesters. This designation should be reserved for human pregnancies unless the species happens to have a nine month pregnancy or a pregnancy which lasts for any period of time divisible by a factor of three.

Why Attempt Standardization of Perfusion Techniques?

Placental perfusion techniques can be powerful tools for answering specific questions on the nature of the manner in which materials reach the fetus and the physiology of pregnancy. If perfusion techniques are to realize their full potential for the study of effects of pharmacological agents on fetal development, greater effort must be made to supply data which can be compared among 1) different laboratories, 2) different studies in the same laboratory on the same materials, and 3) studies on the same laboratory on different compounds. This can only be accomplished by including more specific detail about techniques and expressing data in terms of reference materials and conditions.

REFERENCES

Bailey, D.J., Bradbury, M.W., France, U.M., Hedley, R., Naik, S., and Parry, H. (1979) Cation transport across the guinea-pig placenta perfused in situ. *J. Physiol. (Lond.)* 287, 45-56.

Baker, E. and Morgan, E.H. (1970) Iron transfer across the perfused rabbit placenta. *Life Sci.* 9, 773-779.

Bissonnette, J.M. (1975) Control of vascular volume in sheep umbilical circulation. *J. Appl. Physiol.* 38, 1057-1061.

Bissonnette, J.M. and Farrell, R.C. (1973) Pressure-flow and pressure-volume relationships in the fetal placental circulation. *J. Appl. Physiol.* 35, 355-360.

Bissonnette, J.M. and Gurtner, G.H. (1972) Transfer of inert gases and tritiated water across the sheep placenta. *J. Appl. Physiol.* 32, 64-69.

Bloxam, D.L., Tyler, C.F., and Young, M. (1981) Foetal glutamate as a possible precursor of placental glutamine in the guinea pig. *Biochem. J.* 198, 397-401.

Bradbury, M.W., France, U.M., Hedley, R., and Parry, H. (1977) Transport of cations by the guinea-pig perfused in situ. *J. Physiol. (Lond.)* 272, 22P-23P.

Carroll, M.J. (1981) Short communication: The influence of perfusion in situ on lactate and pyruvate levels in guinea pig placenta. *Placenta* 2, 271-272.

Carstensen, M.H., Leichtweiss, H.P., and Schröder, H. (1982a) The metabolism of the isolated artificially perfused guinea pig placenta. II. Difference of excretion of hydrogen ions, ammonia, carbon dioxide, and lactate into maternal and fetal veins. *J. Perinat. Med.* 10, 154-160.

Carstensen, M.H., Leichtweiss, H.P., and Schröder, H. (1982b) The metabolism of the isolated artificially perfused guinea pig placenta. I. Excretion of hydrogen ions, ammonia, carbon dioxide, and lactate, and the consumption of oxygen and glucose. *J. Perinat. Med.* 10, 147-153.

Eaton, B.M. and Yudilevich, D.L. (1981) Uptake and asymmetric efflux of amino acids at maternal and fetal sides of placenta. *Am. Physiol. Soc.* 241, C106-C112.

Faber, J.J., Hart, F.M., and Poutala, A.C. (1968) Diffusional resistance of the innermost layer of the placental barrier of the rabbit. *J. Physiol. (Lond.)* 197, 381-393.

Fenton, E. (1977) Technique for in situ perfusion of the guinea pig placenta. *Biol. Neonate* 32, 1-4.

Fenton, E., Britton, H.G., and Nixon, D.A. (1976) A study of the permeability of the guinea pig placenta to citrate using recirculating placental perfusion technique. *Biol. Neonate* 29, 299-305.

Forsling, M.L. and Fenton, E. (1977) Permeability of the foetal guinea-pig placental to arginine-vasopressin. *J. Endocrinol.* 72, 409-410.

Hanshaw-Thomas, A., Roserson, N., and Reynolds, F. (1984) Transfer of bupivacaine, lignocaine, and pethidine across the rabbit placenta: Influence of maternal protein binding and fetal flow. *Placenta* 5, 61-70.

Hedley, R. and Bradbury, M.W.B. (1980) Transport of polar non-electrolytes across the intact and perfused guinea-pig placenta. *Placenta* 1, 277-285.

Herberger, J. and Moll, W. (1976) The flow resistance of the material placental vascular bed of anesthetized guinea pigs. *Z. Geburtshilfe. Perinatol.* 180, 61-66.

Kastendieck, E., Kuenzel, W., and Kurz, C.S. (1980) Placental clearance of lactate and bicarbonate in sheep. *Gynecol. Obstet. Invest.* 10, 9-22.

Kastendieck, E. and Moll, W. (1977) The placental transfer of lactate and biocarbonate in the guinea-pig. *Pflüegers Arch.* 370, 165-171.

Kelman, B.J. (1977) Inorganic mercury movements across the perfused guinea pig placenta in late gestation. *Toxicol. Appl. Pharmacol.* 41, 659-665.

Kelman, B.J. (1979) Effects of toxic agents on movements of materials across the placenta. *Fed. Proc.* 38, 2246-2250.

Kelman, B.J. and Sasser, L.B. (1977) Methylmercury movements across the perfused guinea pig placenta in late gestation. *Toxicol. Appl. Pharmacol.* 39, 119-127.

Kelman, B.J. and Sikov, M.R. (1981) Plutonium movements across the hemochorial placenta of the guinea pig. *Placenta* 3, 319-326.

Kelman, B.J. and Springer, D.L. (1982) Movements of benzo(a)pyrene across the hemochorial placenta of the guinea pig. *Proc. Soc. Exp. Biol. Med.* 169, 58-62.

Kelman, B.J. and Sikov, M.R. (1986) Effects of ultrasound on placental function. *UFFC* 2, 218-224.

Kelman, B.J. and Walter, B.K. (1977) Passage of cadmium across the perfused guinea pig placenta. *Proc. Soc. Exp. Biol. Med.* 156, 68-71.

Kelman, B.J. and Walter, B.K. (1980a) Fetal to maternal cadmium movements across the perfused hemochorial placenta of the guinea pig. *Toxicol. Appl. Pharmacol.* 52, 440-406.

Kelman, B.J. and Walter, B.K. (1980b) Transplacental movements of inorganic lead from mother to fetus. *Proc. Soc. Exp. Biol. Med.* 163, 278-282.

Kilstrom, I. (1983) Placental transfer of diethylhexyl phathalate in the guinea-pig placenta perfused in situ. *Acta Pharmacol. Toxicol.* 53, 23-27.

Leichtweiss, H.P. and Schröder, H. (1971) Investigations of glucose transport across the isolated artificially perfused placenta of the guinea-pig. *Pflüegers Arch. Eur. J. Physiol.* 325, 139-148.

Leichtweiss, H.P. and Schröder, H. (1971) Glucose transport through the isolated, bilaterally artifically perfused placenta of the guinea pig. *Arch Gynecol.* 211, 146-147.

McKercher, H.G., Derewlany, L.O., and Radde, I.C. (1982) Free calcium concentrations in Krebs-Ringer biocarbonate buffer: effects on ^{45}Ca and ^{32}P-transport across the perfused guinea pig placenta. *Biochem. Biophys. Res. Comm.* 105, 841-846.

Moll, W. and Kastendieck, E. (1977) Transfor of N_2O, CO and HTO in the artifically perfused guinea-pig placenta. *Respir. Physiol.* 29, 283-302.

Nixon, D.A. (1972) Carbohydrate metabolism in fetal life. *Phys. Biochem. of the Fetus Symposium XIV*, 96-116.

Nixon, D.A., Alexander, D.P., and Hugget, A.S. (1966) Influence of umbilical blood fructose on the production of fructose by the perfused placenta of sheep. *Nature (Lond).* 209, 918-919.

Powers, G.G. and Jenkins, F. (1975) Factors affecting O_2 transfer in sheep and rabbit placenta perfused in situ. *Am. J. Physiol.* 229, 1147-1153.

Reynolds, M. and Young, M. (1968) Transfer of amino-acids across the guinea-pig placenta perfused in situ. *Proc. Int. Union Physiol. Sci.* 7, 365.

Rubio, V., Tan, C.B., Andrews, W.H., Nixon, D.A., Alexander, D.P., and Britton, H.G. (1983) Nitrogen metabolism in the sheep fetus. Observations on the liver and placenta. *Biol. Neonate* 43, 86-91.

Schröder, H. and Leichtweiss, H.P. (1977) Perfusion rates and the transfer of water across isolated guinea pig placenta. *Am. J. Physiol.* 232, 666-670.

Schröder, H., Paul, W., and Leichtweiss, H.P. (1981) Vascular volumes in isolated perfused guinea pig placenta. *Am. Physiol. Soc.* 241, H73-H77.

Schröder, H., Stolp, W., and Leichtweiss, H.P. (1972) Measurements of Na^+ transport in the isolated, artificially perfused guinea pig placenta. *Am. J. Obstet. Gynecol.* 114, 51-57.

Sibley, C.P., Bauman, K.F., and Firth, J.A. (1981) Ultrastructural study of the permeability of the guinea-pig placenta to horseradish peroxidase. *Cell Tissue Res.* 219, 637-647.

Stulc, J. and Svihovec, B. (1979) Lack of effect of ouabain on sodium transport across the rat placenta. *Pflüegers Arch.* 381, 79-81.

Stulc, J. and Svihovec, B. (1984) Transport of inorganic phosphate by the fetal border of the guinea-pig placenta perfused in situ. *Placenta* 5, 9-19.

Stulc, J. and Svihovec, J. (1973) Effects of potassium cyanide, strophanthin or sodium-free perfusion fluid on the electrical potential difference across the guinea-pig placenta perfused in situ. *J. Physiol. (Lond)* 231, 403-415.

Stulc, J. and Svihovec, J. (1977) Placental transport of sodium in the guinea-pig. *J. Physiol. (Lond)* 265, 691-703.

Svihovec, J. and Stulc, J. (1972) Electrical potential difference between maternal and fetal side on the guinea pig placenta perfused in utero. *Cesk. Fysiol.* 21.

Sweiry, J.H. and Yudilevich, D.L. (1984) Assymetric calcium influx and efflux at maternal and fetal sides of the guinea-pig placenta: Kinetics and specificity. *J. Physiol. (Lond)* 355, 295-311.

Thomsen, K., Schniewind, H., Schultze-Mosgau, H., Kramer, M., and Born, H. (1967) Effect and passage of vasoactive drugs in the in vitro perfused placenta of guinea pigs. *Arch. Gynecol.* 204, 51-58.

van Dijk, J.P. and van Kreel, B.K. (1978) Transport and accumulation of a-aminoisobutyric acid (AIB) in the guinea pig placenta. *Pflüegers Arch.* 377, 217-224.

van Dijk, J.P. and van Kreel, B.K. (1982) Electric potential differences in the isolated guinea-pig placenta. *J. Dev. Physiol.* (Oxf.) 4, 23-28.

van Dijk, J.P., van Kreel, B.K., and Heeren, J.W. (1985) Studies on the mechanisms involved in iron transfer across the isolated guinea pig placenta by means of bolus experiments. *J. Dev. Physiol. (Oxf.)* 7, 1-16.

van Kreel, B.K. and van Dijk, J.P. (1977a) Transport of uric acid and hypoxanthine across the isolated guinea pig placenta. *Biol. Neonate* 32, 260-265.

van Kreel, B.K. and van Dijk, J.P. (1977b) The influence of flow rates and flow distributions on diffusion controlled transport across the isolated guinea pig placenta. *J. Comp. Physiol.* B115, 255-264.

Weatherley, A.J., Ross, R., Pickard, D.W., and Care, A.D. (1983) The transfer of calcium during perfusion of the placenta in intact and thyroparathyroidectomized sheep. *Placenta* 4, 271-278.

Wong, C.T. and Morgan, E.H. (1973) Placental transfer of iron in the guinea pig. *J. Exp. Physiol.* 58, 47-58.

HUMAN PLACENTAL
PERFUSIONS

Trophoblast Research 2:525-533, 1987

TRANSFER OF ANTIPYRINE, H₂O, L-GLUCOSE AND α-AMINOISOBUTYRIC ACID ACROSS THE IN VITRO PERFUSED HUMAN PLACENTAL LOBE

Henning Schneider, Mariette Proegler, and Reto Duft

Division of Perinatal Physiology
Department of Obstetrics
University of Zuerich
Frauenklinik-Str. 10
CH-8091 Zuerich, Switzerland

INTRODUCTION

The dual circuit in vitro perfusion of an isolated lobe of human placenta is a useful experimental technique to study transfer of material across the placenta and to identify factors influencing exchange between maternal and fetal circulation. It has repeatedly been noted that there is considerable variation in results among experiments which are not fully explained by differences in size of the perfused lobe (Schneider et al., 1972; Welsch, 1980). While flow rates in the two perfusion circuits can easily be controlled, the surface area of exchange, shunts, and degree of overlap of the two circulations remain major factors causing interexperimental variability. Antipyrine which rapidly equilibrates across most biological membranes has been proposed as a reference for normalization of experimental data on transfer (Dancis et al., 1973). After this correction parameters like membrane permeability or specific transport mechanisms can be explored. This concept should be valid for the study of test compounds which are similarly affected in transfer by the uncontrollable variables as is antipyrine.

In a series of experiments with similar flow rates, we have measured the transfer of labeled water, L-glucose, and the non-metabolized alpha-aminoisobutyric acid (AIB) with antipyrine as reference. Antipyrine transfer served as baseline for all experiments and was measured under steady state conditions of continuous infusion into the maternal circulation. For the three test compounds transfer was explored during steady state and also after a single bolus injection. Labeled water like antipyrine is considered as flow-limited in its diffusion while transfer of L-glucose and AIB is less dependent on flow. To test the usefulness of antipyrine as a reference, the variation of the results of each substance was determined with and without normalization and the correlation coefficient with antipyrine was calculated using regression analysis.

Received: 10 October 1985; Accepted: 5 May 1986

MATERIALS AND METHODS

Perfusion Technique

All placentae were obtained immediately after spontaneous vaginal deliveries or elective Cesarean sections performed at term after uncomplicated pregnancies. The perfusion preparation was quickly established by cannulating a chorionic artery and vein supplying an intact lobe of placental tissue and the villous vascular space (fetal compartment) was rinsed with perfusion medium. The lobe with the catheterized vessels was fixed in a special ring, and the preparation was placed with the decidual surface facing upwards into a perfusion chamber heated by a warm-water circuit to 37°C. The area of villous vasculature which is perfused was recognized by its light color and five metal cannulae were inserted into the intervillous space in this region (maternal compartment). Effluent from the intervillous space collecting on the decidual surface was continously drained using a syphon. The mean flow-rate was 9.33 ± 1.59 (SD) in the fetal circulation and 13.43 ± 1.66 ml/min in the maternal circulation for a placental lobe of 35.57 ± 9.54 g mean wet weight. Details of this perfusion set up have been described before (Schneider et al., 1972; Schneider and Huch, 1985).

Temperature and perfusion pressure were continuously monitored in both circuits next to the chamber, and the metabolic activity of the tissue as monitored by the consumption of glucose and the production of lactate and pyruvate was similar to previously published values (Sodha et al., 1984).

Perfusate

The medium was prepared from two parts NCTC 135 (National Culture and Tissue Collection, from Difco) and one part Earle's bicarbonate buffer with 1 g/l human serum albumen, 100 mg% glucose and 2 mg % AIB (Alpha-aminoisobuytric acid). Antipyrine, because of its flow-limited diffusion characteristics, was used as reference in all perfusions and was added to the maternal perfusate at a concentration of 8 mg%. The medium was prepared fresh prior to each perfusion with addition of 1 g/l of amoxycillin and was equilibrated with 95% oxygen and 5% carbon dioxide in a water bath at 38°C. ^3H_2O (4 µCi/mmol), ^{3}H-L-Glucose (10.7 Ci/mmol) and ^{14}C-AIB (19.9 mCi/mmol), were purchased from New England Nuclear. Antipyrine was obtained from Fluka AG, Switzerland.

Analytical Methods

Antipyrine was measured colorimetrically (Brodie et al., 1949) using a Technicon automated analyzer. Radioactivity was determined in 1.0 ml samples after addition of 15 ml scintillant (Instagel, Packard Instruments) using standard double isotope liquid scintillation techniques with channel cross over correction. In all experiments the wet weight of the perfused tissue was recorded after blotting of excess fluid and in some of them a piece of perfused tissue was excised and homogenized in 3 vol of distilled water and placed in a boiling water bath for 3 min. After centrifugation, radioactivity in 1 ml of the supernatant was counted.

For statistical evaluation comparison by paired t-test was used. Regression analysis using a linear model was applied to test for correlation between test and reference compound.

Experimental Design

Both fetal and maternal compartments were perfused in an open system. After an initial period of stabilization three one minute collections of fetal and maternal venous return were taken and the measured volume was compared to the preset constant flow delivered by the perfusion pump and with any signs of fluid shift from the fetal to the maternal circulation the preparation was discarded. Transfer of ^3H_2O, ^{14}C-AIB, and ^3H-L-glucose was studied from maternal to fetal side after rapid injection of a 300 µl bolus (20 µCi ^3H_2O, or 20 µCi ^3H-L-glucose and 5 µCi ^{14}C-AIB) into the maternal circuit next to the perfusion chamber. Disposable 1 ml hypodermic syringes were used and weighed before and after the injection. Maternal and fetal effluent were separately collected for 4 min following the bolus injection and the measured volume was used for calculation of the flow rates. The total transfer of radioactivity was expressed in percent of the bolus. Antipyrine concentration was also measured in the collection and transfer was expressed in percent of the amount delivered to the intervillous space over the 4 min period. ^{14}C-AIB transfer was studied either together with ^3H_2O or with ^3H-L-glucose as a second isotope. For the second half of the experiment 5 µCi of AIB and 10 µCi of ^3H_2O or ^3H-L-glucose was added to 800 ml perfusate in the maternal reservoir so that the fluid circulating through the intervillous space had a constant concentration of antipyrine and the labeled test compounds. After 40 min of perfusion, three successive samples were collected for one minute each from the maternal and fetal vein. The mean of the flow rate and the concentration of antipyrine as well as radioactivity of the three samples was used to calculate transfer which was converted into clearance values by dividing it by the constant concentration in the maternal perfusate.

RESULTS

The transfer of the four compounds as measured under steady state conditions and expressed as clearance is shown in Table 1. In 19 experiments, transfer of antipyrine and ^3H_2O were measured together, and almost identical values were found as indicated by a clearance index of close to 1. The lower clearance indices for L-glucose and AIB of 0.35 ± 0.03 and 0.47 ± 0.15 are consistent with some limitation of transfer when compared to the flow dependent compounds, antipyrine and ^3H_2O. When the clearance values for ^3H_2O and L-glucose are compared with the clearance indices, the coefficient of variation is considerably reduced. It was expected that similar variables are responsible for the variation of ^3H_2O and antipyrine transfer so that the variation is reduced by formation of the index. The same variables to a significant extent also affect the transfer of L-glucose as shown by a reduced coefficient of variation for the index. For AIB, the formation of the clearance index does not decrease the variation. Therefore, different factors must lead to the variability in AIB transfer. Using regression analyses, the correlation coefficient between the test substances and antipyrine as a reference was calculated. With a coefficient of 0.95, the paired values of antipyrine and ^3H_2O transfers were correlated at a highly significant level ($p < 0.001$). For L-glucose ($R = 0.75$) and AIB ($R = 0.50$), the correlation with antipyrine was less strong ($p < 0.05$).

Table 1

Placental transfer of Antipyrine, ^3H_2O, L-Glucose and AIB

	Antipyrine	^3H_2O	^3H-L-Glucose	^{14}C-AIB
n =	25	19	5	12
Clearance ml/min	2.79 ± 0.84	3.03 ± 0.74	0.72 ± 0.13	1.11 ± 0.32
V	0.30	0.24	0.18	0.29
Clearance Index	–	1.01 ± 0.09	0.35 ± 0.03	0.47 ± 0.15
V	–	0.09	0.09	0.32

mean ± SD,. V = coefficient of variation   Clearance Index: $\dfrac{\text{test substance}}{\text{antipyrine}}$

n = number of experiments

Clearance measurements were performed after 40 min of continuous infusion of test substance and antipyrine into the intervillous space at mean flow-rates in the maternal and fetal perfusion circuit of $Q_M = 13.43 \pm 1.66$ and $Q_F = 9.33 \pm 1.59$. Clearance index was calculated from paired values obtained from the same sample.

Because of the similarity in molecular weight as well as solubility characteristics L-glucose is frequently used as a reference in transport studies of amino acids. The results of five experiments in which AIB transfer was compared directly with L-glucose are shown in Table 2 together with tissue levels determined at the end of the experiment. Transfer of AIB was significantly higher than of L-glucose when compared by paired t-test and relative to L-glucose there was also accumulation of AIB by the tissue.

When the data on placental transfer obtained under steady state conditions of flow and transplacental gradient were compared with transfer measured after bolus injection, interesting differences among the three test compounds became apparent. For ^3H_2O there was no difference between bolus and steady state transfer (Table 3) while L-glucose was somewhat higher during steady state transfer. With AIB, transfer almost doubled.

DISCUSSION

The complexity of the two circulations in the bilaterally in vitro perfused isolated human placental lobe is poorly understood. The difference in the anatomical design of the fetal capillary network and the intervillous space, shunts in both compartments and uneven distribution of flow will influence experimental results. Placental transfer by passive diffusion is determined by flow rates and membrane factors. The clearance of antipyrine and ^3H_2O predominantly is a function of the two

Table 2

Placental transfer of L-Glucose and AIB

	3H-L-Glucose		14C-AIB		AIB : L-Glucose	
	Clearance ml/min	Tissue %	Clearance	Tissue %	Clearance	Tissue
G-78	0.81	37	0.99	57	1.22	1.54
G-80	0.84	33	1.09	48	1.29	1.46
G-84	0.76	35	0.93	47	1.22	1.34
G-85	0.69	33	0.83	48	1.20	1.46
G-86	0.52	14	0.58	20	1.12	1.43

mean ± SD 0.72 ± 0.13   0.88 ± 0.19

└────────$p < 0.01$────────┘

Transfer was measured from the maternal to the fetal side in a steady state of flow-rates and of the transplacental gradient. Tissue levels correspond to radioactivity in the extract of 1 g wet weight and are given in % of the concentration in the maternal perfusate. Ratio of AIB over L-Glucose is shown for clearance and tissue concentration.

flows. Although in this series, perfusion flows were maintained within a narrow range considerable variation in clearance of the two flow dependent substances was observed. It is unlikely that this is explained by differences in exchange area since these would be compensated by flow. With a smaller lobe, there will be higher flows to the exchange area with increased transfer of a flow dependent substance. Shunts as a result of inadequate overlap of maternal and fetal perfused area will reduce the effective flow rates and provide a reasonable explanation for the observed variation in clearance. The close agreement in the transfer of antipyrine and $3H_2O$ confirms previously published results of Challier et al. (1983) and ourselves (1985). Also in animal experiments, under steady state conditions, there was no difference in placental transfer of $3H_2O$ and antipyrine (Meschia et al., 1967). Antipyrine is a useful reference for the study of other mostly flow-dependent substances and interplacental variability is reduced by more than half when transfer is presented as a ratio of test over reference compound.

The observed correlation between L-glucose and antipyrine and the reduction in interexperimental variability in L-glucose transfer is also related to flows. Yudilevich et al. (1981) have interpreted the L-glucose permeability of the isolated guinea pig placenta as "leakiness" which develops as a result of in vitro perfusion. This is supported by a very low L-glucose transfer described by Bissonnette (1981) for the in situ guinea pig placenta. On the other hand, Hedley and Bradbury (1981) have

Table 3

Placental transfer of ^3H_2O, L-Glucose and AIB
during steady state and after bolus injection

	^3H_2O Bolus	^3H_2O Steady state	^{3}H-L-Glucose Bolus	^{3}H-L-Glucose Steady state	^{14}C-AIB Bolus	^{14}C-AIB Steady state
n =	7		5		12	
mean ± SD (%)	21.81 ± 3.59	21.05 ± 5.12	4.28 ± 1.45	5.70 ± 0.72	4.36 ± 1.38	8.17 ± 2.29
	n.s.		$p < 0.05$		$p < 0.001$	

n = number of experiments ·
Transfer was measured from the maternal to the fetal side in a steady state of flow
rate and of the transplacental gradient and after injection of a 300 µl bolus into the
intervillous space with collection of the effluent over 4 min. Transfer is expressed in
% of the substrate delivered to the placenta.

found a physiological leakiness to L-glucose and larger hydrophilic molecules which
increases in the isolated placental preparation. It must be recognized that the
clearance index based on a purely flow-dependent reference, is of limited value for the
study of compounds where some of the variability is due to other factors like
membrane area. The index will not allow comparison of experiments with grossly
different flow rates since the effect of flow on test and reference compound will differ.
On the other hand, L-glucose could serve as a reference for partly membrane limited
compounds like AIB.

For AIB, the correlation with antipyrine showed only borderline significance
and the coefficient of variation was not improved by using the clearance index.
However, when L-glucose is used as reference, the coefficient of variation for AIB
transport is reduced from 0.22 for the absolute values to 0.06 for the index.
Comparison with L-glucose as a marker for extracellular diffusion reveals that under
the given experimental conditions, barely 20% of total AIB transport is carrier-
mediated, and the rest is by passive diffusion. Similar results on AIB transfer relative
to ^3H_2O were found by Kelman and Sikov (1983) in the in situ perfused guinea pig
placenta. They described a good correlation with the flow-dependent marker and also
found a large component of diffusional transfer. It is obvious that with the
physiological concentration gradient between fetal and maternal blood, passive
diffusion under in vivo conditions has no role in transport of amino acids from the
maternal to the fetal side. High transfer rates of AIB across the in situ as well as
isolated perfused guinea pig placenta were found by van Dijk and van Kreel (1978)
against the physiological gradient showing an efficient transport system which
provides adequate supply of amino acids to the fetus. It is unclear how much of the
diffusion of AIB seen in our experiments is an artifact of the in vitro perfusion set up
which tends to exaggerate diffusion through extracellular channels. It will lead to a

distortion of the carrier-mediated portion of placental transport and will interfere with the study of saturation kinetics as was shown by Bissonnette (1981) for glucose transport. We also found little retention of AIB by the tissue and radioactivity per gram wet weight barely reached 50% of the concentration in the maternal perfusate. In the guinea pig placenta, the tissue level was around 10 times the maternal level (van Dijk and van Kreel, 1978) and a similar concentration factor was seen by Kelman and Sikov (1983) with an open circuit with no radioactivity in the fetal perfusate. Higher tissue levels with a concentration factor of close to 3 are also seen for human placenta when the fetal side is perfused for 2-3 hr in a closed system (Schneider, unpublished) or with tissue slices (Schneider and Dancis, 1974). This would suggest that under the conditions of in vitro perfusion with open circuits passive diffusion accounts for most of the transfer of AIB across the human placenta.

Bolus injections of a pair of differently labeled substances have been mainly used to measure the kinetics of uptake of a test substance by the syncytiotrophoblast or the fetal surface of the placental barrier. For the measurement of transplacental transport, the amount of tracer recovered from both circuits must be taken into consideration. With almost total recovery transfer should be equivalent to transfer estimations under steady state conditions (Lassen and Perl, 1979). With a sample collection period limited to 4 min after bolus injection the mean recovery was 81% for ^3H_2O, 84% for ^3H-L-glucose and 76% for ^{14}C-AIB. Although recovery of ^3H_2O was incomplete, there was no difference between transfer after bolus injection and under steady state conditions. This differs from the intact sheep where transfer with bolus injection of antipyrine was only 60% of ^3H_2O measured under steady state conditions (Bissonnette et al., 1979). While for both passively diffusing markers there is little difference in transfer after bolus injection compared to the steady state, AIB transfer is considerably reduced with bolus injection, which most likely reflects tissue uptake. The early recovered fraction of transfer represents extracellular diffusion while the active component is related to the equilibration of the labeled amino acid with the tissue pool.

SUMMARY

In an open circuit in vitro perfusion technique of the intervillous space and the fetal villous capillaries of an isolated lobe of human placenta mean clearance values of 2.79 ± 0.84, 3.03 ± 0.74, 0.72 ± 0.13, and 1.11 ± 0.32 ml/min were measured for the transfer from maternal to fetal side for antipyrine, ^3H_2O, ^3H-L- glucose, and ^{14}C-AIB. There was a close correlation between the transfer of the reference substance antipyrine and ^3H_2O while for L-glucose and AIB transfer showed only a weak correlation with antipyrine. The use of the ratio between test substance over antipyrine as reference as an index reduces interexperimental variation for ^3H_2O and L-glucose. Comparison of AIB transfer with L-glucose reveals that diffusion is the major component of transport under the presented experimental conditions.

When placental transfer is measured after bolus injection no difference was seen for the rapidly diffusing ^3H_2O, for L-glucose there is slight reduction, and AIB transfer was cut in half.

ACKNOWLEDGEMENTS

The authors are very grateful to Mrs. B. Benz for her expert technical assistance. This work was supported by a grant from the Schweizerischer Nationalfonds.

REFERENCES

Bissonnette, J.M., Cronau, J.Z., Richards, L.L., and Wickham, W.K. (1979) Placental transfer of water and nonelectrolytes during a single circulatory passage. *Am. J. Physiol.* 236 (1), 47-52.

Bissonnette, J.M. (1981) Studies in vivo of glucose transfer across the guinea pig placenta. *Placenta Suppl.* 2, 155-162.

Brodie, B.B., Axelrod, J., Soberman, R., and Levy, B.B. (1949) The estimation of antipyrine in biological materials. *J. Biol. Chem.* 179, 25-29.

Challier, J.C., D'Athis, P., Guerre-Millo, M., and Nandakumaran, M. (1983) Flow-dependent transfer of antipyrine in the human placenta in vitro. *Reprod. Nutr. Develop.* 23, 41-50.

Dancis, J., Jansen, V., Kayden, H.J., Schneider, H., and Levitz, M. (1973) Transfer across perfused human placenta. II. Fatty acids. *Ped. Res.* 7, 192-197.

Hedley, R. and Bradbury, M.W.B. (1980) Transport of polar nonelectrolytes across the intact and perfused guinea pig placenta. *Placenta* 1, 277-285.

Kelman, B.J. and Sikov, M.R. (1983) Clearance of α-aminoisobutyric acid during in situ perfusion of the guinea pig placenta. *Trophoblast Res.* 1, 71-80.

Lassen, N.A. and Perl, W. (1979) *Tracer Methods in Medical Physiology.* Raven Press, New York.

Meschia, G., Battaglia, F.C., and Bruns, P.D. (1967) Theoretical and experimental study of transplacental diffusion. *J. Appl. Physiol.* 22, 1171-1178.

Schneider, H., Panigel, M., and Dancis, J. (1972) Transfer across the perfused human placenta of antipyrine, sodium and leucine. *Am. J. Obstet. Gynecol.* 114, 822-828.

Schneider, H. and Dancis, J. (1974) Amino acid transport in human placental slices. *Am. J. Obstet. Gynecol.* 120, 1092-1098.

Schneider, H., Moehlen, K.H., and Dancis, J. (1979) Transfer of amino acids across the in vitro perfused human placenta. *Ped. Res.* 13, 236-240.

Schneider, H. and Huch, A. (1985) Dual in vitro perfusion of an isolated lobe of human placenta: method and instrumentation. In: *In Vitro Perfusion of Human Placental Tissue*, (eds.), H. Schneider and J. Dancis, Basel, Karger, pp. 40-47.

Schneider, H., Proegler, M., and Sodha, R.J. (1985) Effect of flow rate ratios on the diffusion of antipyrine and ^3H_2O in the isolated dually in vitro perfused lobe of the human placenta. In: *In Vitro Profusion of Human Placental Tissues*, (eds.), H. Schneider and F. Dancis, Basel, Karger, pp. 114-123.

Sodha, R.J., Proegler, M., and Schneider, H. (1984) Transfer and metabolism of norepinephrine studied from maternal-to-fetal and fetal-to-maternal sides in the in vitro perfused human placental lobe. *Am. J. Obstet. Gynecol.* 148, 474-484.

van Dijk, J.P. and van Kreel, B.K. (1978) Transport and accumulation of α-aminoisobutyric acid (A.I.B.) in the guinea pig placenta. *Pflüegers Arch.* 377, 217-224.

Welsch, F. (1980) Effects of acetylcholine on the clearance of ^3H-antipyrine in bilaterally perfused lobules of human term placenta. *Gynecol. Obstet. Invest.* II, 49-55.

Yudilevich, D.L., Eaton, B.M., and Mann, G.E. (1981) Carriers and receptors at the maternal and fetal sides of the placenta studied by a single-circulation paired-tracer dilution technique. *Placenta Suppl.* 2, 139-149.

Trophoblast Research 2:535-544, 1987

THE MODULATION OF GLUCOSE TRANSFER ACROSS THE HUMAN PLACENTA BY INTERVILLOUS FLOW RATES: AN IN VITRO PERFUSION STUDY

Nicholas P. Illsley[1], S. Hall, and T.E. Stacey

Section of Perinatal and Child Health
MRC Clinical Research Centre
Watford Road
Harrow, Middlesex HA1 3UJ, United Kingdom

INTRODUCTION

There is substantial evidence that a high proportion of fetal growth retardation in man is the result of uteroplacental vascular insufficiency. The association of growth retardation with maternal hypertensive disease, cyanotic heart disease and the relationship between growth retardation and low prepregnancy blood volume support the suggestion that chronic flow restriction may result in intrauterine growth retardation (Seeds, 1984). At the same time, in cases of essential hypertension and pre-eclampsia, a reduction in uteroplacental blood flow of 50-60% has been demonstrated (Brown and Veall, 1953; Lunell et al., 1979). Blood flow in growth retarded pregnancies can be reduced by over 50%, even in those women who do not display signs of hypertensive disorders or fetal malformations (Nylund et al., 1983).

Fetal weight has been shown to be closely related to placental size and uteroplacental blood flow (or maternal blood flow; MBF) in a number of species (Bruce and Abdul-Karim, 1973; Wootton et al., 1977; Myers et al., 1982; Garris, 1983; Jones and Parer, 1983). Two methods have commonly been used to examine the interaction of these variables. The first technique is the correlation of spontaneously occurring differences in fetal weight with both placental weight and MBF (Wootton et al., 1977; Myers et al., 1982; Garris, 1983). In this type of investigation, however, an element of ambiguity remains with regard to the primary mechanism regulating fetal weight. Does placental size (and transport capacity) determine nutrient transfer and fetal growth (Wootton et al., 1977; Simmons et al., 1979; Santonge and Ross, 1983) or does MBF regulate fetal (and placental) growth (Myers et al., 1982; Garris, 1983)?

The second technique, described initially by Wigglesworth (1964), involves the ligation of a uterine artery to reduce MBF. Fetal and placental weights, nutrient transfer, and various biochemical markers are measured to determine the effects of reduced MBF on growth (Wigglesworth, 1964; Kollee et al., 1979; Jones and Parer, 1983; Garris, 1984). This technique suffers from two problems. The first is the

[1]To whom correspondence should be addressed: Laboratory of Renal Biophysics, Cardiovascular Research Institute, University of California, San Francisco, California 94143 USA
Submitted: 24 September 1985; Accepted: 25 March 1986

difficulty encountered in determining whether there are acute and/or chronic adaptations to the procedures. The second is in determining the point at which maternal and fetoplacental systems have reached a new homeostatic equilibrium.

This investigation was designed to examine the links between MBF and the transfer of glucose across the term human placenta. We used an in vitro placental perfusion technique which eliminated the problem of extrapolating animal data. By perfusing the placenta over a short time span, the longer term adaptations to reduced MBF involving protein synthesis or structural alterations were minimised. Furthermore, perfusing placentae in isolation permitted determination of the relationship between MBF and nutrient transport, free from intervening maternal and fetal factors.

METHODS

The placentae used in this investigation were from normal (non-pathological), term (39-41 wk) pregnancies obtained after vaginal delivery or elective Cesarean section. In vitro placental perfusion was conducted as described previously (Illsley et al., 1984) with a number of minor modifications. Fetal and maternal circulations were maintained in a single-pass, open-circuit configuration for all experiments. The perfusion medium was a Krebs-Ringer phosphate/bicarbonate buffer gassed with 95% oxygen/5% carbon dioxide (maternal) or 95% nitrogen/5% carbon dioxide (fetal), and containing 1.0 g/l D-glucose and 7.5 g/l dextran (average mol. wt. 17,900; Sigma Chemical Company).

Fetal arterial pressure was monitored continuously using a pressure transducer (Druck PDCR 10/DPI 201) sited close to the fetal arterial cannulation point. Fetal flow rate was measured continuously using a liquid flowmeter (Phase Separations Ltd.) on the fetal venous outlow (sensitivity of $\pm$ 0.02 ml/min). Pressure and flow signals were acquired via an analogue to digital converter (PCI 1001; CIL Electronics) and collected by an Apple Europlus microcomputer.

In experiments where flow rates were not varied, the mean fetal flow rate was 6.9 ± 0.2 ml/min ($\pm$ s.d.) and the mean maternal flow rate was 20.7 ± 1.2 ml/min ($\pm$ s.d.) (standard flow rates). In experiments where flow rates varied, the maximum fetal flow rate was 12.2 ml/min and the maximum maternal flow rates was 22.0 ml/min. Tissue weight of the perfused lobule was the weight of that volume of tissue which appeared blanched after dual perfusion at the maximum maternal flow rate. Using the mean perfused tissue weight for these experiments (21.1 ± 4.4 g; $\pm$ s.d.), the maximum flow rates were calculated as equivalent to approximately 300 ml/min for the fetal flow and 550 ml/min for the maternal flow in a 500 g (term) placenta. These rates correspond to the physiologic umbilical and uteroplacental flow rates found in vivo (Jouppila and Kirkinen, 1984; Metcalfe et al., 1955).

The umbilical arterial pressure range, in experiments where fetal flow was not altered, was 29 ± 4 to 34 ± 5 mm Hg (mean $\pm$ s.d.). In experiments where fetal flow rate was altered, the maximum fetal arterial pressures were 65 and 72 mm Hg. Perfusion time was approximately 150 min from the start of fetal perfusion. The pH of perfusate samples from maternal and fetal arterial inflows was checked periodically and found to be in the range 7.34 - 7.42. Perfusions in which fetal-to-maternal leakage of perfusate consistently exceeded 0.15 ml/min (2.0 - 2.5% of total fetal flow) were not used.

Two min samples were taken from maternal and fetal outflows for measurement of radioactivity. Samples containing [3H] were mixed with scintillant (0.5 ml sample plus 4.5 ml NE-262; Nuclear Enterprises Ltd), and counted on an LKB-Wallac 1216 RACKBETA II Liquid Scintillation Counter. Samples containing [125I] or [51Cr] were counted on an LKB-Wallac 80000 Gamma Sample Counter. [51Cr] EDTA and [3H] 3-O-methylglucose ([3H] MG; 2-5 Ci/mmol) were obtained from Amersham International plc. [125I] antipyrine ([125] AP) was synthesized by the Division of Radiochemistry, CRC.

Clearance values for each tracer were calculated as the product of flow rate and venous tracer concentration in the acceptor circulation divided by the arterial tracer concentration in the donor circulation. The clearance index (CI) was defined as the ratio of [3H] MG clearance to [125] AP clearance. The transfer index (TI) was defined as the ratio of [3H] MG clearance to that of [51Cr] EDTA.

Experimental Procedures

(a)   Permeability

The maternal to fetal permeability of [3H] MG was measured by determining its clearance index during steady state infusion of [3H] MG and [125I] AP into the maternal circulation. A constant concentration of the tracers was infused for 20 min prior to sampling the fetal outflow.

(b)   Modulation of Transfer by Fetal Flow

A constant concentration of [3H] MG was infused into the maternal arterial circulation and maternal to fetal clearance rates were determined at a series of fetal flow rates. Fetal flow was increased or decreased in a series of 20 min steps, the clearance being calculated from samples taken during the latter 10 min of a flow step, once fetal arterial pressure had stabilized. The fetal flow rate was increased to its maximum value prior to initiating the clearance experiments. This was to ensure that if an increasing set of fetal flow rate steps were used, later increases in flow rate would not open up new areas in the fetal capillary bed. Fetal arterial pressure was a linear function of fetal flow rate over the flow rate range used, indicating that there was no change in the perfused fetal bed volume during the experiment.

(c)   Modulation of Transfer by Maternal Flow

During these experiments constant concentrations of [3H] MG and [51Cr] EDTA were infused via the maternal circulation. Maternal to fetal clearance rates for the tracers were calculated at maternal flow rates consisting of four ascending or descending steps, each lasting 30 min. Samples for radiolabel assay were taken from the last 15 min of each flow rate step.

RESULTS

(a)   [3H] MG Permeability

The [3H] MG transfer rate was compared to the maximum, flow-limited ([125I] AP) transfer rate by measuring its clearance index at standard flow rates. The CI for [3H] MG was also compared to the CI determined previously for a series of substances which cross the placenta by simple diffusion (Illsley et al., 1985). The CI for [3H] MG determined from 5 perfusions (0.91 $\pm$ 0.16; mean $\pm$ s.d.) is plotted in Figure 1, along with the previously determined CI for urea, creatinine, mannitol, CrEDTA and raffinose, against molecular weight. It is apparent that the CI for [3H] MG is well in excess of that predicted solely on the basis of molecular weight (0.35), and is not significantly different from the value of 1.00 (p $<$ 0.01; Student's t-test), representing the CI for flow-limited transfer.

(b)   Modulation of [3H] MG Transfer by Fetal Flow Rate

The rapid rate of [3H] MG maternofetal transfer apparent from the previous result raises the question of whether its transfer, like that of antipyrine, is modulated by flow rate. This was investigated initially by measuring [3H] MG clearance as fetal flow rate was varied. Figure 2 illustrates the results from two such perfusions in which [3H] MG clearance is plotted against fetal flow rate. Two samples were measured at each flow rate step. Both curves show a decrease in [3H] MG clearance below a fetal flow rate of 4-6 ml/min.

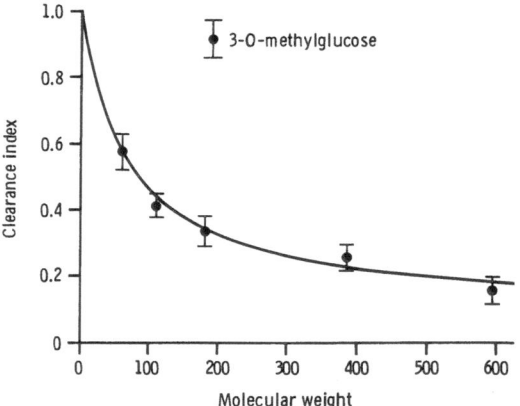

Figure 1. Plot of clearance index against molecular weight showing the clearance index for [3H] MG for five perfusions (mean $\pm$ s.e.m.) compared to the clearance indices for (left to right; mean $\pm$ s.e.m.; n = 6) urea, creatinine, mannitol, CrEDTA and raffinose. Data for points other than [3H] MG taken from Illsley et al. (1985).

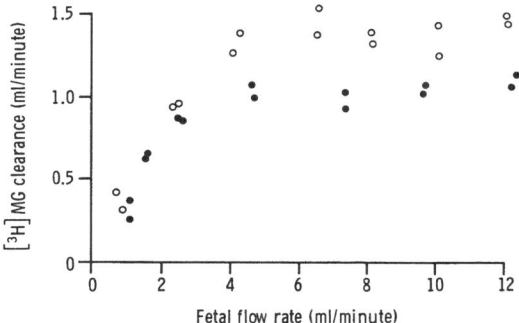

Figure 2. Plot of [3H] MG clearance (ml/min) against fetal flow rate (ml/min) for two perfusions.

(c)   Modulation of [3H] MG Transfer by Maternal Flow Rate

Changes in fetal flow do not cause alterations in the placental exchange area since the fetal circulation flows through a discrete vascular bed. Alterations in the maternal flow rate however, will produce some variation in the exchange area due to the non-vascular, pool-flow nature of the intervillous circulation. Corrections for alterations in the exchange area were made by normalizing [3H] MG clearance to the simultaneously measured clearance of [51Cr] EDTA. Unlike antipyrine, the transfer of [51Cr] EDTA is substantially diffusion-limited (Illsley et al., 1985), making it a better marker of exchange area.

The transfer of [3H] MG and [51Cr] EDTA was measured in 5 perfusions, where the flow rate was varied between 4.0 and 22.0 ml/min. The results of these experiments are shown in Figure 3. The TI is expressed as a percentage of its value at the maximum maternal flow rate. The transfer index (percent TI) for [3H] MG is plotted against the percentage of the maximum flow rate (22.0 ml/min). The results demonstrate a decrease in [3H] MG clearance as maternal flow rate is reduced.

Umbilical arterial pressure and fetal flow rate were measured throughout these experiments. There were no changes in either of these variables as a result of alterations in the maternal flow rate. The results shown in Figure 3 are not therefore due to the effects of changing intervillous flow on the fetal capillary bed.

DISCUSSION

The transplacental passage of glucose or glucose analogues has been documented in a number of species (Widdas, 1962; Stacey et al., 1977; Yudilevich et al., 1979; Bissonnette, 1981) including man (Rice et al., 1976; Carstensen et al., 1977; Johnson and Smith, 1985). This evidence shows that hexoses, including 3-O-methylglucose, are transported across the placenta by a facilitated diffusion

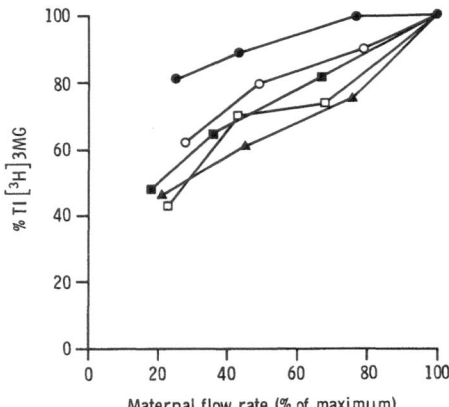

Figure 3. Transfer index for [3H] MG ([3H MG clearance/[51Cr] EDTA clearance) plotted against maternal flow rate for 5 perfusions. Transfer index is plotted as a percentage of its value at a maximum maternal flow rate. Maternal flow rate is plotted as a percentage of its maximum value (22.0 ml/min). Maximum TI values were - 2.61 (closed circles), 3.76 (open circles), 4.73 (closed squares), 5.50 (open squares), and 3.39 (triangles).

mechanism. The clearance index for [3H] MG calculated here shows that the rate of [3H] MG transfer is rapid, and close to that for [125I] AP. Since the transfer of antipyrine is known to be flow-limited, this observation raises the possibility that [3H] MG transfer may also be modulated by flow rate. Investigation of this possibility, by following changes in [3H] MG clearance subsequent to alterations in fetal flow rate showed that below 4 ml/min, [3H] MG clearance dropped markedly.

The more interesting question perhaps is whether [3H] MG transfer is modulated by changes in maternal blood flow (MBF). The problem encountered in trying to quantify this is that changes in the maternal flow rate also cause changes in the placental exchange area. An exchange area marker was therefore required to which [3H] MG data could be normalized so as to correct for this problem. Previously, antipyrine or oxygen have been used as indicators of exchange area (Schneider et al., 1985; Wier et al., 1985). Antipyrine suffers from the problem that because its transfer is flow-limited, variations in flow rate, flow pattern, and flow homogeneity may be reflected in the exchanged area measurement. We chose therefore to use [51Cr] EDTA as an exchange marker since it is nonmetabolisable and its transfer is substantially diffusion-limited (Illsley et al., 1985), eliminating the problems likely to be encountered using oxygen or antipyrine. When [3H] MG clearance is normalized to [51Cr] EDTA clearance, it is apparent that [3H] MG transfer is regulated, at least in part, by MBF. A 50% reduction in MBF reduces [3H] MG transfer by between 9 and 36%.

Saintonge and Rosso (1981) have reported that in the guinea pig, spontaneous fetal growth retardation was associated with reduced placental weight, reduced MBF and a reduced transfer of [3H] MG to the fetus. They later concluded that fetal growth was primarily controlled by placental size (Saintonge and Rosso, 1983). The data described here suggests first, that changes in MBF modulate placental glucose transfer, separate from the question of placental size. Secondly, a reduction in MBF is directly responsible for the decrease in [3H] MG transfer observed here, as opposed to a mechanism which operates via modification of maternal or fetoplacental metabolic or hormonal responses.

Reductions in umbilical blood flow are found as a result of MBF reductions in sheep (Clapp et al., 1980; Block et al., 1984) and also as a consequence of chronic fetal hypoxia in man (Jouppila and Kirkinen, 1984). Our data indicate that glucose transfer is sensitive to changes in umbilical flow rate. As noted above, [3H] MG transfer decreased markedly below a fetal flow rate of 4 ml/min. As normal umbilical flow rates are, however, equivalent to approximately 12 ml/min in the perfusion, it might be expected that a drop in the umbilical flow rate of some two thirds would be required before such effects would be seen. Oh et al. (1975) have observed that a drop of a third in the umbilical blood flow of the rat does not cause a significant change in fetal glucose uptake.

The data presented here suggest that reductions in MBF can contribute to fetal growth retardation through the mechanism of reduced placental transfer of nutrients such as glucose. It seems possible that a further contribution might stem from reductions in umbilical blood flow (resulting from decreased MBF) which could lead to further reductions in transfer.

SUMMARY

Intervillous flow rates have long been suspected of affecting the materno-fetal transfer of nutrients. In this study, the effect of intervillous and umbilical flow rates on the transfer of 3-O-methylglucose was investigated, using the in vitro dually-perfused human placenta.

Below a fetal (umbilical) flow rate of 4-6 ml/min, the maternofetal clearance of 3-O-methylglucose was markedly decreased by the reductions in fetal flow rate. The maternofetal clearance of 3-O-methylglucose was also decreased by reductions in maternal (intervillous) flow rate. The clearance of 3-O-methylglucose was normalized to that of CrEDTA to correct for alterations in exchange area as the maternal flow rate was changed.

The decrease of 3-O-methylglucose with reductions in the intervillous flow rate suggests a mechanism whereby the reduced uteroplacental blood flow might contribute to fetal growth retardation.

REFERENCES

Bissonnette, J.M. (1981) Studies in vivo of glucose transfer across the guinea-pig placenta. *Placenta* (Suppl. 2), 155-162.

Block, B.S., Llanos, A.J., and Creasey, R.K. (1984) Responses of the growth retarded fetus to acute hypoxemia. *Am. J. Obstet. Gynecol.* 148, 878-885.

Brown, J.C.M. and Veall, N. (1953) The maternal placental blood flow in normotensive and hypertensive women. *J. Obstet. Gynaecol. Br. Cmwlth.* 60, 141-147.

Bruce, N.W. and Abdul-Karim, R. (1975) Relationships between fetal weight, placental weight and maternal circulation in the rabbit at different stages of gestation. *J. Reprod. Fertil.* 32, 15-24.

Cartensen, M., Leichtweiss, H-P., Molson, G., and Schröder, H. (1977) Evidence for a specific transport of D-hexoses across the human term placenta in vitro. *Arch. Gynakol.* 222, 187-196.

Clapp, J.F., Szeto, H.H., Larrow, R., Hewitt, J., and Mann, L.I. (1980) Umbilical blood flow response to embolization of the uterine circulation. *Am. J. Obstet. Gynecol.* 138, 60-67.

Garris, D. (1983) Regional variations in guinea pig uterine blood flow during pregnancy: relationship to intrauterine growth of the fetal-placental unit. *Teratol.* 27, 101-107.

Garris, D. (1984) Intrauterine growth of the guinea pig fetal-placental unit throughout pregnancy: regulation by uteroplacental blood flow. *Teratol.* 29, 93- 99.

Illsley, N.P., Aarnoudse, J.G., Penfold, P., Bardsley, S.E., Coade, S.B., Stacey, T.E., and Hytten, F.E. (1984) Mechanical and metabolic viability of a placental perfusion system in vitro under oxygenated and anoxic conditions. *Placenta* 5, 213-225.

Illsley, N.P., Hall, S., Penfold, P., and Stacey, T.E. (1985) Diffusional permeability of the human placenta. *Contrib. Gynecol. Obstet.* 13, 92-97.

Johnson, L.W. and Smith, C.H. (1985) Glucose transport across the basal plasma membrane of human syncytiotrophoblast. *Biochim. Biophys. Acta* 815, 44-50.

Jouppila, P. and Kirkinen, P. (1984) Umbilical vein blood flow as an indicator of fetal hypoxia. *Br. J. Obstet. Gynaecol.* 91, 107-110.

Kollee, L.A.A., Monnens, L.A.H., Trijbels, J.M.F., Veerkamp, J.H., and Jansen, A.J.M. (1979) Experimental intrauterine growth retardation in the rat. Evaluation of the Wigglesworth model. *Early Humn. Develop.* 3, 295-300.

Lunell, N-O., Sarby, B., Lewander, R., and Nylund, L. (1979) Comparison of uteroplacental blood flow in normal and intrauterine growth retarded pregnancy. Measurements with Indium 113m and a computer linked gamma camera. *Gynecol. Obstet. Invest.* 10, 106-118.

Metcalfe, J., Romney, S.L., Ramsey, L.H., Reid, D.E., and Burwell, C.S. (1955) Estimation of uterine blood flow in normal human pregnancies at term. *J. Clin. Invest.* 34, 1632-1640.

Myers, S.A., Sparks, J.W., Makowski, E.L., Meschia, G., and Battaglia, F.C. (1982) Relationship between placental blood flow and placental and fetal size in guinea pig. *Am. J. Physiol.* 243, H404-H409.

Nylund, L., Lunnell, N-O., Lewander, R., and Sarby, B. (1983) Uteroplacental blood flow index in uterine growth retardation of fetal or maternal origin. *Br. J. Obstet. Gynaecol.* 90, 16-20.

Oh, W., Omori, K., Hobel, C.J., Erenberg, A., and Emmanoulidies, G.C. (1975) Umbilical blood flow and glucose uptake in lamb fetus following single umbilical artery ligation. *Biol. Neonate* 26, 291-299.

Rice, P.A., Rourke, J.E., and Nesbitt, R.E.L. (1976) In vitro perfusion studies of the human placenta. IV. Some characteristics of the glucose transport system in the human placenta. *Gynecol. Obstet. Invest.* 7, 213-221.

Saintonge, J. and Rosso, P. (1981) Placental blood flow and transfer of nutrient analogs in large, average and small guinea pig littermates. *Ped. Res.* 15, 152-156.

Saintonge, J. and Rosso, P. (1973) Placental blood flow and transfer of nutrient analogues during normal gestation in the guinea pig. *Placenta* 4, 31-40.

Schneider, H., Progler, M., and Sodha, R.J. (1985) Effect of flow rate ratio on the diffusion of antipyrine and [^{3}H] H_2O in the isolated dually in vitro perfused lobe of the human placenta. *Contrib. Gynecol. Obstet.* 13, 114-123.

Seeds, J.W. (1984) Impaired fetal growth: definition and clinical diagnosis. *Obstet. Gynecol.* 64, 303-310.

Simmons, M.A., Battaglia, F.C., and Meschia, G. (1979) Placental transfer of glucose. *J. Dev. Physiol.* 1, 227-243.

Stacey, T.E., Boyd, R.D.H., Ward, R.H.T., and Weedon, A.P. (1977) Placental permeability in the sheep. *Annal. de Recherches Vet.* 8, 345-352.

Widdas, W.F. (1952) Inability of diffusion to account for placental glucose transfer in the sheep and consideration of the kinetics of a possible carrier transfer. *J. Physiol.* 118, 23-39.

Wier, P.J. and Miller, R.K. (1985) Oxygen transfer as an indicator of perfusion viability in the isolated human placental lobule. *Contr. Gynecol. Obstet.* 13, 127-131.

Wigglesworth, J.C. (1964) Experimental growth retardation in the fetal rat. *J. Pathol. Bacteriol.* 88, 1-13.

Wootton, R., McFadyen, I.R., and Cooper, J.E. (1977) Measurement of placental blood flow in the pig and its relation to placental and fetal weight. *Biol. Neonate* 31, 333-339.

Yudilevich, D.L., Eaton, B.M., Short, A.H., and Leichtweiss, H-P (1979) Glucose carriers at maternal and fetal sides of the trophoblast in guinea pig placenta. *Am. J. Physiol.* 236, C205-C212.

Trophoblast Research 2:545-555, 1987

LONG TERM HUMAN PLACENTAL LOBULE PERFUSION - AN ULTRASTRUCTURAL STUDY

P. Anthony di Sant'Agnese[1,4], Karen L. de Mesy Jensen[1],
Patrick J. Wier[2,3], Debabrata Maulik[2], and Richard K. Miller[2,3]

Departments of Pathology and Laboratory Medicine[1],
Obstetrics and Gynecology[2], and Radiation Biology and Biophysics[3]
University of Rochester Medical Center
Rochester, New York 14642 USA

INTRODUCTION

The importance of the ultrastructural evaluation of perfused placental tissue was first emphasized by Panigel (1965, 1971). Ultrastructural study of perfused tissue is now generally accepted as mandatory. Several recent publications have addressed the issue of placental ultrastructural changes after in vitro dual perfusion (Contractor et al., 1984; Illsley et al., 1985; Kaufmann, 1985). These studies involved placental tissue evaluated after only 2-4 hr of perfusion. In this paper we present high resolution light microscopic and electron microscopic findings of placental lobules dually perfused for up to 12 hr. There was excellent overall preservation of morphology. Biochemical and physiological evaluations were also performed demonstrating viability throughout the perfusion period.

MATERIALS AND METHODS

The details of the placental perfusion technique have been previously published (Miller et al., 1985). In brief, dual perfusion of isolated term human placental lobules in a closed recirculating system was performed using a modified tissue culture medium 199 (Difco, Detroit) with heparin, glucose, dextran 40, and gentamicin. The maternal perfusate was gassed with 95% O_2/5% CO_2 and the fetal perfusate was gassed with 95% N_2/5% CO_2. The perfusates were exchanged for fresh ones every 4 hr.

Parameters which were monitored during the perfusions have been detailed elsewhere (Wier et al., 1983; Miller et al., 1985; Wier and Miller, 1985) and included fetal capillary pressure and permeability, fetal volume stability, oxygen transfer and consumption, and the metabolism of carbohydrates and protein hormones.

Multiple tissue samples from 11 lobule perfusions (7 perfusions of 12 hr duration; 4 perfusions of 8 hr duration) were embedded in large plastic blocks and studied by high resolution light microscopy. Multiple tissue samples of preperfusion controls from adjacent lobules were also evaluated in this manner. One sample from

[4]To whom correspondence should be addressed.
Received: 10 October 1985; Accepted: 15 May 1986

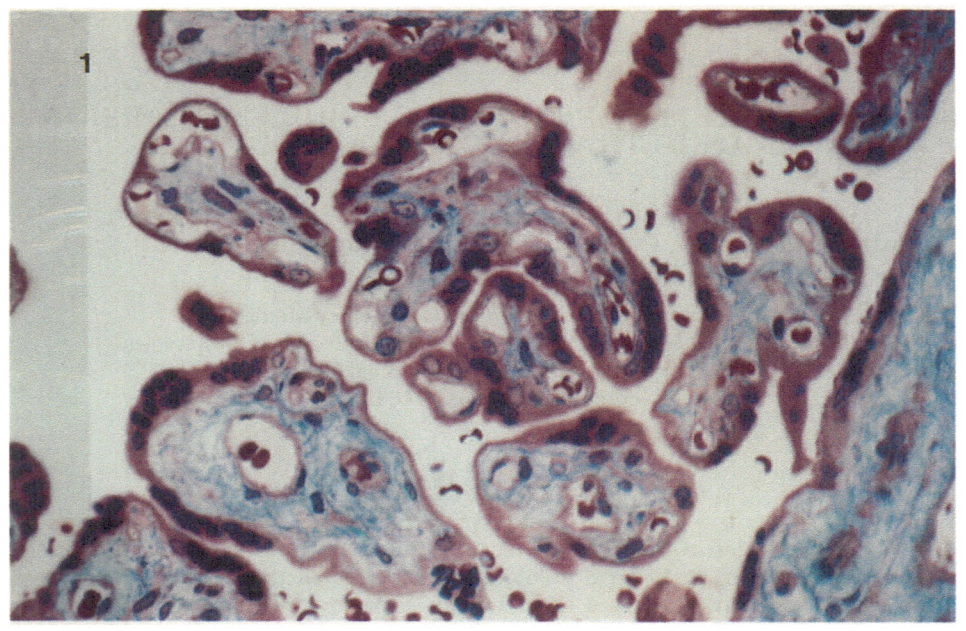

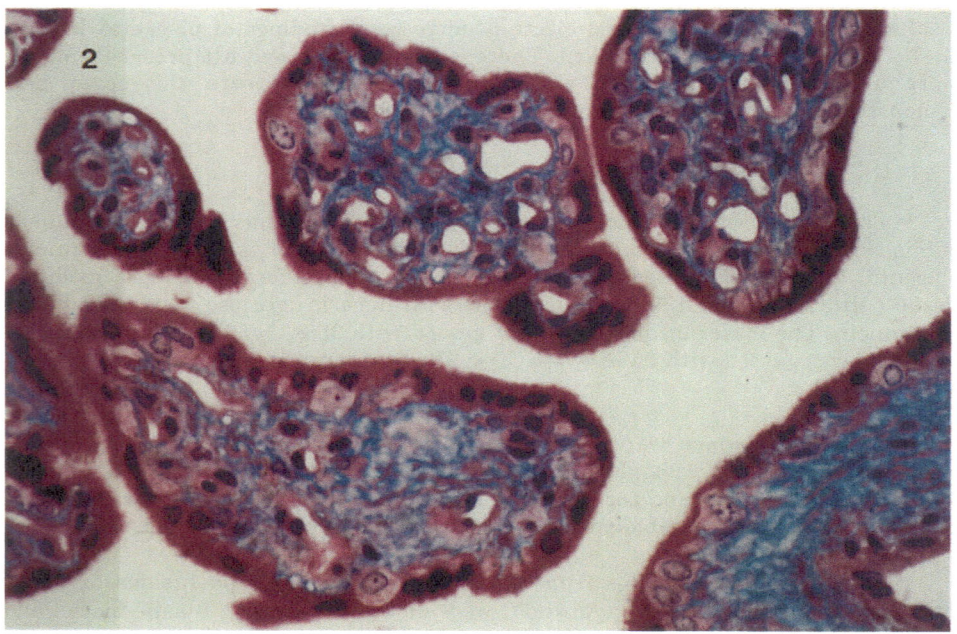

perfused lobules was taken from the area between the two maternal perfusion cannulae and a similar area was taken from control lobules. A total of 10 perfusion specimens were further studied by electron microscopy along with selected nonperfused controls. The details of the specimen processing and special techniques used have been previously published (di Sant'Agnese and de Mesy Jensen, 1984a, b) and are presented in condensed form. The tissues samples measured 1.5 x 0.5 x 0.3 cm and were fixed at least 24 hr in a 4°C 4% paraformaldehyde/1% glutaraldehyde phosphate buffered solution at pH 7.3. The tissues were postfixed in 1% osmium tetroxide, dehydrated and embedded in Spurr epoxy resin in large JB-4 molding trays (Polysciences). Large 2 1/2 μ thick sections were stained with a new versatile H & E-like dibasic stain (di Sant'Agnese and de Mesy Jensen, 1984a) and evaluated to select an area for ultrastructural study. Tissue from this area was directly removed from the slide by an inverted BEEM capsule re-embedding "pop-off" technique (di Sant'Agnese and de Mesy Jensen, 1984b) and thin sectioned. The thin sections were stained with alcoholic uranyl acetate and lead citrate and examined with an Hitachi HS-8 electron microscope.

RESULTS

Light Microscopy

(Figures 1-3) Eight of eleven perfusion specimens studied by high resolution light microscopy of plastic embedded tissue appeared indistinguishable from pre-perfusion controls except for the obvious absence of red blood cells. There was a generalized hypertrophy and hyperplasia of cytotrophoblast in all perfused specimens. Occasional mitotic figures were noted in the cytotrophoblast of the perfused specimens and the ratio of cytotrophoblast to intermediate type cells was increased. Three specimens exhibited moderate focal villus edema which was sometimes associated with coarse vacuolization of Hofbauer cells and confluent subsynctiotrophoblastic vesiculation as well as syncytiotrophoblastic vacuolization (particularly on the microvillus side of the cell). A similar degree of villus edema was noted in the preperfusion controls of one of these three specimens but without the other associated changes in the Hofbauer cells and syncytiotrophoblast. Focal trophoblastic necrosis was noted in two specimens and in one of these, areas of necrosis were seen to partially overlap villus edema. In areas of necrosis, the syncytiotrophoblast appeared darker staining and contracted, with clubbing of microvilli. Nuclear pyknosis was also seen. In places the necrotic syncytium was detached from the villus.

Figure 1. High resolution plastic section of preperfused control placenta stained with a versatile dibasic stain (Spurr plastic, basic fuchsin-methylene blue X300).

Figure 2. Twelve hr perfused placental lobule with no significant morphologic alterations except for increased numbers of cytotrophoblast. Note absence of red blood cells (Spurr plastic, basic fuchsin-methylene blue X300).

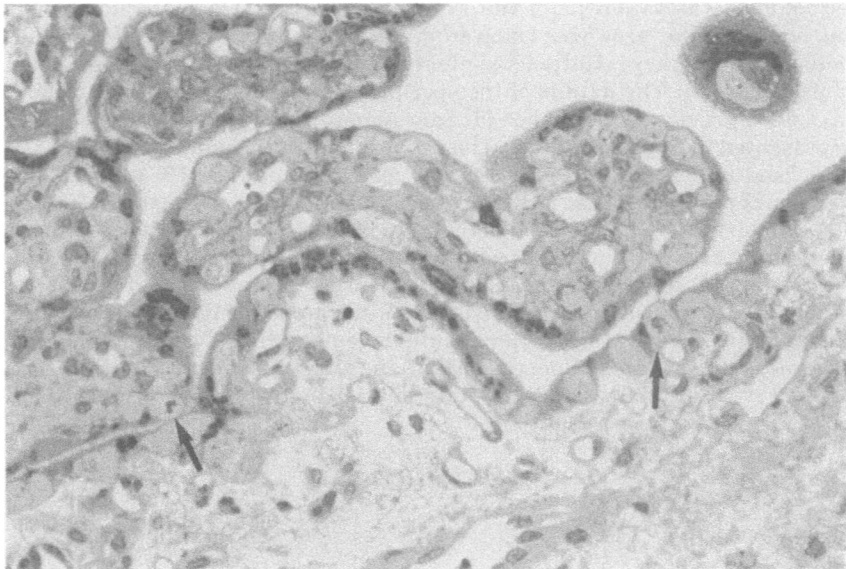

Figure 3. Eight hr perfused placental lobule lightly stained for increased cytologic detail demonstrating hyperplasia and hypertrophy of cytotrophoblast. Two mitoses are seen in cytotrophoblast cells (arrows). Villus edema is focally present and associated with vacuolated Hofbauer cells. Note vacuoles in outer third of syncytiotrophoblast (Spurr plastic, basic fuchsin-methylene blue X300).

Electron Microscopy

Most specimens studied showed excellent preservation of the cellular ultrastructure of all cell types present in the placental villus. The findings were comparable to preperfusion controls except for the presence of widely scattered highly electron dense lipid droplets in the syncytiotrophoblast of the perfused specimens. Fine particulate electron dense material approximately 5-30 nm in size and presumed to be dextran particles were seen in some fetal capillaries and occasionally were attached to the syncytiotrophoblast microvilli (Figures 4-9).

In specimens where edema was present, the villus stromal collagen fibers were widely displaced by electron lucent "empty" zones which presumably had contained fluid. Long dendritic processes of stromal mesenchymal cells (reticular cells) were associated with these zones. Hofbauer cells often contained large electron lucent vacuoles, some of which contained presumed dextran in either a dispersed or concentrated state. Large confluent subsyncytiotrophoblastic spaces were noted in some areas as were smaller rounded vacuoles usually present in the upper third of the syncytiotrophoblast. Some of these altered vacuoles appear to contain dextran. More rarely, presumptive transsyncytiotrophoblastic channels were noted.

In areas where focal necrosis was present, the necrosis was confined to the trophoblast with syncytiotrophoblast usually showing more advanced necrosis than

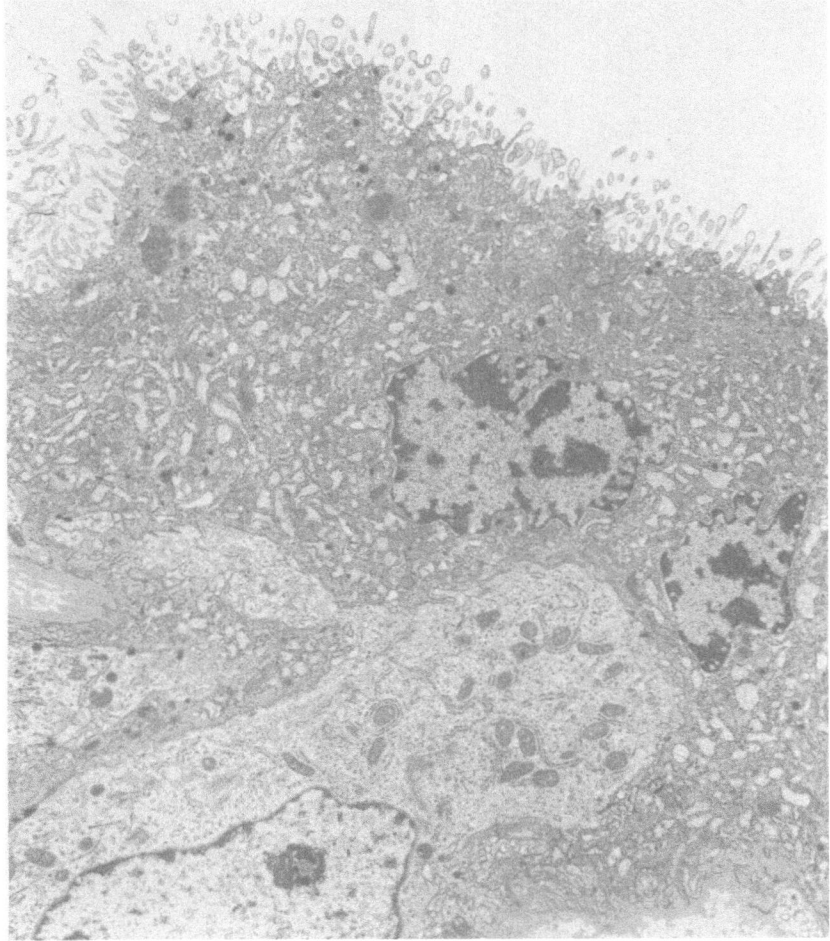

Figure 4. Preperfused control placenta (X5800).

the cytotrophoblast. Necrosis of the syncytiotrophoblast was characterized by diffuse cytoplasmic microvesiculation. Microvilli were lost and club-like cytoplasmic projections were seen in their place. Nuclear chromatin was condensed into large irregular coarse clumps. Necrosis of the cytotrophoblast was accompanied by similar changes except of course the alterations of the cell surface.

DISCUSSION

The results of the study indicate that excellent ultrastructural preservation of human placental tissues can be obtained after long term (up to 12 hr) single lobule perfusion. In the long term perfusions, 73%, showed no evidence of morphologic changes over the perfusion period. Biochemical and physiologic parameters measured over this period also indicated good viability of the tissue (Wier et al., 1983; Miller

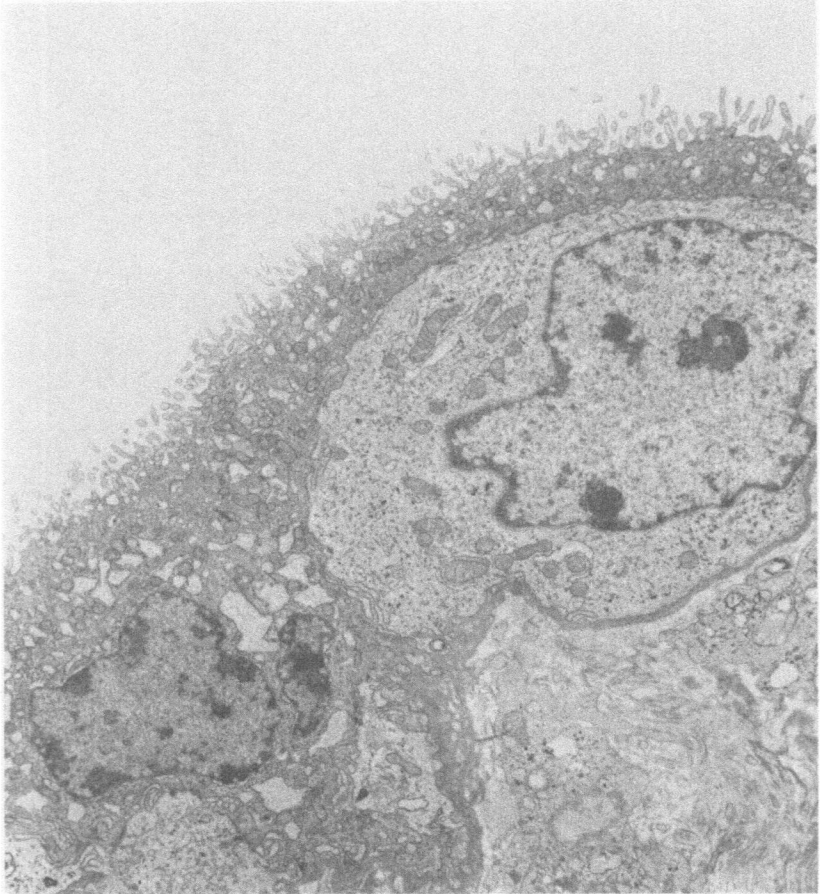

Figure 5. Twelve hr perfused placenta lobule which is very similar in appearance to preperfused control specimen (X6700).

et al., 1985; Wier and Miller, 1985; Wier et al., 1985). Fetal arterial pressure was stable with no significant fetal volume loss. Neither O_2 nor glucose consumption by the tissue were significantly reduced. The appearance rate of human chorionic gonadotrophin (HCG) in the maternal perfusate indicated HCG synthesis throughout the perfusion period. Similar in vitro perfusion systems recently described Contractor et al. (1984); Illsley et al. (1985); and Kaufmann (1985) have achieved ultrastructural preservation for periods of 2-4 hr. The major detectable morphologic difference from preperfusion controls was a hyperplasia and hypertrophy of cytotrophoblast cells, some of which are seen in mitosis in the perfused specimens. These changes have been ascribed to nonspecific injury to the syncytiotrophoblast and a repair phenomenon (Fox, 1970). Despite the presence of cytotrophoblast hyperplasia and hypertrophy, in most areas the syncytiotrophoblast was ultrastructurally similar to

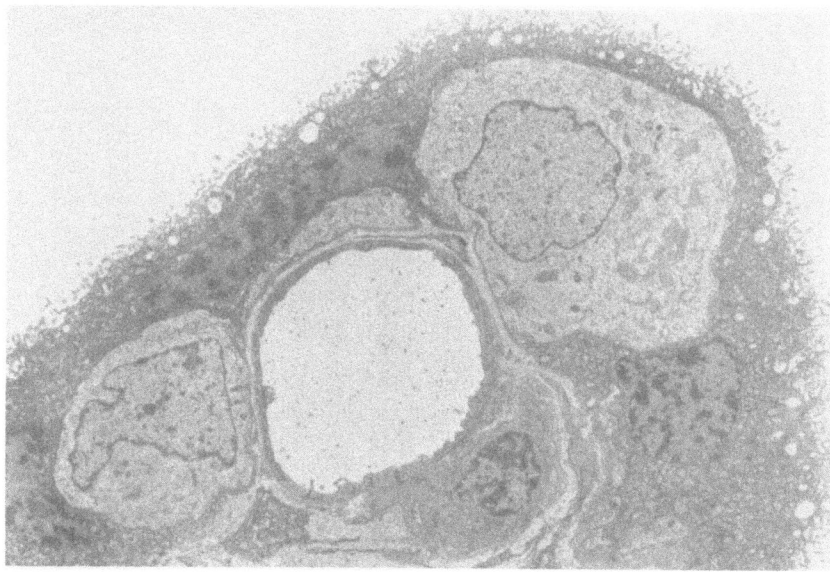

Figure 6. Eight hr perfused placental lobule with electron-dense particular material (probably dextran) in fetal blood vessels. Note excellent preservation of cells (X3700).

syncytiotrophoblast in preperfused placentae. In two cases focal trophoblastic necrosis was seen. Ultrastructurally the necrotic trophoblast resembles that which has been described in a variety of ischemic or anoxic experimental conditions including in vivo (Sheppard and Bonnar, 1980; Panigel, 1974), organ culture (MacLennan et al., 1972), and placental perfusion (Panigel, 1971). The necrosis of the underlying cytotrophoblast as has been reported by Panigel (1974). The syncytiotrophoblast appears to be relatively more vulnerable to a variety of insults than does the cytotrophoblast. The necrosis may reflect zone of inadequate perfusion. While no necrosis was noted in control tissue, there exists the possibility that the necrosis was a result of placental injury during delivery which became manifest after long term perfusion.

The specimens from three perfusions showed moderate focal villus edema which was variably associated with coarse vacuolization of Hofbauer cells, marked confluent subsyncytiotrophoblast vesiculation, vacuoles and rare dilated transcyto-plasmic channels in the syncytiotrophoblast. This constellation of findings could represent imbalances in fetal-maternal hydrostatic, osmotic, and oncotic forces which are of importance in placental fluid dynamics (Schroeder et al., 1982). It has been postulated that Hofbauer cells regulate fluid and protein balance in the placental villi (Demir and Erbengi, 1984; Enders and King, 1970). When this system becomes overloaded transsyncytiotrophoblastic channels for fluid and protein transport become dilated (Kaufmann et al., 1982; Kaufmann, 1985). The subsyncytio-trophoblastic vesiculation may be part of this channel system. Kaufmann (1985) apparently noted edema in most of the perfused placentae he studied. Villus edema can also occur in assocation with ischemia (Kaufmann, 1985). These ischemic

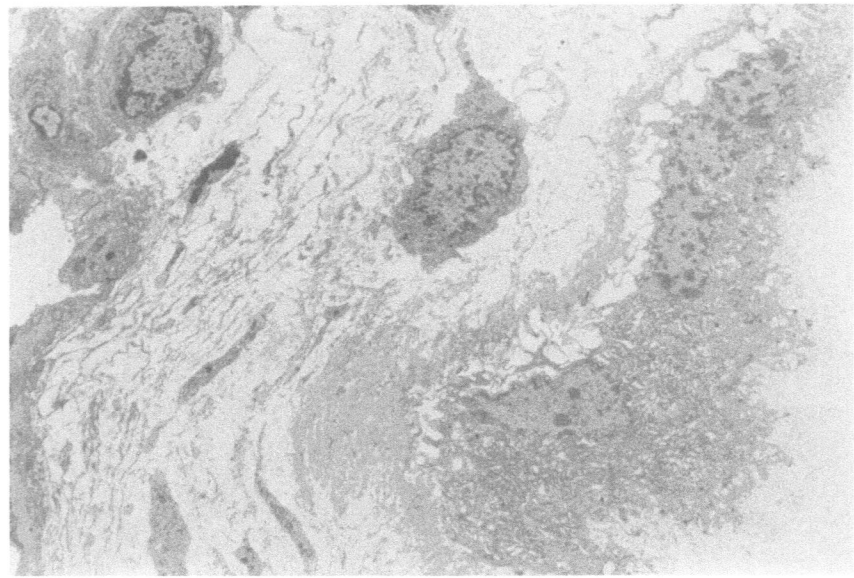

Figure 7. Twelve hr perfused placental lobule with villus stromal edema with subsyncytiotrophoblast vesiculation. Note preservation of cytologic ultrastructure (X2000).

related fluid shifts may have occurred in the few areas of our perfusion specimens where trophoblast necrosis was associated with edema. As was evident in one case, edema may also be related to pre-existing pathology since edema was also seen in the preperfusion specimen.

SUMMARY

Eleven human placental lobules from term placentae were dually perfused for 8 hr (4 specimens) and 12 hr (7 specimens) with a modified tissue culture media 199. Eight of these eleven specimens (73%) that were studied by high resoltion light microscopy of large plastic sections were indistinguishable from preperfusion controls except for hyperplasia and hypertrophy of the cytotrophoblast. This finding may reflect some mild injury to the syncytiotrophoblast. Three specimens exhibited focal villus edema with associated vacuolization of Hofbauer cells, subsyncytio-trophoblastic vesiculation and vacuoles in the outer third of the syncytiotrophoblast. This edema and associated changes could represent imbalances in fetal-maternal hydrostatic, osmotic, and oncotic forces. Two specimens exhibited focal trophoblast necrosis which partially overlapped with some of the areas of edema. The necrosis may indicate zones of inadequate perfusion or placental injury during delivery. Ultrastructural study confirmed the excellent cytologic preservation suggested by light microscopy. This work along with the biochemical and physiologic parameters measured indicate that viable placental tissue can be maintained by long term perfusion for at least 12 hr.

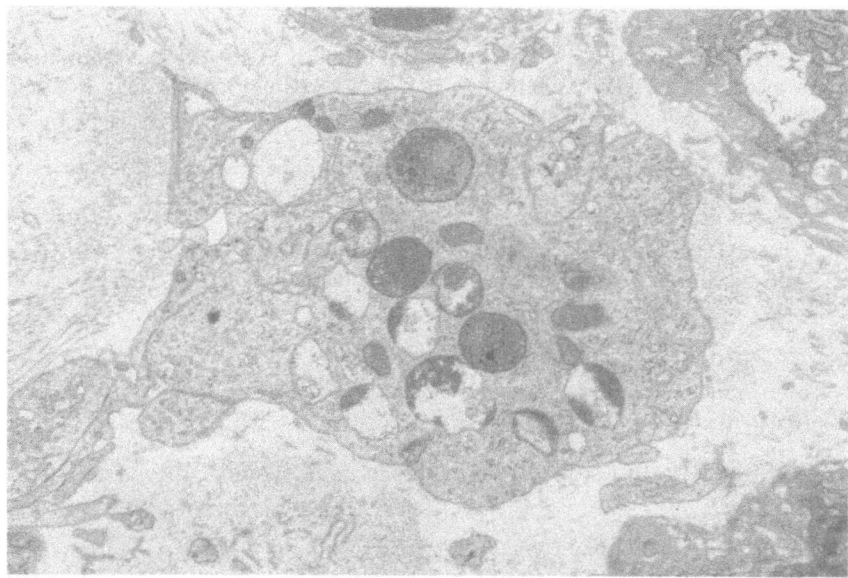

Figure 8. Perfused placental lobule with Hofbauer cell containing vacuoles some of which are filled with finely particulate electron-dense material (probably dextran) (X9600).

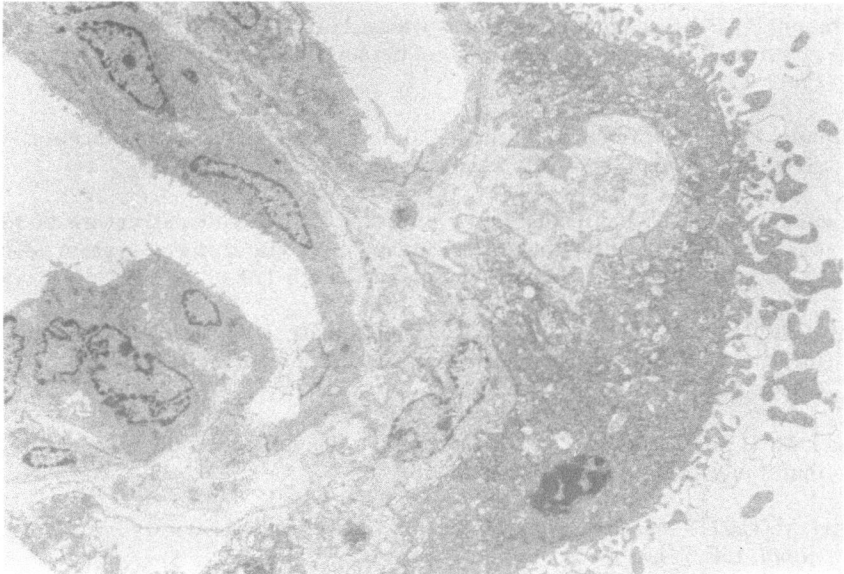

Figure 9. Twelve hr perfused placental lobule with necrosis of syncytiotrophoblast and less advanced changes (early necrosis) of cytotrophoblast. Note preservation of stromal cytologic elements (X2000).

REFERENCES

Contractor, S.F., Easton, B.M., Firth, J.A., and Bauman, K.F. (1984) A comparison of the effects of different perfusion regimes on the structure of the isolated placental lobule. *Cell Tissue Res.* 237, 609-617.

Demir, R. and Erbengi, T. (1984) Some new findings about Hofbauer cells in the chorionic villi of the human placenta. *Acta Anat.* 119, 18-26.

di Sant'Agnese, P.A. and de Mesy Jensen, K.L. (1984a) Dibasic staining of large epoxy tissue sections and applications to surgical pathology. *Am. J. Clin. Pathol.* 81, 25-29.

di Sant'Agnese, P.A. and de Mesy Jensen, K.L. (1984b) Diagnostic electron microscopy on reembedded ("popped-off") areas of large Spurr epoxy sections. *Ultrastruc. Pathol.* 6, 247-253.

Enders, A.C. and King, B.F. (1970) The cytology of Hofbauer cells. *Anat. Rec.* 167, 231-252.

Fox, H. (1978) *Pathology of the Placenta.* (Major Problems in Pathology Series, Vol. VII), (ed.), J.L. Bennington, Philadelphia, W.B. Saunders, Ltd., pp. 158-161.

Illsley, N.P., Fox, H., Van der Veen, L., Chawner, L., and Penfold, P. (1985) Human placental ultrastructure after in vitro dual perfusion. *Placenta* 6, 23-32.

Kaufmann, P., Schröeder, H., and Leichtweiss, H.-P. (1982) Fluid shift across the placenta. II. Fetomaternal transfer of horseradish peroxidase in the guinea pig. *Placenta* 3, 339-348.

Kaufmann, P. (1985) Influence of ischemia and artificial perfusion on placental ultrastructure and morphometry. *Contr. Gynecol. Obstet.* 13, 18-26.

MacLennan, A.H., Sharp, F., and Shaw-Dunn, J. (1972) The ultrastructure of human trophoblast in spontaneous and induced hypoxia using a system of organ culture. *J. Obstet. Gynaecol. Brit. Comm.* 79, 113-121.

Miller, R.K., Wier, P.J., Maulik, D., and di Sant'Agnese, P.A. (1985) Human placenta in vitro: Characterization during 12 hr of dual perfusion. *Contr. Gynecol. Obstet.* 13, 77-84.

Panigel, M. (1965) Evolution de l'ultrastructure des villosites du placenta humain maintenic en survie. *J. Micros.* 4, 158.

Panigel, M. (1971) Morphological evaluation of perfused tissues. *Acta Endocrinol. Suppl.* 158, 74-94.

Panigel, M. (1974) Electron microscopic studies on placental function and maternal fetal exchange. In: *Modern Perinatal Medicine*, (ed.), L. Gluck, Chicago, Year Book, pp 37-65.

Schroeder, H., Nelson, P., and Power, G. (1982) Fluid shift across the placenta: I. The effect of dextran T40 in the isolated guinea-pig placenta. *Placenta* 3, 327- 338.

Sheppard, B.L. and Bonnar, J. (1980) Ultrastructural abnormalities of placental villi in placentae from pregnancies complicated by intrauterine fetal growth retardation: Their relationship to decidual spiral arteries lesions. *Placenta* 1, 145-156.

Wier, P.J., Miller, R.K., Maulik, D., and di Sant'Agnese, P.A. (1983) Bidirectional transfer of alpha aminoisobutyric acid by the perfused human lobule. *Trophoblast Res.* 1, 37-54.

Wier, P.J. and Miller, R.K. (1985) Oxygen transfer as an indicator of perfusion variability in the isolated human placental lobule. *Contr. Gynecol. Obstet.* 13, 127-131.

Wier, P.J., Miller, R.K., Maulik, D., di Sant'Agnese, P.A., and Shah, Y. (1985) Dual perfusion of the human placental lobule: A model for placental toxicology. *Teratol.* 31, 60A.

Trophoblast Research 2:557-571, 1987

ARE THERE MEMBRANE-LINED CHANNELS THROUGH THE TROPHOBLAST?
A STUDY WITH LANTHANUM HYDROXIDE

Peter Kaufmann[1], Hobe Schroeder[2], Heinz-Peter Leichtweiss[2], and Elke Winterhager[1]

[1] Abt. Anatomie, RWTH Aachen
Melatener Str. 211, D 5100 Aachen
West Germany

[2] Abt. experimentelle Medizin
Universitaets-Frauenklinik
Martinistr. 52, D 2000 Hamburg 20
West Germany

INTRODUCTION

Transfer experiments by Stulc et al. (1969) gave evidence for the existence of water-filled routes, so-called pores or channels, across the rabbit placenta. Their radius was calculated to be approximately 10 nm. These findings have been confirmed for the guinea pig placenta by Thornburg and Faber (1977) as well as by Hedley and Bradbury (1980). Morphological studies of both placentae, however, failed to demonstrate structural correlates, i.e., membrane-lined pathways across the trophoblast from the maternal lacunar surface to the basal trophoblastic surface that faces the fetal capillaries (Enders, 1965; Firth and Farr, 1977; Kaufmann and Davidoff, 1977). Among large numbers of syncytiotrophoblastic intracellular membrane systems like endoplasmic reticulum, Golgi cisternae, and absorptive tubules, structures with apparent luminal contacts to either surface are missing. In a recent publication, Gammal (1985) described so-called syncytial channels in the rhesus monkey placenta. These structures reported by Gammal are not comparable with those discussed in the present publication for several reasons: the "channels" described by Gammal are lamellar spaces rather than tubular channels, their luminal diameters range from 50 to 100 nm, and these "channels" are restricted to the superficial zone of the syncytiotrophoblast and never reach the basal trophoblastic surface.

In the fully isolated, artificially perfused guinea pig placenta, hydrostatic and colloid-osmotic forces may cause feto-maternal fluid shifts. Under these conditions, the syncytiotrophoblastic membrane-lined channels open with diameters from 40 nm to about 5 μm (cf. Figure 1b) (Kaufmann et al., 1982). These channels are the routes for the fluid shift which drags large amounts of albumin and horseradish peroxidase from the fetal into the maternal circulation. When normal pressure conditions are produced, the original transfer rates are re-established and, the channels disappear

Received: 10 October 1985; Accepted: 6 May 1986

within a few minutes. These phenomena are not related to rupture of the trophoblast since the pathways are membrane-lined, and their formation is reversible. It has been proposed that these are pre-existing but artificially dilated channels being identical with those postulated by the physiologists.

In preliminary experiments, tracer proteins like ferritin, horseradish peroxidase, hemoglobin, cytochrome, and myoglobin were too large to pass such undilated channels under normal pressure conditions. The cytochemical reactivity of microperoxidase (M.W. 1,880), which is small enough for the transcanalicular transfer, was too sparse to identify the postulated 10 nm channels. Therefore, lanthanum hydroxide was chosen as an extracellular marker to give proof for such transtrophoblastic pathways.

MATERIALS AND METHODS

Eleven guinea pigs were used at approximately day 60 of pregnancy. One placenta from each animal was isolated and prepared for double sided artificial perfusion as described by Leichtweiss and Schroeder (1971). To remove the fetal and maternal blood as well as to sustain placental nutrition before the start of the experiment, both placental vessel systems were perfused with an oxygenated fluid (TC 199, Difco, with 30 g/l dextran T40 and 1 g/l albumin).

To demonstrate extracellular pathways across the placental barrier, lanthanum hydroxide was used as a tracer. It was prepared following the methods of Revel and Karnovsky (1967): a 4% solution of lanthanum-nitrate was brought to pH 7.6 with 0.01 N NaOH. The resulting lanthanum hydroxide solution was mixed either with collidine buffer or with 2% OsO_4 in collidine buffer (pH 7.2) in equal parts so that both final solutions contained 2% lanthanum hydroxide.

Five types of experiments were performed:

A) Perfusion of the fetal vascular system with lanthanum hydroxide was performed, while the perfusion of the maternal circulation with TC 199 was maintained.

A1) Following 10 min of perfusion, fixation with 2.2% glutaraldehyde in phosphate buffer (340 mosm), and 5 min perfusion with collidine buffer to remove phosphate buffer and excess fixative. Lanthanum collidine OsO_4 medium was then perfused for 20 to 30 min. Perfusion was stopped as soon as a pressure exceeding 20 mm Hg in the tube system indicated the blockage of microvessels by high molecular weight lanthanum complexes.

A2) Identical lanthanum collidine OsO_4 perfusion as described in A1 was performed without prefixation with glutaraldehyde and without rinsing with collidine buffer.

A3) A 20 min perfusion with lanthanum hydroxide in collidine buffer was performed followed by 10 min perfusion with the lanthanum collidine OsO_4 medium.

B)   Perfusion of the maternal vessel system of the placenta with lanthanum hydroxide was established while the perfusion of the fetal vessels with TC 199 was maintained.

B1)   A 10 min prefixation with 2.2% glutaraldehyde in phosphate buffer, 5 min perfusion with colidine buffer and a 20 to 30 min perfusion with the lanthanum collidine OsO_4 medium was performed .

B2)   A perfusion with lanthanum collidine OsO_4 for 20 to 30 min without prefixation or rinsing was also done.

Another series of experiments, was to perfuse the isolated placenta via both the fetal as well as the maternal circulation simultaneously with lanthanum hydroxide. This possibility has been refuted because of two reasons: Due to the method of ultrathin sectioning, the postulated long, branched, and winding channels cannot be demonstrated in full length, reaching from the maternal to the fetal trophoblastic surface, even if a complete filling of the channels by simultaneous tracer application could be obtained. When compared with the experiments described above, it is not possible to determine, for each single segment, whether it has been filled from the fetal or from the maternal side. Moreover, lanthanum precipitates in the circulation not perfused with the tracer could no longer be regarded as a sign of transtrophoblastic passage for the latter. Thus, the results would become even more confusing. Furthermore, from pilot experiments, it became clear that lanthanum application in combination with fixation from both sides is obsolete, since simultaneous perfusion fixation via both vasculature beds severely damages the organ because of dramatic shrinkage. Thus, we restricted the studies to separate lanthanum perfusion experiments via the fetal or maternal circulation.

Detailed data for pressures as well as flow recording during the experiments are given in Schroeder et al. (1982). Following cessation of perfusion on both sides, the placenta was cut into pieces of approximately 3 x 2 x 0.5 mm, which were immersed for 2 hr in phosphate buffered glutaraldehyde, postfixed for 2 hr in phosphate buffered OsO_4, dehydrated in graded series of ethanol, and embedded in Epon. Ultrathin sections were examined with a Philips 300 electron microscope.

RESULTS

Lanthanum Application via the Fetal Circulation

Perfusing with lanthanum hydroxide solution via the fetal vessel system results in intensely marked electron opaque lanthanum precipitation in the luminal endothelium, the basement membrane between endothelium and trophoblast, as well as the basal plasmalemma of the trophoblast (Figures 2, 3b, and 4). In particular the focal basal invaginations of the trophoblastic plasmalemma are stained by this method (Figures 2 and 3b). In accordance with previous results (Kaufmann et al., 1982) (cf. Figure 1a), there is no structural evidence for a vesicular uptake of lanthanum precipitates from the basal plasmalemma into the trophoblast. The above findings do not depend on whether the material was prefixed, fixed in parallel, or only postfixed (cf. Methods A1 to A3). Within the syncytiotrophoblast, slender, tortuous, sometimes branching, membrane-lined, lanthanum-filled tubuli could be observed,

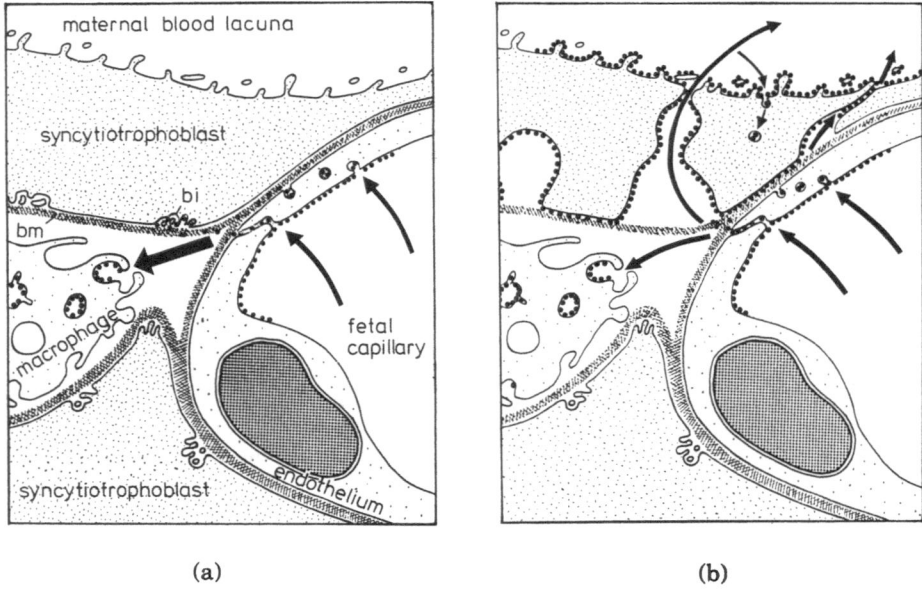

(a)   (b)

Figure 1. Schematic summary of previous experiments by Kaufmann et al., (1982) suggesting the existence of transtrophoblastic channels. a) After injection of horseradish peroxidase into the fetal perfusion fluid the tracer passes the fetal endothelium, and then spreads in the basement membranes (bm). Vesicular uptake into the trophoblast from the basal surface has never been observed. All of the tracer protein is phagocytized by macrophages. There is neither morphological nor physiological evidence for a feto-maternal protein transfer in the isolated guinea- pig main placenta under normal pressure conditons. b) When the fetal venous hydrostatic pressure is raised to about 20 mm Hg a fluid shift from the fetal to the maternal circulation occurs. It is accompanied by a marked feto-maternal protein transfer. Morphologically the syncytiotrophoblast is passed by bag-like, membrane-lined channels, the surfaces of which are labeled with the tracer protein. Some of the tracer is back transferred from the lacunar surface into the trophoblast by micropinocytosis, indicating a materno-fetal vesicular transport. The basal membrane invaginations (bi), visible in Figure 1a, disappear as soon as such bag-like channels are formed (modified from Kaufmann et al., 1982).

some of which are in direct continuity with the basal membrane infoldings. Bag-like invaginations as demonstrated under the conditions of bulk flow from the fetal to the maternal circulation are the exception (Kaufmann et al., 1982).

The results of the first series of fetal perfusion experiments (A1), using glutaraldehyde prefixation, followed by simultaneous application of lanthanum hydroxide and OsO_4 are ultrastructurally the best ones obtained. Channel filling, however, is less prominent. Shorter, narrow segments (10 to 20 nm) are visible in the basal parts of the trophoblast, commonly connected with the basal trophoblastic

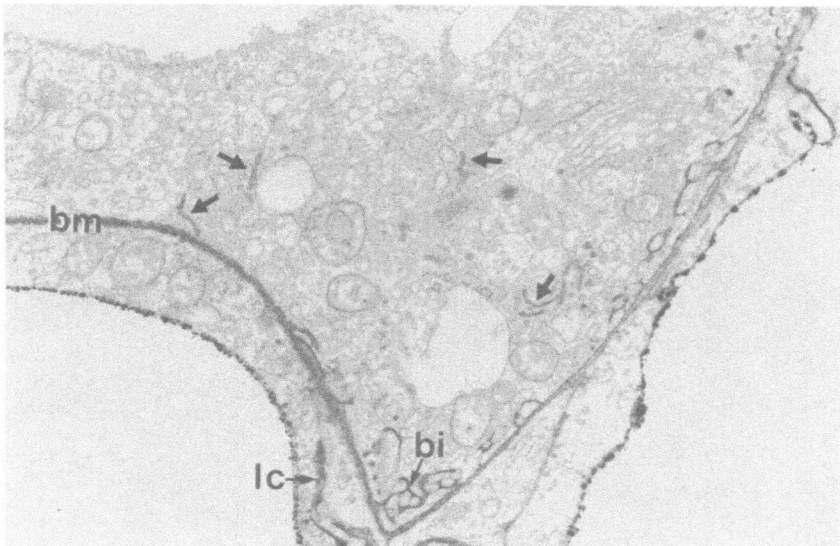

Figure 2. Perfusion of the fetal circulation with a lanthanum OsO$_4$ collidine medium, following glutaraldehyde prefixation. The precipitates of lanthanum hydroxide are clearly visible along the endothelial surface, in the lateral intercellular clefts of the endothelium (lc), in the basement membranes (bm), in the basal membrane invaginations of the trophoblast (bi), as well as in some tubular structures inside the trophoblast (arrows). Because of the labeling, the latter must be continuous with the basal trophoblastic surface. X14,500

invaginations (Figure 2). Continuation of channel segments with the maternal blood lacunae are not seen.

When perfusing directly with lanthanum collidine OsO$_4$ (no prefixiation with glutaraldehyde, series A2) the channels are mostly restricted to the basal half of the trophoblast (Figure 3a). Superficially positioned channel segments filled with the tracer are less common (Figure 3b). Continuation of the channel segments with the maternal blood lacunae could not be demonstrated. Some occasional lanthanum precipitations within the maternal blood lacunae may be a faint hint in this direction (Figure 3a).

As soon as the lanthanum collidine medium is applied without prefixation and without OsO$_4$ (series A3), the placental ultrastructure is not well preserved. Unit membranes are almost invisible. However, the narrow branching channels can be visualized in all parts of the trophoblast (Figure 4), not only connected to the basal invaginations, but sometimes even short segments in continuation to the apical trophoblastic surface. Distinct lanthanum precipitation inside the maternal blood lacunae (Figure 4) points to a free feto-maternal parasyncytiotrophoblastic passage of the material under those experimental conditions. The channels exhibit varying diameters ranging from 15 to 30 nm.

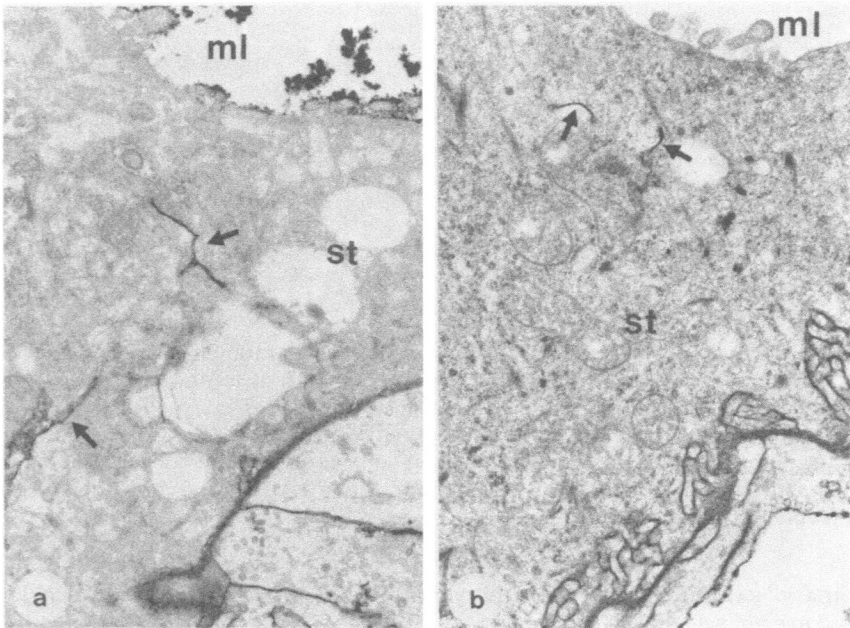

Figure 3. Perfusion of the fetal vessels with lanthanum OsO₄ collidine medium without prefixation results in a rather poor ultrastructural tissue preservation. a) Different from the foregoing condition the traced tubular segments (arrows) are longer. Some of the lanthanum hydroxide is visible in the maternal lacuna (ml), indicating an extracellular passage of the tracer across the syncytiotrophoblast (st). X17,500 b) In another experiment of the same type, the basal membrane invaginations are perfectly traced. Inside the syncytiotrophoblast two short lanthanum-filled channel segments (arrows) are visible, only. No lanthanum transfer into the maternal blood lacuna (ml) can be seen. X17,500

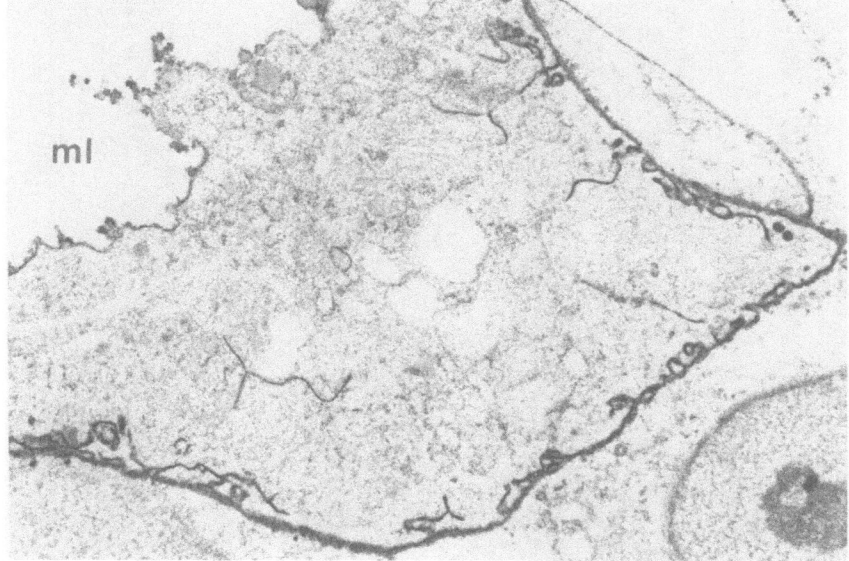

Figure 4. Perfusion with lanthanum hydroxide without prefixation or even simultaneous fixation with OsO_4 is followed by a very poor ultrastructural preservation. Under this condition, however, numerous slender, branching channels with about 15 nm luminal diameter, filled with the tracer can be demonstrated. Again, ample precipitates can be seen in the maternal lacuna. Inspite of the poor ultrastructure, this picture seems to represent the most natural configuration of the transtrophoblastic channels. X14,500

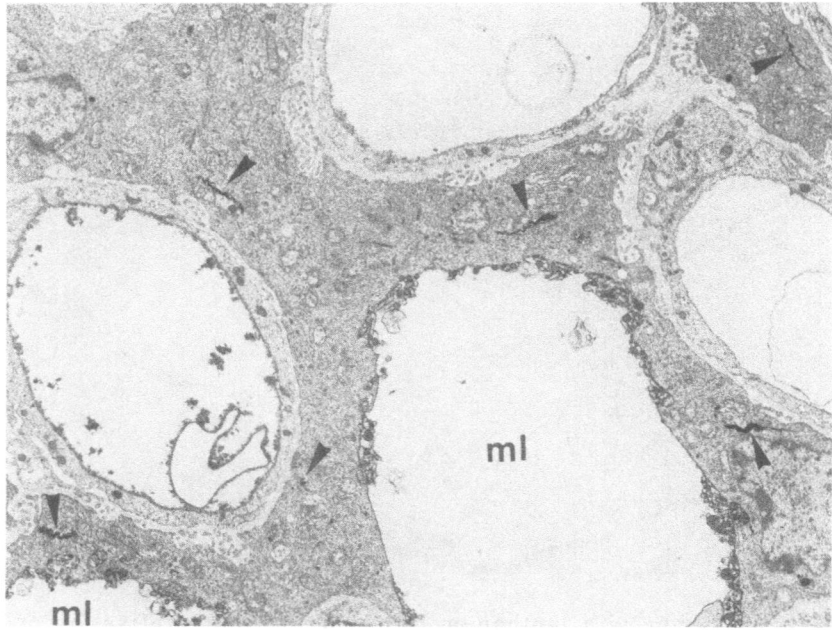

Figure 5. Lanthanum OsO$_4$ collidine application via the maternal blood lacunae (ml) results in much less lanthanum filled channel-like structures than in the fetal perfusion. Following prefixation with glutaraldehyde, the ultrastructural appearance of the trophoblast is sufficient. Evenly distributed lanthanum-filled short channel segments (arrowheads) are demonstrable in the trophoblast. The spiral-like precipitation of the tracer is characteristic for this type of experiment. X6,000

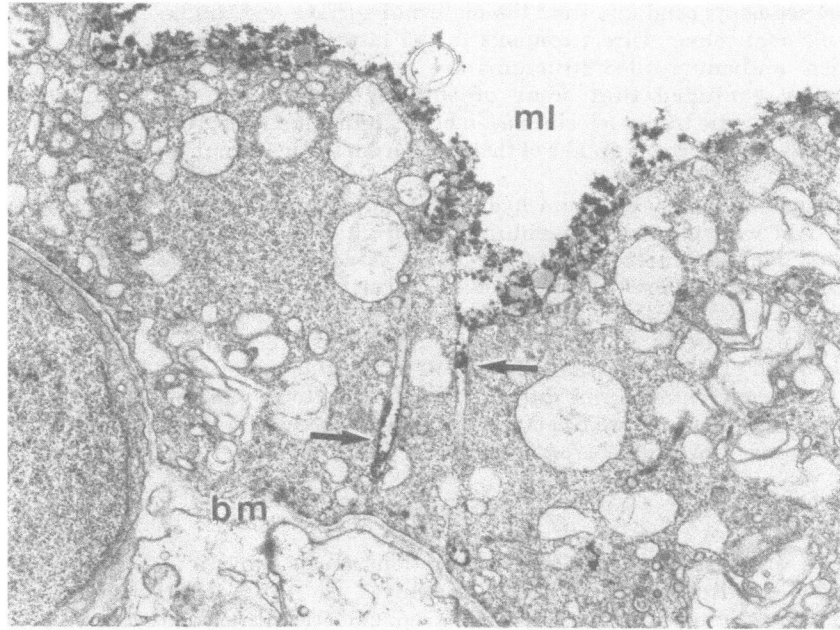

Figure 6. Following maternal lanthanum perfusion without prefixation the trophoblastic organelles are dilated vesicularly pointing to ischaemic artifacts. Some long, slightly dilated channels extending from the maternal lacuna (ml) deeply into the trophoblast (arrows) are visible. The plasma membrane surrounding these channels can clearly be seen. There are no signs of tracer transfer across the trophoblast into the basement membranes (bm) in the interstitial space. X16,000

Lanthanum Application via the Maternal Circulation

Tracing the postulated transtrophoblastic channels with lanthanum hydroxide from the maternal side proved to be more difficult than from the fetal side. In series B1 - glutaraldehyde prefixation followed by rinsing with collidine buffer and perfusion with lanthanum collidine OsO_4 medium, the diameters of the channels are difficult to estimate. Spiral-like precipitations of the lanthanum hydroxide and poor membrane preservation cause considerable problems. Short channel segments are evenly distributed throughout the material (Figure 5). Following the majority of electron micrographs of this group, the diameters range from 15 to 25 nm. Some of the channel segments originate from the maternal surface and can be traced to near the basal plasmalemma. Direct contacts to the latter are not demonstrable. Not all roundish lanthanum filled structures are necessarily channel cross sections. It cannot be excluded that some of the sections are a result of vesicular transtrophoblastic transport, since even brief glutaraldehyde prefixation did not fully prevent micropinocytotic uptake of the tracer from the maternal surface.

Application of lanthanum hydroxide simultaneously with OsO_4 (series B2) resulted in well preserved membrane-lined channels which extended from the maternal lacunar surface up to the basal third of the trophoblast (Figure 6). Connections to the basal plasmalemma could not be identified, nor did lanthanum occur in the interstitial space and in the fetal vessel lumina. Different from the fetal perfusion experiments, in this series, the channels have a stretched shape, are less branched and exhibit large diameters of 50 to 100 nm (Figure 6). The evaluation of the electron micrographs was difficult, since pinocytotic vesicles pinched off at the maternal plasmalemma are filled with lanthanum precipitates too and can hardly be discriminated from channel cross sections.

DISCUSSION

As was discussed in earlier publications (Kaufmann and Schroeder, 1980; Schroeder and Kaufmann, 1980; Kaufmann et al., 1982), there is no macro- or micropinocytotic uptake from the basal trophoblastic surface into the trophoblast (Figure 1a). The only contradictory observation has been published by Sibley et al. (1981). The authors described small amounts of horseradish peroxidase in syncytiotrophoblastic vesicles following administration of the tracer protein via the fetal circulation which occur only when a Krebs Ringer bicarbonate glucose (KRBG) solution was used but not TC 199. Since the respective plates of the KRBG experiments reveal typical signs of feto-maternal bulk flow with dilated channels, as described by Kaufmann et al. (1982), these "vesicles" are interpreted as cross sections of such dilated channels rather than as signs of vesicular uptake. This interpretation is in agreement with the current views of the authors (Firth and Sibley, personal communication, 1984). Thus, pinocytosis into and across the guinea pig syncytiotrophoblast, first described by King and Enders (1971), is a one-way system in the materno-fetal direction. This fact is important for the interpretation of our results. Moreover, as will be discussed later, this view may be of interest for the functional implication of the channel system. If this concept of a one-way micropinocytosis is considered to be correct, then each membrane lined structure within the trophoblast filled with lanthanum hydroxide from the fetal vessels has to be interpreted as part of a system continuous with the basal trophoblastic surface. As

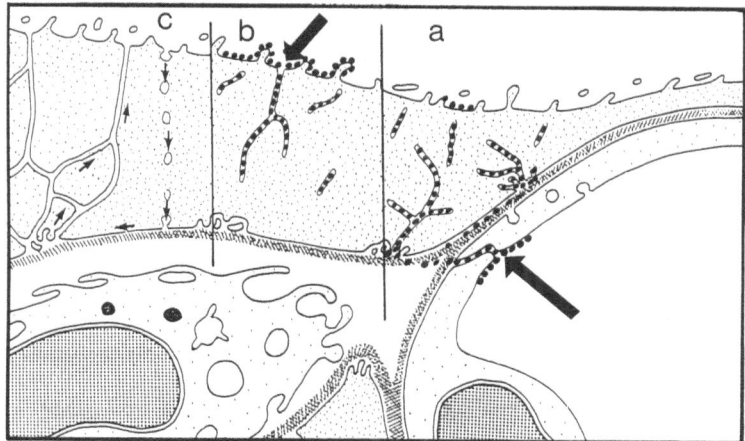

Figure 7. Schematic summary of the experimental results: a) Tracer application from the fetal side results in a well developed channel system extending from the basal surface deeply into the trophoblast. Connections with the maternal surface are uncertain. Tracer transfer into the lacunar lumina, however, points to the existence of a complete transtrophoblastic channel system as depited in part c of this diagram. b) Maternal application of the tracer yields an apical filling of the channel system, only. In this case there are no obvious signs for a continuous materno-fetal channel system. c) Comprising these results, a complete transtrophoblastic channel system with about 15 nm luminar diameter is highly probable. It may act as a way for plasma membrane reflux from the basal towards the apical trophoblastic surface to reduce a basal membrane surplus as a result of the unidirectional materno-fetal vesicular transport.

described above, such structures are common in all parts of the trophoblastic layer (Figure 7a).

As soon as such lanthanum filled invaginations of the plasma membrane are in open connection with the maternal blood lacunae, a transtrophoblastic channel system would have been proven. This channel system would be in agreement with transport calculations presented for the guinea pig placenta by Thornburg and Faber (1977), Hedley and Bradbury (1980), Stulc (personal communication, 1984), and Stulc et al. (1969) for the rabbit placenta. These authors described a free paracellular diffusional transfer between maternal and fetal blood of hydrophilic substances which only can diffuse in a water-filled system but hardly pass the plasmalemma.

From the current results, there is strong evidence but not complete proof, for such a continuous transtrophoblastic system. All experiments described demonstrated the existence of membrane-lined, lanthanum-filled, channel-like structures, which were detectable as shorter or longer segments in all parts of the syncytiotrophoblast. Since the diameter of these structures (about 15 nm) was narrow as compared to the thickness of the ultrathin section (50 to 100 nm), even longer channels could be followed in one section for a limited length only. Thus, the complete arrangement of the postulated channel system could only be deduced, but never directly visualized. According to Revel and Karnovsky (1967), lanthanum compounds are not known to penetrate into cells, except if a rupture of the plasma membranes occurs. This fact makes it an useful tracer for the extracellular space and its intracellular connections. The same substance, however, is known to be cytotoxic. Therefore, application in the unfixed or not yet perfectly fixed tissue may cause artifacts. Since all intratrophoblastic precipitations of lanthanum hydroxide observed so far are surrounded by plasma membranes these cannot be derived by artifacts. One may argue that lanthanum hydroxide transfer takes place across an incomplete channel system (Figure 7a) to the superficial zone of the trophoblast and then is completed either by artificial rupture of the superficial lamella of trophoblast or by a kind of vesicular shuttle between channel end and lacuna. A similar process has been discussed by Hedley and Bradbury (1980) as an alternative to a channel system. Simionescu (1981) described comparable vesicle shuttles for the capillary endothelium. For both phenomena, our current studies did not reveal structural correlates.

Lanthanum application via the maternal circulation results in a channel system which is connected to the lacunar lumina and deeply extends into the trophoblast. Connections to the basal plasmalemma were not demonstrable (Figure 7b). If this system is continuous with the one demonstrated by fetal lanthanum application, then a channel connection between both trophoblastic surfaces would exist (Figure 7c). According to our results it seems reasonable that both systems are one and the same.

Even if some uncertainties concerning the morphological results must be admitted, the physiological evidence (Stulc et al., 1969; Thornburg and Faber, 1977; Hedley and Bradbury, 1980) supports our interpretation of the morphological results. The assumption that the structures postulated by the physiologists and described by the morphologists are identical, becomes probable by the fact that the diameters described by both disciplines fit nicely together. Following Thornburg and Faber (1977) free paracellular diffusion by hydrophilic, lipid-insoluble substances with a molecular weight of 5,500 daltons, e.g., inulin takes place in the guinea pig chorioallantoic placenta. This diffusion demands the existence of fluid filled routes of 10 to 20 nm in diameter (Faber, personal communcation, 1982). Hedley and Bradbury (1980) came to similar conclusions, postulating at least "transient channels of say confluent vesicles" across the trophoblast with a radius of 10 nm. The current results confirm that at least 3/4 of the trophoblastic layer from both sides are penetrated by persistent channels (Figures 7a and 7b). As discussed above, transport across the remaining lamella of about 1/4 of the trophoblastic thickness is still uncertain and may be bridged either by such confluent vesicles or by stable channel connections. Faber's rough estimation (personal communication, 1982), that the cross sectional area of the channels (i.e., total pore area) amounts to 0.1 to 0.01% of the lacunar trophoblastic surface, corresponds to the current results. Since the thickness of the ultrathin sections varies (50 to 100 nm) and only parts of the channel system are filled with lanthanum precipitates, the corresponding morphological data cannot be presented. However, demonstrating one channel segment near the maternal surface every 1 to 10 μm (Figure 5) is approximately in the same range.

Last but not least, previous experimental findings (Kaufmann et al., 1982) support the concept of transtrophoblastic channels under the conditions of feto-maternal fluid filtration: the trophoblast is traversed by a system of large channels, measuring 1 to 5 μm in diameter (Figure 1b). These channels are formed within a few minutes after pressure elevation in the umbilical vein and are involuted again within 10 min after pressure normalization. Bulk flow of water with simultaneous transfer of albumin and horseradish peroxidase, as long as the channels are open, as well as the full reversibility of this process with disappearance of the large channels, indicate that the trophoblast does not merely disintegrate. These phenomena make a system of dilating and narrowing channels highly probable. The amounts of membrane used for channel dilation could be mobilized from the intensive basal membrane invaginations of the trophoblast, indicated by the disappearance of the invaginations with increased channel enlargement (Figure 1b) (Kaufmann et al., 1982). Furthermore these folded membrane areas are the regular points of origin of the nondilated channels (Figures 4, 7a, and 7c).

Ultrastructurally, there are no signs that the channels are continuous to any of the syncytiotrophoblastic organelles of similar membrane-lined structure. They seem to establish a totally separate system. It may be argued that they are remainders of lateral intercellular clefts derived from former cellular stages of the trophoblast. This idea cannot be refuted; however, it is less probable since the trophoblast is losing its single cell nature already during implantation. In all later stages, only occasionally single cytotrophoblastic cells are incorporated into the syncytium from its basal side. The corresponding remainder of cell membranes would not transverse the trophoblastic layer from the lacunar to the basal surface.

When the results of our experimental protocols are compared with each other, one has the impression that fixation of the tissue with glutaraldehyde or OsO_4 prior to, or simultaneously with, lanthanum application hinders the tracer transfer along the channels in feto-maternal direction, but supports the same transfer in materno-fetal direction. As mentioned above, micropinocytosis takes place in the materno-fetal direction only. Thus the trophoblast as a continuous layer will suffer a loss of membrane at the maternal (apical) surface resulting in an excess of membrane at the basal side. This imbalance can be equilibrated theoretically by an undirectional membrane flux back in baso-superficial direction via the transtrophoblastic channels (Figure 7c). Tight junctions which act as a barrier against such a membrane flux in epithelia consisting of cells (Dragsten et al., 1981) could not be observed along the channels. According to the results of King and Enders (1971), membrane turnover and recycling of the apical membrane will occur primarily as a result of vesicular endocytotic processes. Transcytosis causing an apico-basal imbalance is the less common phenomenon. Freeze fracture investigations of Sideri (1983), however, point out membrane differences in respect to particle density between the apical and the basal membrane of the syncytiotrophoblast. This membrane polarity, typical for epithelial cells, could not be compensated by the hypothetical membrane recycling mechanism, due to the paucity of baso-apical channel connections and their restricted efficiency to transport membrane material.

If an unidirectional membrane flux along the transtrophoblastic channels takes place, the asymmetric influence of fixation on lanthanum transfer across the channels can be explained. A still active membrane flux would increase feto-maternal lanthanum transport but would hinder the opposite transport. Fixation will decrease or stop membrane flux and thus would guarantee comparable lanthanum transfer in both directions.

In summary we are aware that we have proven channels passing 3/4 of the trophoblast only (Figure 7). However, comparing morphological and physiological results, it is probable that a complete channel system connecting maternal blood lacuna with the interstitial space between trophoblast and fetal capillaries exists in the guinea pig placenta. Experimental results presented for the human placenta (Kaufmann, 1984) as well as physiological data obtained in the rabbit (Stulc et al., 1969) indicate that these channels are also valid for other placentae.

SUMMARY

Application of lanthanum hydroxide as an extracellular tracer via the maternal or fetal circulation of the isolated guinea pig placenta results in the proposed appearance of membrane-lined channels extending from the apical or basal trophoblastic surface into the syncytiotrophoblast. The luminal diameter of these channels in most experiments ranges from 15 to 25 nm. In the basal half of the trophoblast, the channels form a rather dense meshwork with few extensions towards the microvillous surface of the maternal lacunae only. Since these channels are winding and branching structures which usually are not completely filled with lanthanum hydroxide, it is impossible to trace the channels in full length. There is indirect evidence only that these channels really pass the trophoblast completely from the superficial (maternal) towards the basal (fetal) surface. These morphological results fit into the corresponding physiological concepts requiring 10 to 20 nm wide water filled routes across the syncytiotrophoblast which allow unrestricted materno-fetal diffusion of water soluble, lipid insoluble molecules with an effective molecular diameter of about 1.5 nm.

REFERENCES

Dragsten, P.R., Blumenthal, R., and Handler, J.S. (1981) Membrane asymmetry in epithelia: is the tight junction a barrier to diffusion in the plasma membrane? *Nature* 294, 718-722.

Enders, A.C. (1965) A comparative study of the fine structure of the trophoblast in several hemochorial placentas. *Am. J. Anat.* 116, 29-68.

Firth, J.A. and Farr, A. (1977) Structural features and quantitative age-dependent changes in the intervascular barrier of the guinea-pig haemochorial placenta. *Cell Tissue Res.* 184, 507-516.

Gammal, E.B. (1985) Syncytial channels in the villous trophoblast of the macaque. *J. Anat.* 141, 181-191.

Hedley, R. and Bradbury, M.B.W. (1980) Transport of polar non-electrolytes across the intact and perfused guinea-pig placenta. *Placenta* 1, 277-285.

Kaufmann, P. (1985) Influence of ischemia and artificial perfusion on placental ultrastructure and morphometry. *Contr. Gynecol. Obstet.* 13, 18-26.

Kaufmann, P. and Davidoff, M. (1977) The guinea-pig placenta. *Adv. Anat. Embryol. Cell Biol.* 53(2), 1-92.

Kaufmann, P. and Schroeder, H. (1980) *Protein transport across the guinea pig chorioallantoic placenta.* XIth Int. Cong. Anat., Mexico City.

Kaufmann, P., Schroeder, H., and Leichtweiss, H.-P. (1982) Fluid shift across the placenta: II. Fetomaternal transfer of horseradish peroxidase in the guinea pig. *Placenta* 3, 339-348.

King, B.F. and Enders, A.C. (1971) Protein absorption by the guinea pig chorioallantoic placenta. *Am. J. Anat.* 130, 409-430.

Leichtweiss, H.-P. and Schroeder, H. (1971) Untersuchungen ueber den Glucosetransport durch die isolierte, beiderseits kuenstlich perfundierte meerschweinchenplacenta. *Pflüegers Arch.* 325, 139-148.

Revel, J.P. and Karnovsky, M.J. (1967) Hexagonal array of subunits in intercellular junctions of the mouse heart and liver. *J. Cell Biol.* 33, C7-C12.

Schroeder, H. and Kaufmann, P. (1980) Do transtrophoblastic channels exist in the guinea-pig placenta. *Int. Congr. Physiol. Sci.*, Szeged.

Schroeder, H., Nelson, P., and Power, B. (1982) Fluid shift across the placenta. I. The effect of dextran T40 in the isolated guinea pig placenta. *Placenta* 3, 327- 338.

Sibley, C.P., Bauman, K.F., and Firth, J.A. (1981) Ultrastructural study of the permeability of the guinea-pig placenta to horseradish peroxidase. *Cell Tissue Res.* 219, 637-647.

Sideri, M., de Virgiliis, G., Rainoldi, R., and Remotti, G. (1983) The ultrastructural basis of the nutritional transfer: evidence of different patterns in the plasma membranes of the multilayered placental barrier. *Trophoblast Res.* 1, 15-26.

Simionescu, N. (1981) Transcytosis and traffic of membranes in the endothelial cell. In: *International Cell Biology*, (ed.), G. Schweiger, Berlin, Springer Publishers.

Stulc, J., Friederich, R., and Jiricka, Z. (1969) Estimation of the equivalent pore dimensions in the rabbit placenta. *Life Sci.* 8, 167-180.

Thornburg, K. and Faber, J.J. (1977) Transfer of hydrophilic molecules by placenta and yolk sac of the guinea pig. *Am. J. Physiol.* 233, C111-C124.

Trophoblast Research 2:573-584, 1987

TRANSFER OF HORSERADISH PEROXIDASE ACROSS THE HUMAN PLACENTAL COTYLEDON PERFUSED IN VITRO

Mario Sideri[1,3], Elena Zannoni[1], and Jean-Claude Challier[2]

[1]First Clinic of Obstetrics and Gynecology
Section of Pathological Obstetrics and Gynecology
CNR Centre of Cytopharmacology
University of Milan
Milan, Italy

[2]Université Pierre et Marie Curie
Department of Biology of Reproduction
Paris, France

INTRODUCTION

Physiological studies have indicated that the human placenta is to some degree permeable to macromolecules (Challier et al., 1985). The pathway and the cellular mechanisms by which this passage occurs are still not identified. Our previous investigations on the junctional pattern of the human villous trophoblast have suggested that the syncytium and the fetal capillary endothelium are the most important structures involved in placental transport (De Virgiliis et al., 1982; Sideri et al., 1983). Conflicting results have been reported by in vitro tracer studies either at capillary or at syncytial levels. In the human placenta incubated in vitro with HRP conjugated IgG, Lin (1980) detected the protein in the inter-endothelial space and in the transcytotic vesicles of the fetal capillary, while King (1982) observed the reaction product only in transcytotic vesicles. Both authors were unable to follow the endocytotic vesicles containing HRP conjugated IgG in their trophoblastic pathways. In animal studies, Kaufmann et al. (1982) have suggested that transtrophoblastic channels, detected under perfusing conditions of fluid-shift, are involved in the transfer of HRP. Sibley et al. (1981, 1983) failed to demonstrate any uptake of HRP into the syncytium. These two authors showed a passage of HRP through the fetal capillary endothelium via the inter-endothelial spaces. In addition, Firth et al. (1983) indicated that the junctional complexes between two adjacent endothelial cells are capable of some selection, at least for the charged molecules.

In order to define the pathway and the cellular mechanisms of the placental macromolecular transfer, we investigated the passage of cationic and anionic HRP across the human placental lobule perfused in vitro.

[3]To whom corespondence should be addressed: Prima Clinica Ostetrica e Ginecologica dell'Università di Milano, v. Commenda, 12, 20122 Milano, Italy

Received: 1 February 1986; Accepted: 29 July 1986

MATERIALS AND METHODS

Placental Perfusion

Nineteen human placentae at term, from normal vaginal deliveries, were perfused in vitro according to the technique described by Schneider et al. (1972). After the introduction of a cannula into a small fetal artery, a placental lobule was perfused with an Earle's solution, containing serum albumin 1%, heated at 37°C and oxygenated with a mixture of 95% O_2 and 5% CO_2. Pressures in the arterial and venous fetal circuits were kept within the physiological ranges (Table 1). The maternal circuit was realized by the insertion of two cannulas into the intervillous space close to the perfused cotyledon.

HRP Infusion Into the Fetal Circuit

After 20 min perfusion of the fetal circuit only, an Evan's blue bolus (1 ml of a 0.42 mM solution) was injected into the fetal artery, in order to evaluate the size of fetal or maternal vascular compartment (appearance time). The appearance time is a delay between the time of injection and its appearance in the outflow. It includes a few seconds of passage in the catheter. Then a 1 ml bolus of cationic (type VI Sigma; lots 121 F9605 and 31 F95205) or anionic (type VIII Sigma; lot 12 F9850) HRP was performed into the fetal circuit. HRP type VI contained 330 purpurogallin units per mg solid and type VIII 100 U/mg. Most of the experiments were run using 2 mg/ml of type VI HRP or 6.6 mg/ml of type VIII HRP, except the dose dependent study in which the concentrations of type VI varied from 0.2 to 20 mg/ml. The perfusion was stopped after a time corresponding to the appearance time and the tissue was left to incubate with the tracer until fixation. The latter was carried out by perfusing the fixative through the maternal circulation after a time which varied from 1 to 15 min from the HRP bolus injection.

HRP Infusion Into the Maternal Circuit

After 20 min perfusion of the placental cotyledon on both the placental sides, an Evan's blue bolus was injected into the maternal circuit in order to measure the appearance time. Then a bolus of cationic (Type VI Sigma) or anionic (type VIII Sigma) HRP in concentrations varying from 2 to 20 mg/ml was injected into the maternal circuit. The perfusion was then stopped until fixation, which was carried out by perfusing the fixative through the maternal circulation after a time which varied from 1 to 15 min from the HRP bolus injection.

Electron Microscopy

Pre-fixation was carried out by perfusing the maternal circuit with a solution of gultaraldehyde (2%) in cacodylate buffer 0.1 M, pH 7.35 at 4°C for 15 min. Then, samples of terminal villi were excised under a stereomicroscope and immersed in the same fixative for 2 hr at 4°C. The reaction for precipitating HRP was performed by preincubating the tissue in a 1% DAB (Polysciences) solution, in TRIS buffer, 0.2 M, pH 7.4, for 30 min and then incubating the specimens in a solution containing 1% DAB in the same buffer, added with H_2O_2 0.01%, according to the technique

Table 1

Parameters of perfusion

Side perfused	N	Fa (ml/min)	D (ml/mn)	Pf (mmHg)	Qm (ml/min)	Pm (mmHg)	AT (sec)	LE (min)
Cationic HRP								
Fetal	7	9.8** ±2.6	0.32 ±0.31	48 ±12.3	–	–	10.8 ±3.0	37 ±14.7
Maternal and fetal	7	9.8 ±2.0	0.57 ±0.34	48 ±10.8	20.4 ±4.9	N=3 36/62*§ N=3 2/4*§	11.1 ±4.9	43.8 ±13.4
Anionic HRP								
Fetal	3	9.2 ±2.3	0.67 ±0.50	41.3 ±15.0	–	–	8.3 ±1.0	34 ±20
Maternal and fetal	2	9.8/12.5*	08/0.9*	19/22*	19/22.4*	2/4*§	7.5/13*	47/13*

N = number of experiments; Fa = fetal flow rate; D = artero-venous difference of fetal flow rates; Pf = fetal pressure; Qm = maternal flow rate; Pm = maternal pressure; A.T. = appearance time; L.E. = duration of experiment from the start of perfusion to the end of fixation; * = range; ** = mean ± SD; § = difference due to a change in maternal cannula.

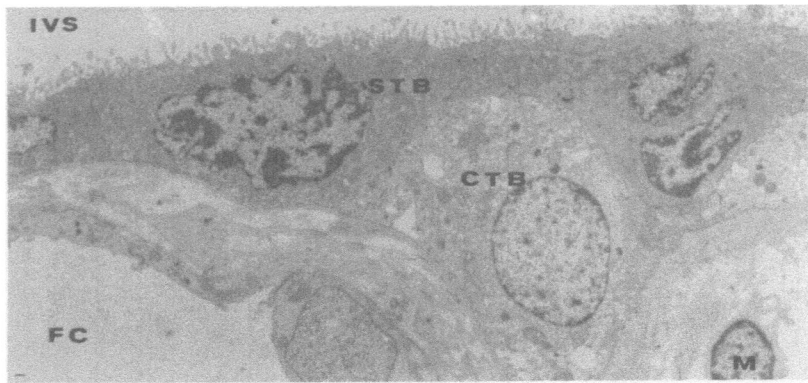

Figure 1. The figure shows the morphology of the placental villus after 20 min of perfusion. Note the good preservation of the tissue. IVS = Intervillous Space; STB = Syncytiotrophoblast; CTB = Cytotrophoblast; M = Macrophage; FC = Fetal Capillary. Magnification 3500X.

described by Graham and Karnowksy (1966). The specimens were then postfixed in cacodylate buffered osmium tetroxyde solution, dehydrated in a graded series of alcohols and embedded in epon. Thin sections were cut with an ultramicrotom Porter Mt-2B and examined with a Philips EM 200 electron microscope, without additional staining.

RESULTS

Morphology of the Perfused In Vitro Placenta

The fine morphology of the in vitro perfused placenta was nicely preserved in these short term experiments (Figure 1). Most of the fetal capillaries appeared clear from the red blood cells and slightly distended. A lesser number of capillaries (<10%) remained filled by the red blood cells and showed some degree of vasoconstriction, indicating that they were in poorly perfused areas. The interendothelial spaces and the junctional complexes appeared as in the control specimen. The villous stroma showed in some experiments various degrees of edema. In the syncytium, the rough endoplasmic reticulum and the mitochondria remained unchanged. The microvillar surface, however, did not show the peculiar aspect due to a layer of glycocalix.

Figures 2A, B, and C. The figures show the morphology of the fetal capillary endothelium after 15 min incubation with a cationic HRP bolus injected into the fetal circuit. Note the tracer on the luminal endothelial surface and in the macrophage (Figure 2A). In the interendothelial space, the tracer is present along the whole length of the junctional area (Figure 2B) or it shows a sharp stop (Figure 2C). FC = Fetal Capillary; M = Macrophage. Magnification for Figure 2A = 2700X, Figure 2B = 31500X, and Figure 2C = 26000X.

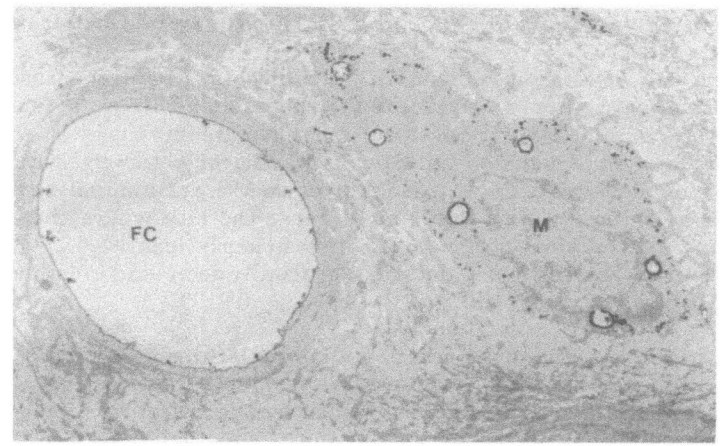

2A

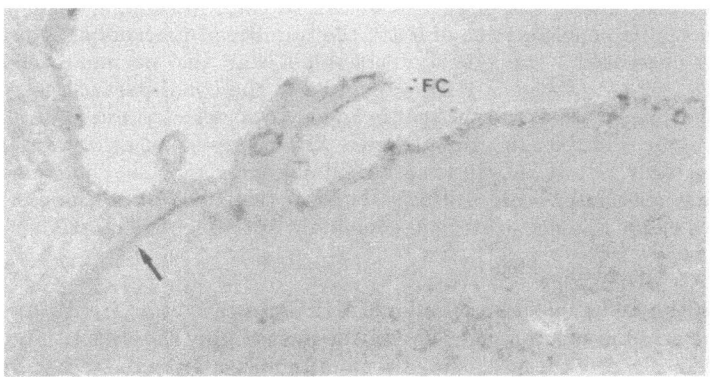

2B

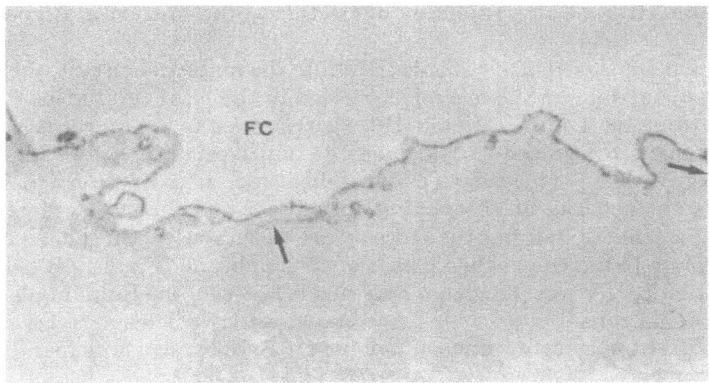

2C

Observations After a Bolus Injection of HRP Into the Fetal Circuit

After a bolus injection of cationic HRP into the fetal circuit, the tracer formed a continuous layer on the luminal surface of the endothelial cells of the fetal capillary (Figures 2A, 2B, 2C, and 3A). Cationic HRP did penetrate into the intercellular space between two adjacent endothelial cells, but it was never observed to permeate the endothelium completely and to form a dense layer on the antiluminal surface of the endothelium in continuity with the luminal one. The site where the tracer was stopped in the intercellular space was sometimes clearly identified by a sharp line (Figure 2C), but in other instances, the signal gradually decreased in intensity until it disappeared completely (Figure 2B).

The tracer was also observed in numerous transcytotic vesicles of the endothelial cells (Figures 2B, 2C, and 3A). However, only the vesicles closely located on the luminal endothelial surface were loaded by the tracer, and never a vesicle containing cationic HRP, opened on the abluminal endothelial surface. The signal decreased in intensity as the vesicle was closer to the abluminal endothelial surface. Cationic HRP was also observed in the stromal macrophages (Figure 2A). These cells showed the tracer after 5 min of incubation. However, after longer incubation time or when using higher concentration of HRP, the number of macrophages present in the stroma was decreased. The syncytiotrophoblast was also permeated by the tracer (Figures 3A and 3B). Cationic HRP was found in the basal part of the syncytium as well as in the apical one. It was present in endocytotic vesicles and sometimes in short tubular structures of the syncytium (Figure 3B). Only some areas of the syncytium showed the tracer. However, these areas did not show any morphological pecularity nor were they associated with synthesis (beta) or transfer (alpha) zones. In addition, the presence or the absence of stromal edema did not affect the location of HRP in the tissue.

After the bolus injection of anionic HRP into the fetal circuit, the tracer was found in the stromal macrophages and in the syncytium (Figures 4A and 4B). Some transcytotic vesicles of the fetal capillary endothelium also were stained by the tracer. In the syncytium anionic HRP was present in endocytotic vesicles of either the basal or the apical zone (Figure 4B).

Observations After a Bolus Injection of HRP Into the Maternal Circuit

After a bolus injection of cationic HRP into the maternal circuit, the tracer was found in the syncytium, in the macrophages and in the fetal capillaries (Figure 5A). In the syncytium, most of the cationic HRP was confined to the apical part, contained in large vacuoles, recognized as lysosomes or multivesicular bodies, and in small vesicles (Figure 5C). These latter were observed in close relationship to the microvillous surface and often opening into the lysosomes. In some instances endocytotic vesicles containing the tracer were observed in the basal part of the syncytium, close to the trophoblastic basement membrane (Figure 5B). In the fetal capillary cationic HRP was found as a continuous layer on the luminal plasmalemma of the endothelial cells (Figure 5A). Some transcytotic vesicles also were stained by the tracer. After the bolus injection of anionic HRP, the tracer was found only in the intervillous space.

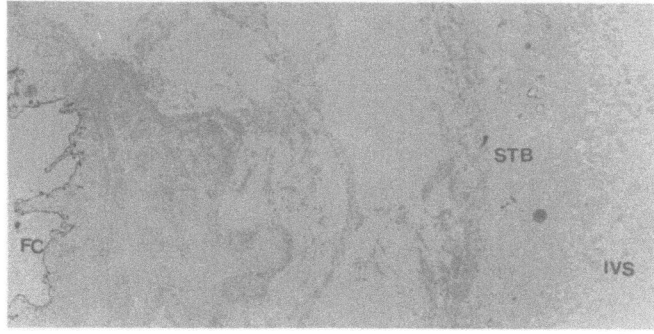

3A

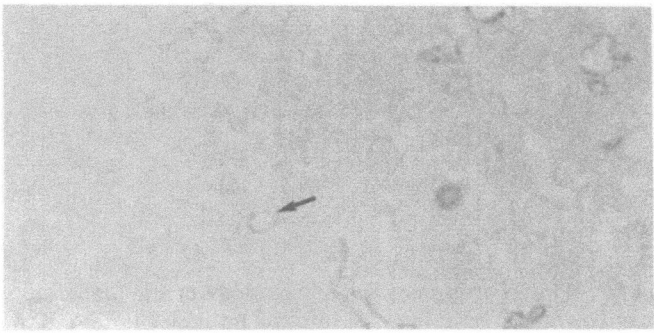

3B

Figures 3A and 3B. The figures show the morphology of the placental villus after 15 min incubation with cationic HRP injected into the fetal circuit. Note the tracer on the luminal surface of the fetal capillary endothelium (Figure 3A) in the syncytial endocytotic vesicles and a weak staining of the trophoblastic basement membrane. In Figure 3B, note the presence of the tracer in endocytotic vesicles and tubular structures of the syncytium. The arrow indicates a vesicle close to the intervillous space. IVS = Intervillous Space; STB = Syncytiotrophoblast; FC = Fetal Capillary. Magnification for Figure 3A = 1850X and 9000X for Figure 3B.

DISCUSSION

The aim of this study was the direct visualization of the macromolecular transfer across the trophoblastic barrier. The perfusion technique used in these experiments provided this opportunity with the trophoblastic morphology well preserved in each experiment. However, it should be noted that in these experiments, the perfusion technique was utilized to get the substance into the tissue, but the permeation of the latter was carried out in experimental condition of "incubation", due to the fact that the perfusion was not running. This allows the observation of the

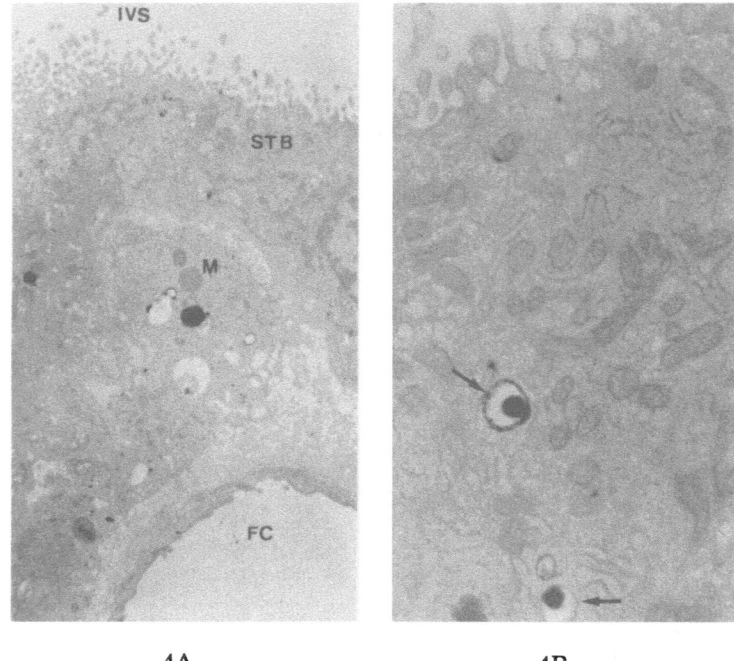

4A   4B

Figures 4A and 4B. The figures show the morphology of the placental villus after 15 min incubation with anionic HRP injected into the fetal circuit. Note the tracer in the macrophage, and in the syncytial endocytotic vesicles. In the latter anionic HRP is contained in the center of the vesicle (Figure 4B) as expected for a fluid phase marker. IVS = Intervillous Space; STB = Syncytiotrophoblast; M = Macrophage; FC = Fetal Capillary. Magnification of Figure 4A = 3000X and Figure 4B = 7500X.

cellular mechanisms of HRP transfer. However, after stopping the perfusion, some degree of hypoxia might occur; it is known that this latter affects the endo/exocytotic processes as well as the structure of the junctional complexes present between endothelial cells. We rule out this possibility on the basis of three considerations: a) all the experiments were performed in a short time (max 15 min); b) the syncytium undergoes high degree of structural changes when exposed to hypoxia (Schweikhart and Kaufmann, 1977), and the present results show a very well preserved syncytial structure; c) in two experiments we continuously perfused HRP on the fetal side (without stopping the double side perfusion) and had results similar to the perfusion/incubation technique. Two types of HRP, cationic and anionic, which varied by their isoelectric points, were used. The first one binds to the plasma membranes and it is therefore a marker of the adsorptive endocytosis (Silverstein et al., 1977; Steinman et al., 1983; Stenseth et al., 1983). The second one is repulsed from the plasma membranes and it is a marker of the fluid phase endocytosis. In these experiments Sigma type VI HRP (cationic) and Sigma type VIII (anionic) were used which were probably not completely pure preparations (Sibley et al., 1983). In fact, using the anionic HRP, also a discrete staining of the plasma membrane was observed.

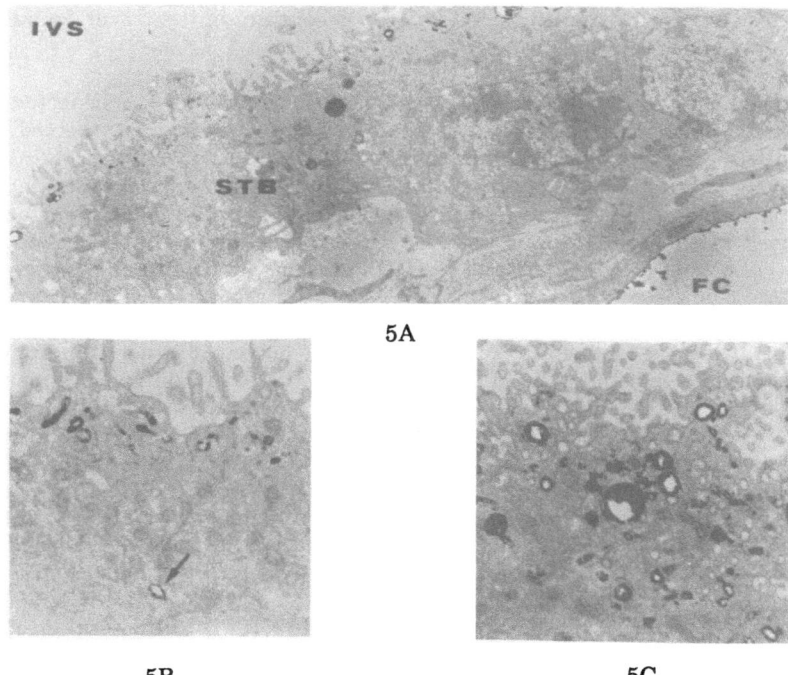

Figures 5A, 5B, and 5C. The figures show the morphology of the placental villus after 15 min incubation with cationic HRP injected into the maternal circuit. Note the presence of the tracer in the apical area of the syncytium and on the luminal surface of the fetal capillary endothelium. In Figure 5B, the arrow indicates a solitary vesicle close to the trophoblastic basement membrane. In Figure 5C, note the number of vesicles stained by the tracer. Most of the coated vesicles do not contain HRP.

This study demonstrates that in the human placenta, proteins of high molecular weight (at least 40000 daltons) can pass through the fetal capillary endothelium and the syncytium in both directions. In the fetal capillary, the passage occurs through the interendothelial space and/or via the transcytotic vesicles. These results do not demonstrate which pathway is involved, nor can they exclude one of the two routes. The problem certainly lies in the fact that the HRP concentrations at these sites are too low to allow for the visualization of the tracer.

The results at the syncytial levels are more interesting. In fact, cationic HRP can cross the syncytium in both directions. In addition, the pathways from fetus to mother and mother to fetus are different. After the bolus injection of HRP into the fetal circuit, the tracer is found in restricted areas of the syncytium, contained in endocytotic vesicles and cysternae, located along the whole thickness of this layer, from the basement membrane to the microvillar surface. On the contrary, after a bolus injection of HRP into the maternal circuit, the tracer is detected at the apical part of the syncytium, contained in endocytotic vesicles and multivesicular bodies.

Only in few cases are some endocytotic vesicles containing the tracer found close to the syncytial basement membrane.

In addition, while anionic HRP does cross the syncytium when injected into the fetal circuit, the same tracer is not detectable in the tissue when injected into the maternal circuit. These observations indicate that the pathway of protein transfer from mother to fetus is more selective than from the fetus to mother, and that the site of selection (at least for the charged proteins) is located at the syncytial surface. Also, considering the mother to fetus pathway, it is noteworthy that there is a large disproportion between the amount of HRP present in the apical portion of the syncytium, compared with the basal part. These findings are in agreement with previous studies (Lin, 1980; King, 1982) that failed to demonstrate the pathway by which IgG (which are known to cross the placenta) leave the syncytium and suggest the presence of a selection site, in the mother to fetus pathway, which appears to be located at the lysosomes. Since cationic HRP covalently binds to the plasma membranes, it is also possible that these results are simply demonstrating a membrane flow, instead of protein transfer (Steinman et al., 1983). In the guinea pig placenta, Kaufmann et al. (1982) have hypothesized the presence of transtrophoblastic network for the transport of macromolecules. The present results do not support that hypothesis since they indicate that the pathway from mother to fetus is different and more selective than the pathway from fetus to mother.

Although after the bolus injection of cationic HRP into the fetal circuit short tubular structures in the syncytium are stained by the tracer, their distribution and morphology do not fit the channel hypothesis. Finally, the stromal macrophages contained HRP in both fetus to mother and mother to fetus experiments. However, the role played by these cells in placental protein transfer still remains to be clarified.

In conclusion, the study of the macromolecular transfer across the human placental cotyledon perfused in vitro indicates that the pathway of the transfer from mother to fetus is different and more selective than the fetus to mother. It is likely that classes of macromolecules cross the human placenta by using the same nonspecific pathways of HRP and, depending on their isoelectric point, bind to the cell membrane or enter the bulk flow of fluid.

SUMMARY

Physiological studies have indicated that the human placenta is permeable to macromolecules. However, the routes of this passage have never been investigated previously. In this study we have investigated the passage of horseradish peroxidase (HRP) across the trophoblast membrane and the fetal capillary endothelium of the human placental cotyledon perfused in vitro. The results showed that a) the in vitro placental perfusion technique is a valid method to study directly the route of macromolecular passage; b) fetal capillary is a highly permeable structure in both the directions (mother to fetus and fetus to mother) and the passage occurs through the interendothelial space and/or via transcytotic vesicles; c) the syncytium is permeable to HRP in both directions, but the pathways from fetus to mother and from mother to fetus are different. In addition, anionic HRP does not enter into the syncytium when injected into the maternal circuit. These results confirm that the trophoblastic membrane and the fetal capillary endothelium in the human placental cotyledon perfused in vitro are permeable to macromolecules in both the directions and suggest that the mother to fetus pathway is different and more selective than the fetus to mother pathway.

REFERENCES

Challier, J.C., Guerre-Millo, M., Nandakumaran, M., Gerbaut, L., d'Athis, P. (1985) Clearance of compounds of different molecular size in the human placenta in vitro. *Biol. Neonate* 48, 143-148.

De Virgiliis, G., Sideri, M., Fumagalli, G., and Remotti, G. (1982) The junctional pattern of the human villous trophoblast (freeze fracture study). *Gynecol. Obstet. Invest.* 14, 263-272.

Firth, J.A., Bauman, K.F., and Sibley, C.P. (1983) The intercellular junctions of guinea pig placental capillaries: a possible structural basis for endothelial solute permeability. *J. Ultrastr. Res.* 85, 45-57.

Graham, R.C. and Karnovsky, M.J. (1966) The early stage of adsorption of injected horseradish peroxidase in the proximal tubule of the mouse kidney. Ultrastructural cytochemistry by a new technique. *J. Histochem. Cytochem.* 14, 291-302.

Kaufmann, P., Schroder, H., and Leichtweiss, P.H. (1982) Fluid shift across the placenta: Feto-maternal transfer of horseradish peroxidase in the guinea pig. *Placenta* 3, 339-348.

King, B.F. (1982) Absorption of peroxidase conjugated immunoglobulin G by human placenta: An in vitro study. *Placenta* 3, 395-406.

Lin, C.T. (1980) Immunoelectron microscopic localization of immunoglobulin G in human placenta. *J. Histochem. Cytochem.* 28, 339-346.

Schneider, H., Panigel, M., and Dancis, J. (1972) Transfer across the human placental of antipyrine, sodium and leucine. *Am. J. Obstet. Gynecol.* 114, 822-828.

Schweikhart, G. and Kaufmann, P. (1977) Zur Abgrenzung normaler, artefizieller and pathologischer Strukturen in reifen menschlichen Plazentasotten. I. Ultrastruktur des Syncytiotrophoblasten. *Arch. Gynak.* 222, 213-230.

Sibley, C.P., Baumann, K.F., and Firth, J.A. (1981) Ultrastructural study of the permeability of the guinea pig placenta to horseradish peroxidase. *Cell Tissue Res.* 219, 637-647.

Sibley, C.P., Bauman, K.F., and Firth, J.A. (1983) Molecular charge as determinant of macromolecule permeability across the fetal capillary endothelium of the guinea-pig placenta. *Cell Tissue Res.* 229, 365-377.

Sideri, M., De Virgiliis, G., Rainoldi, R., and Remotti, G. (1983) The ultrastructural basis of the nutritional transfer: Evidence of different patterns in the plasma membranes of the multilayered placental barrier. *Trophoblast Res.* 1, 15-26.

Silverstein, S.C., Steinman, R.M., and Cohn, Z.A. (1977) Endocytosis. *Ann. Rev. Biochem.* 46, 669-722.

Steinman, R.M., Mellman, I.S., Muller, W.A., and Cohn, Z.A. (1983) Endocytosis and recycling of plasma membranes. *J. Cell Biol.* 96, 1-27.

Stenseth, K., Hedin, U., and Thyberg, U. (1983) Endocytosis, intracellular transport and turnover of anionic and cationic proteins in cultured mouse peritoneal macrophages. *Eur. J. Cell Biol.* 31, 15-25.

Trophoblast Research 2:585-596, 1987

INTERRELATIONSHIPS OF PERFUSION PARAMETERS IN THE DUAL-PERFUSED HUMAN PLACENTAL COTYLEDON

Randy B. Howard, Janet Levy, Tomokazu Hosokawa, and M. Helen Maguire[1]

Department of Pharmacology, Toxicology and Therapeutics and Ralph L. Smith Research Center University of Kansas Medical Center Kansas City, Kansas 66103

INTRODUCTION

The technique of in vitro dual-perfusion of an isolated human term placental cotyledon, in which intervillous space as well as fetal cotyledonary vasculture is perfused (Panigel, 1968; Schneider et al., 1972), has been widely used for study of placental uptake, metabolism and transfer (cf. Schneider and Dancis, 1985). We used this preparation to determine effects of vasoactive agents and drugs on resistance of the human fetoplacental vascular bed, carrying out pharmacological dose-response studies in which fetoplacental perfusion pressure was monitored (Hosokawa et al., 1985; Howard et al., 1986; Howard et al., in preparation). The preparation does not appear to have been used for physiological study of the regulation of human fetoplacental vascular resistance. With the objective of improving the understanding of hemodynamic characteristics of the dual-perfused placental cotyledon, we perfused the fetal circuit of preparations with a salt solution which had arterial gas and pH values similar to those reported for human umbilical arterial blood (Longo, 1972). Perfusion was standardized with the aid of fetal venous gas and pH analysis, and the relationship of the different perfusion parameters to each other were analyzed statistically. In particular, it was anticipated that the flow rate of maternal perfusate might be an important determinant of other perfusion parameters, because maternal blood is the source of oxygen and nutrients for placenta and fetus in vivo. Moreover, it was considered that examination of the interrelationship of perfusion parameters with fetal vascular resistance may indicate which parameters contribute to fetal vascular resistance and which may be indicators of the level of fetal resistance in the preparation.

MATERIALS AND METHODS

Materials

Dextran (MW 37,000-43,000) was supplied by Chemical Dynamics Corporation, South Plains, NJ or ICN, Cleveland, OH. Sodium chloride was Fisher biological

[1]To whom correspondence should be addressed.

Submitted: 10 October 1985; Accepted: 8 August 1986

grade; all other salts and reagents were the best grade commercially available. Coomassie Blue G was obtained from Sigma Chemical Company, St. Louis, MO.

Dual Perfused Placental Cotyledon

Human placentae from normal full term pregnancies were obtained immediately after spontaneous delivery or delivery by cesarean section. Constant flow perfusion of the fetal circulation and intervillous space of a single cotyledon from each placenta was performed essentially as described by Schneider et al. (1972) with Earle's salt solution (Earle, 1943) containing 4 g% of dextran. Perfusate was gassed with 6% carbon dioxide and 94% nitrogen for delivery to the fetal circuit or with 95% oxygen and 5% carbon dioxide for delivery to the maternal circuit; pH and gas tensions of perfusate sampled close to the arterial cannulas were: fetal, pH 7.343 $\pm$ 0.016, pO_2 27 $\pm$ 3 mm Hg, pCO_2 43.9 $\pm$ 1.4 mm Hg; maternal, pH 7.392 $\pm$ 0.019, pO_2 540 $\pm$ 19 mm Hg, pCO_2 35.0 $\pm$ 0.9 mm Hg (means $\pm$ SD; n = 6-12 cotyledons). Preparations and perfusates were maintained at 37°C. Fetal and maternal arterial perfusion pressures were monitored with Statham P23D transducers and recorded on a Gilson ICT-2H Duograph. Gilmont #12 flowmeters were used to monitor perfusate flow rates. Gas and pH values of perfusates were determined using a Radiometer BMS blood gas analyzer.

When setting up the preparation, perfusion of the fetal circuit of a single cotyledon was commenced, keeping perfusion pressure below 60 mm Hg. Perfusion of the maternal circuit at a flow rate of 12-24 ml per min was then begun. A drop in fetal perfusion pressure usually ensued and flow rate of the fetal perfusate was increased to raise fetal perfusion pressure to 40-50 mm Hg. Maternal and fetal flow rates were further adjusted so that fetal venous gas values were within the range: pH, 7.32-7.39 units, pCO_2, 38-47 mm Hg and pO_2, 150-260 mm Hg. Fetal flow rate was typically 3-8 ml per min and fetal perfusion pressure was 40-50 mm Hg. The preparation was then allowed to stabilize under constant flow conditions so that fetal perfusion pressure was stable at 40-50 mm Hg.

Perfusion Parameters

Forty-one dual-perfused cotyledons were set up under the "steady state" conditions described above. Samples of fetal venous perfusates were then taken for measurement of fetal venous pH (F_V pH), fetal venous pCO_2 (F_V pCO_2), and fetal venous pO_2 (F_V pO_2). Maternal and fetal flow rates were noted (Q_M, Q_F, ml/min). Cotyledon weight (Wt, g) was determined at the end of experiment using fetal arterial injection of Coomassie Blue G solution for demarcation of the cotyledon (Hosokawa et al., 1985). Range of cotyledon weights was 8 to 27 g. Values of Q_M/g and Q_F/g (ml/min/g) and fetal vascular resistance (R_F, mm Hg x min x g/ml) were calculated. Mean values ($\pm$ S.D.) of perfusion parameters and cotyledon weight are listed in Table 1.

Statistical Methods

Simple linear regression was used to study the relationships of maternal flow per g, weight and fetal vascular resistance with the fetal venous gas and pH values

Table 1

Parameters of steady state perfusion of isolated placental cotyledon[a]

	FvpH —	$FvpO_2$ (mmHg)	$FvpCO_2$ (mmHg)	Q_M/g (ml/min/g)	Q_F/g (ml/min/g)	PP^b (mmHg)	R_F (mmHg.min.g/ml)	Wt (g)
$\bar{X}$	7.356	198	42.6	1.32	0.41	44.5	115	15.1
S.D.	0.021	35	2.3	0.34	0.15	3.9	40	4.7

[a] Results from 38-41 cotyledons, each from a different placenta.
[b] Baseline fetal perfusion pressure.

Table 2

Simple correlation coefficients of maternal flow per g (Q_M/g) with other perfusion parameters[a]

	$F_{Vp}H$	$F_{Vp}CO_2$	$F_{Vp}O_2$	Q_F/g	R_F	Wt
r	.238	-.490**	.373*	.643***	-.572***	-.791***

a   N = 38-41 cotyledons.   *   p < .05   **   p < .01   ***   p < .001

and with fetal flow per g. The statistical significance of the resulting simple correlation coefficients was computed using a t-test (Hays, 1973). Multiple linear regression was used to investigate the relationship between resistance and two sets of parameters which correlated with resistance. Separate analyses were performed; in the first, fetal venous gas values and pH served as predictor variables and in the second maternal flow per g, fetal venous pCO_2 and cotyledon weight were the predictor variables. A series of partial correlation coefficients was calculated with each multiple regression analysis. These were used to determine, within a given set, which predictor variables were most closely related to fetal vascular resistance independent of their relationships to each other.

Tests of significance reported with the partial correlation coefficients were based on partial F statistics (Draper and Smith, 1981a). Analyses were performed using BMDP statistical software (Dixon, 1982): BMDP6D for simple linear regression and BMDP2R for multiple regression.

RESULTS

Relationships between steady state perfusion parameters and maternal flow per g, as measured in 41 cotyledons were initially studied individually via simple linear regression analysis. All perfusion parameters except fetal venous pH were significantly related to maternal flow per g; cotyledon weight was also significnatly related to maternal flow per g (Table 2). Five of these relationships are presented graphically in Figures 1 and 2. Especially evident from the figures is the varied nature of these relationships, both in terms of strength and direction. For example, maternal flow per g is seen to be moderately related to both fetal venous pCO_2 and fetal venous pO_2, but in opposite directions. Fetal flow per g and fetal vascular resistance are both highly related to maternal flow per g, but in opposite directions, indicating that higher maternal flow per g was associated with higher fetal flow per g and lower fetal vascular resistance.

Further statistical analysis focused on the relationships between each of the perfusion parameters and cotyledon weight, again via simple linear regression. The resulting correlation coefficients are presented in Table 3. These coefficients indicate that both maternal flow per g and fetal flow per g have strong, negatively directed linear relationships with cotyledon weight. Fetal vascular resistance and fetal venous pCO_2 are also highly related to weight, but in a positive direction.

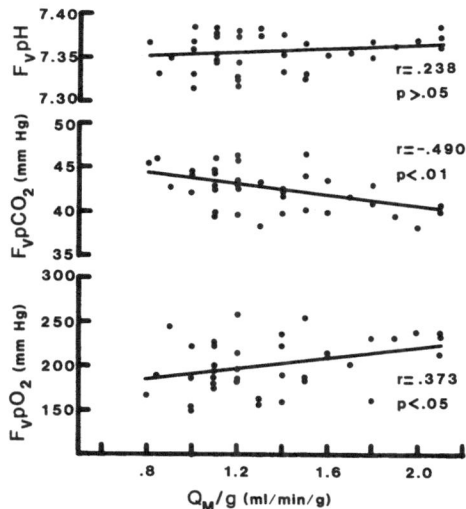

Figure 1. Scatter plots showing the relationship between maternal flow per g (Q_M/g) and the steady state fetal venous perfusate gas and pH values in each of 39 different dual-perfused cotyledons. Maternal perfusate was gassed with 95% oxygen - 5% carbon dioxide. Fetal perfusate was gassed with 6% carbon dioxide - 94% nitrogen. Top panel: $F_{Vp}H$ vs. Q_M/g; middle panel: $F_{Vp}CO_2$ vs. Q_M/g; bottom panel: $F_{Vp}O_2$ vs. Q_M/g.

Table 3

Simple correlation coefficients of cotyledon weight with other perfusion parameters[a]

	$F_{Vp}H$	$F_{Vp}CO_2$	$F_{Vp}O_2$	Q_F/g	R_F	Q_M/g
r	.383*	.599***	-.280	-.686***	.663***	-.791***

[a]   N = 38-41 cotyledons.
*   p < .05
***   p < .001

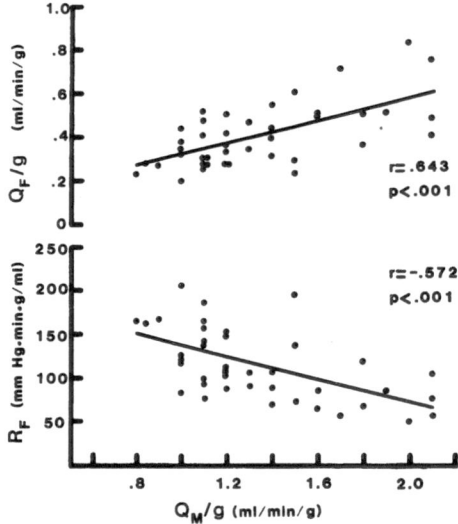

Figure 2. Scatter plots showing the relationship between maternal flow per g (Q_M/g) and steady state fetal flow per g (Q_F/g) (upper panel) and between maternal flow per g (Q_M/g) and calculated fetal vascular resistance (R_F) (lower panel) in each of 41 different dual-perfused cotyledons.

The results of a third set of regression analyses, assessing the linear associations between each perfusion parameter and resistance, are shown in Table 4. The correlations shown there suggest that resistance is highly associated with maternal flow rate, and fetal venous pCO_2. Although statistically significant, the correlation coefficients between resistance and fetal venous pH and pO_2, suggest weaker relationships between these parameters and resistance.

Table 4

Simple correlation coefficients of fetal vascular resistance (R_F) with other perfusion parameters[a]

	$F_{Vp}H$	$F_{Vp}CO_2$	$F_{Vp}O_2$	Q_F/g	Q_M/g	Wt
r	-.349*	.576***	-.334*	-.882***	-.572***	.663***

[a]   N = 38-41 cotyledons.
*   p<.05
***   p<.001

Two separate multiple regression analyses further examined the way in which perfusion parameters and cotyledon weight interact to influence measures of fetal vascular resistance. One analysis focused on the fetal venous gas and pH values. Three partial correlation coefficients were computed and are presented in Table 5. Partial correlation coefficients assess the degree or extent of relationship between two sets of measurements, while holding constant the influence of a third or fourth related measurement. For example, as shown in Table 5, the partial correlation of resistance and fetal venous pH is close to zero (-0.082) when the influence of both fetal venous pCO_2 and pO_2 are statistically removed. Fetal venous pCO_2, however, remains related to resistance (partial r = 0.448) when fetal venous pH and fetal venous pO_2 are statistically controlled. The second multiple regression analysis examined the combined influence of maternal flow per g, cotyledon weight and fetal venous pCO_2 on resistance. Resulting partial correlations are presented in Table 6, and suggest that none of these three parameters is related to resistance independent of its relationship to the other two.

Table 5

Simple and partial correlation coefficients of fetal vascular resistance (R_F) with fetal venous gas and pH values[a]

	$F_{Vp}H$	$F_{Vp}CO_2$	F_VO_2
simple r	-.337*	.573***	-.314*
partial r[b]	-.082	.448**	-.100

a   N = 36 cotyledons.
b   Two parameters were partialled out.
*   p<.05
**   p<.01
*** p<.001

Table 6

Simple and partial correlation coefficients of fetal vascular resistance (R_F) with three perfusion parameters[a]

	$F_{Vp}CO_2$	Q_M/g	Wt
simple r	.576***	-.598***	.661***
partial r[b]	.298	-.159	.266

a   N = 36 cotyledons.
b   Two parameters were partialled out.
*** p<.001

DISCUSSION

This study has compared interrelationships of the perfusion parameters pertaining under steady state conditions of dual-perfused human placental cotyledon in which the fetal circuit is perfused with a low oxygen perfusate. Primary goals of the study were to determine the relationship of maternal flow per g to other perfusion parameters, and to delineate those parameters which are most important in determining fetal vascular resistance in the perfused cotyledon.

The correlation of perfusion parameters with maternal flow per g (Figures 1 and 2 and Table 2) showed that higher fetal venous pO_2 (an index of maternal to fetal oxygen transfer) and lower fetal venous pCO_2 (a measure of fetal to maternal CO_2 transfer and metabolic waste removal), were associated with higher maternal flow rates, indicating that the level of maternal flow per g may be an important determinant of fetal perfusate gas parameters. The lack of significant correlation between fetal venous pH and maternal flow per g (Table 2) may mean that, under the steady state conditions of perfusion, lactate production is not important in determining fetal venous pH, and that the small increase in pH associated with higher maternal flow per g reflects only the decrease in fetal venous pCO_2, which is itself not large enough to result in major changes in pH.

The strong positive correlation between maternal flow per g and fetal flow per g and the inverse relationship between maternal flow per g and fetal vascular resistance (Figure 2, Table 2) suggest that fetal flow and resistance are matched to and perhaps determined by maternal flow in the in vitro preparation of human placental cotyledon. It is true that both maternal flow and fetal flow were set by the experimenter; each preparation was standardized by adjusting both maternal and fetal flow rates so that fetal venous gas and pH values fell within the desired range and fetal perfusion pressure was between 40 and 50 mm Hg. Initially fetal flow was established; commencement of maternal flow was usually followed by a drop in fetal perfusion pressure and fetal flow rate was then increased to raise fetal perfusion pressure to the 40-50 mm Hg. From the experimental protocol it can be seen that the establishment of maternal flow determines fetal perfusion pressure which thereby determines the adjustment of fetal flow rates required to maintain fetal perfusion pressure in the 40-50 mm Hg range. Similarly, resistance has also been determined by maternal flow rate since resistance is equal to the fetal perfusion pressure divided by fetal flow per g which was set in response to maternal flow rate.

The strong negative relationship between maternal flow per g and cotyledon weight (Table 2) indicates that, although the maternal flow was increased in large cotyledons to achieve fetal gas values and perfusion pressure within the desired range, it was not increased in proportion to cotyledon size with the result that large cotyledons were less adequately perfused. In effect, the size (weight) of the cotyledon actually determined, to a large extent, maternal flow per g, so that there was a trend when setting up preparations, towards use of lower maternal flow per g with large cotyledons. Thus, size and maternal flow per g of the cotyledons were very closely related (r = -.791). Any perfusion parameter closely related to maternal flow per g would also be expected to be closely (albeit inversely) to cotyledon weight. This is seen to be the case as measured parameters which showed strong correlations with maternal flow per g (for example, fetal venous pCO_2, fetal flow per g, and fetal resistance) also correlated strongly with weight (Table 3).

While it is impossible that cotyledon weight alone could influence any of the parameters measured, it appears likely that weight primarily influences maternal flow per g which then largely determines all other parameters. Correlation coefficients demonstrated that fetal venous gas and pH values moved in less favorable directions with increasing cotyledon weight, i.e. pH and fetal venous pO_2 tended to become lower and fetal venous pCO_2 higher (Table 3). This would be expected if fetal venous gas and pH values were determined by maternal flow per g which was found to decrease as cotyledon size (weight) increased.

Fetal vascular resistance similarly correlated with perfusate gas values and flow rates (Table 4), and therefore could have been determined by more than one variable or, alternatively, by a single variable (such as maternal flow per g) which in turn may have largely determined the values of all the other variables. In an attempt to dissect out the important interrelationships, multiple regression analysis was used to investigate the relationship of resistance to two subsets of parameters. Results showed that, of the fetal venous gas and pH values, pCO_2 was the only parameter related to resistance exclusive of its relationship to the other variables (Table 5). A similar analysis of the relationship of resistance to maternal flow per g, cotyledon weight and fetal venous pCO_2, failed to indicate that any one of these parameters was related significantly to resistance over and above its relationship to the other two (Table 6). This finding was, in part, a consequence of the strong correlations among the three predictor variables. As shown in Table 7, maternal flow per g and weight are more closely related to each other than either is to resistance. Statistical control of any two of these predictors removes much of the influence of the remaining variable on resistance. If, for example, both weight and maternal flow per g are statistically controlled, their influences on fetal venous pCO_2 are removed. The partial correlation coefficient then reflects the extent to which fetal venous pCO_2 is related to resistance excluding any mediating effects of either weight or maternal flow per g. As shown in Table 6, this correlation coefficient is quite low (.298). Similarly, if the influence of fetal venous pCO_2 and weight are controlled, maternal flow per g is virtually unrelated to resistance ($r = -.159$). Therefore, none of these three parameters is individually related to resistance when the effects of the other two are statistically controlled.

Although maternal flow per g and weight are so highly correlated that the influence of each alone on resistance could not be statistically separated, it was possible to assess their combined influence. The results of a multiple regression analysis where fetal vascular resistance was predicted from weight and maternal flow per g yielded a squared multiple correlation coefficient (R^2) (Draper and Smith, 1981b) of .44. This finding indicates that 44% of the variability in resistance measures, across cotyledons, is associated with variations in cotyledon weight and maternal flow per g. The addition of fetal venous pCO_2 to the regression equation increases R^2 to only 0.50. These results suggest that maternal flow per g and weight, in combination, are predictive of resistance. It seems likely, however, that maternal flow per g is the primary factor and interacts with weight to determine fetal venous pCO_2 and resistance. Since larger cotyledons tended to be set up with lower maternal flow per g (relative under perfusion), and to have a higher fetal venous pCO_2

Table 7

Correlation matrix for fetal vascular resistance (R_F) with three other perfusion parameters[a]

	Q_M/g	Wt	$F_{VP}CO_2$	R_F
Q_M/g	----	-.791***	-.490**	-.572***
Wt	----	----	.599***	.663***
$F_{VP}CO_2$	----	----	----	.576***

[a]   Simple correlation coefficients form 38-41 different cotyledons.
** $p < .01$
*** $p < .001$

(associated with relative under perfusion), weight, maternal flow per g and fetal venous pCO_2 would be expected to be related to each other; if maternal and fetal flows are indeed matched, as suggested earlier, these three variables would therefore be related to fetal flow per g and resistance as well. It is possible that large cotyledons have intrinsically higher fetal vascular resistances, but it is also possible that such a relationship in vitro is an artifact of the technique. Since the maternal perfusate is delivered through randomly placed glass cannulas instead of through the spiral arteries which deliver maternal flow in vivo, it may be the case that in larger cotyledons, perfusate delivery is not as well distributed in the intervillous space and therefore results in a lower real maternal flow per g and higher resistance.

The interrelationships we report here of perfusion parameters measured in 41 cotyledons perfused under steady state conditions, indicate that the flow rate per g of maternal perfusate determines gas values of the fetal venous perfusate, flow rate per g of fetal perfusate and fetal vascular resistance, although an influence of cotyledon weight on these parameters cannot be ruled out. We believe these findings comprise the first indication in human placenta of a relationship between fetal and maternal flows at the level of the cotyledon. Such a correlation appears consistent with the concept of a local mechanism for regulation of fetoplacental vascular resistance in response to changes in maternal perfusion of the in situ placenta; a primary factor in such a mechanism may be change in uteroplacental oxygen delivery resulting from reduced maternal perfusion. Further to these findings we have shown that cessation of maternal flow of the dual-perfused cotyledon results in an increase in fetoplacental perfusion pressure (unpublished results). Moreover, we have also demonstrated that acute reduction in oxygen tension in the maternal perfusate of dual-perfused cotyledons elicits a prompt reversible fetoplacental vasoconstriction (Howard et al., in preparation), observations which provide further support for the existence of local mechanisms for regulation of fetoplacental vascular resistance.

SUMMARY

Human term placental cotyledons were set up for in vitro dual perfusion using Earle's salt solution containing 4 g% dextran (MW approx. 40,000), which was gassed with 95% O_2 - 5% CO_2 for delivery to the maternal circuit, and with 94% N_2 - 6% CO_2 for delivery to the fetal circuit. Maternal and fetal flow rates were adjusted so that fetal venous gas values and pH were within acceptable ranges and fetal perfusion pressure was 40-50 mm Hg. Interrelationships of the following perfusion parameters, fetal venous gas and pH values (F_V pO_2, F_V pCO_2 and F_V pH), maternal and fetal flow rates per g, fetal vascular resistance (R_F) and cotyledon weight, were studied in 41 cotyledons. Simple linear regression analysis between pairs of parameters showed that maternal flow per g correlated negatively with F_V pCO_2 and positively with F_V pO_2. Fetal flow per g and R_F were highly related to maternal flow per g, fetal flow in a positive, and R_F in a negative direction. Maternal and fetal flows per g had strong negatively directed relationships with cotyledon weight; R_F and F_V pCO_2 were highly, positively, related to weight. Multiple regression analysis showed that the combination of maternal flow per g and weight was a significant determinant of R_F. The analysis indicates that the level of maternal flow may determine fetal venous gas values and fetoplacental vascular resistance, although an effect of cotyledon weight on these parameters cannot be ruled out, and suggests that in human placenta there is a correlation between fetal and maternal flows at the levels of the cotyledon.

ACKNOWLEDGEMENTS

The authors thank Dr. C.R. King and the staff of the delivery room of Bell Memorial Hospital for help in obtaining placentae. This work was supported by NIH Grants HD 14888 and HD 02528.

REFERENCES

Dixon, W.J. (1983) *BMDP Statistical Software*. Berkeley: University of California Press.

Draper, N.R. and Smith, H. (1981a) *Applied Regression Analysis*, second ed., New York, John Wiley and Sons, p. 102.

Draper, N.R. and Smith, H. (1981b) *Applied Regression Analysis*, second ed., New York, John Wiley and Sons, p. 33.

Earle, W.R. (1943) Production of malignancy in vitro iv. The mouse fibroblast cultures and changes seen in the living cells. *J. Natl. Cancer Inst.* 4, 165-212.

Hays, W.L. (1973) *Statistics for the Social Sciences*, second ed., New York, Holt, Rinehart and Winston, Inc., p. 661.

Hosokawa, T., Howard, R.B., and Maguire, M.H. (1985) Conversion of angiotensin I to angiotensin II in the human fetoplacental vascular bed. *Brit. J. Pharmacol.* 84, 237-241.

Howard, R.B., Hosokawa, T., and Maguire, M.H. (1986) Pressor and depressor actions of prostanoids in the intact human fetoplacental vascular bed. *Prost. Leuk. Med.* 21, 323-330.

Howard, R.B., Hosokawa, T., and Maguire, M.H. Responses of the human feto-placental vascular bed to vasoactive agents: Studies in the dual-perfused human placental cotyledon. In preparation.

Howard, R.B., Hosokawa, T., and Maguire, M.H. Hypoxia-induced fetoplacental vasoconstriction in perfused human placental cotyledons. In preparation.

Longo, L (1972) Disorders of placental transfer. In: *Pathophysiology of Gestation*, eds., N.S. Assali and C.R. Brinkmann, New York, Academic Press, pp. 1-65.

Panigel, M. (1968) Placental perfusion. In: *Fetal Homeostasis*, vol. 4, ed., R. Wynn, New York, Appleton Century Crofts, pp. 15-25.

Schneider, H., Panigel, M., and Dancis, J. (1972) Transfer across the perfused human placenta of antipyrine, sodium and leucine. *Am. J. Obstet. Gynecol.* 113, 822-828.

Schneider, H. and Dancis, J. (1985) In vitro perfusion of human placental tissue. *Contrib. Gynecol. Obstet.*, Vol. 13.

Trophoblast Research 2:597-605, 1987

IN VITRO PERFUSION OF HUMAN PLACENTA

- A Workshop Report -

Henning Schneider[1] and Joseph Dancis[2]

[1]Division of Perinatal Physiology
Department of Obstretrics
University of Zuerich
Franenklinik - Str. 10
CH - 8091 Zuerich, Switzerland

[2]Department of Pediatrics
New York University
School of Medicine
550 First Avenue
New York, New York 10016 USA

The workshop on in vitro perfusion of human placenta concentrated on recent advances that might influence the use of the techniques. Presentations were made by Henning Schneider, Nicholas Illsley, and Richard Miller. Maurice Panigel and Heinz-Peter Leichtweiss introduced the discussion The allotted time passed quickly, with questions and conversations overflowing into the corridors during the rest of the conference. In the following, we shall review briefly the workshop and attempt to summarize the ensuing conversations in a form that we hope will be useful to the reader.

ISCHEMIC CHANGES DEVELOPING IN HUMAN PLACENTAL TISSUE BEFORE THE START OF IN VITRO PERFUSION - Henning Schneider, Zuerich

Different from perfusion of animal placentae, the in vitro perfusion of isolated human placental tissue is preceded by a significant period of partial or total ischemia which begins with the moment of cord clamping. The perfusionist has only limited control over the length of the time period before in vitro perfusion of the fetal and maternal compartments can be started. The extent of tissue changes due to ischemia may influence the result of the perfusion experiment and tissue changes may develop at the time of reperfusion because of the toxic effect of oxygen in ischemic tissue.

Measurement of tissue lactate shows a continuous rise with time after cord-clamping and differences between placentae from vaginal deliveries and from elective cesarean section can be explained by differences in duration of ischemia. The

Submitted: 10 December 1985; Accepted: 16 April 1986

constant lactate production calculated from the rise in tissue levels up to 30 min after cord-clamping gives the mean value of 0.35 μmoles/g/min (Duft and Schneider, 1985) which is in the same order as the lactate production during well oxygenated perfusion (Sodha et al., 1984). There is no indication of increased lactate production as a result of accelerated glycolysis during ischemia. Tissue ATP levels show a gradual decline with length of ischemia. A possible rapid initial drop in tissue ATP is difficult to demonstrate because no samples during the early part of ischemia can be obtained (Bloxam and Bobinski, 1984). There is recent evidence of increased resistance of fetal tissues including placenta to ischemia as shown by a better preservation of tissue ATP levels when compared to adult tissues (Harkness et al., 1985). When samples are taken from different sites in the same placenta, considerable differences are seen in ATP levels while lactate concentrations show less variation. Hypoxanthine and adenosine gave similar variations (Maguire et al., 1985), and different degrees of hypoxia at different sites in the same placenta must be considered.

The toxicity of oxygen during reperfusion of ischemic tissue is explained by the accumulation of "oxygen radicals" which develops from oxidation of metabolites. These radicals if not rapidly removed, react with polyunsaturated fatty acids of membrane lipids leading to disruption of membranous structures (Halliwell, 1978). Protective measures like perfusion with medium with low oxygen content containing neutralizing enzymes like catalase and superoxide dismutase as well as antioxidants before the tissue is exposed to high oxygen concentrations should be tried.

EXTENDED PERFUSIONS OF THE HUMAN PLACENTAL LOBULE AND PERFUSION/PERFUSION OVERLAP - Richard K. Miller, Rochester

The approach to long term perfusion developed in Rochester has been successful in maintaining the preparation for a period of 12 hr with the placenta retaining the capacity for establishing a transplacental gradient and EM morphology remaining quite acceptable. The limit, so far, is the stamina of the investigators!

Both maternal and fetal perfusates are recirculated, conserving materials. Precautions for sterility are stringent and antibiotics are added to the perfusates. Enriched media, such as those used for tissue culture, are used as perfusates. The perfusates are replaced every 4 hrs with fresh medium to replenish nutrients and remove metabolites. Details of the technique have been published before (Miller et al., 1985).

In isolating the lobule for perfusion, most of the placenta is trimmed away. To avoid damaging the lobule and to obtain secure fixation in the perfusion chamber, a significant amount of nonperfused placenta is retained in the chamber. Under the conditions of long term perfusion with recirculated maternal and fetal perfusates, the system is "closed" and substrate diffuses from the perfused area into the nonperfused placenta. The magnitude of this "4th compartment" depends on the diffusion characteristics of the perfused materials. This phenomenon probably pertains to a lesser degree to all perfusion preparations but is greatly exaggerated in the closed system with prolonged perfusion.

Dr. Miller also addressed the question of the correct placement of the maternal cannulae. Most perfusionists attempt to define the fetally perfused area by searching for blanching on the maternal surface, into which are placed the maternal cannulae. Placement may be made more precise by measuring oxygen transfer. An increase in pO_2 following insertion of the maternal cannula will indicate proper alignment (Wier and Miller, 1985).

Dr. Schneider has used another approach to this problem. An intermittent injection of perfusate into the fetal artery with increase in pressure causes detectable movement in the decidual plate, defining the perfused area.

DIFFERENT EXPERIMENTAL APPROACHES TO THE MEASUREMENT OF TRANSPLACENTAL FLUX - Nicholas Illsley, San Francisco

The measurement of transplacental flux is complicated by the rapid occurrence of backflux as well as uptake and release of substrate on both sides of the tissue barrier interposed between the two blood streams. In addition to the constant substrate infusion for flux measurements under steady state conditions, various bolus techniques have been proposed. Bissonnette (1981) studied the transfer of glucose and other hexoses across the guinea pig placenta in vivo by injecting ^{14}C-labeled glucose as a bolus through a catheter into the descending aorta of the sow and a single fetal sample was taken by continuously withdrawing blood from the umbilical vein over 20-25 sec starting with injection. Tritiated water was injected as a second isotope together with ^{14}C-glucose and the placental transfer index was calculated from the ratio of ^{14}C-hexose: ^3HOH in fetal whole blood over ^{14}C-hexose: ^3HOH in maternal plasma. The paired tracer dilution technique was first applied to the placenta by Yudilevich (1979). It is mainly used for measurement of cellular uptake either from the maternal or fetal compartment and not so much for transplacental flux. Total uptake is corrected for diffusion into the extracellular space by use of an extracellular marker for a second tracer.

While rapid back flux of tracer interferes with unidirectional flux determinations in all of these methods the use of the in vitro perfusion technique of human placental tissue for flux determination is further complicated by the pool type flow behavior on the maternal side.

One of the basic assumptions for the use of bolus injection to measure tissue uptake or transplacental flux is that it is injected into a perfusion stream moving along parallel-sided, long capillaries. To solve some of these problems, a new method was recently described which requires two bolus injections of labeled substrate, first into the donor side with fractionated outflow collection on the acceptor side and subsequently a bolus injection on the acceptor side with outflow collection on the same side (Wooton et al., 1985). Complex mathematical data treatment was applied to deconvolve both efflux curves which permitted calculation of true placental flux. A hydraulic two pool perfusion model with preset flows between the two pools was built for verification of the method. Good agreement was found between the flux derived from the preset flow and that obtained with dye bolus injections into the perfusion streams and deconvolution of the two efflux curves. To avoid the complex data treatment the venous outflow after the two bolus injections may also be collected as a pooled sample and the fraction of the injected bolus transferred unidirectionally from one side to the other side is equivalent to the ratio of the total radioactivity in the venous outflow after the first bolus over that recovered from the venous outflow following the second bolus. This rather simple approach was also verified using the described model (Illsley, 1985).

The questions raised during the workshop and after fell into two broad categories: the pros and cons of the several experimental designs and the value of marker molecules such as antipyrine and L-glucose. Both are complex questions that could be discussed at length. The following brief comments are intended as useful introductory guidelines.

Bolus Technique

By presenting substrate, briefly as a bolus, it is hoped to obtain information on events at the membrane surface, particularly uptake rates, avoiding other complicating factors. Yudilevich's mathematical treatment of data is commonly used. The brevity of the experiments makes possible sequential studies on one placenta. The application of the technique to guinea pig placenta has provided interesting information. In the hands of one of us (Schneider, 1983), the consistency of results in sequential boli given in the same experiment has been disappointing.

The bolus technique offers an additional advantage in the study of unstable materials. This has been demonstrated in a paper during this conference on vitamin C transport. Dehydroascorbic acid is rapidly reduced to L-ascorbic acid under the usual perfusion conditions. By using the bolus technique Leichtweiss et al. were able to study the uptake of L-ascorbic acid by human placenta, avoiding the confusion of contaminating dehydroascrobic acid.

The use of the bolus technique to study transfer across the placenta exposes the substrate to the full range of dynamic events within the placenta and at the second membrane. Despite this, Illsley believes that useful data may be obtained on selected materials, as he described at this workshop.

The "Australian" Technique

In the original technical description, the perfused placenta is fixed with the maternal surface up. The group from Harrow, England, has preferred the reverse to facilitate collection of maternal effluent without pooling (Penfold et al., 1981). The two approaches have not been compared in one laboratory so that it remains moot as to whether either presents distinct advantages.

Open:Open Perfusions

The perfusates are not recirculated and the materials under study are presented in either the maternal or fetal perfusate and the rate of appearance in the recipient perfusate is measured. By not recirculating the perfusates, a constant transplacental gradient is maintained providing data on transfer across the placenta under relatively steady-state conditions. The kinetic factors that are emphasized are uptake from the donor circuit and release into the recipient. Backflow does not appear to be a significant complicating factor. The importance of intraplacental events in modifying transfer rates is reduced by preventing the build up of large placental concentrations. It is, however, not eliminated and can be very significant.

Open:Closed Perfusions

The non-recirculated perfusate maintains a constant concentration of the compound under study. Recirculating the recipient circuit closes the system so that it proceeds over time towards an equilibrium against the non-recirculated perfusate concentration. There is a progressive decrease in gradients with increased prominence in backflow. The effects of intraplacental events (mixing with placental pools, metabolism) become more evident.

There are several possible advantages to closing one circulation. Perfusion materials are conserved. Slowly transferred materials may accumulate to detectable levels. The most common application has been the investigation of the establishment of transplacental gradients. By leaving one circuit open, a constant supply of nutrients is maintained and an accumulation of metabolites is prevented or reduced.

Closed:Closed Perfusions

The most interesting application of this approach has been to the long term perfusions, as described above. The method is still too new to discuss its usefulness.

The preceding cursory review emphasizes current experimental designs. The flexibility of the perfusion technique permits further modifications applicable to particular questions. The emphasis so far has been on placental transfer. Protein synthesis, another major placental function, has received scant attention. Interest in pharmacology and toxicology has just begun. Two papers in the present conference on the control of the fetal circulation suggest additional avenues of study.

The Use of Marker Substances in Placental Transfer

A good portion of the general discussion was devoted to the proper choice of a reference or marker substance to monitor the transfer potential of a given perfusion preparation.

First the question must be raised why marker substances are necessary and then criteria for the selection of the proper marker can be discussed. Transfer by passive diffusion is dependent on a number of variables of which some are known to the investigator. It is one of the advantages of the isolated in vitro perfusion that the concentration difference between maternal and fetal sides can be controlled and diffusion rates can be expressed relative to the transplacental gradient either as clearance or permeability. Other variables however remain unknown. These are related either to perfusion flow for rapidly diffusing substances ("flow limited") or to membrane characteristics for slowly diffusing compounds ("membrane limited"). While the isolated perfusion preparation allows determination of the flow-rate in the two circuits, the presence of shunts and poor overlap leads to considerable differences between measured flow and effective flow. The fraction of the total flow which actually reaches the part of the membrane area which is equally well perfused on the other side is not easily determined (Schneider et al., 1986; Schroeder et al., 1985).

The tissue weight measured at the end of the experiment provides only a very crude estimate of the extent of the functioning exchange area, so that the membrane surface also remains unknown. A control for differences in flow and membrane however, is desirable to reduce the considerable degree of variability in diffusion found among experiments under otherwise standardized conditions. Furthermore, if the effect of either flow or membrane on transfer is to be studied, a correction for the variability introduced by the other parameter is required. The use of different flow rates within one experiment may at the same time lead to changes in perfused area and a correction for the membrane related change would be needed to properly study the effect of flow changes (Illsley et al., 1986). In certain protocols, a control for both flow and membrane related variables may be needed.

Since the variables cannot be measured directly, marker substances which are predominantly affected in diffusion by either one of the variables are used together with the compound under study. The transfer or clearance obtained for a test compound can then be normalized for flow or membrane related variables by forming a ratio with the transfer or clearance of the appropriate marker, giving a transfer or clearance index (Dancis et al., 1973). When active or carrier mediated transfer is studied, a marker is used to correct for passive diffusion which may account for a substantial portion of total transfer.

It is clear that the appropriate marker must be chosen for a given experimental design. Antipyrine and ^3HOH are the two classical flow limited markers, both being practically inert, a basic requirement for a marker substance. The rate of placental diffusion of these substances is directly proportional to the flow delivering the substance on one side of the membrane and the receiving flow on the other side. Antipyrine was used in the in situ perfusion of the guinea pig placenta, where only one flow can be easily measured and controlled. Changes in antipyrine transfer were taken as evidence for variations in maternal blood flow to the in situ placenta (Kelman and Sasser, 1977). In the isolated human placental preparation, antipyrine has proved to be a good marker to correct for differences in effective flow rates among experiments and also for possible changes within a single experiment (Schneider et al., 1972). The ratio of the transfer or clearance of the test substance over the flow limited marker, usually called index, provides normalized results. In a strict sense this way of data correction is only applicable to test substances with the same degree of flow limitation as the marker and it may be misleading when it is used for slowly diffusing, i.e., "membrane limited", material. A drop in flow rate would result in a greater decline of diffusion of the marker than of the test compound resulting in a rise in the index which may be incorrectly interpreted.

The interexperimental variation in diffusion of an only partially flow-limited compound is therefore not completely corrected by use of the clearance index with a flow limited marker (Schneider et al., 1985). For compounds which are predominantly "membrane limited" in diffusion a marker from this group of substances is required. ^{51}Cr-EDTA as well as creatinine have been used for this purpose (Illsley et al., 1985; Eaton and Contractor, 1985). Creatinine may be preferable since like antipyrine a simple method for determination is available so that isotopes can be reserved for the test compound. However its predominantly "membrane limited" diffusion under in vitro perfusion conditions remains to be verified. Membrane markers are indicators of the exchange area since their diffusion rate is directly proportional to the membrane surface. With flow limited markers differences in exchange area may be compensated by differences in effective flows.

However it must be recognized that surface area is not the only membrane factor determining diffusion. There is good evidence that in vitro perfusion may result in changes in membrane permeability (Hedley and Bradbury, 1980) and the membrane marker will not differentiate between differences in surface area and differences in permeability. A membrane marker because of its slow transfer may be useful in a recirculating system to monitor possible spontaneous changes in permeability particularly during long term perfusion. In recirculating systems the classic "flow limited" markers are not very useful to monitor changes during the experiment since equilibration between the two circuits is rapidly achieved. This is avoided by the measurement of oxygen transfer which has been recommended as a flow marker for recirculating systems (Wier and Miller, 1985).

In studies of carrier mediated transport stereoisomers of the compound under investigation are ideal as diffusion markers. Because of identical physio-chemical characteristics the stereoisomer will diffuse at the same rate as the test molecule. Since most of the carrier transport systems are stereospecific, the stereoisomer usually is not subject to mediated transport. Since the appropriate stereoisomers may not be easily available extracellular markers like L-glucose have been used to correct for diffusion (Eaton and Yudilevich, 1981). However, it should be noted that such a marker only allows correction for extracellular diffusion while differences in physico-chemical characteristics between marker and test molecule may give different rates of transcellular diffusion which would lead to either over or under correction.

REFERENCES

Bissonnette, J.M. (1981) Studies in vivo on glucose transfer across the guinea pig placenta. *Placenta Suppl.* 2, 155-162.

Bloxam, D.L. and Bobinski, P.M. (1984) Energy metabolism and glycolysis in the human placenta during ischemia and in normal labor. *Placenta* 5, 381-394.

Dancis, J., Jansen, V., Kayden, H.J., Schneider, H., and Levitz, M. (1973) Transfer across perfused human placenta. II. Free fatty acids. *Pediatr. Res.* 7, 192-197.

Duft, R. and Schneider, H. (1985) Ischemic postpartum changes in human placental tissue. Unpublished observations.

Eaton, B.M. and Yudilevich, D.L. (1981) Uptake and asymmetric efflux of amino acids at maternal and fetal sides of placenta. *Am. J. Physiol.* 241, C106-C112.

Eaton, B.M. and Contractor, S.F. (1985) Maternal to fetal movement of creatinine as a measure of perfusion efficiency and diffusional transfer in the isolated human placental lobule. *Placenta* 6, 341-346.

Harkness, R.A., Coade, S.B., Simmonds, R.J., and Duffy, S. (1985) Effect of a failure of energy supply on adenine nucleotide breakdown in placentae and other fetal tissues from rat and guinea pig. *Placenta* 6, 199-216.

Halliwell, B. (1978) Biochemical mechanisms accounting for the toxic action of oxygen on living organisms: the key role of superoxide dismutase. *Cell Biol. Int. Rep.* 2, 113-128.

Hedley, R. and Bradbury, M.W.B. (1980) Transport of polar non-electrolytes across the intact and perfused guinea pig-placenta. *Placenta* 1, 227-285.

Illsley, N. (1985) A simplified method for measurement of unidirectional transplacental flux. Unpublished observation.

Illsley, N.P., Hall, S., and Stacey, T.E. (1987) The modulation of glucose and oxygen transfer across the human placenta by intervillous flow rates: an in vitro perfusion study. *Trophoblast Res.* 2, 537-546.

Kelman, B.J. and Sasser, L.B. (1977) Methylmercury movements across the perfused guinea pig placenta in late gestation. *Toxicol. Appl. Pharmacol.* 39, 119-127.

Maguire, M.H., Westermeyer, F.A., and King, Ch.R. (1985) Adenosine, inosine and hypoxanthine levels in human term placenta. *Abst. 10th Rochester Trophoblast Conference*, 34.

Miller, R.K., Wier, P.J., Maulik, D., and di Sant'Agnese, P.A. (1985) Human placenta in vitro: Characterization during 12 hr of dual perfusion. *Contrib. Gyn. Obstet.* 13, 77-84.

Penfold, P., Drury, L., Simmonds, R.J., and Hytten, F.E. (1981) Studies of a single placental cotyledon in vitro: I. The preparation and its viability. *Placenta* 2, 149-154.

Schneider H. (1983) Placental transfer of D- and L-glucose studies in the in vitro perfused lobule using bolus injections. Unpublished.

Schneider, H., Pangiel, M., and Dancis, J. (1972) Transfer across the perfused human placenta of antipyrine, sodium and leucine. *Am. J. Obstet. Gynecol.* 114, 822-828.

Schneider, H., Proegler, M., and Duft, R. (1986) Transfer of antipyrine, ^3H_2O, L-glucose and α-aminoisobutyric acid across the in vitro perfused human placental lobe. *Trophoblast Res.* 2, 481-489.

Schneider, H., Proegler, M., and Sodha, R.J. (1985) Effect of flow rate ratio on the diffusion of antipyrine and ^3H_2O in the isolated dually in vitro perfused lobe of human placenta. *Contrib. Gyn. Obstet.* 13, 114-123.

Schröeder, H., Leichtweiss, H.-P., and Rachor, D. (1985) Passive exchange and the distribution of flows in the isolated human placenta. *Contrib. Gyn. Obstet.* 13, 106-113.

Sodha, R.J., Proegler, M., and Schneider, H., (1984) Transfer and metabolism of norepinephrine studied from maternal-to-fetal and fetal-to-maternal sides in the in vitro perfused human placenta lobe. *Am. J. Obstet. Gynecol.* 148, 474-481.

Wier, P.J. and Miller, R.K. (1985) Oxygen transfer as an indicator of perfusion variability in the isolated human placental lobule. *Contrib. Gyn. Obstet.* 13, 127-131.

Wooton, R., Illsley, N., and Hall, S., (1985) A new method for measuring unidirectional transplacental flux. *Clin. Phys. Physiol. Meas.* 6, 47-57.

Yudilevich, D.L., Eaton, B.M., Short, A.H., and Leichtweiss, H-P., (1979) Glucose carriers at maternal and fetal sides of the trophoblast in guinea pig placenta. *Am. J. Physiol.* 237, C205-C212.

LIST OF CONTRIBUTORS AND ATTENDEES

Promila Agrawal
Department of Pharmacology
University of Kansas
Kansas City, Kansas 66103

Mahmond Ahmed
Department of Obstetrics/Gynecology
University of Missouri
Kansas City, Missouri 64108

Michelle Allen
March of Dimes
11 State Street
Pittsford, New York 14534

William Allen
Thoroughbred Breeders' Association
Animal Research Station
307 Huntingdon Road
Cambridge, BC3 0JQ, United Kingdom

Eliane Alsat
U166 INSERM
Maternite Baudelocque
123 Bd de Port Royal
75014 Paris, France

Zelda Althaus
Pharmacodynamics Branch
Reproductive and Developmental
  Toxicology
National Center for Toxicology Research
Jefferson Arkansas 72079

Marvin Amstey
Department of Obstetrics/Gynecology
University of Rochester
601 Elmwood Avenue
Rochester, New York 14642

Deborah Anderson
Department of Obstetrics/Gynecology
Brigham and Women's Hospital
75 Francis Street
Boston, Massachusetts 02115

Teresita Angtuaco
Department of Radiology
University of Arkansas
4301 West Markham
Little Rock, Arkansas 72205

Douglas Antczak
New York State College of Veterinary
  Medicine
Cornell University
Ithaca, New York 14853

Birget Bader
Department of Obstetrics/Gynecology
University of Rochester
601 Elmwood Avenue
Rochester, New York 14642

Raymond Baggs
Dept. of Laboratory Animal Medicine
University of Rochester
601 Elmwood Avenue
Rochester, New York 14642

John R. Bailey
Pharmacodynamics Branch
Reproductive and Developmental
Toxicology
National Center for Toxicology Research
Jefferson Arkansas 72079

Eytan Barnea
Department of Obstetrics/Gynecology
Yale University
333 Cedar Street
New Haven, Connecticut 06510

Stanley L. Barnwell
Department of Pharmacology
Vanderbilt University
Nashville, Tennessee 37232

Rebecca Beaconsfield
SCIP Research Unit
University of London
Regent's Park
London, N.W.L., England

Jackson Beecham
Department of Obstetrics/Gynecology
University of Rochester
Rochester, New York 14642

Alan Beer
L2021-Women's Hospital
University of Michigan
Ann Arbor, Michigan 48109

Frank Bellino
Department of Biological Sciences
109 Cooke
SUNY/Buffalo
Buffalo, New York 14260

Robert Benveniste
814 Dreyfus Laboratories Building
Michael Reese Hospital
Lake Shore Drive at 31st Street
Chicago, Illinois 60616

Ross Berkowitz
Department of Obstetrics/Gynecology
Brigham and Women's Hospital
75 Francis Street
Boston, Massachusetts 02115

Lorelle Bestervelt
Toxicology Program
University of Michigan
Ann Arbor, Michigan 48109

Kathy Birk
Department of Obstetrics/Gynecology
University of Rochester
601 Elmwood Avenue
Rochester, New York 14642

Donna Biscardi
Department of Obstetrics/Gynecology
University of Rochester
601 Elmwood Avenue
Rochester, New York 14642

Charles Blomquist
Department of Obstetrics/Gynecology
St. Paul-Ramsey Medical Center
640 Jackson Street
St. Paul, Minnesota 55101

Irving Boime
Department of Pharmacology
Washington University
St. Louis, Missouri 63110

Robert D.H. Boyd
Department of Child Health
University of Manchester
St. Mary's Hospital
Manchester, M13 OJH, United Kingdom

Robert Brent
Stein Research Center
Thomas Jefferson University
920 Chancellor Street
Philadelphia, Pennsylvania 19107

William Buhi
Box J294
Department of Obstetrics/Gynecology
University of Florida
Gainesville, Florida 32610

Bridget Butterworth
Department of Pathology
Cambridge University
Tennis Court Road
Cambridge, CB2 1QP, United Kingdom

Jack Butler
Hoffmann LaRoche, Inc.
Nutley, New Jersey 07110

John Carbone
Stein Research Center
Thomas Jefferson University
920 Chancellor Street
Philadelphia, Pennsylvania 19107

Patrick Carmody
Department of Obstetrics/Gynecology
Children's Hospital of Buffalo
140 Hodge Street
Buffalo, New York 14222

Anna Cavinato
Chemistry Department
Memphis State University
Memphis, Tennessee 38163

Lise Cedard
INSERM U166
Maternite Baudelocque
123 Bd de Port Royal
75014 Paris, France

Bojana Cemerikic
INEP Department of Endocrinology,
  Immunology and Nutrition
11080 Zemun, Banatska
31b Yugoslavia

Jean-Claude Challier
Department of Biology of Reproduction
University of Paris
Paris, France

John Challis
Department of Obstetrics/Gynecology
St. Joseph's Hospital
268 Grosvenor Street
London, Ontario
NGA 4V2, Canada

Kenneth Chepernik
Department of Anatomy
Jefferson Medical College
1020 Locust Street
Philadelphia, Pennsylvania 19107

John Choate
Department of Obstetrics/Gynecology
University of Rochester
601 Elmwood Avenue
Rochester, New York 14642

H. Howard Cockrill
National Center for Toxicological
  Research
Jefferson, Arkansas

Laurence Cole
Department of Obstetrics/Gynecology
Yale University
333 Cedar Street
New Haven, Connecticut 06510

Carolyn Coulam
Methodist Hospital of Indiana
1604 North Capitol Avenue
Indianapolis, Indiana 46202

Steward Cramer
Department of Pathology
Rochester General Hospital
1425 Portland Avenue
Rochester, New York 14621

Jeanne Cullinan
Department of Obstetrics/Gynecology
University of Rochester
601 Elmwood Avenue
Rochester, New York 14642

Joseph Dancis
Department of Pediatrics
New York University
550 First Avenue
New York, New York 10016

Kathleen Day
Dermatology Unit
University of Rochester
601 Elmwood Avenue
Rochester, New York 14642

Karen de Mesy Jensen
Department of Pathology and
  Laboratory Medicine
University of Rochester
601 Elmwood Avenue
Rochester, New York 14642

Edgard Delvin
Shriners Hospital
1529 Cedar Avenue
Montreal, Quebec
H3G 1A6, Canada

Lidia Derewlany
Hospital for Sick Children
555 University Avenue
Toronto, Ontario
M5G 1X8, Canada

Gernot Desoye
Department of Obstetrics/Gynecology
University of Graz
A8036 Graz, Austria

P. Anthony di Sant'Agnese
Department of Pathology and
  Laboratory Medicine
University of Rochester
601 Elmwood Avenue
Rochester, New York 14642

Richard Doherty
Department of Pediatrics
University of Rochester
601 Elmwood Avenue
Rochester, New York 14642

Reto Duft
Department of Obstetrics
University of Zuerich
Frauenklinik Strasse 10
CH 8091 Zuerich, Switzerland

Murdoch Elder
Institute of Obstetrics/Gynecology
Hammersmith Hospital
DuCane Road
London, W12 0HS, United Kingdom

Jane Erway
March of Dimes
11 State Street
Pittsford, New York 14534

Will Faber
Division of Toxicology
University of Rochester
601 Elmwood Avenue
Rochester, New York 14642

W. Page Faulk
Methodist Hospital of Indiana
1604 North Capitol Avenue
Indianapolis, Indiana 46202

Michael Feinman
Obstetrics/Gynecology
University of Pennsylvania
3400 Spruce Street
Philadelphia, Pennsylvania 19104

Dorothy Feldman
Department of Toxicology and Pathology
Hoffmann LaRoche, Inc.
Nutley, New Jersey 07110

Charles Fisher
Department of Pediatrics
Cleveland Metropolatin General
  Hospital
Cleveland, Ohio 44109

John Fisher
Department of Pathology
Children's Hospital of Buffalo
219 Bryant Street
Buffalo, New York 14222

Stanley Fisher
North Shore University Hospital
Manhasset, New York 11030

Stephen Fortunato
Department of Obstetrics/Gynecology
University of Texas at Dallas
5323 Harry Hines Blvd.
Dallas, Texas 75235

Fred Fumia
St. Francis Hospital
114 Woodland Street
Hartford, Connecticut 06105

Ulysse Gaspard
Department of Obstetrics/Gynecology
State University of Liege
81 Blvd de la Constitution
B4020 Liege, Belgium

Olga Genbacev
INEP Institute of Endocrinology,
  Immunology, and Nutrition
Banatska 31b
11080 Zemun, Yugoslavia

Mary Giknis
Ciba-Geigy, Inc.
556 Morris Avenue
Summit, New Jersey 07901

David Glance
Department of Obstetrics/Gynecology
Hammersmith Hospital
DuCane Road
London, W12 0HS, United Kingdom

J.D. Glazier
Department of Child Health
University of Manchester
St. Mary's Hospital
Manchester, M13 OJH, United Kingdom

Donald Goldstein
Department of Obstetrics/Gynecology
Brigham and Women's Hospital
75 Francis Street
Boston, Massachusetts 02115

Jill Goldstein
Department of Pharmacy
Children's Hospital of Buffalo
219 Bryant Street

Sonia Goldstein
Hopital St-Antoine
CNRSU A 524
27 rue Chaligny
75012 Paris, France

Carll Goodpasture
Vivigen, Inc.
550 St. Michaels Drive
Sante Fe, New Mexico 87502

Barbara Gordon
Nutrition and Food Science Program
Hunter College, CUNY
425 East 25th Street
New York, New York

S. Hall
Section of Perinatal and Child Health
MRC Clinical Research Centre
Watford Road
Harrow, Middlesex,
HA1 3UJ, United Kingdom

Orville Hinsvark
Pennwalt Pharmaceuticals
775 Jefferson Road
Rochester, New York 14623

Nicholas Hole
Department of Immunology
University of Liverpool
PO Box 147
Liverpool, L69 3BX, United Kingdom

Tomokazu Hosokawa
Department of Pharmacology
University of Kansas
Kansas City, Kansas   66103

Randy Howard
Cleveland Clinic Foundation
Box 251
9500 Euclid Avenue
Cleveland, Ohio 44106

Bae-Li Hsi
INSERM U210
Faculte de Medecine
Avenue de Vallombrose
06034 Nice, France

Joan Hunt
Department of Pathology and Oncology
University of Kansas
39th Street and Rainbow Blvd.
Kansas City, Kansas 66103

Robert Hussa
Department of Obstetrics/Gynecology
Medical College of Wisconsin
8700 West Wisconsin Avenue
Milwaukee, Wisconsin 53226

Nicholas Illsley
Laboratory of Renal Biophysics
Cardiovascular Research Institute
University of California
San Francisco, California 94143

Laird Jackson
Division of Medical Genetics
Jefferson Medical College
1025 Walnut Street
Philadelphia, Pennsylvania 19107

Marcela Jensen
Stein Research Center
Thomas Jefferson University
920 Chancellor Street
Philadelphia, Pennsylvania 19107

Peter Johnson
Department of Immunology
University of Liverpool
PO Box 147
Liverpool, L69 3BX, United Kingdom

Janet Jordan
Department of Obstetrics/Gynecology
University of Arkansas for Medical
Sciences
4301 West Markham
Little Rock, Arkansas 72205

Sharad Joshi
Department of Obstetrics/Gynecology
Albany Medical College at Union
  University
Albany, New York 12208

John Josimovich
Department of Obstetrics/Gynecology
New Jersey Medical School
100 Bergen Street
Newark, New York 07103

Mont Juchau
Department of Pharmacology
University of Washington
Seattle, Washington 98195

Peter Karl
North Shore University Hospital
Manhasset, New York 11030

Peter Kaufmann
Department of Anatomy
RWTH Aachen
Melatener Strasse 211
D5100 Aachen, FR Germany

John Kelly
Loon Lake Road
Hornell, New York 14843

Bruce Kelman
Biology and Chemistry Department
Battelle Pacific Northwest Laboratories
Richland, Washington 99352

Peter Keng
Cancer Center
University of Rochester
601 Elmwood Avenue
Rochester, New York 14642

Robert Kilpper
Xerox Corporation
Xerox Square
Rochester, New York 14614

Tai Kim
Department of Chemical Engineering
  and Materials Science
Syracuse University
Syracuse, New York

Harvey Kliman
Department of Pathology and
Laboratory Medicine
University of Pennsylvania School of
Medicine
3400 Spruce Street
Philadelphia, Pennsylvania 19104

Arnold Klopper
Department of Obstetrics/Gynecology
University of Aberdeen
Royal Infirmary
Aberdeen AB9 2ZB, Scotland

Ernest Kohorn
F322 Box 3333
Yale University
New Haven, Connecticut 06510

Thomas Koszalka
Stein Research Center
Thomas Jefferson University
920 Chancellor Street
Philadelphia, Pennsylvania 19107

Alexander Krichevsky
Department of Medicine
Columbia University
College of Physicians and Surgeons
New York, New York 10021

Douglas Kuhn
Department of Pediatrics
Syracuse University
750 East Adams Street
Syracuse, New York 13210

Arun Kulkarni
Toxicology Program
University of Michigan
Ann Arbor, Michigan 48109-2029

Janice Lage
Department of Pathology
Brigham and Women's Hospital
75 Francis Street
Boston, Massachusetts 02215

Julian Leakey
Pharmacodynamics Branch
Reproductive and Developmental
Toxicology
National Center for Toxicology Research
Jefferson Arkansas 72079

Mary Lee
Department of Physiology
Columbia University
630 West 168th Street
New York, New York 10032

Heinz-Peter Leichtweiss
Universitatsfrauenklinik
Abtlg. Exper. Med.
Martinistr. 52
D2000 Hamburg 20, FR Germany

Amol Lele
Children's Hospital of Buffalo
130 Hodge Avenue
Buffalo, New York 14222

Y. Leung
219 Bryant Street
Buffalo, New York 14222

Janet Levy
Department of Pharmacology
University of Kansas
Kansas City, Kansas 66103

George Lipe
Pharmacodynamics Branch
Reproductive and Developmental
Toxicology
National Center for Toxicology Research
Jefferson Arkansas 72079

Belisario Lisboa
Universitatsfrauenklinik
Abtlg. Exper. Med.
Martinistr. 52
D2000 Hamburg 20, FR Germany

Juliet Lobo
Department of Biological Sciences
109 Cooke
SUNY/Buffalo
Buffalo, New York 14260

Y.W. (Charles) Loke
Department of Pathology
Cambridge University
Tennis Court Road
Cambridge, CB2 1QP, England

Christopher Lu
Renal Division
Brigham and Women's Hospital
75 Francis Street
Boston, Massachusetts 02115

M. Helen Maguire
Department of Pharmacology
University of Kansas
Rainbow Blvd. at 39th
Kansas City, Kansas 66103

Elliott Main
Department of Obstetrics/Gynecology
University of Pennsylvania
3400 Spruce Street
Philadelphia, Pennsylvania 19104

Ahad Makarachi
Albany Medical College
Albany, New York

Andre Malassine
Maternite Baudelcoque
INSERM U166
123 Bld de Port Royal
75014 Paris, France

David Manchester
Department of Pediatrics and
Pharmacology
University of Colorado
4200 East 9th Street
Denver, Colorado 80262

Donald Mattison
Department of Obstetrics/Gynecology
University of Arkansas
4301 West Markham
Little Rock, Arkansas 72205

Debabrata Maulik
Department of Obstetrics/Gynecology
Truman Medical Center
2301 Holmes Street
Kansas City, Missouri 64108

Peter McConnachie
Methodist Hospital of Indiana
1604 North Capitol Avenue
Indianapolis, Indiana 46202

John McCoshen
Department of Obstetrics/Gynecology
University of Manitoba
59 Emily Street
Winnipeg, Manitoba
R3E 0W3 Canada

John McIntyre
Methodist Hospital of Indiana
1604 North Capitol Avenue
Indianapolis, Indiana 46202

P. Jeremy McLaughlin
Department of Immunology
University of Liverpool
PO Box 147
Liverpool, L69 3BX, United Kingdom

Don McNellis
National Institutes of Health
Bethesda, Maryland

Susan Daniels McQueen
Department of Pharmacology
Washington University
4566 Scott
St. Louis, Missouri 63110

Leon Metlay
Department of Pathology
University of Rochester
601 Elmwood Avenue
Rochester, New York 14642

Harold Miller
647 Sunset Lane
East Lansing, Michigan 48823

Richard Miller
Department of Obstetrics/Gynecology
University of Rochester
601 Elmwood Avenue
Rochester, New York 14642

Bryan Mitchell
Department of Obstetrics/Gynecology
St. Joseph's Hospital
London, Ontario
N6A 4V2, Canada

Francoise Mondon
Maternite Baudelcoque
INSERM U166
123 Bld de Port Royal
75014 Paris, France

Robert Moore
443 Richmond Park West - 506D
Richmond Heights, Ohio 44143

John Moore
3697 Townley Road
Shaker Heights, Ohio 44122

W.M.O. Moore
Department of Obstetrics/Gynecology
University of Manchester
St. Mary's Hospital
Manchester, M13 OJH, United Kingdom

Donald Morrish
Department of Medicine
7-117 Clinical Sciences Building
University of Alberta
Edmonton, Alberta
T6G 2G3, Canada

Frank Morriss
Department of Pediatrics
University of Texas at Houston
6431 Fannin Street
Houston, Texas 77030

Eberhard Muechler
Department of Obstetrics/Gynecology
University of Rochester
601 Elmwood Avenue
Rochester, New York 14642

Zulf Mughal
Department of Child Health
St. Mary's Hospital
Manchester, M13 OJH, England

Leslie Myatt
Department of Obstetrics/Gynecology
Hammersmith Hospital
DuCane Road
London, W12 0HS, United Kingdom

Moses Namkung
Department of Pharmacology
SJ-30
University of Washington
Seattle, Washington 98195

Peter Nathanielsz
Cornell University
Ithaca, New York 14853

Michael Nelson
Department of Obstetrics/Gynecology
The Jewish Hospital of St. Louis
216 South Kingshighway
St. Louis, Missouri 63110

Wendy Ng
Warner Lambert
Ann Arbor, Michigan

Allen Okey
The Hospital for Sick Children
555 University Avenue
Toronto, Ontario
M5G 1X8, Canada

L.K. Owens
Department of Pharmacology
Vanderbilt University
Nashville, Tennessee 37232

Maurice Panigel
University of Paris
4 Villa Patric Boudard
75016 Paris, France

Rogelio Perez-D'Gregorio
Department of Obstetrics/Gynecology
University of Rochester
601 Elmwood Avenue
Rochester, New York 14642

A.M.C.M. Pijnenburg
Department of Chemical Pathology
Erasmus University
3000 DR Rotterdam, The Netherlands

Alan Poisner
Department of Pharmacology
University of Kansas
Kansas City, Kansas 66103

Roselle Poisner
Department of Pharmacology
University of Kansas
Kansas City, Kansas 66103

J. Clare Poleman
James A. Baker Institute for Animal
   Health
NYS College of Veterinary Medicine
Cornell University
Ithaca, New York 14853

Jainine Polifka
Stein Research Center
Thomas Jefferson University
920 Chancellor Street
Philadelphia, Pennsylvania 19107

Mariette Proegler
Department of Obstetrics
University of Zuerich
Frauenklinik Strasse 10
CH 8091 Zuerich, Switzerland

Thomas Putman
March of Dimes
11 State Street
Pittsford, New York 14534

James Quebbeman
Department of Obstetrics/Gynecology
University of Michigan
Ann Arbor, Michigan 48109

Ingeborg Radde
The Hospital for Sick Children
555 University Avenue
Toronto, Ontario
M5G 1X8, Canada

Elizabeth Ramsey
3420 Que Street NW
Washington, DC 20007

Regis Rebourcet
Hopital St-Antoine
CNRSUA 524
27 rue Chaligny
75012 Paris, France

Allan Rettie
Department of Pharmacology
SJ-30
University of Washington
Seattle, Washington 98195

Marie Rex
American Cancer Society
1400 North Winton Road
Rochester, New York 14609

Philip Rice
Department of Chemical Engineering
   and Materials Science
Syracuse University
320 Hinds Hall
Syracuse, New York 13210

Matthew Rose
Department of Obstetrics/Gynecology
Hammersmith Hospital
DuCane Road
London, W12 0HS, United Kingdom

Richardus Ross
Department of Pediatrics
University of Cincinnati
231 Bethesda Avenue
Cincinnati, Ohio

Anne Rueter
Department 47V
Abbott Laboratories
Building AP-10
Abbott Park, Illinois 60064

Carolyn Salafia
Department of Laboratory Medicine
Danbury Hospital
Danbury, Connecticut 06810

Risa Saltzman
Division of Toxicology
University of Rochester
601 Elmwood Avenue
Rochester, New York 14642

Steve Sanko
Department of Obstetrics/Gynecology
University of Rochester
601 Elmwood Avenue
Rochester, New York 14642

B.V. Rama Sastry
Department of Pharmacology
Vanderbilt University
Nashville, Tennessee 37232

Henning Schneider
Division of Perinatal Physiology
Department of Obstetrics
University of Zuerich
Frauenklinik - Strasse 10
CH 8091 Zuerich, Switzerland

Peter Schochet
Department of Obstetrics/Gynecology
University of Pennsylvania
3400 Spruce Street
Philadelphia, Pennsylvania 19104

Hobe Schroeder
Abt. Experimentelle Medizin
Universitaets Frauenklinik
Martinistr. 52
D2000 Hamburg, 20 FR Germany

A. Enrico Semprini
University of Milano
Milano, Italy

Yogesh Shah
Department of Obstetrics/Gynecology
University of Rochester
601 Elmwood Avenue
Rochester, New York 14642

Shashi Sharma
Department of Obstetrics/Gynecology
University of Rochester
601 Elmwood Avenue
Rochester, New York 14642

Kathleen Shiverick
Department of Pharmacology and
  Therapeutics
Box J267
University of Florida
Gainesville, Florida 32610

C.P. Sibley
Department of Physiology
University of Manchester
St. Mary's Hospital
Manchester, M13 OJH, United Kingdom

Mario Sideri
Department of Obstetrics/Gynecology
University of Milano
v. Commenda 12
20122 Milano, Italy

Alan Siegel
Department of Biological Sciences
SUNY/Buffalo
109 Cooke Hall
Buffalo, New York 14260

Roy Simmons
Pennwalt Pharmaceuticals
775 Jefferson Road
Rochester, New York 14623

William Slikker
Pharmacodynamics Branch
Reproductive and Developmental
  Toxicology
National Center for Toxicology Research
Jefferson Arkansas 72079

Lisa Smith
Department of Obstetrics/Gynecology
University of Rochester
601 Elmwood Avenue
Rochester, New York 14642

Thomas Smith
Department of Pathology
Children's Hospital of Buffalo
219 Bryant Street
Buffalo, New York 14222

R.N. Sreenathan
Department of Anatomy
Mysore 570001, India

T.E. Stacey
Section of Perinatal and Child Health
MRC Clinical Research Centre
Watford Road
Harrow, Middlesex,
HA1 3UJ, United Kingdom

Christiane Steinborn
Universitatsfrauenklinik
Abtlg. Exper. Med.
Martinistr. 52
D2000 Hamburg 20, FR Germany

Laura Stenzler
James A. Baker Institute
        for Animal Health
NYS College of Veterinary Medicine
Cornell University
Ithaca, New York 14853

Janice Stern
Plenum Publishing Company
233 Spring Street
New York, New York 10013

Peter Stern
Department of Immunology
University of Liverpool
PO Box 147
Liverpool, L69 3BX, United Kingdom

Daniel Stetka
Division of Human Genetics
Children's Hospital of Buffalo
219 Bryant Street
Buffalo, New York 14222

Jerome Strauss, III
Obstetrics/Gynecology
University of Pennsylvania
3400 Spruce Street
Philadelphia, Pennsylvania 19104

Julie Strizki
Department of Obstetrics/Gynecology
University of Pennsylvania
3400 Spruce Street
Philadelphia, Pennsylvania 19104

Vojin Sulovic
INEP Department of Endocrinology,
    Immunology and Nutrition
11080 Zemun, Banatska
31b Yugoslavia

Urvashi Surti
Department of Pathology
University of Pittsburgh
Pittsburgh, Pennsylvania 15213

Howard Sussman
Stanford University
Stanford, California

Ossama Tawfik
Department of Pathology and Oncology
University of Kansas
39th Street and Rainbow Blvd.
Kansas City, Kansas 66103

Francois Teasdale
Hopital Sainte-Justine
3175 Chemin Ste-Catherine
Montreal, H3T 1C5, Quebec, Canada

Henry Thiede
Department of Obstetrics/Gynecology
University of Rochester
601 Elmwood Avenue
Rochester, New York 14642

Peter J. Thomford
Department of Obstetrics/Gynecology
University of Arkansas
Slot 518
4301 West Markham
Little Rock, Arkansas 72205

Donald Torry
Methodist Hospital of Indiana
1604 North Capitol Avenue
Indianapolis, Indiana 46202

Kimberly Treinen
Toxicology Program
University of Michigan
Ann Arbor, Michigan 48109

J. P. van Dijk
Department of Chemical Pathology
Erasmus University
3000 DR Rotterdam, The Netherlands

Bernard van Kreel
Department of Chemical Pathology
Erasmus University
3000 DR Rotterdam, The Netherlands

Ljiljana Vicovac
INEP - Institute of Endocrinology,
Immunology, and Nutrition
Banatska 31b
11080 Zemun, Yugoslavia

Stephen Volsen
Thoroughbred Breeders' Association
Equine Fertility Unit
Animal Research Station
307 Huntingdon Road
Cambridge, CG3 0JQ, United Kingdom

Mirjana Vuckovic
INEP - Institute of Endocrinology,
Immunology, and Nutrition
Banatska 31b
11080 Zemun, Yugoslavia

Laurene Wang
Department of Pharmaceutical
Chemistry
University of California
San Francisco, California 94143

B. Stuart Ward
Department of Obstetrics/Gynecology
University of Manchester
St. Mary's Hospital
Manchester, M13 0JH, United Kingdom

Terri Wasmoen
Department of Immunology
Mayo Clinic
Rochester, Minnesota 55905

Paul Webb
Department of Immunology
University of Liverpool
PO Box 147
Liverpool, L69 3BX, United Kingdom

Peter Weiss
Department of Obstetrics/Gynecology
University of Graz
8036 Graz, Austria

Lowell Weitkamp
Division of Genetics
University of Rochester
601 Elmwood Avenue
Rochester, New York 14642

Jackie White
Department of Obstetrics/Gynecology
University of Rochester
601 Elmwood Avenue
Rochester, New York 14642

Michael White
Gannett Rochester Newspapers
55 Exchange Blvd.
Rochester, New York 14614

Tacey White
Division of Toxicology
University of Rochester
601 Elmwood Avenue
Rochester, New York 14642

Patrick Wier
Medicine and Environmental
      Health Department
Exxon Corporation
PO Box 235
East Millstone, New Jersey 08873

Elke Winterhager
Abt. Anatomie, RWTH Aachen
Melatener Str. 211
D5100 Aachen, FR Germany

Gary Wood
Department of Pathology and Oncology
University of Kansas
39th Street and Rainbow Blvd.
Kansas City, Kansas 66103

James Woods
Department of Obstetrics/Gynecology
University of Cincinnati
231 Bethesda Avenue
Cincinnati, Ohio 45367

Ralph Wynn
1510 West 25th Street
Sunset Island #2
Miami Beach, Florida 33140

Katsumi Yazaki
Department of Obstetrics/Gynecology
Gumma University
Maebashi, Gumma-ken 371 Japan

Chang Jing Yeh
INSERM U210
Faculte de Medecine
Avenue de Vallombrose
06034 Nice, France

Ming Neg Yeh
80 Dyer Court
Norwood, New Jersey 07648

Frank Young
Food and Drug Administrion
1471 Parklawn Building
5600 Fishers Lane
Rockville, Maryland 20857

John Young
National Center for Toxicology Research
Jefferson, Arkansas 72079

Zhu Xiaoyu
Wuhan, Hubei
Peoples Republic of China

Elena Zannoni
First Clinic of Obstetrics and
  Gynecology
University of Milan
Milan, Italy

Jeanette Zavislan
Department of Obstetrics/Gynecology
University of Rochester
601 Elmwood Avenue
Rochester, New York 14642

John Zongrone
Department of Obstetrics/Gynecology
University of Rochester
601 Elmwood Avenue
Rochester, New York 14642

INDEX

All page numbers listed below represent the first page of each chapter,
where the subject is located.